## TO THE STUDENT

This Solutions Manual contains answers to all the exercises and solutions to all problems at the end of each chapter in the text.

This book will be valuable to you if you use it properly. It is not intended to be a substitute for answering the exercises by yourself. It is important that you go through the process of writing down the answers to the exercises and the solutions to problems before you see them in the solutions manual. Once you have seen the answer, much of the value of whether or not you have learned the material is lost. This does not mean that you cannot learn from the answers. If you absolutely cannot answer a question or problem, study the answer carefully to see where you are having difficulty. You should spend sufficient time answering each assigned exercise. Only after you have made a serious attempt to answer a question or problem should you resort to the solutions manual.

Chemistry is one of the most challenging courses of study you will encounter. The core of the study of chemistry is problem solving. Skill at solving problems is achieved through effective and consistent practice. Watching and listening to others solve problems may be of some use but will not result in facility with chemistry problems. The methods used for solving problems in this manual are essentially the same as those used in the text. The basic steps in problem solving are fairly universal:

1. Read the problem carefully and determine the type of problem.
2. Develop a plan for solving the problem.
3. Write down the given information, including units, in an organized fashion.
4. Write a complete set-up and solution for the problem.
5. Use the solutions manual to check the answer and solution.

Your solution might be correct and yet will vary from the manual. Usually, these differences are the result of completing operations in a different order, or using separate steps instead of a single line approach. If you make an error, take the time to analyze what went wrong. An error caused by an improper calculator entry can be frustrating, but it is not as serious as the inability to properly set up the problem. Do not become discouraged. Once you understand the steps in a problem, go back and rework it without looking at the answer.

The areas of organic and biochemistry present a different type of learning. Learning organic chemistry takes a great amount of practice. You need to put in a lot of time writing formulas, names, and reaction equations if you are to be successful. Initially there is a certain amount of memorization of material. For example, you must learn the names and formulas of the first ten aliphatic hydrocarbons: methane, ethane, propane, butane, pentane, hexane, heptane, octane, nonane, and decane. Say them to yourself several times daily; write the names together with their formulas over and over again, and soon you will know them very well. The names of many other classes of organic compounds are derived from the names of the aliphatic hydrocarbons. Remember, **practice** is the key word.

We have made every attempt to produce a manual with as few errors as possible. Please let us know of errors that you encounter so that they can be corrected. Allow this manual to help you have success as you begin the great adventure of studying chemistry.

Morris Hein
Scott Pattison
Susan Arena

# CONTENTS

# STANDARDS FOR MEASUREMENT

1. 100 cm = 1 m
   1000 m = 1 km
   100,000 cm = 1 km

   $$(1\ \text{km})\left(\frac{1000\ \text{m}}{\text{km}}\right)\left(\frac{100\ \text{cm}}{\text{m}}\right) = 100{,}000\ \text{cm}$$

2. 7.6 cm

3. The volumetric flask is a more precise measuring instrument than the graduated cylinder. This is because the narrow opening at the point of measurement (calibration mark in the neck of the flask) means that a small error made in not filling the flask exactly to the mark will be a smaller percentage error in total volume than will a similar error with the cylinder.

4. The three materials would sort out according to their densities with the most dense (mercury) at the bottom and the least dense (glycerin) at the top. In the cylinder, the solid magnesium would sink in the glycerin and float on the liquid mercury.

5. Order of increasing density: ethyl alcohol, vegetable oil, salt, lead.

6. The density of ice must be less than 0.91 g/mL and greater than 0.789 g/mL.

7. Heat is a form of energy, while temperature is a measure of the intensity of heat (how hot the system is).

8. Density is the ratio of the mass of a substance to the volume occupied by that mass. Density has the units of mass over volume. Specific gravity is the ratio (no units) of the density of a substance to the density of a reference substance (usually water at a specific temperature for solids and liquids). Specific gravity has no units.

9. Rule 1. When the first digit after those you want to retain is 4 or less, that digit and all others to its right are dropped. The last digit retained is not changed.

   Rule 2. When the first digit after those you want to retain is 5 or greater, that digit and all others to the right of it are dropped and the last digit retained is increased by one.

10. The number of degrees between the freezing and boiling point of water are
    Fahrenheit 180°F
    Celsius 100°C
    Kelvin 100 K

11. (a) gram = g
    (b) microgram = μg
    (c) centimeter = cm
    (d) micrometer = μm
    (e) milliliter = mL
    (f) deciliter = dL

12. (a) milligram = mg
    (b) kilogram = kg
    (c) meter = m
    (d) nanometer = nm
    (e) angstrom = Å
    (f) microliter = μL

13. (a) 503 zero is significant
    (b) 0.007 zeros are not significant
    (c) 4200 zeros are not significant
    (d) 3.0030 zeros are significant
    (e) 100.00 zeros are significant
    (f) $8.00 \times 10^2$ zeros are significant

14. (a) 63,000 zeros are not significant
    (b) 6.004 zeros are significant
    (c) 0.00543 zeros are not significant
    (d) 8.3090 zeros are significant
    (e) 60. zero is significant
    (f) $5.0 \times 10^{-4}$ zero is significant

15. Significant figures
    (a) 0.025 (2)
    (b) 22.4 (3)
    (c) 0.0404 (3)
    (d) $5.50 \times 10^3$ (3)

16. Significant figures
    (a) 40.0 (3)
    (b) 0.081 (2)
    (c) 129,042 (6)
    (d) $4.090 \times 10^{-3}$ (4)

17. Round to three significant figures
    (a) 93.2
    (b) 0.0286
    (c) 4.64
    (d) 34.3

18. Round to three significant figures
    (a) 8.87
    (b) 21.3
    (c) 130. $(1.30 \times 10^2)$
    (d) $2.00 \times 10^6$

19. Exponential notation
    (a) $2.9 \times 10^6$
    (b) $5.87 \times 10^{-1}$
    (c) $8.40 \times 10^{-3}$
    (d) $5.5 \times 10^{-6}$

20. Exponential notation
    (a) $4.56 \times 10^{-2}$
    (b) $4.0822 \times 10^3$
    (c) $4.030 \times 10^1$
    (d) $1.2 \times 10^7$

21. (a)
$$\begin{array}{r} 12.62 \\ 1.5 \\ \underline{0.25} \\ 14.37 = 14.4 \end{array}$$

(b) $(2.25 \times 10^3)(4.80 \times 10^4) = 10.8 \times 10^7 = 1.08 \times 10^8$

(c) $\dfrac{(452)(6.2)}{14.3} = 195.97 = 2.0 \times 10^2$

(d) $(0.0394)(12.8) = 0.504$

(e) $\dfrac{0.4278}{59.6} = 0.00718 = 7.18 \times 10^{-3}$

(f) $10.4 + (3.75)(1.5 \times 10^4) = 5.6 \times 10^4$

22. (a)
$$\begin{array}{r} 15.2 \\ -2.75 \\ \underline{15.67} \\ 28.1 \end{array}$$

(b) $(4.68)(12.5) = 58.5$

(c) $\dfrac{182.6}{4.6} = 4.0 \times 10^1$ or $40.$

(d)
$$\begin{array}{r} 1986 \\ 23.84 \\ \underline{0.012} \\ 2009.852 = 2010. = 2.010 \times 10^3 \end{array}$$

(e) $\dfrac{29.3}{(284)(415)} = 2.49 \times 10^{-4}$

(f) $(2.92 \times 10^{-3})(6.14 \times 10^5) = 1.79 \times 10^3$

23. Fractions to decimals (3 significant figures)

(a) $\dfrac{5}{6} = 0.833$ (c) $\dfrac{12}{16} = 0.750$

(b) $\dfrac{3}{7} = 0.429$ (d) $\dfrac{9}{18} = 0.500$

24. Decimals to fractions

(a) $0.25 = \dfrac{1}{4}$ (c) $1.67 = 1\dfrac{2}{3}$ or $\dfrac{5}{3}$

(b) $0.625 = \dfrac{5}{8}$ (d) $0.888 = \dfrac{8}{9}$

25. (a) $3.42x = 6.5$

$$\frac{3.42x}{3.42} = \frac{6.5}{3.42}$$

$$x = \frac{6.5}{3.42} = 1.9$$

(b) $\frac{x}{12.3} = 7.05$

$$x = (7.05)(12.3) = 86.7$$

(c) $\frac{0.525}{x} = 0.25$

$$0.525 = 0.25x$$

$$x = \frac{0.525}{0.25} = 2.1$$

26. (a) $x = \frac{212-32}{1.8}$

$$x = 1.0 \times 10^2$$

(b) $8.9\ \frac{\text{g}}{\text{mL}} = \frac{40.90\ \text{g}}{x}$

$$\left(8.9\frac{\text{g}}{\text{mL}}\right)x = 40.90\ \text{g}$$

$$x = \frac{40.90\ \text{g}}{8.9\frac{\text{g}}{\text{mL}}} = 4.6\ \text{mL}$$

(c) $72 = 1.8x + 32$

$$72 - 32 = 1.8x$$

$$40. = 1.8x$$

$$\frac{40.}{1.8} = x$$

$$22 = x$$

27. (a) $(28.0\ \text{cm})\left(\frac{1\ \text{m}}{100\ \text{cm}}\right) = 0.280\ \text{m}$

(b) $(1000.\ \text{m})\left(\frac{1\ \text{km}}{1000\ \text{m}}\right) = 1.000\ \text{km}$

(c) $(9.28\ \text{cm})\left(\frac{10\ \text{mm}}{1\ \text{cm}}\right) = 92.8\ \text{mm}$

(d) $(10.68\ \text{g})\left(\frac{1000\ \text{mg}}{1\ \text{g}}\right) = 1.068 \times 10^4\ \text{mg}$

(e) $(6.8 \times 10^4\ \text{mg})\left(\frac{1\ \text{g}}{1000\ \text{mg}}\right)\left(\frac{1\ \text{kg}}{1000\ \text{g}}\right) = 6.8 \times 10^{-2}\ \text{kg}$

(f) $(8.54\ \text{g})\left(\frac{1\ \text{kg}}{1000\ \text{g}}\right) = 0.00854\ \text{kg}$

(g) $(25.0 \text{ mL})\left(\dfrac{1 \text{ L}}{1000 \text{ mL}}\right) = 2.50 \times 10^{-2} \text{ L}$

(h) $(22.4 \text{ L})\left(\dfrac{10^6 \ \mu\text{L}}{1 \text{ L}}\right) = 2.24 \times 10^7 \ \mu\text{L}$

28. (a) $(4.5 \text{ cm})\left(\dfrac{1 \text{ m}}{100 \text{ cm}}\right)\left(\dfrac{1 \text{ Å}}{10^{-10} \text{ m}}\right) = 4.5 \times 10^8 \text{ Å}$

(b) $(12 \text{ nm})\left(\dfrac{10^{-9} \text{ m}}{1 \text{ nm}}\right)\left(\dfrac{100 \text{ cm}}{1 \text{ m}}\right) = 1.2 \times 10^{-6} \text{ cm}$

(c) $(8.0 \text{ km})\left(\dfrac{1000 \text{ m}}{1 \text{ km}}\right)\left(\dfrac{1000 \text{ mm}}{1 \text{ m}}\right) = 8.0 \times 10^6 \text{ mm}$

(d) $(164 \text{ mg})\left(\dfrac{1 \text{ g}}{1000 \text{ mg}}\right) = 0.164 \text{ g}$

(e) $(0.65 \text{ kg})\left(\dfrac{1000 \text{ g}}{1 \text{ kg}}\right)\left(\dfrac{1000 \text{ mg}}{1 \text{ g}}\right) = 6.5 \times 10^5 \text{ mg}$

(f) $(5.5 \text{ kg})\left(\dfrac{1000 \text{ g}}{1 \text{ kg}}\right) = 5.5 \times 10^3 \text{ g}$

(g) $(0.468 \text{ L})\left(\dfrac{1000 \text{ mL}}{1 \text{ L}}\right) = 468 \text{ mL}$

(h) $(9.0 \ \mu\text{L})\left(\dfrac{1 \text{ L}}{10^6 \ \mu\text{L}}\right)\left(\dfrac{1000 \text{ mL}}{1 \text{ L}}\right) = 9.0 \times 10^{-3} \text{ mL}$

29. (a) $(42.2 \text{ in.})\left(\dfrac{2.54 \text{ cm}}{1 \text{ in.}}\right) = 107 \text{ cm}$

(b) $(0.64 \text{ mi})\left(\dfrac{5280 \text{ ft}}{1 \text{ mi}}\right)\left(\dfrac{12 \text{ in.}}{1 \text{ ft}}\right) = 4.1 \times 10^4 \text{ in.}$

(c) $(2.00 \text{ in.}^2)\left(\dfrac{2.54 \text{ cm}}{1 \text{ in.}}\right)^2 = 12.9 \text{ cm}^2$

(d) $(42.8 \text{ kg})\left(\dfrac{2.205 \text{ lb}}{\text{kg}}\right) = 94.4 \text{ lb}$

(e) $(3.5 \text{ qt})\left(\dfrac{946 \text{ mL}}{1 \text{ qt}}\right) = 3.3 \times 10^3 \text{ mL}$

(f) $(20.0 \text{ gal})\left(\dfrac{4 \text{ qt}}{1 \text{ gal}}\right)\left(\dfrac{0.946 \text{ L}}{1 \text{ qt}}\right) = 75.7 \text{ L}$

30. (a) The conversion is: m → cm → in. → ft

$$(35.6 \text{ m})\left(\frac{100 \text{ cm}}{1 \text{ m}}\right)\left(\frac{1 \text{ in.}}{2.54 \text{ cm}}\right)\left(\frac{1 \text{ ft}}{12 \text{ in.}}\right) = 117 \text{ ft}$$

(b) $(16.5 \text{ km})\left(\dfrac{1 \text{ mi}}{1.609 \text{ km}}\right) = 10.3 \text{ mi}$

(c) $(4.5 \text{ in.}^3)\left(\dfrac{2.54 \text{ cm}}{1 \text{ in.}}\right)^3\left(\dfrac{10 \text{ mm}}{1 \text{ cm}}\right)^3 = 7.4 \times 10^4 \text{ mm}^3$

(d) $(95 \text{ lb})\left(\dfrac{453.6 \text{ g}}{1 \text{ lb}}\right) = 4.3 \times 10^4 \text{ g}$

(e) $(20.0 \text{ gal})\left(\dfrac{4 \text{ qt}}{1 \text{ gal}}\right)\left(\dfrac{0.946 \text{ L}}{1 \text{ qt}}\right) = 75.7 \text{ L}$

(f) The conversion is: $\text{ft}^3 \rightarrow \text{in.}^3 \rightarrow \text{cm}^3 \rightarrow \text{m}^3$

$$(4.5 \times 10^4 \text{ ft}^3)\left(\frac{12 \text{ in.}}{1 \text{ ft}}\right)^3\left(\frac{2.54 \text{ cm}}{1 \text{ in.}}\right)^3\left(\frac{1 \text{ m}}{1000 \text{ cm}}\right)^3 = 1.3 \times 10^3 \text{ m}^3$$

31. $\left(55 \dfrac{\text{mi}}{\text{hr}}\right)\left(1.609 \dfrac{\text{km}}{1 \text{ mi}}\right) = 88 \dfrac{\text{km}}{\text{hr}}$

32. The conversion is: $\dfrac{\text{km}}{\text{hr}} \rightarrow \dfrac{\text{mi}}{\text{hr}} \rightarrow \dfrac{\text{ft}}{\text{hr}} \rightarrow \dfrac{\text{ft}}{\text{s}}$

$$\left(55 \frac{\text{km}}{\text{hr}}\right)\left(\frac{1 \text{ mi}}{1.609 \text{ km}}\right)\left(\frac{5280 \text{ ft}}{1 \text{ mi}}\right)\left(\frac{1 \text{ hr}}{3600 \text{ s}}\right) = 50. \frac{\text{ft}}{\text{s}}$$

33. The conversion is: $\dfrac{\text{m}}{\text{s}} \rightarrow \dfrac{\text{cm}}{\text{s}} \rightarrow \dfrac{\text{in.}}{\text{s}} \rightarrow \dfrac{\text{ft}}{\text{s}}$

$$\left(\frac{100. \text{ m}}{9.92 \text{ s}}\right)\left(\frac{100 \text{ cm}}{1 \text{ m}}\right)\left(\frac{1 \text{ in.}}{2.54 \text{ cm}}\right)\left(\frac{1 \text{ ft}}{12 \text{ in.}}\right) = 33.1 \frac{\text{ft}}{\text{s}}$$

34. The conversion is: $\frac{\text{mi}}{\text{hr}} \rightarrow \frac{\text{km}}{\text{hr}} \rightarrow \frac{\text{km}}{\text{s}}$

$$\left(\frac{229\ \text{mi}}{1\ \text{hr}}\right)\left(\frac{1.609\ \text{km}}{\text{mi}}\right)\left(\frac{1\ \text{hr}}{3600\ \text{s}}\right) = 0.102\ \frac{\text{km}}{\text{s}}$$

35. The conversion is: $\frac{\text{mi}}{\text{hr}} \rightarrow \frac{\text{km}}{\text{hr}} \rightarrow \frac{\text{km}}{\text{s}}$

$$\left(\frac{27{,}000\ \text{mi}}{\text{hr}}\right)\left(\frac{1.609\ \text{km}}{\text{mi}}\right)\left(\frac{1\ \text{hr}}{3600\ \text{s}}\right) = 12\ \frac{\text{km}}{\text{s}}$$

36. The conversion is: mi $\rightarrow$ km $\rightarrow$ m $\rightarrow$ s
93 million miles = $9.3 \times 10^7$ mi

$$(9.3 \times 10^7\ \text{mi})\left(\frac{1.609\ \text{km}}{\text{mi}}\right)\left(\frac{1000\ \text{m}}{1\ \text{km}}\right)\left(\frac{1\ \text{s}}{3.00 \times 10^8\ \text{m}}\right) = 5.0 \times 10^2\ \text{s}$$

37. $(176\ \text{lb})\left(\frac{453.6\ \text{g}}{1\ \text{lb}}\right)\left(\frac{1\ \text{kg}}{1000\ \text{g}}\right) = 79.8\ \text{kg}$

38. The conversion is: oz $\rightarrow$ lb $\rightarrow$ g $\rightarrow$ mg

$$(1\ \text{oz})\left(\frac{1\ \text{lb}}{16\ \text{oz}}\right)\left(\frac{453.6\ \text{g}}{1\ \text{lb}}\right)\left(\frac{1000\ \text{mg}}{1\ \text{g}}\right) = 3 \times 10^4\ \text{mg}$$

39. $(5.0\ \text{grains})\left(\frac{1\ \text{lb}}{7000.\ \text{grains}}\right)\left(\frac{453.6\ \text{g}}{1\ \text{lb}}\right) = 0.32\ \text{g}$

40. $(21\ \text{lb})\left(\frac{453.6\ \text{g}}{1\ \text{lb}}\right) = 9.5 \times 10^3\ \text{g} = \text{mass condor}$

$$\left(\frac{9.5 \times 10^3\ \text{g}/(\text{condor})}{3.2\ \text{g}/(\text{hummingbird})}\right) = 3.0 \times 10^3 \text{ hummingbirds to equal the mass of one (1) condor}$$

41. $\left(\frac{\$1.49}{283.5\ \text{g}}\right)\left(\frac{453.6\ \text{g}}{1\ \text{lb}}\right) = \frac{\$2.38}{\text{lb}}$

42. The conversion is: $\frac{\$}{\text{oz}} \rightarrow \frac{\$}{\text{lb}} \rightarrow \frac{\$}{\text{g}} \rightarrow \$$

$$\left(\frac{\$350}{1\ \text{oz}}\right)\left(\frac{14.58\ \text{oz}}{1\ \text{lb}}\right)\left(\frac{1\ \text{lb}}{453.6\ \text{g}}\right)(250\ \text{g}) = \$2800$$

43. The conversion is: $\frac{\$}{\text{L}} \rightarrow \frac{\$}{\text{qt}} \rightarrow \frac{\$}{\text{gal}} \rightarrow \$$

$$\left(\frac{\$0.35}{1\text{ L}}\right)\left(\frac{0.946\text{ L}}{1\text{ qt}}\right)\left(\frac{4\text{ qt}}{1\text{ gal}}\right)(15.8\text{ gal}) = \$21$$

44. The conversion is: mi → gal → qt → L

$$(525\text{ mi})\left(\frac{1\text{ gal}}{35\text{ mi}}\right)\left(\frac{4\text{ qt}}{1\text{ gal}}\right)\left(\frac{0.946\text{ L}}{1\text{ qt}}\right) = 57\text{ L}$$

45. The conversion is: $\frac{\text{drops}}{\text{mL}} \rightarrow \frac{\text{drops}}{\text{qt}} \rightarrow \frac{\text{drops}}{\text{gal}} \rightarrow \text{drops}$

$$\left(\frac{20.\text{ drops}}{\text{mL}}\right)\left(\frac{946\text{ mL}}{\text{qt}}\right)\left(\frac{4\text{ qt}}{\text{gal}}\right)(1.0\text{ gal}) = 7.6 \times 10^4\text{ drops}$$

46. $(42\text{ gal})\left(\frac{4\text{ qt}}{\text{gal}}\right)\left(\frac{0.946\text{ L}}{\text{qt}}\right) = 160\text{ L}$

47. The conversion is: $\text{ft}^3 \rightarrow \text{in.}^3 \rightarrow \text{cm}^3 \rightarrow \text{mL}$

$$(1.00\text{ ft}^3)\left(\frac{12\text{ in.}}{\text{ft}}\right)^3\left(\frac{2.54\text{ cm}}{1\text{ in.}}\right)^3\left(\frac{1\text{ mL}}{1\text{ cm}^3}\right) = 2.83 \times 10^4\text{ mL}$$

48. $V = A \times h$ $\quad$ A = area $\quad$ h = height

The conversion is: $\frac{\text{cm}^3}{\text{nm}} \rightarrow \frac{\text{cm}^3}{\text{m}} \rightarrow \text{m}^2$

$$A = \frac{V}{h} = \left(\frac{200\text{ cm}^3}{0.5\text{ nm}}\right)\left(\frac{1\text{ nm}}{10^{-9}\text{ m}}\right)\left(\frac{1\text{ m}}{100\text{ cm}}\right)^3 = 4 \times 10^5\text{ m}^2$$

49. (a) $(27\text{ cm})(21\text{ cm})(4.4\text{ cm}) = 2.5 \times 10^3\text{ cm}^3$

(b) $2.5 \times 10^3\text{ cm}^3$ is $2.5 \times 10^3\text{ mL}\left(\frac{1\text{ L}}{1000\text{ mL}}\right) = 2.5\text{ L}$

(c) $(2.5 \times 10^3\text{ cm}^3)\left(\frac{1\text{ in.}}{2.54\text{ cm}}\right)^3 = 1.5 \times 10^2\text{ in.}^3$

50. $(16 \text{ in.})(8 \text{ in.})(10 \text{ in.})\left(\frac{2.54 \text{ cm}}{1 \text{ in.}}\right)^3 = 2 \times 10^4 \text{ cm}^3 = 2 \times 10^4 \text{ mL}$

$(2 \times 10^4 \text{ mL})\left(\frac{1 \text{ L}}{1000 \text{ mL}}\right) = 2 \times 10^1 \text{ L}$

$(2 \times 10^1 \text{ L})\left(\frac{1 \text{ qt}}{0.946 \text{ L}}\right)\left(\frac{1 \text{ gal}}{4 \text{ qt}}\right) = 5 \text{ gal}$

51. $°C = \frac{°F - 32}{1.8}$ $\quad \frac{98.6 - 32}{1.8} = 37.0°C$

52. $°F = 1.8°C + 32$ $\quad (1.8)(45) + 32 = 113°F$ Summer!

53. (a) $\frac{162 - 32}{1.8} = 72.2°C$ Remember to express the answer to the same precision as the original measurement.

(b) $°C + 273 = K$ $\quad \frac{0.0 - 32}{1.8} + 273 = 255.2 \text{ K}$

(c) $1.8(-18) + 32 = -0.40°F$

(d) $212 - 273 = -61°C$

54. (a) $1.8(32) + 32 = 90.°F$

(b) $\frac{-8.6 - 32}{1.8} = -22.6°C$

(c) $273 + 273 = 546 \text{ K}$

(d) $°C = 100 - 273 = -173°C$
$(-173)(1.8) + 32 = -279°F = -300°F$ (1 significant figure in 100 K)

55. $°F = °C$

$°F = 1.8(°C) + 32$ substitute °F for °C

$°F = 1.8(°F) + 32$

$-32 = 0.8\ (°F)$

$\frac{-32}{0.8} = °F$

$-40 = °F$

$-40°F = -40°C$

56. $°F = -°C$

$°F = 1.8(°C) + 32$ substitute $-°C$ for $°F$

$-°C = 1.8(°C) + 32$

$2.8(°C) = -32$

$°C = \frac{-32}{2.8}$

$°C = -11.4$

$-11.4°C = 11.4°F$

57. $d = \frac{m}{V} = \frac{78.26\ g}{50.00\ mL} = 1.565\ \frac{g}{mL}$

58. $d = \frac{m}{V} = \frac{39.9\ g}{12.8\ mL} = 3.12\ \frac{g}{mL}$

59. $29.6\ mL - 25.0\ mL = 4.6\ mL$ (volume of chromium)

$d = \frac{m}{V} = \frac{32.7\ g}{4.6\ mL} = 7.1\ \frac{g}{mL}$

60. $106.773\ g - 42.817\ g = 63.956\ g$ (mass of the liquid)

$d = \frac{m}{V} = \frac{63.956\ g}{50.0\ mL} = 1.28\ \frac{g}{mL}$

61. $d = \frac{m}{V}$

$m = d\,V = \left(1.19\ \frac{g}{mL}\right)(250.0\ mL) = 298\ g$

62. $d = \frac{m}{V}$

$m = d\,V = \left(13.6\ \frac{g}{mL}\right)(25.0\ mL) = 3.40 \times 10^2\ g$

63. $d = \left(\frac{1032\text{ g}}{1\text{ L}}\right)\left(\frac{1\text{ L}}{1000\text{ mL}}\right) = 1.032\ \frac{\text{g}}{\text{mL}}$

$d = \left(\frac{1032\text{ g}}{1\text{ L}}\right)\left(\frac{1\text{ kg}}{1000\text{ g}}\right) = 1.032\ \frac{\text{kg}}{\text{L}}$

64. The conversion is: $\text{L} \rightarrow \text{cm}^3 \rightarrow \text{g} \rightarrow \text{lb}$

$(3.1\text{ L})\left(\frac{1000\text{ cm}^3}{1\text{ L}}\right)\left(1.03\frac{\text{g}}{\text{cm}^3}\right)\left(\frac{1\text{ lb}}{453.6\text{ g}}\right) = 7.0\text{ lbs}$

65. area per lane marker $= (2.5\text{ ft})(4.0\text{ in.})\left(\frac{1\text{ ft}}{12\text{ in.}}\right) = 0.83\text{ ft}^2$

The conversion is: $\frac{\text{lane marker}}{\text{ft}^2} \rightarrow \frac{\text{lane}}{\text{qt}} \rightarrow \frac{\text{lane}}{\text{gal}} \rightarrow \text{lane}$

$\left(\frac{1\text{ lane marker}}{0.83\text{ ft}^2}\right)\left(\frac{43\text{ ft}^2}{1.0\text{ qt}}\right)\left(\frac{4\text{ qt}}{1.0\text{ gal}}\right)(15\text{ gal}) = 3100\text{ lane markers}$

66. $V = \text{side}^3 = (0.50\text{ m})^3 = 0.125\text{ m}^3$

$(0.125\text{ m}^3)\left(\frac{100.\text{ cm}}{\text{m}}\right)^3\left(\frac{1\text{ L}}{1000\text{ cm}^3}\right) = 125\text{ L}$

Yes, the cube will hold the solution. 125 L − 8.5 L = 116.5 L additional solution is necessary to fill the container.

67. The conversion is: $\frac{\mu\text{g}}{\text{m}^3} \rightarrow \frac{\mu\text{g}}{\text{L}} \rightarrow \frac{\mu\text{g}}{\text{day}}$

$\left(\frac{180\ \mu\text{g}}{1\text{ m}^3}\right)\left(\frac{1\text{ m}^3}{1000\text{ L}}\right)\left(2 \times 10^4\ \frac{\text{L}}{\text{day}}\right) = 4000\ \mu\text{g/day ingested}$

Yes, the technician is at risk. This is well over the toxic limit.

68. $\frac{°\text{F} - 32}{1.8} = °\text{C}$

$\frac{(4.5 - 32)}{1.8} = -15.3°\text{C} \equiv 4.5°\text{F}$

$-15°\text{C} > -15.3°\text{C}$

$-15°\text{C} > 4.5°\text{F}$

therefore, $-15°\text{C}$ is the higher temperature

69. $m_{pan\ 1} = m_{pan\ 2}$ (when balanced)

$m_{pan\ 1} = m_{flask} + m_{alcohol} = m_{flask} + (100.\ mL)(0.789\ g/mL)$

$= m_{flask} + 78.9\ g$

let $x$ = volume of turpentine

$$m_{pan\ 2} = (m_{flask} + 11.0\ g) + m_{turpentine} = (m_{flask} + 11.0\ g) + x\left(0.87\ \frac{g}{mL}\right)$$

Since $m_{pan\ 1} = m_{pan\ 2}$

$$m_{flask} + 78.9\ g = (m_{flask} + 11.0\ g) + x\left(0.87\ \frac{g}{mL}\right)$$

$$78.9\ g = 11.0\ g + x\left(0.87\ \frac{g}{mL}\right)$$

$$67.9 = x\left(0.87\ \frac{g}{mL}\right)$$

$x$ = 78 mL turpentine

70. $d = \frac{m}{V}$ $\qquad$ $V = \frac{m}{d}$

$$V_a = \frac{25\ g}{10.\ \frac{g}{mL}} = 2.5\ mL$$

$$V_B = \frac{65\ g}{4.0\ \frac{g}{mL}} = 16\ mL$$

B occupies the larger volume

71.

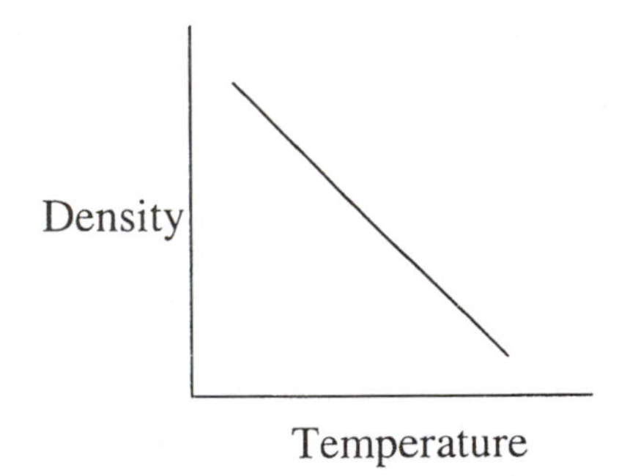


Since $d = \frac{m}{V}$, as the volume increases, the density decreases. As solids are heated the density decreases due to an increase in the volume of the solid.

72. $m = d\,V = (0.789\ g/mL)(35.0\ mL) = 27.6\ g$ ethyl alcohol
$27.6\ g + 49.28\ g = 76.9\ g$ (mass of cylinder and alcohol)

73. $d = \frac{m}{V}$ The cube with the largest V has the lowest density. Use Table 2.5.

| | |
|---|---|
| Cube A - lowest density | 1.74 g/mL - magnesium |
| Cube B | 2.70 g/mL - aluminum |
| Cube C - highest density | 10.5 g/mL - silver |

74. The volume of the aluminum cube is:

$$V = \frac{m}{d} = \frac{500.\ g}{2.70\frac{g}{mL}} = 185\ mL$$

This is the same volume as the gold cube thus:

$m = d\,V = (185\ mL)(19.3\ g/mL) = 3.57 \times 10^3\ g$ of gold

75. $$d = \frac{m}{V} = \frac{24.12\ g}{25.0\ mL} = \frac{0.965\ g}{mL}$$

76. $150.50\ g - 88.25\ g = 62.25\ g$ (mass of liquid)

$$d = \frac{m}{V} \text{ thus } V = \frac{m}{d} = \frac{62.25\ g}{1.25\frac{g}{mL}} = 49.8\ mL \quad \text{(volume of liquid)}$$

The container must hold at least 50 mL.

77. $H_2O$ $$\frac{50\ g}{1.0\frac{g}{mL}} = 50\ mL$$

alcohol $$\frac{50\ g}{0.789\frac{g}{mL}} = 60\ mL$$

Ethyl alcohol has the greater volume due to its lower density.

78. $V = (2.00\text{ cm})(15.0\text{ cm})\ (6.00\text{ cm}) \left(\frac{1\text{ mL}}{1\text{ cm}^3}\right) = 180.\text{ mL}$

$$d = \frac{\text{m}}{\text{V}} = 3300\text{ g}/180.\text{ mL} = 18.3\text{ g/mL}$$

The density of pure gold is 19.3 g/mL (from Table 2.5), therefore, the gold bar is not pure gold, since its density is only 18.3 g/mL, or it is hollow inside.

79. $93.3\text{ kg} = 9.33 \times 10^4\text{ g}$ $\quad d(\text{gold}) = 19.3\text{ g/mL}$

$$V = \left(\frac{9.33 \times 10^4\text{ g}}{19.3\text{ g/mL}}\right) = 4.83 \times 10^3\text{ mL} = 4.83 \times 10^3\text{ cm}^3$$

convert to ounces

$$(9.33 \times 10^4\text{ g})\left(\frac{1\text{ lb}}{453.6\text{ g}}\right)\left(\frac{14.58\text{ oz}}{1\text{ lb}}\right) = 3.0 \times 10^3\text{ oz gold}$$

$$(3.0 \times 10^3\text{ oz})(\$345/\text{oz}) = \$1.04 \times 10^6$$

80. Volume of slug $\quad 30.7\text{ mL} - 25.0\text{ mL} = 5.7\text{ mL}$

Density of slug $\quad d = \frac{\text{m}}{\text{V}} = \frac{15.454\text{ g}}{5.7\text{ mL}} = 2.7\text{ g/mL}$

| | |
|---|---|
| Mass of liquid, cylinder, and slug | 125.934 g |
| Mass of slug (subtract) | -15.454 g |
| Mass of cylinder (subtract) | -89.450 g |
| Mass of the liquid | 21.030 g |

Density of liquid $d = \frac{\text{m}}{\text{V}} = \frac{21.030\text{ g}}{25.0\text{ mL}} = 0.841\text{ g/mL}$

CHAPTER 3

# CLASSIFICATION OF MATTER

1. Answers will vary
   Solids: sugar, salt, copper, silver
   Liquids: water, sulfuric acid, ethyl alcohol, benzene
   Gases: hydrogen, oxygen, nitrogen, ammonia

2. (a) The attractive forces among the ultimate particles of a solid (atoms, ions, or molecules) are strong enough to hold these particles in a fixed position within the solid and thus maintain the solid in a definite shape. The attractive forces among the ultimate particles of a liquid (usually molecules) are sufficiently strong enough to hold them together (preventing the liquid from rapidly becoming gas) but are not strong enough to hold the particles in fixed positions (as in a solid).

   (b) The ultimate particles in a liquid are quite closely packed (essentially in contact with each other) and thus the volume of the liquid is fixed at a given temperature. But, the ultimate particles in a gas are relatively far apart and essentially independent of each other. Consequently, the gas does not have a definite volume.

   (c) In a gas the particles are relatively far apart and are easily compressed, but in a solid the particles are closely packed together and are virtually incompressible.

3. The water in the beaker does not fill the test tube. Since the tube is filled with air (a gas) and two objects cannot occupy the same space at the same time, the gas is shown to occupy space.

4. Mercury and water are the only liquids in the table which are not mixtures.

5. Air is the only gas mixture found in the table.

6. Three phases are present within the bottle; solid and liquid are observed visually, while gas is detected by the immediate odor.

7. The system is heterogeneous as multiple phases are present.

8. A system containing only one substance is not necessarily homogeneous. Two phases may be present. Example: ice in water.

9. A system containing two or more substances is not necessarily heterogeneous. In a solution only one phase is present. Examples: sugar dissolved in water, dilute sulfuric acid.

10. Silicon 25.67% Hydrogen 0.87%

In 100 g $\frac{25.67 \text{ g Si}}{0.87 \text{ g H}} = 30 \text{ g Si/1 g H}$

Si is 28 times heavier than H, thus since 30 > 28, there are more Si atoms than H atoms.

11. The symbol of an element represents the element itself. It may stand for a single atom or a given quantity of the element.

12.

| | | | |
|---|---|---|---|
| phosphorus | P | sodium | Na |
| aluminum | Al | nitrogen | N |
| hydrogen | H | nickel | Ni |
| potassium | K | silver | Ag |
| magnesium | Mg | plutonium | Pu |

13. (a) Si – 1 atom silicon SI – System International or 1 atom sulfur
1 atom iodine

(b) Pb – 1 atom lead PB – 1 atom phosphorus 1 atom boron

(c) 4P – 4 atoms phosphorus $P_4$ – 1 molecule phosphorus
(made of 4 phosphorus atoms)

14.

| | | | |
|---|---|---|---|
| Na | sodium | Ag | silver |
| K | potassium | W | tungsten |
| Fe | iron | Au | gold |
| Sb | antimony | Hg | mercury |
| Sn | tin | Pb | lead |

15.

| | | | |
|---|---|---|---|
| H | hydrogen | S | sulfur |
| B | boron | K | potassium |
| C | carbon | V | vanadium |
| N | nitrogen | Y | yttrium |
| O | oxygen | I | iodine |
| F | fluorine | W | tungsten |
| P | phosphorus | U | uranium |

16. In an element all atoms are alike, while a compound contains two or more elements (different atoms) which are chemically combined. Compounds may be decomposed into simpler substances while elements cannot.

17. 84 metals 7 metalloids 18 nonmetals

18. 7 metals  1 metalloid  2 nonmetals

19. 1 metal  0 metalloids  5 nonmetals

20. The symbol for gold is based upon the Latin word for gold, aurum.

21. (a) iodine
    (b) bromine

22. A compound is composed of two or more elements which are chemically combined in a definite proportion by mass. Its properties differ from those of its components. A mixture is the physical combining of two or more substances (not necessarily elements). The composition may vary, the substances retain their properties, and they may be separated by physical means.

23. Molecular compounds exist as molecules formed from two or more elements bonded together. Ionic compounds exist as cations and anions held together by electrical attractions.

24. Compounds are distinguished from one another by their characteristic physical and chemical properties.

25. (a) $H_2$ – 2 atoms

    (b) $H_2O$ – 3 atoms

    (c) $H_2SO_4$ – 7 atoms

26. Cations are positively charged, while anions are charged negatively.

27. $H_2$ – hydrogen  $Cl_2$ – chlorine

    $N_2$ – nitrogen  $Br_2$ – bromine

    $O_2$ – oxygen  $I_2$ – iodine

    $F_2$ – fluorine

28. Homogeneous mixtures contain only one phase, while heterogeneous mixtures contain two or more phases.

29.

| Metals | Nonmetals |
|---|---|
| solid at room T (except Hg) | solids, liquids or gas at room T |
| luster | dull (no luster) |
| conduct heat & electricity | insulator (does not conduct electricity) |
| malleable | react with each other forming compounds. |
| react with nonmetals to form compounds | |

30. diatomic molecules (a) $H_2$, (c) HCl, (e) NO

31. (a) Potassium, iodine
 (b) Sodium, carbon, oxygen
 (c) Aluminum, oxygen
 (d) Calcium, bromine
 (e) Hydrogen, carbon, oxygen

32. (a) Magnesium, bromine
 (b) Carbon, chlorine
 (c) Hydrogen, nitrogen, oxygen
 (d) Barium, sulfur, oxygen
 (e) Aluminum, phosphorus, oxygen

33. (a) ZnO
 (b) $KClO_3$
 (c) NaOH
 (d) $C_2H_6O$

34. (a) $AlBr_3$
 (b) $CaF_2$
 (c) $PbCrO_4$
 (d) $C_6H_6$

35. (a) 2 atoms H, 1 atom O
 (b) 2 atoms Na, 1 atom S, 4 atoms O
 (c) 4 atoms H, 2 atoms C, 2 atoms O

36. (a) 1 atom Al, 3 atoms Br
 (b) 1 atom Ni, 2 atoms N, 6 atoms O
 (c) 12 atoms C, 22 atoms H, 11 atoms O

37. (a) 2 atoms
 (b) 5 atoms
 (c) 11 atoms
 (d) 8 atoms
 (e) 16 atoms

38. (a) 2 atoms
 (b) 2 atoms
 (c) 9 atoms
 (d) 5 atoms
 (e) 17 atoms

39. (a) 1 atom O
 (b) 4 atoms O
 (c) 2 atoms O
 (d) 3 atoms O
 (e) 9 atoms O

40. (a) 2 atoms H (d) 4 atoms H
(b) 6 atoms H (e) 8 atoms H
(c) 12 atoms H

41. (a) mixture (c) compound
(b) element (d) mixture

42. (a) element (c) element
(b) compound (d) mixture

43. (a) mixture (c) element
(b) compound (d) mixture

44. (a) mixture (c) mixture
(b) element (d) compound

45. (a) $CH_2O$ (c) $C_{25}H_{52}$
(b) $C_4H_9$

46. (a) HO (c) $Na_2Cr_2O_7$
(b) $C_2H_6O$

47. Yes. The gaseous elements are all found on the extreme right of the periodic table. They are the entire last column and in the upper right corner of the table. Hydrogen is the exceptions and located at the upper left of the table.

48. No. The only common liquid elements (at room temperature) are mercury and bromine.

49. $\dfrac{18 \text{ metals}}{36 \text{ elements}} \times 100 = 50\%$ metals

50. $\dfrac{26 \text{ solids}}{36 \text{ elements}} \times 100 = 72\%$ solids

51. (a) 181 atoms/molecule

63 C
88 H
1 Co
14 N
14 O
1 P
181 atoms

(b) $\frac{63\text{ C}}{181\text{ atoms}} \times 100 = 35\%$ C atoms

(c) $\frac{1\text{ CO}}{181\text{ atoms}} = \frac{1}{181}$ metals

52. $HNO_3$ has 5 atoms/molecule

7 dozen = 84

(84 molecules)(5 atoms/molecule) = 420 atoms

or $(7\text{ dz})\left(\frac{12\text{ molecules}}{\text{dz}}\right)\left(\frac{5\text{ atoms}}{\text{molecule}}\right) = 420\text{ atoms}$

53. Each represents eight units of sulfur. In 8 S the atoms are separate and distinct. In $S_8$ the atoms are joined as a unit (molecule).

54. $Ca(H_2PO_4)_2$

$(10\text{ formula units})\left(\frac{4\text{ atoms H}}{\text{formula unit}}\right) = 40\text{ atoms H}$

55. $C_{145}H_{293}O_{168}$

145 C
293 H
168 O
606 atoms/molecule

56. (a) magnesium, manganese, molybdenum, mendeleevium, mercury
(b) carbon, phosphorus, sulfur, selenium, iodine, astatine, boron
(c) sodium, potassium, iron, silver, tin, antimony

57. The conversion is: $\text{cm}^3 \rightarrow \text{L} \rightarrow \text{mg} \rightarrow \text{g} \rightarrow \$$

$$(1 \times 10^{15}\text{ cm}^3)\left(\frac{1\text{ L}}{1000\text{ cm}^3}\right)\left(\frac{4 \times 10^{-4}\text{ mg}}{\text{L}}\right)\left(\frac{1\text{ g}}{1000\text{ mg}}\right)\left(\frac{\$19.40}{\text{g}}\right) = \$8 \times 10^6$$

58.

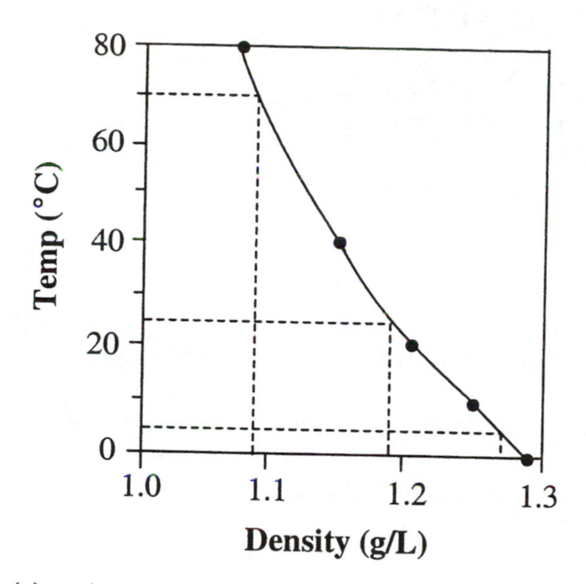


(a) As temperatures decreases, density increases.

(b) approximately 1.28 g/L 5°C
approximately 1.19 g/L 25°C
approximately 1.09 g/L 70°C

CHAPTER 4

# PROPERTIES OF MATTER

1. Solid (melting point of acetic acid is 16.7°C)

2. Solid    102 K = −171°C    melting point of chlorine is −101.6°C

3. Small bubbles appear at each electrode. Gas collects above the electrodes. The system now contains water and gas.

4. Water disappears. Gas appears above each electrode and as bubbles in solution.

5. Physical properties are characteristics which may be determined without altering the composition of the substance. Chemical properties describe the ability of a substance to form new substances by chemical reaction or decomposition.

6. A new substance is always formed during a chemical change, but never formed during physical changes.

7. Potential energy is the energy of position. By the position of an object, it has the potential of movement to a lower energy state. Kinetic energy is the energy matter possesses due to its motion.

8. (a) $118.0°C + 273 = 391.0\ K$
   (b) $(118.0°C)\ 1.8 + 32 = 244.4°F$

9. (a) physical    (d) chemical
   (b) physical    (e) chemical
   (c) physical    (f) chemical

10. (a) chemical    (d) chemical
    (b) physical    (e) chemical
    (c) physical    (f) physical

11. Although the appearance of the platinum wire changed during the heating, the original appearance was restored when the wire cooled. No change in the composition of the platinum could be detected.

12. The copper wire, like the platinum wire, changed to a glowing red color when heated (physical change). Upon cooling, the original appearance of the copper wire was not restored, but a new substance, black copper(II) oxide, had appeared (chemical change).

13. Reactants: copper, oxygen
    Product: copper(II) oxide

14. Reactant: water
    Product: hydrogen, oxygen

15. The kinetic energy is converted to thermal energy (heat), chiefly in the brake system, and eventually dissipated into the atmosphere.

16. The transformation of kinetic energy to thermal energy (heat) is responsible for the fiery reentry of a space vehicle.

17. (a) + (d) +
    (b) – (e) –
    (c) +

18. (a) + (d) +
    (b) – (e) –
    (c) –

19. $E = (m)\ (\text{specific heat}))\ (\Delta t)$
    $= (75\ g)(4.184\ J/g°C)(70.0°C - 20.0°C)$
    $= 1.6 \times 10^4\ J$

20. $E = (m)\ (\text{specific heat}))\ (\Delta t)$
    $= (65\ g)(0.473\ J/g°C)(95°C - 25°C)$
    $= 2.2 \times 10^3\ J$

21. $E = (m)(\text{specific heat})\ (\Delta t)$; change kJ to J

$$\text{specific heat} = \frac{E}{m(\Delta t)} = \frac{5.866 \times 10^3\ J}{(250.0\ g)(100.0°C - 22°C)} = 0.30\ J/g°C$$

22. $E = (m)(\text{specific heat})\ (\Delta t)$; change kg to g and kJ to J

$$\text{specific heat} = \frac{E}{m(\Delta t)} = \frac{3.07 \times 10^4\ J}{(1.00 \times 10^3\ J)(630.0°C - 20.0°C)} = 5.03 \times 10^{-2}\ J/g°C$$

23. heat lost by gold = heat gained by water     $x$ = final temperature

$$(m)(\text{specific heat})(\Delta t) = (m)(\text{specific heat})(\Delta t)$$

$$(325\ g)(0.131\ J/g°C)(427°C - x) = (200.0\ g)(4.184\ J/g°C)(x - 22.0°C)$$

$$18180 \text{ J} - 42.575x \text{ J/°C} = 836.8x \text{ J/°C} - 18410 \text{ J}$$
$$18180 \text{ J} + 18410 \text{ J} = 836.8x + 42.575x$$
$$36590 \text{ J} = 879.4x \text{ J/°C}$$
$$41.6\text{°C} = x$$

24. heat lost by iron = heat gained by water
($x$ = final temperature; $\Delta t$ = change in temperature)

$$m = Vd = (2.0 \text{ L})\left(\frac{1000 \text{ mL}}{1 \text{ L}}\right)\left(1.0 \frac{\text{g}}{\text{mL}}\right) = 2.0 \times 10^3 \text{ g } H_2O$$

$$(\text{m})(\text{specific heat})(\Delta t) = (\text{m})(\text{specific heat})(\Delta t)$$
$$(500.0 \text{ g})(0.473 \text{ J/g°C})(212\text{°C} - x) = (2.0 \times 10^3 \text{ g})(4.184 \text{ J/g°C})(x - 24.0\text{°C})$$
$$50138 \text{ J} - 236.5x \text{ J/°C} = 8368x \text{ J/°C} - 200832 \text{ J}$$
$$18179 \text{ J} + 18409.6 \text{ J} = 836.8x + 42.575x$$
$$250970 \text{ J} = 8604.5x \frac{\text{J}}{\text{°C}}$$
$$x = 29\text{°C}$$
$$\Delta t = 29\text{°C} - 24\text{°C} = 5\text{°C}$$

25. E = (m)(specific heat)($\Delta t$)
= (250. g)(0.096 cal/g°C)(150.0°C − 24°C)
= $3.0 \times 10^3$ cal

26. E = (m)(specific heat)($\Delta t$) change kJ to J $x$ = final temperature
$4.00 \times 10^4$ J = (500.0 g)(4.184 J/g°C)($x$ − 10.0°C)
$4.00 \times 10^4$ J = 2092$x$ J/°C − 20920 J
60,920 J = 2092$x$ J/°C
60,920°C = 2092$x$
29.1°C = $x$

27. E = (m)(specific heat)($\Delta t$)
heat lost by coal = heat gained by water $x$ = mass of coal in g
(5500 cal/g)$x$ = (500.0 g)(1.00 cal/g°C)(90.0°C − 20.0°C) = 35,000 cal

$$x = \frac{35{,}000 \text{ cal}}{5500 \text{ cal/g}} = 6.36 \text{ g coal}$$

28. (7000. cal)(4.184 J/cal) = 29290 J
heat lost by coal = heat gained by water $x$ = mass of coal in g

$4.0\ \text{L}\ H_2O = 4.0 \times 10^3\ \text{g}\ H_2O$

$$\left(2.929 \times 10^4\ \frac{\text{J}}{\text{g}}\right) x = (4.0 \times 10^3\ \text{g}) \left(4.184\ \frac{\text{J}}{\text{g}^\circ\text{C}}\right) (100.0^\circ\text{C} - 20.0^\circ\text{C})$$

$$\left(2.929 \times 10^4\ \frac{\text{J}}{\text{g}}\right) x = 1.3 \times 10^6\ \text{J}$$

$x = 44$ g coal

29. (a) E = (m)(specific heat)(Δt)
(100.0 g)(0.0921 cal/g°C)(100.0°C − 10.0°C) = 829 cal to heat Cu

(b) let $x$ = temperature of Al after adding 829 cal
829 cal = (100.0 g)(0.215 cal/g°C)($x$ − 10.0°C)
829 cal = (21.5 cal/°C)$x$ − 215 cal
$x$ = 48.6°C (final temperature for aluminum)
Therefore the copper gets hotter since it ended up at 100.0°C.
Note: You can figure this out without calculation if you consider the specific heats of the metals. Since the specific heat of copper is much less than aluminum the copper heats more easily.

30. heat lost by iron = heat gained by water $\qquad$ $x$ = initial temperature of iron

$$(\text{m})\ (\text{specific heat})(\Delta t) = (\text{m})(\text{specific heat})(\Delta t)$$

$$(500.0\ \text{g})(0.473\ \text{J/g}^\circ\text{C})(x - 90.0^\circ\text{C}) = (400.\ \text{g})(4.184\ \text{J/g}^\circ\text{C})(90.0^\circ\text{C} - 10.0^\circ\text{C})$$

$$\left(237\ \frac{\text{J}}{^\circ\text{C}}\right) x - 2.13 \times 10^4\ \text{J} = 1.34 \times 10^5\ \text{J}$$

$$\left(237\ \frac{\text{J}}{^\circ\text{C}}\right) x = 1.55 \times 10^5\ \text{J}$$

$$x = \frac{1.55\times10^5\ \text{J}^\circ\text{C}}{237\ \text{J}}$$

$$x = 654^\circ\text{C}$$

31. heat lost by metal = heat gained by water $\qquad$ $x$ = specific heat of metal

$$(\text{m})\ (\text{specific heat})(\Delta t) = (\text{m})(\text{specific heat})(\Delta t)$$

$$(20.0\ \text{g})(x)(2.03^\circ\text{C} - 29.0^\circ\text{C}) = (100.0\ \text{g})(4.184\ \text{J/g}^\circ\text{C})(29.0^\circ\text{C} - 25.0^\circ\text{C})$$

$$4060x\ \text{g}^\circ\text{C} - 580x\ \text{g}^\circ\text{C} = 12134\ \text{J} - 10460\ \text{J}$$

$$(3480\ \text{g}^\circ\text{C})x = 1674\ \text{J}$$

$$x = 0.481\ \text{J/g}^\circ\text{C}$$

32. heat lost = heat gained $\qquad$ $x$ = final temperature

(m) (specific heat)(Δt) = (m)(specific heat)(Δt) $\qquad$ (specific heats are the same)

$$(10.0\ \text{g})\left(4.184\ \frac{\text{J}}{\text{g°C}}\right)(50.0\ \text{°C} - x) = (50.0\ \text{g})\left(4.184\ \frac{\text{J}}{\text{g°C}}\right)(x - 10.0\text{°C})$$

$$500. - 10.0x = 50.0x - 500.$$

$$1.00 \times 10^3\ \text{J} = 60.0x$$

$$16.7\text{°C} = x$$

33. Specific heats for the metals are Fe: 0.473 J/g°C; Cu: 0.385 J/g°C; Al: 0.900 J/g°C. The metal with the lowest specific heat will warm most quickly, therefore, the copper pan heats fastest, and fries the egg fastest.

34. In order for the water to boil both the pan and water must reach 100.0°C. Specific heat for copper is 0.385 J/g °C.

$$(300.0\ \text{g})\left(0.385\ \frac{\text{J}}{\text{g°C}}\right)(100.\text{°C} - 25\text{°C}) + (800.0\ \text{g})\left(4.184\ \frac{\text{J}}{\text{g°C}}\right)(100.\ \text{°C} - 25\text{°C})$$

$= 8.7 \times 10^3\ \text{J} + 2.5 \times 10^5\ \text{J}$

$= 2.6 \times 10^5$ J needed to heat the pan and water

$$(2.6 \times 10^5\ \text{J})\left(\frac{1\ \text{s}}{628\ \text{cal}}\right) = 414\ \text{s} = 6.9\ \text{min} = 6\ \text{min} + 54\ \text{s}$$

The water will boil at 6:06 and 54 s p.m.

35. Heat is transferred from the molecules of coffee on the surface to the air above them. As you blow you move the warmed air molecules away from the surface replacing them with cooler ones which are warmed by the coffee and cool it in a repeating cycle. Inserting a spoon into hot coffee cools the coffee by heat transfer as well. Heat is transferred from the coffee to the spoon lowering the temperature of the coffee and raising the temperature of the spoon.

36. The potatoes will cook at the same rate whether the water boils vigorously or slowly. Once the boiling point is reached the water temperature remains constant. The energy available is the same so the cooking time should be equal.

37. (250 mL)(0.04) = 10 mL fat
(10 mL)(0.8 g/mL) = 8 g fat in a glass of milk

38. mercury + sulfur → compound
The mercury and sulfur react to form a compound since the properties of the product are

different from the properties of either reactant.

$$(100.0 \text{ mL})\left(13.6 \frac{\text{g}}{\text{mL}}\right) = 1.36 \times 10^3 \text{ g mercury}$$

| 1360 g | + | 100.0 g | | 1460 g |
|---|---|---|---|---|
| mercury | + | sulfur | → | compound |

This supports the Law of Conservation of Matter since the mass of the product is equal to the mass of the reactants.

CHAPTER 5

# EARLY ATOMIC THEORY AND STRUCTURE

1.

| | Element | Atomic number |
|---|---|---|
| (a) | copper | 29 |
| (b) | nitrogen | 7 |
| (c) | phosphorus | 15 |
| (d) | radium | 88 |
| (e) | zinc | 30 |

2. The neutron is about 1840 times heavier than an electron.

3.

| **Particle** | **charge** | **mass** |
|---|---|---|
| proton | +1 | 1 amu |
| neutron | 0 | 1 amu |
| electron | −1 | 0 |

4. An atom is electrically neutral, containing equal numbers of protons and electrons. An ion has a charge resulting from an imbalance between the numbers of protons and electrons.

5. Isotopic notation $^{A}_{Z}X$

   Z represents the atomic number
   A represents the mass number

6. Isotopes contain the same number of protons and the same number of electrons. Isotopes have different numbers of neutrons and thus different atomic masses.

7. Gold nuclei are very massive (compared to an alpha particle) and have a large positive charge. As the alpha particles approach the atom, some are deflected by this positive charge. Those approaching a gold nucleus directly are deflected backwards by the massive positive nucleus.

8. (a) The nucleus of the atom contains most of the mass since only a collision with a very dense, massive object would cause an alpha particle to be deflected back towards the source.

(b) The deflection of the alpha particles from their initial flight indicates the nucleus of the atom is also positively charged.

(c) Most alpha particles pass through the fold foil undeflected leading to the conclusion that the atom is mostly empty space.

9. In the atom, protons and neutrons are found within the nucleus. Electrons occupy the remaining space within the atom outside the nucleus.

10. The nucleus of an atom contains nearly all of its mass.

11. (a) Dalton contributed the concept that each element is composed of atoms which are unique, and can combine in ratios of small whole numbers.

(b) Thomson discovered the electron, determined its properties, and found that the mass of a proton is 1840 times the mass of the electron. He developed the Thomson model of the atom.

(c) Rutherford devised the model of a nuclear atom with a positive charge and mass concentrated in the nucleus. Most of the atom is empty space.

12. Electrons: Dalton – electrons are not part of his model
Thomson – electrons are scattered throughout the positive mass of matter in the atom
Rutherford – electrons are located out in space away from the central positive mass

Positive matter: Dalton – no positive matter in his model
Thomson – positive matter is distributed throughout the atom
Rutherford – positive matter is concentrated in a small central nucleus

13. Atomic masses are not whole numbers because:

(a) the neutron and proton do not have identical masses and neither is exactly 1 amu.

(b) most elements exist in nature as a mixture of isotopes with different numbers of neutrons. The atomic mass is the average of all these isotopes.

14. The isotope of C with a mass of 12 is an exact number by definition. The mass of other isotopes such as $^{63}_{29}Cu$ will not be an exact number for reasons given in Exercise 13.

15. The isotopes of hydrogen are protium, deuterium, and tritium.

16. All three isotopes of hydrogen have the same number of protons (1) and electrons (1). They differ in the number of neutrons (0, 1, and 2).

17. $^{52}_{24}Cr$ chromium-52

18. (a) 201 - 121 = 80 protons; electrical charge of the nucleus is +80.

    (b) Hg, mercury

19. All six isotopes have 20 protons and 20 electrons. The number of neutrons are

| Isotope mass number | Neutrons |
|---|---|
| 40 | 20 |
| 42 | 22 |
| 43 | 23 |
| 44 | 24 |
| 46 | 26 |
| 48 | 28 |

20. The most abundant Ca isotope has a mass number of 40. This is certain because 40 is about the average of all the isotopes and has the lowest mass number on the list. An arithmetic average would be between 40 and 48. Since the atomic mass is 40.08, there must be only small amounts of the other isotopes.

21. (a) $^{55}_{26}Fe$ (c) $^{6}_{3}Li$

    (b) $^{26}_{12}Mg$ (d) $^{188}_{79}Au$

22. (a) $^{59}_{27}Co$ Nucleus contains 27 protons and 32 neutrons

    (b) $^{31}_{15}P$ Nucleus contains 15 protons and 16 neutrons

    (c) $^{184}_{74}W$ Nucleus contains 74 protons and 110 neutrons

    (d) $^{235}_{92}U$ Nucleus contains 92 protons and 143 neutrons

23. For each isotope:
(%)(amu) = that portion of the average atomic mass for that isotope.
Add together to obtain the average atomic mass.
(0.2360)(205.9745 amu) + (0.2260)(206.9759 amu) +
(0.5230)(207.9766 amu) + (0.01480)(203.973 amu)
= 48.61 amu + 46.78 amu + 108.8 amu + 3.019 amu
= 207.2 amu = average atomic mass Pb

24. (0.7899)(23.985 amu) + (0.1000)(24.986) amu) + (0.1101)(25.983 amu)
= 18.95 amu + 2.500 amu + 2.861 amu
= 24.31 amu = average atomic mass Mg

25. (0.604)(68.9257 amu) + (1.00 − 0.604)(70.9249 amu)
= 41.6 amu + 28.1 amu
= 69.7 amu = average atomic mass
The element is gallium (see periodic table).

26. (0.300)(6.015 amu) + (0.7000)(7.016 amu)
= 1.805 amu + 4.911 amu = 6.716 amu = average atomic mass of Li sample

27. $V_{sphere} = \frac{4}{3}\pi r^3$ $\quad$ $r_A$ = radius of atom, $r_N$ = radius of nucleus

$$\frac{V_{atom}}{V_{nucleus}} = \frac{\frac{4}{3}\pi r_A^3}{\frac{4}{3}\pi r_N^3} = \frac{r_A^3}{r_N^3} = \frac{(1.0 \times 10^{-8})^3}{(1.0 \times 10^{-13})^3} = 1.0 \times 10^{15} : 1.0$$

(ratio of atomic volume to nuclear volume)

28. $\frac{3.0 \times 10^{-8}\ \text{cm}}{2.0 \times 10^{-13}\ \text{cm}} = 1.5 \times 10^5 : 1.0$ (ratio of the diameter of an Al atom to its nucleus diameter)

29. (a) In Rutherford's experiment the majority of alpha particles passed through the gold foil without deflection. This shows that the atom is mostly empty space and the nucleus is very small.

(b) In Thomson's experiments with the cathode ray tube rays were observed coming from both the anode and the cathode.

(c) In Rutherford's experiment an alpha particle was occasionally dramatically deflected by the nucleus of a gold atom. The direction of deflection showed the nucleus to be positive.

30. (a) These atoms are isotopes.

(b) These atoms are adjacent to each other on the periodic table. The atoms have the same mass.

31. The nucleus of the atoms will have 2 less protons (lowering the nuclear charge by 2) and 2 less neutrons. The nuclear mass will be reduced by 4 amu.

32. $\frac{1.5\ \text{cm}}{0.77 \times 10^{-8}\ \text{cm}} = 1.9 \times 10^8 : 1.0$ $\quad$ ($1.9 \times 10^8$ enlargement)

33. The properties of an element are related to the number of protons and electrons. If the number of neutrons differs, isotopes result. Isotopes of the same element are still the same element even though the nuclear composition of the atoms are different.

34. $^{210}$Bi has 210 − 83 = 127 neutrons → largest number of neutrons/atom
$^{210}$Po has 210 − 84 = 126 neutrons
$^{210}$At has 210 − 85 = 125 neutrons
$^{211}$At has 211 − 85 = 126 neutrons

35. percent of sample $^{60}$Q = $x$
percent of sample $^{63}$Q = $1 - x$

$$\begin{aligned}(x)(60.\text{ amu}) + (1 - x)(63\text{ amu}) &= 61.5\text{ amu}\\ 60.\,x\text{ amu} + 63\text{ amu} - 63x\text{ amu} &= 61.5\text{ amu}\\ 63\text{ amu} - 61.5\text{ amu} &= 63x\text{ amu} - 60x\text{ amu}\\ 1.5 &= 3x\\ 0.50 &= x\\ {}^{60}\text{Q} &= 50\%\\ {}^{63}\text{Q} &= 50\%\end{aligned}$$

36. Compare the mass of the unknown atom to the mass of carbon-12 ($1.9927 \times 10^{-23}$ g)

$$\left(\frac{2.18 \times 10^{-22}\text{ g}}{1.9927 \times 10^{-23}\text{ g C}}\right)(12.0\text{ g C}) = 131\text{ g} \qquad \text{(atomic mass of unknown element)}$$

37. $$(40.0\text{ g})\left(\frac{1\text{ atom}}{6.63 \times 10^{-24}\text{ g}}\right) = 6.03 \times 10^{24}\text{ atoms Ar}$$

38.

| | protons | neutrons | electrons |
|---|---|---|---|
| He | 2 | 2 | 2 |
| C | 6 | 6 | 6 |
| N | 7 | 7 | 7 |
| O | 8 | 8 | 8 |
| Ne | 10 | 10 | 10 |
| Mg | 12 | 12 | 12 |
| Si | 14 | 14 | 14 |
| S | 16 | 16 | 16 |
| Ca | 20 | 20 | 20 |

39.

| | Atomic Number | Mass Number | Symbol | Protons | Neutrons |
|---|---|---|---|---|---|
| (a) | 8 | 16 | O | 8 | 8 |
| (b) | 28 | 58 | Ni | 28 | 30 |
| (c) | 80 | 199 | Hg | 80 | 119 |

40.

| | Element | Symbol | Atomic # | Protons | Neutrons | Electrons |
|---|---|---|---|---|---|---|
| (a) | platinum | $^{195}Pt$ | 78 | 78 | 117 | 78 |
| (b) | phosphorus | $^{30}P$ | 15 | 15 | 15 | 15 |
| (c) | iodine | $^{127}I$ | 53 | 53 | 74 | 53 |
| (d) | krypton | $^{84}Kr$ | 36 | 36 | 48 | 36 |
| (e) | selenium | $^{79}Se$ | 34 | 34 | 45 | 34 |
| (f) | calcium | $^{40}Ca$ | 20 | 20 | 20 | 20 |

CHAPTER 6

# NOMENCLATURE OF INORGANIC COMPOUNDS

1. (a) $NaClO_3$ (d) $Cu_2O$
   (b) $H_2SO_4$ (e) $Zn(HCO_3)_2$
   (c) $Sn(C_2H_3O_2)_2$ (f) $Fe_2(CO_3)_3$

2. No, if elements combine in a one-to-one ratio the charges on their ions must be equal and opposite in sign. They could be +1, −1, or +2, −2 or +3, −3 etc.

3. (a) HBrO hypobromous acid
   $HBrO_2$ bromous acid
   $HBrO_3$ bromic acid
   $HBrO_4$ perbromic acid

   (b) HIO hypoiodous acid
   $HIO_2$ iodous acid
   $HIO_3$ iodic acid
   $HIO_4$ periodic acid

4. The system for naming binary compounds composed of two nonmetals uses the stem of the second element in the formula plus the suffix ide. A prefix is attached to each element indicating the number of atoms of that element in the formula. Thus $N_2O_5$ is named dinitrogen pentoxide and $PCl_3$ is named phosphorus trichloride.

5. Chromium(III) compounds

   (a) $Cr(OH)_3$ (d) $Cr(HCO_3)_3$ (g) $CrPO_4$ (j) $CrF_3$
   (b) $Cr(NO_3)_3$ (e) $Cr_2(CO_3)_3$ (h) $Cr_2(C_2O_4)_3$
   (c) $Cr(NO_2)_3$ (f) $Cr_2(Cr_2O_7)_3$ (i) $Cr_2O_3$

6. Magnesium forms one series of compounds in which the cation is $Mg^{2+}$. Thus the name for $MgCl_2$ (magnesium chloride) does not need to be distinguished from any other compound. Copper forms two series of compounds in which the copper ion is $Cu^+$ and $Cu^{2+}$. Thus the name copper chloride does not indicate which compound is in question. Therefore, $CuCl_2$ is called copper(II) chloride to indicate that the compound contains the $Cu^{2+}$ ion.

7. Formulas of compounds.

   (a) Na and I NaI (d) K and S $K_2S$
   (b) Ba and F $BaF_2$ (e) Cs and Cl CsCl
   (c) Al and O $Al_2O_3$ (f) Sr and Br $SrBr_2$

8. Formulas of compounds.

| | | | | | |
|---|---|---|---|---|---|
| (a) | Ba and O | $BaO$ | (d) | Be and Br | $BeBr_2$ |
| (b) | H and S | $H_2S$ | (e) | Li and Si | $Li_4Si$ |
| (c) | Al and Cl | $AlCl_3$ | (f) | Mg and P | $Mg_3P_2$ |

9.

| | | | |
|---|---|---|---|
| sodium | $Na^+$ | cobalt(II) | $Co^{2+}$ |
| magnesium | $Mg^{2+}$ | barium | $Ba^{2+}$ |
| aluminum | $Al^{3+}$ | hydrogen | $H^+$ |
| copper(II) | $Cu^{2+}$ | mercury(II) | $Hg^{2+}$ |
| iron(II) | $Fe^{2+}$ | tin(II) | $Sn^{2+}$ |
| iron(III) | $Fe^{3+}$ | chromium(III) | $Cr^{3+}$ |
| lead(II) | $Pb^{2+}$ | tin(IV) | $Sn^{4+}$ |
| silver | $Ag^+$ | manganese(II) | $Mn^{2+}$ |
| | | bismuth(III) | $Bi^{3+}$ |

10.

| | | | |
|---|---|---|---|
| chloride | $Cl^-$ | hydrogen sulfate | $HSO_4^-$ |
| bromide | $Br^-$ | hydrogen sulfite | $HSO_3^-$ |
| fluoride | $F^-$ | chromate | $CrO_4^{2-}$ |
| iodide | $I^-$ | carbonate | $CO_3^{2-}$ |
| cyanide | $CN^-$ | hydrogen carbonate | $HCO_3^-$ |
| oxide | $O^{2-}$ | acetate | $C_2H_3O_2^-$ |
| hydroxide | $OH^-$ | chlorate | $ClO_3^-$ |
| sulfide | $S^{2-}$ | permanganate | $MnO_4^-$ |
| sulfate | $SO_4^{2-}$ | oxalate | $C_2O_4^{2-}$ |

11.

| Ion | $Br^-$ | $O^{2-}$ | $NO_3^-$ | $PO_4^{3-}$ | $CO_3^{2-}$ |
|---|---|---|---|---|---|
| $K^+$ | $KBr$ | $K_2O$ | $KNO_3$ | $K_3PO_4$ | $K_2CO_3$ |
| $Mg^{2+}$ | $MgBr_2$ | $MgO$ | $Mg(NO_3)_2$ | $Mg_3(PO_4)_2$ | $MgCO_3$ |
| $Al^{3+}$ | $AlBr_3$ | $Al_2O_3$ | $Al(NO_3)_3$ | $AlPO_4$ | $Al_2(CO_3)_3$ |
| $Zn^{2+}$ | $ZnBr_2$ | $ZnO$ | $Zn(NO_3)_2$ | $Zn_3(PO_4)_2$ | $ZnCO_3$ |
| $H^+$ | $HBr$ | $H_2O$ | $HNO_3$ | $H_3PO_4$ | $H_2CO_3$ |

12.

| Ion | $SO_4^{2-}$ | $Cl^-$ | $AsO_4^{3-}$ | $C_2H_3O_2^-$ | $CrO_4^{2-}$ |
|---|---|---|---|---|---|
| $NH_4^+$ | $(NH_4)_2SO_4$ | $NH_4Cl$ | $(NH_4)_3AsO_4$ | $NH_4C_2H_3O_2$ | $(NH_4)_2CrO_4$ |
| $Ca^{2+}$ | $CaSO_4$ | $CaCl_2$ | $Ca_3(AsO_4)_2$ | $Ca(C_2H_3O_2)_2$ | $CaCrO_4$ |
| $Fe^{3+}$ | $Fe_2(SO_4)_3$ | $FeCl_3$ | $FeAsO_4$ | $Fe(C_2H_3O_2)_3$ | $Fe_2(CrO_4)_3$ |
| $Ag^+$ | $Ag_2SO_4$ | $AgCl$ | $Ag_3AsO_4$ | $AgC_2H_3O_2$ | $Ag_2CrO_4$ |
| $Cu^{2+}$ | $CuSO_4$ | $CuCl_2$ | $Cu_3(AsO_4)_2$ | $Cu(C_2H_3O_2)_2$ | $CuCrO_4$ |

13. Nonmetal binary compound formulas

(a) carbon monoxide, $CO$
(b) sulfur trioxide, $SO_3$
(c) carbon tetrabromide, $CBr_4$
(d) phosphorus trichloride, $PCl_3$
(e) nitrogen dioxide, $NO_2$
(f) dinitrogen pentoxide, $N_2O_5$
(g) iodine monobromide, $IBr$
(h) silicon tetrachloride, $SiCl_4$
(i) phosphorus pentiodide, $PI_5$
(j) diboron trioxide, $B_2O_3$

14. Naming binary nonmetal compounds:

(a) $CO_2$ carbon dioxide
(b) $N_2O$ dinitrogen oxide
(c) $PCl_5$ phosphorus pentachloride
(d) $CCl_4$ carbon tetrachloride
(e) $SO_2$ sulfur dioxide
(f) $N_2O_4$ dinitrogen tetroxide
(g) $P_2O_5$ diphosphorus pentoxide
(h) $OF_2$ oxygen difluoride
(i) $NF_3$ nitrogen trifluoride
(j) $CS_2$ carbon disulfide

15. (a) sodium nitrate, $NaNO_3$
(b) magnesium fluoride, $MgF_2$
(c) barium hydroxide, $Ba(OH)_2$
(d) ammonium sulfate, $(NH_4)_2SO_4$
(e) silver carbonate, $Ag_2CO_3$
(f) calcium phosphate, $Ca_3(PO_4)_2$
(g) potassium nitrite, $KNO_2$
(h) strontium oxide, $SrO$

16. (a) $K_2O$, potassium oxide
(b) $NH_4Br$, ammonium bromide
(c) $CaI_2$, calcium iodide
(d) $BaCO_3$, barium carbonate

(e) $Na_3PO_4$, sodium phosphate
(f) $Al_2O_3$, aluminum oxide
(g) $Zn(NO_3)_2$, zinc nitrate
(h) $Ag_2SO_4$, silver sulfate

17.
(a) $CuCl_2$ copper(II) chloride
(b) $CuBr$ copper(I) bromide
(c) $Fe(NO_3)_2$ iron(II) nitrate
(d) $FeCl_3$ iron(III) chloride
(e) $SnF_2$ tin(II) fluoride
(f) $HgCO_3$ mercury(II) carbonate

18. Formulas:

(a) tin(IV) bromide $SnBr_4$
(b) copper(I) sulfate $Cu_2SO_4$
(c) iron(III) carbonate $Fe_2(CO_3)_3$
(d) mercury(II) nitrite $Hg(NO_2)_2$
(e) titanium(IV) sulfide $TiS_2$
(f) iron(II) acetate $Fe(C_2H_3O_2)_2$

19. Acid formulas:

(a) hydrochloric acid, $HCl$
(b) chloric acid, $HClO_3$
(c) nitric acid, $HNO_3$
(d) carbonic acid, $H_2CO_3$
(e) sulfurous acid, $H_2SO_3$
(f) phosphoric acid, $H_3PO_4$

20. Formulas of acids:

(a) acetic acid, $HC_2H_3O_2$
(b) hydrofluoric acid, $HF$
(c) hypochlorous acid, $HClO$
(d) boric acid, $H_3BO_3$
(e) nitrous acid, $HNO_2$
(f) hydrosulfuric acid, $H_2S$

21. Naming acids:

(a) $HNO_2$, nitrous acid
(b) $H_2SO_4$, sulfuric acid
(c) $H_2C_2O_4$, oxalic acid
(d) $HBr$, hydrobromic acid
(e) $H_3PO_3$, phosphorous acid
(f) $HC_2H_3O_2$, acetic acid
(g) $HF$, hydrofluoric acid
(h) $HBrO_3$, bromic acid

22. Naming acids:

(a) $H_3PO_4$, phosphoric acid
(b) $H_2CO_3$ carbonic acid
(c) $HIO_3$, iodic acid
(d) HCl, hydrochloric acid
(e) HClO, hypochlorous acid
(f) $HNO_3$, nitric acid
(g) HI, hydroiodic acid
(h) $HClO_4$ perchloric acid

23. Formulas for:

(a) silver sulfite $Ag_2SO_3$
(b) cobalt(II) bromide $CoBr_2$
(c) tin(II) hydroxide $Sn(OH)_2$
(d) aluminum sulfate $Al_2(SO_4)_3$
(e) manganese(II) fluoride $MnF_2$
(f) ammonium carbonate $(NH_4)_2CO_3$
(g) chromium(III) oxide $Cr_2O_3$
(h) cupric chloride $CuCl_2$
(i) potassium permanganate $KMnO_4$
(j) barium nitrite $Ba(NO_2)_2$
(k) sodium peroxide $Na_2O_2$
(l) iron(II) sulfate $FeSO_4$
(m) potassium dichromate $K_2Cr_2O_7$
(n) bismuth(III) chromate $Bi_2(CrO_4)_3$

24. Formulas for:

(a) sodium chromate $Na_2CrO_4$
(b) magnesium hydride $MgH_2$
(c) nickel(II) acetate $Ni(C_2H_3O_2)_2$
(d) calcium chlorate $Ca(ClO_3)_2$
(e) lead(II) nitrate $Pb(NO_3)_2$
(f) potassium dihydrogen phosphate $KH_2PO_4$
(g) manganese(II) hydroxide $Mn(OH)_2$
(h) cobalt(II) hydrogen carbonate $Co(HCO_3)_2$
(i) sodium hypochlorite NaClO

| | | |
|---|---|---|
| (j) | arsenic(V) carbonate | $As_2(CO_3)_5$ |
| (k) | chromium(III) sulfite | $Cr_2(SO_3)_3$ |
| (l) | antimony(III) sulfate | $Sb_2(SO_4)_3$ |
| (m) | sodium oxalate | $Na_2C_2O_4$ |
| (n) | potassium thiocyanate | KSCN |

25.

| | Formula | Name |
|---|---|---|
| (a) | $ZnSO_4$ | zinc sulfate |
| (b) | $HgCl_2$ | mercury(II) chloride |
| (c) | $CuCO_3$ | copper(II) carbonate |
| (d) | $Cd(NO_3)_2$ | cadmium nitrate |
| (e) | $Al(C_2H_3O_2)_3$ | aluminum acetate |
| (f) | $CoF_2$ | cobalt(II) fluoride |
| (g) | $Cr(ClO_3)_3$ | chromium(III) chlorate |
| (h) | $Ag_3PO_4$ | silver phosphate |
| (i) | NiS | nickel(II) sulfide |
| (j) | $BaCrO_4$ | barium chromate |

26.

| | Formula | Name |
|---|---|---|
| (a) | $Ca(HSO_4)_2$ | calcium hydrogen sulfate |
| (b) | $As_2(SO_3)_3$ | arsenic(III) sulfite |
| (c) | $Sn(NO_2)_2$ | tin(II) nitrite |
| (d) | $FeBr_3$ | iron(III) bromide |
| (e) | $KHCO_3$ | potassium hydrogen carbonate |
| (f) | $BiAsO_4$ | bismuth(III) arsenate |
| (g) | $Fe(BrO_3)_2$ | iron(II) bromate |
| (h) | $(NH_4)_2HPO_4$ | ammonium monohydrogen phosphate |
| (i) | NaClO | sodium hypochlorite |
| (j) | $KMnO_4$ | potassium permanganate |

27. Formulas for:

| | | |
|---|---|---|
| (a) | baking soda | $NaHCO_3$ |
| (b) | lime | CaO |
| (c) | Epsom salts | $MgSO_4 \cdot 7\ H_2O$ |

(d) muriatic acid $HCl$
(e) vinegar $HC_2H_3O_2$
(f) potash $K_2CO_3$
(g) lye $NaOH$

28. Formulas for:

(a) fool's gold $FeS_2$
(b) saltpeter $NaNO_3$
(c) limestone $CaCO_3$
(d) cane sugar $C_{12}H_{22}O_{11}$
(e) milk of magnesia $Mg(OH)_2$
(f) washing soda $Na_2CO_3 \bullet 10\ H_2O$
(g) grain alcohol $C_2H_5OH$

29. Naming compounds

(a) $Ba(NO_3)_2$, barium nitrate
(b) $NaC_2H_3O_2$, sodium acetate
(c) $PbI_2$, lead(II) iodide
(d) $MgSO_4$, magnesium sulfate
(e) $CdCrO_4$, cadmium chromate
(f) $BiCl_3$, bismuth(III) chloride
(g) NiS, nickel(II) suflide
(h) $Sn(NO_3)_4$, tin(IV) nitrate
(i) $Ca(OH)_2$, calcium hydroxide

30. ide: suffix is used to indicate a binary compound except for hydroxides, cyanides, and ammonium compounds.

ous: used as a suffix to name an acid that has a lower oxygen content than the -ic acid (e.g. $HNO_2$, nitrous acid and $HNO_3$, nitric acid); also used as a suffix to name the lower ionic charge of a multivalent metal (e.g. $Fe^{2+}$, ferrous and $Fe^{3+}$, ferric).

hypo: used as a prefix in naming an acid that has a lower oxygen content that the -ous acid when there are more than two oxyacids with the same elements (e.g. HClO, hypochlorous acid and $HClO_2$, chlorous acid).

per: used as a prefix in naming an acid that has a higher oxygen content than the -ic acid when there are more than two oxyacids with the same elements (e.g. $HClO_4$, perchloric acid and $HClO_3$, chloric acid).

ite: the suffix of a salt derived from an -ous acid.

ate: the suffix of a salt derived from an -ic acid.

Roman numerals: In the Stock System Roman numerals are used in naming compounds that contain metals that may exist in more than one type of cation. The charge of a metal is indicated by a Roman numeral written in parenthesis immediately after the name of the metal.

31. (a) $AgNO_3 + NaCl \rightarrow AgCl + NaNO_3$
    (b) $Fe_2(SO_4)_3 + Ca(OH)_2 \rightarrow Fe(OH)_3 + CaSO_4$
    (c) $KOH + H_2SO_4 \rightarrow K_2SO_4 + H_2O$

32. (a) 50 $e^-$, 50 p (b) 48 $e^-$, 50 p (c) 46 $e^-$, 50 p

33. The formula for a compound must be electrically neutral. Therefore $X = +3$ and $Y = -2$ since in $X_2Y_3$ this would give $2(+3) + 3(-2) = 0$.

34. $Li_3Fe(CN)_6$
    $AlFe(CN)_6$
    $Zn_3[Fe(CN)_6]_2$

35. (a) $N^{3-}$ nitride, $NO_2^-$ nitrite — One has oxygen the other does not, charges on the ions differ.

    (b) $NO_2^-$ nitrite, $NO_3^-$ nitrate — The number of oxygens differ, but the charge is the same.

    (c) $HNO_2$ nitrous acid, $HNO_3$ nitric acid — The number of oxygens in the compounds differ but they both have only one hydrogen.

36.

| | |
|---|---|
| $(NH_4)_2O$ | ammonium oxide |
| $(NH_4)_2CO_3$ | ammonium carbonate |
| $NH_4Cl$ | ammonium chloride |
| $NH_4C_2H_3O_2$ | ammonium acetate |
| $ZnO$ | zinc oxide |
| $ZnCO_3$ | zinc carbonate |
| $ZnCl_2$ | zinc chloride |
| $Zn(C_2H_3O_2)_2$ | zinc acetate |

| | |
|---|---|
| $H_2CO_3$ | carbonic acid |
| $HC_2H_3O_2$ | acetic acid |
| $HCl$ | hydrochloric acid |
| $H_2O$ | water |

CHAPTER 7

# QUANTITATIVE COMPOSITION OF COMPOUNDS

1. A mole is an amount of substance containing the same number of particles as there are atoms in exactly 12 g of carbon-12.

   It is Avogadro's number ($6.022 \times 10^{23}$) of anything (atoms, molecules, ping-pong balls, etc.).

2. A mole of gold (197.0 g) has a higher mass than a mole of potassium (39.10 g).

3. Both samples (Au and K) contain the same number of atoms. ($6.022 \times 10^{23}$).

4. A mole of gold atoms contains more electrons than a mole of potassium atoms, as each Au atom has 79 $e^-$, while each K atom has only 19 $e^-$.

5. No. Avogadro's number is a constant. The mole is defined as Avogadro's number of C-12 atoms. Changing the atomic mass to 50 amu would change only the size of the atomic mass unit, not Avogadro's number.

6. $6.022 \times 10^{23}$

7. There are Avogadro's number of particles in one mole of substance.

8. (a) A mole of oxygen atoms (O) contains $\mathbf{6.022 \times 10^{23}}$ atoms.

   (b) A mole of oxygen molecules ($O_2$) contains $\mathbf{6.022 \times 10^{23}}$ molecules.

   (c) A mole of oxygen molecules ($O_2$) contains $\mathbf{1.204 \times 10^{24}}$ atoms.

   (d) A mole of oxygen atoms (O) has a mass of **16.00** g.

   (e) A mole of oxygen molecules ($O_2$) has a mass of **32.00** g.

9. $6.022 \times 10^{23}$ molecules in one molar mass of $H_2SO_4$.
   $4.215 \times 10^{24}$ atoms in one molar mass of $H_2SO_4$.

10. Choosing 100.0 g of a compound allows us to simply drop the % sign and use grams for each percent.

11. Molar masses

| | | | |
|---|---|---|---|
| (a) KBr | 1 | K | 39.10 g |
| | 1 | Br | 79.90 g |
| | | | 119.0 g |
| (b) $Na_2SO_4$ | 2 | Na | 45.98 g |
| | 1 | S | 32.07 g |
| | 4 | O | 64.00 g |
| | | | 142.1 g |
| (c) $Pb(NO_3)_2$ | 1 | Pb | 207.2 g |
| | 2 | N | 28.02 g |
| | 6 | O | 96.00 g |
| | | | 331.2 g |
| (d) $C_2H_5OH$ | 2 | C | 24.02 g |
| | 6 | H | 6.048 g |
| | 1 | O | 16.00 g |
| | | | 46.07 g |
| (e) $HC_2H_3O_2$ | 4 | H | 4.032 g |
| | 2 | C | 24.02 g |
| | 2 | O | 32.00 g |
| | | | 60.05 g |
| (f) $Fe_3O_4$ | 3 | Fe | 167.6 g |
| | 4 | O | 64.00 g |
| | | | 231.6 g |
| (g) $C_{12}H_{22}O_{11}$ | 12 | C | 144.1 g |
| | 22 | H | 22.18 g |
| | 11 | O | 176.0 g |
| | | | 342.3 g |
| (h) $Al_2(SO_4)_3$ | 2 | Al | 53.96 g |
| | 3 | S | 96.21 g |
| | 12 | O | 192.0 g |
| | | | 342.2 g |

(i) $(NH_4)_2HPO_4$

| | | |
|---|---|---|
| 9 | H | 9.072 g |
| 2 | N | 28.02 g |
| 1 | P | 30.97 g |
| 4 | O | 64.00 g |
| | | 132.1 g |

12. Molar masses

(a) $NaOH$

| | | |
|---|---|---|
| 1 | Na | 22.99 g |
| 1 | O | 16.00 g |
| 1 | H | 1.008 g |
| | | 40.00 g |

(b) $Ag_2CO_3$

| | | |
|---|---|---|
| 2 | Ag | 215.8 g |
| 1 | C | 12.01 g |
| 3 | O | 48.00 g |
| | | 275.8 g |

(c) $Cr_2O_3$

| | | |
|---|---|---|
| 2 | Cr | 104.0 g |
| 3 | O | 48.00 g |
| | | 152.0 g |

(d) $(NH_4)_2CO_3$

| | | |
|---|---|---|
| 2 | N | 28.02 g |
| 8 | H | 8.064 g |
| 1 | C | 12.01 g |
| 4 | O | 48.00 g |
| | | 96.09 g |

(e) $Mg(HCO_3)_2$

| | | |
|---|---|---|
| 1 | Mg | 24.31 g |
| 2 | H | 2.016 g |
| 2 | C | 24.02 g |
| 6 | O | 96.00 g |
| | | 146.3 g |

(f) $C_6H_5COOH$

| | | |
|---|---|---|
| 7 | C | 84.07 g |
| 6 | H | 6.048 g |
| 2 | O | 32.00 g |
| | | 122.1 g |

(g) $C_6H_{12}O_6$

| | | |
|---|---|---|
| 6 | C | 72.06 g |
| 12 | H | 12.10 g |
| 6 | O | 96.00 g |
| | | 180.2 g |

(h) $K_4Fe(CN)_6$

| | | |
|---|---|---|
| 4 | K | 156.4 g |
| 1 | Fe | 55.85 g |
| 6 | C | 72.06 g |
| 6 | N | 84.06 g |
| | | 368.4 g |

(i) $BaCl_2 \bullet 2\,H_2O$

| | | |
|---|---|---|
| 1 | Ba | 137.3 g |
| 2 | Cl | 70.90 g |
| 4 | H | 4.032 g |
| 2 | O | 32.00 g |
| | | 244.2 g |

13. Moles of atoms.

(a) $(22.5\text{ g Zn})\left(\dfrac{1\text{ mol}}{65.39\text{ g}}\right) = 0.344\text{ mol Zn}$

(b) $(0.688\text{ g Mg})\left(\dfrac{1\text{ mol}}{24.31\text{ g}}\right) = 2.83 \times 10^{-2}\text{ mol Mg}$

(c) $(4.5 \times 10^{22}\text{ atoms Cu})\left(\dfrac{1\text{ mol}}{6.022 \times 10^{23}\text{ atoms}}\right) = 7.5 \times 10^{-2}\text{ mol Cu}$

(d) $(382\text{ g Co})\left(\dfrac{1\text{ mol}}{58.93\text{ g}}\right) = 6.48\text{ mol Co}$

(e) $(0.055\text{ g Sn})\left(\dfrac{1\text{ mol}}{118.7\text{ g}}\right) = 4.6 \times 10^{-4}\text{ mol Sn}$

(f) $(8.5 \times 10^{24}\text{ molecules } N_2)\left(\dfrac{2\text{ atoms N}}{1\text{ molecule } N_2}\right)\left(\dfrac{1\text{ mol N atoms}}{6.022 \times 10^{23}\text{ atoms N}}\right) = 28\text{ mol N atoms}$

14. Number of moles.

(a) $(25.0\text{ g NaOH})\left(\dfrac{1\text{ mol}}{40.00\text{ g}}\right) = 0.625\text{ mol NaOH}$

(b) $(44.0\text{ g } Br_2)\left(\dfrac{1\text{ mol}}{159.8\text{ g}}\right) = 0.275\text{ mol } Br_2$

(c) $(0.684\text{ g } MgCl_2)\left(\dfrac{1\text{ mol}}{95.21\text{ g}}\right) = 7.18 \times 10^{-3}\text{ mol } MgCl_2$

(d) $(14.8\text{ g }CH_3OH)\left(\frac{1\text{ mol}}{32.04\text{ g}}\right) = 0.462\text{ mol }CH_3OH$

(e) $(2.88\text{ g }Na_2SO_4)\left(\frac{1\text{ mol}}{142.1\text{ g}}\right) = 2.03 \times 10^{-2}\text{ mol }Na_2SO_4$

(f) $(4.20\text{ lb }ZnI_2)\left(\frac{453.6\text{ g}}{1\text{ lb}}\right)\left(\frac{1\text{ mol}}{319.2\text{ g}}\right) = 5.97\text{ mol }ZnI_2$

15. Number of grams.

(a) $(0.550\text{ mol Au})\left(\frac{197.0\text{ g}}{1\text{ mol}}\right) = 108\text{ g Au}$

(b) $(15.8\text{ mol }H_2O)\left(\frac{18.02\text{ g}}{\text{mol}}\right) = 285\text{ g }H_2O$

(c) $(12.5\text{ mol }Cl_2)\left(\frac{70.90\text{ g}}{\text{mol}}\right) = 886\text{ g }Cl_2$

(d) $(3.15\text{ mol }NH_4NO_3)\left(\frac{80.05\text{ g}}{\text{mol}}\right) = 252\text{ g }NH_4NO_3$

16. Number of grams.

(a) $(4.25 \times 10^{-4}\text{ mol }H_2SO_4)\left(\frac{98.09\text{ g}}{\text{mol}}\right) = 0.0417\text{ g }H_2SO_4$

(b) $(4.5 \times 10^{22}\text{ molecules }CCl_4)\left(\frac{1\text{ mol}}{6.022 \times 10^{23}\text{ molecules}}\right)\left(\frac{153.8\text{ g}}{\text{mol}}\right) = 11\text{ g }CCl_4$

(c) $(0.00255\text{ mol Ti})\left(\frac{47.87\text{ g}}{\text{mol}}\right) = 0.122\text{ g Ti}$

(d) $(1.5 \times 10^{16}\text{ atoms S})\left(\frac{32.07\text{ g}}{6.022 \times 10^{23}\text{ atoms}}\right) = 8.0 \times 10^{-7}\text{ g S}$

17. Number of molecules.

(a) $(1.26\text{ mol }O_2)\left(\frac{6.022 \times 10^{23}\text{ molecules}}{\text{mol}}\right) = 7.59 \times 10^{23}\text{ molecules }O_2$

(b) $(0.56\text{ mol }C_6H_6)\left(\frac{6.022 \times 10^{23}\text{ molecules}}{\text{mol}}\right) = 3.4 \times 10^{23}\text{ molecules }C_6H_6$

(c) $(16.0 \text{ g CH}_4)\left(\dfrac{6.022 \times 10^{23} \text{ molecules}}{16.04 \text{ g}}\right) = 6.01 \times 10^{23} \text{ molecules CH}_4$

(d) $(1000. \text{ g HCl})\left(\dfrac{6.022 \times 10^{23} \text{ molecules}}{36.46 \text{ g}}\right) = 1.652 \times 10^{25} \text{ molecules HCl}$

18. (a) $(1.75 \text{ mol Cl}_2)\left(\dfrac{6.022 \times 10^{23} \text{ molecules}}{\text{mol}}\right) = 1.05 \times 10^{24} \text{ molecules Cl}_2$

(b) $(0.27 \text{ mol C}_2\text{H}_6\text{O})\left(\dfrac{6.022 \times 10^{23} \text{ molecules}}{\text{mol}}\right) = 1.6 \times 10^{23} \text{ molecules C}_2\text{H}_6\text{O}$

(c) $(12.0 \text{ g CO}_2)\left(\dfrac{6.022 \times 10^{23} \text{ molecules}}{44.01 \text{ g}}\right) = 1.64 \times 10^{23} \text{ molecules CO}_2$

(d) $(100. \text{ g CH}_4)\left(\dfrac{6.022 \times 10^{23} \text{ molecules}}{16.04 \text{ g}}\right) = 3.75 \times 10^{24} \text{ molecules CH}_4$

19. Number of grams.

(a) $(1 \text{ atom Pb})\left(\dfrac{207.2 \text{ g}}{6.022 \times 10^{23} \text{ atoms}}\right) = 3.441 \times 10^{-22} \text{ g Pb}$

(b) $(1 \text{ atom Ag})\left(\dfrac{107.9 \text{ g}}{6.022 \times 10^{23} \text{ atoms}}\right) = 1.792 \times 10^{-22} \text{ g Ag}$

(c) $(1 \text{ molecule H}_2\text{O})\left(\dfrac{18.02 \text{ g}}{6.022 \times 10^{23} \text{ molecules}}\right) = 2.992 \times 10^{-23} \text{ g H}_2\text{O}$

(d) $(1 \text{ molecule C}_3\text{H}_5(\text{NO}_3)_3)\left(\dfrac{227.1 \text{ g}}{6.022 \times 10^{23} \text{ molecules}}\right) = 3.771 \times 10^{-22} \text{ g C}_3\text{H}_5(\text{NO}_3)_3$

20. (a) $(1 \text{ atom Au})\left(\dfrac{197.0 \text{ g}}{6.022 \times 10^{23} \text{ atoms}}\right) = 3.271 \times 10^{-22} \text{ g Au}$

(b) $(1 \text{ atom U})\left(\dfrac{238.0 \text{ g}}{6.022 \times 10^{23} \text{ atoms}}\right) = 3.952 \times 10^{-22} \text{ U}$

(c) $(1 \text{ molecule NH}_3)\left(\dfrac{17.03 \text{ g}}{6.022 \times 10^{23} \text{ molecules}}\right) = 2.828 \times 10^{-23} \text{ g NH}_3$

(d) $(1 \text{ molecule C}_6\text{H}_4(\text{NH}_2)_2)\left(\dfrac{108.1 \text{ g}}{6.022 \times 10^{23} \text{ molecules}}\right) = 1.795 \times 10^{-22} \text{ g C}_6\text{H}_4(\text{NH}_2)_2$

21\. (a) $(8.66 \text{ mol Cu})\left(\frac{63.55 \text{ g}}{\text{mol}}\right) = 550. \text{ g Cu}$

(b) $(125 \text{ mol Au})\left(\frac{197.0 \text{ g}}{\text{mol}}\right)\left(\frac{1 \text{ kg}}{1000 \text{ g}}\right) = 24.6 \text{ kg Au}$

(c) $(10 \text{ atoms C})\left(\frac{1 \text{ mol}}{6.022 \times 10^{23} \text{ atoms}}\right) = 2 \times 10^{-23} \text{ mol C}$

(d) $(5000 \text{ molecules } CO_2)\left(\frac{1 \text{ mol}}{6.022 \times 10^{23} \text{ molecules}}\right) = 8 \times 10^{-21} \text{ mol } CO_2$

22\. (a) $(28.4 \text{ g S})\left(\frac{1 \text{ mol}}{32.07 \text{ g}}\right) = 0.886 \text{ mol S}$

(b) $(2.50 \text{ kg NaCl})\left(\frac{1000 \text{ g}}{\text{kg}}\right)\left(\frac{1 \text{ mol}}{58.44 \text{ g}}\right) = 42.8 \text{ mol NaCl}$

(c) $(42.4 \text{ g Mg})\left(\frac{6.022 \times 10^{23} \text{ atoms}}{24.31 \text{ g}}\right) = 1.05 \times 10^{24} \text{ atoms Mg}$

(d) $(485 \text{ mL } Br_2)\left(\frac{3.12 \text{ g}}{\text{mL}}\right)\left(\frac{1 \text{ mol}}{159.8 \text{ g}}\right) = 9.47 \text{ mol } Br_2$

23\. One mole of carbon disulfide ($CS_2$) contains:

(a) $6.022 \times 10^{23}$ molecules of $CS_2$

(b) $(6.022 \times 10^{23} \text{ molecules of } CS_2)\left(\frac{1 \text{ C atom}}{1 \text{ molecule } CS_2}\right) = 6.022 \times 10^{23} \text{ C atoms}$

(c) $(6.022 \times 10^{23} \text{ molecules of } CS_2)\left(\frac{2 \text{ S atoms}}{1 \text{ molecule } CS_2}\right) = 1.204 \times 10^{24} \text{ S atoms}$

(d) $(6.022 \times 10^{23}$ atoms$) + (1.204 \times 10^{24}$ atoms $= 1.806 \times 10^{24}$ atoms

24\. One mole of ammonia ($NH_3$) contains

(a) $6.022 \times 10^{23}$ molecules of $NH_3$

(b) $(6.022 \times 10^{23} \text{ molecules of } NH_3)\left(\frac{1 \text{ N atom}}{\text{molecule } NH_3}\right) = 6.022 \times 10^{23} \text{ N atoms}$

(c) $(6.022 \times 10^{23} \text{ molecules of } NH_3)\left(\frac{3 \text{ H atoms}}{\text{molecule } NH_3}\right) = 1.807 \times 10^{24} \text{ H atoms}$

(d) $(6.022 \times 10^{23} \text{ atoms}) + (1.807 \times 10^{24} \text{ atoms} = 2.409 \times 10^{24} \text{ atoms}$

25. Atoms of oxygen in:

(a) $(16.0 \text{ g } O_2)\left(\frac{1 \text{ mol}}{32.00 \text{ g}}\right)\left(\frac{2 \text{ mol O}}{1 \text{ mol } O_2}\right)\left(\frac{6.022 \times 10^{23} \text{ atoms}}{\text{mol}}\right) = 6.02 \times 10^{23} \text{ atoms O}$

(b) $(0.622 \text{ mol MgO})\left(\frac{1 \text{ mol O}}{\text{mol MgO}}\right)\left(\frac{6.022 \times 10^{23} \text{ atoms}}{\text{mol}}\right) = 3.75 \times 10^{23} \text{ atoms O}$

(c) $(6.00 \times 10^{22} \text{ molecules } C_6H_{12}O_6)\left(\frac{6 \text{ atoms O}}{\text{molecule } C_6H_{12}O_6}\right) = 3.60 \times 10^{23} \text{ atoms O}$

26. Atoms of oxygen in:

(a) $(5.0 \text{ mol } MnO_2)\left(\frac{2 \text{ mol O}}{\text{mol } MnO_2}\right)\left(\frac{6.022 \times 10^{23} \text{ atoms}}{\text{mol}}\right) = 6.0 \times 10^{24} \text{ atoms O}$

(b) $(255 \text{ g } MgCO_3)\left(\frac{1 \text{ mol}}{84.32 \text{ g}}\right)\left(\frac{3 \text{ mol O}}{\text{mol } MgCO_3}\right)\left(\frac{6.022 \times 10^{23} \text{ atoms}}{\text{mol}}\right) = 5.46 \times 10^{24} \text{ atoms O}$

(c) $(5.0 \times 10^{18} \text{ molecules } H_2O)\left(\frac{1 \text{ atom O}}{\text{molecule } H_2O}\right) = 5.0 \times 10^{18} \text{ atoms O}$

27. The number of grams of:

(a) silver in 25.0 g AgBr

$(25.0 \text{ g AgBr})\left(\frac{107.9 \text{ g Ag}}{187.8 \text{ g AgBr}}\right) = 14.4 \text{ g Ag}$

(b) nitrogen in 6.34 mol $(NH_4)_3PO_4$

$(6.34 \text{ mol } (NH_4)_3PO_4)\left(\frac{42.03 \text{ g N}}{\text{mol } (NH_4)_3PO_4}\right) = 266 \text{ g N}$

(c) oxygen in $8.45 \times 10^{22}$ molecules $SO_3$
The conversion is: molecules $SO_3 \rightarrow$ mol $SO_3 \rightarrow$ g O

$(8.45 \times 10^{22} \text{ molecules } SO_3)\left(\frac{1 \text{ mol}}{6.022 \times 10^{23} \text{ molecules}}\right)\left(\frac{48.00 \text{ g O}}{\text{mol } SO_3}\right) = 6.74 \text{ g O}$

28. The number of grams of:

(a) chlorine in 5.00 g $PbCl_2$

$$(5.00 \text{ g } PbCl_2)\left(\frac{70.90 \text{ g Cl}}{278.1 \text{ g } PbCl_2}\right) = 1.27 \text{ g Cl}$$

(b) hydrogen in 4.50 g $H_2SO_4$

$$(4.50 \text{ g } H_2SO_4)\left(\frac{2.016 \text{ g H}}{98.09 \text{ g } H_2SO_4}\right) = 9.25 \times 10^{-2} \text{ g H}$$

(c) Grams of hydrogen in $5.45 \times 10^{22}$ molecules $NH_3$
The conversion is: molecules $NH_3 \rightarrow$ moles $NH_3 \rightarrow$ g H

$$(5.45 \times 10^{22} \text{ molecules } NH_3)\left(\frac{1 \text{ mol}}{6.022 \times 10^{23} \text{ molecules}}\right)\left(\frac{3.024 \text{ g H}}{\text{mol } NH_3}\right) = 2.74 \text{ g H}$$

29. Percent composition

(a) NaBr

| | | |
|---|---|---|
| Na | 22.99 g | |
| Br | 79.90 g | |
| | 102.9 g | |

$$\left(\frac{22.99 \text{ g}}{102.9 \text{ g}}\right)(100) = 22.34\% \text{ Na}$$

$$\left(\frac{79.90 \text{ g}}{102.9 \text{ g}}\right)(100) = 77.65\% \text{ Br}$$

(b) $KHCO_3$

| | |
|---|---|
| K | 39.10 g |
| H | 1.008 g |
| 3 O | 48.00 g |
| C | 12.01 g |
| | 100.1 g |

$$\left(\frac{39.10 \text{ g}}{100.1 \text{ g}}\right)(100) = 39.06\% \text{ K}$$

$$\left(\frac{1.008 \text{ g}}{100.1 \text{ g}}\right)(100) = 1.007\% \text{ H}$$

$$\left(\frac{12.01 \text{ g}}{100.1 \text{ g}}\right)(100) = 12.00\% \text{ C}$$

$$\left(\frac{48.00 \text{ g}}{100.1 \text{ g}}\right)(100) = 47.95\% \text{ O}$$

(c) $FeCl_3$

| | |
|---|---|
| Fe | 55.85 g |
| 3 Cl | 106.4 g |
| | 162.3 g |

$$\left(\frac{55.85 \text{ g}}{162.3 \text{ g}}\right)(100) = 34.41\% \text{ Fe}$$

$$\left(\frac{106.4 \text{ g}}{162.3 \text{ g}}\right)(100) = 65.56\% \text{ Cl}$$

(d) $SiCl_4$

| | | | |
|---|---|---|---|
| Si | 28.09 g | $\left(\frac{28.09\text{ g}}{169.9\text{ g}}\right)(100) = 16.53\%$ | Si |
| 4 Cl | 141.8 g | $\left(\frac{141.8\text{ g}}{169.9\text{ g}}\right)(100) = 83.46\%$ | Cl |
| | 169.9 g | | |

(e) $Al_2(SO_4)_3$

| | | | |
|---|---|---|---|
| 2 Al | 53.96 g | $\left(\frac{53.96\text{ g}}{342.2\text{ g}}\right)(100) = 15.77\%$ | Al |
| 3 S | 96.21 g | $\left(\frac{96.21\text{ g}}{342.2\text{ g}}\right)(100) = 28.12\%$ | S |
| 12 O | 192.0 g | $\left(\frac{192.0\text{ g}}{342.2\text{ g}}\right)(100) = 56.11\%$ | O |
| | 342.2 g | | |

(f) $AgNO_3$

| | | | |
|---|---|---|---|
| Ag | 107.9 g | $\left(\frac{107.9\text{ g}}{169.9\text{ g}}\right)(100) = 63.51\%$ | Ag |
| N | 14.01 g | $\left(\frac{14.01\text{ g}}{169.9\text{ g}}\right)(100) = 8.246\%$ | N |
| 3 O | 48.00 g | $\left(\frac{48.00\text{ g}}{169.9\text{ g}}\right)(100) = 28.25\%$ | O |
| | 169.9 g | | |

30. Percent composition

(a) $ZnCl_2$

| | | | |
|---|---|---|---|
| Zn | 65.39 g | $\left(\frac{65.39\text{ g}}{136.3\text{ g}}\right)(100) = 47.98\%$ | Zn |
| 2 Cl | 70.90 g | $\left(\frac{70.90\text{ g}}{136.3\text{ g}}\right)(100) = 52.02\%$ | Cl |
| | 136.3 g | | |

(b) $NH_4C_2H_3O_2$

| | | | |
|---|---|---|---|
| N | 14.01 g | $\left(\frac{14.01\text{ g}}{77.09\text{ g}}\right)(100) = 18.17\%$ | N |
| 7 H | 7.056 g | $\left(\frac{7.056\text{ g}}{77.09\text{ g}}\right)(100) = 9.153\%$ | H |
| 2 C | 24.02 g | $\left(\frac{24.02\text{ g}}{77.09\text{ g}}\right)(100) = 31.16\%$ | C |
| 2 O | 32.00 g | $\left(\frac{32.00\text{ g}}{77.09\text{ g}}\right)(100) = 41.51\%$ | O |
| | 77.09 g | | |

(c) $MgP_2O_7$

| | | |
|---|---|---|
| Mg | 24.31 g | $\left(\frac{24.31\text{ g}}{198.3\text{ g}}\right)(100) = 12.26\%$ Mg |
| 2 P | 61.94 g | $\left(\frac{61.94\text{ g}}{198.3\text{ g}}\right)(100) = 31.24\%$ P |
| 7 O | 112.0 g | $\left(\frac{112.0\text{ g}}{198.3\text{ g}}\right)(100) = 56.48\%$ O |
| | 198.3 g | |

(d) $(NH_4)_2SO_4$

| | | |
|---|---|---|
| 2 N | 28.02 g | $\left(\frac{28.02\text{ g}}{132.2\text{ g}}\right)(100) = 21.20\%$ N |
| 8 H | 8.064 g | $\left(\frac{8.064\text{ g}}{132.2\text{ g}}\right)(100) = 6.100\%$ H |
| S | 32.07 g | $\left(\frac{32.07\text{ g}}{132.2\text{ g}}\right)(100) = 24.26\%$ S |
| 4 O | 64.00 g | $\left(\frac{64.00\text{ g}}{132.2\text{ g}}\right)(100) = 48.41\%$ O |
| | 132.2 g | |

(e) $Fe(NO_3)_3$

| | | |
|---|---|---|
| Fe | 55.85 g | $\left(\frac{55.85\text{ g}}{241.9\text{ g}}\right)(100) = 23.09\%$ Fe |
| 3 N | 42.03 g | $\left(\frac{42.03\text{ g}}{241.9\text{ g}}\right)(100) = 17.37\%$ N |
| 9 O | 144.0 g | $\left(\frac{144.0\text{ g}}{241.9\text{ g}}\right)(100) = 59.53\%$ O |
| | 241.9 g | |

(f) $ICl_3$

| | | |
|---|---|---|
| I | 126.9 g | $\left(\frac{126.9\text{ g}}{233.3\text{ g}}\right)(100) = 54.39\%$ I |
| 3 Cl | 106.4 g | $\left(\frac{106.4\text{ g}}{233.3\text{ g}}\right)(100) = 45.61\%$ Cl |
| | 233.3 g | |

31. Percent of iron

(a) FeO

| | | |
|---|---|---|
| Fe | 55.85 g | $\left(\frac{55.85\text{ g}}{71.85\text{ g}}\right)(100) = 77.73\%$ Fe |
| O | 16.00 g | |
| | 71.85 g | |

(b) $Fe_2O_3$

| | |
|---|---|
| 2 Fe | 111.7 g |
| 3 O | 48.00 g |
| | 159.7 g |

$$\left(\frac{111.7\ g}{159.7\ g}\right)(100) = 69.94\%\ Fe$$

(c) $Fe_3O_4$

| | |
|---|---|
| 3 Fe | 167.4 g |
| 4 O | 64.00 g |
| | 231.6 g |

$$\left(\frac{167.6\ g}{231.6\ g}\right)(100) = 72.37\%\ Fe$$

(d) $K_4Fe(CN)_6$

| | |
|---|---|
| Fe | 55.85 g |
| 4 K | 156.4 g |
| 6 C | 72.06 g |
| 6 N | 84.06 g |
| | 368.4 g |

$$\left(\frac{55.85\ g}{368.4\ g}\right)(100) = 15.16\%\ Fe$$

32. Percent chlorine

(a) KCl

| | |
|---|---|
| K | 39.10 g |
| Cl | 35.45 g |
| | 74.55 g |

$$\left(\frac{35.45\ g}{74.55\ g}\right)(100) = 47.55\%\ Cl$$

(b) $BaCl_3$

| | |
|---|---|
| Ba | 137.3 g |
| 2 Cl | 70.90 g |
| | 208.2 g |

$$\left(\frac{70.90\ g}{208.2\ g}\right)(100) = 34.05\%\ Cl$$

(c) $SiCl_4$

| | |
|---|---|
| Si | 28.09 g |
| 4 Cl | 141.8 g |
| | 169.9 g |

$$\left(\frac{141.8\ g}{169.9\ g}\right)(100) = 83.46\%\ Cl$$

(d) LiCl

| | |
|---|---|
| Li | 6.941 g |
| Cl | 35.45 g |
| | 42.39 g |

$$\left(\frac{35.45\ g}{42.39\ g}\right)(100) = 83.63\%\ Cl$$

Highest % Cl is in LiCl; lowest % Cl is in $BaCl_2$

33. Percent composition of an oxide

14.20 g oxide
−6.20 g P
8.00 g oxygen

$$\left(\frac{6.20\ g}{14.20\ g}\right)(100) = 43.7\%\ P$$

$$\left(\frac{8.00\ g}{14.20\ g}\right)(100) = 56.3\%\ O$$

34. Percent composition of ethylene chloride

| | |
|---|---|
| 6.00 g C | $\left(\dfrac{6.00\text{ g}}{24.75\text{ g}}\right)(100) = 24.2\%\ \text{C}$ |
| 1.00 g H | $\left(\dfrac{1.00\text{ g}}{24.75\text{ g}}\right)(100) = 4.04\%\ \text{H}$ |
| 17.75 g Cl | $\left(\dfrac{17.75\text{ g}}{24.75\text{ g}}\right)(100) = 71.72\%\ \text{Cl}$ |
| 24.75 g total | |

35. (a) $H_2O$ (by inspection of formulas)
    (b) $N_2O_3$ (by inspection of formulas)
    (c) equal (by inspection of formulas)

36. (a) $KClO_3$ (by inspection of formulas)
    (b) $KHSO_4$ (by inspection of formulas)
    (c) $Na_2CrO_4$ (by inspection of formulas)

37. Empirical formulas from percent composition.

(a) Step 1. Express each element as grams/100 g material.

63.6% N = 63.6 g N/100 g material

36.4% O = 36.4 g O/100 g material

Step 2. Calculate the relative moles of each element.

$$(63.6\text{ g N})\left(\frac{1\text{ mol}}{14.01\text{ g}}\right) = 4.54\text{ mol N}$$

$$(36.4\text{ g O})\left(\frac{1\text{ mol}}{16.00\text{ g}}\right) = 2.28\text{ mol O}$$

Step 3: Change these moles to whole numbers by dividing each by the smaller number.

$$\frac{4.54\text{ mol N}}{2.28} = 1.99\text{ mol N}$$

$$\frac{2.28 \text{ mol O}}{2.28} = 1.00 \text{ mol O}$$

The simplest ratio of N:O is 2:1. The empirical formula, therefore, is $N_2O$.

(b) 46.7% N, 53.3% O

$$(46.7 \text{ g N})\left(\frac{1 \text{ mol}}{14.01 \text{ g}}\right) = 3.33 \text{ mol N} \qquad \frac{3.33}{3.33} = 1.00 \text{ mol N}$$

$$(53.3 \text{ g O})\left(\frac{1 \text{ mol}}{16.00 \text{ g}}\right) = 3.33 \text{ mol O} \qquad \frac{3.33}{3.33} = 1.00 \text{ mol O}$$

The empirical formula is NO.

(c) 25.9% N, 71.4% O

$$(25.9 \text{ g N})\left(\frac{1 \text{ mol}}{14.01 \text{ g}}\right) = 1.85 \text{ mol N} \qquad \frac{1.85}{1.85} = 1.00 \text{ mol N}$$

$$(74.1 \text{ g O})\left(\frac{1 \text{ mol}}{16.00 \text{ g}}\right) = 4.63 \text{ mol O} \qquad \frac{4.63}{1.85} = 2.5 \text{ mol O}$$

Since these values are not whole numbers, multiply each by 2 to change them to whole numbers.

(1.00 mol N)(2) = 2.00 mol N; (2.5 mol O)(2) = 5.00 mol O

The empirical formula is $N_2O_5$.

(d) 43.4% Na, 11.3% C, 45.3% O

$$(43.4 \text{ g Na})\left(\frac{1 \text{ mol}}{22.99 \text{ g}}\right) = 1.89 \text{ mol Na} \qquad \frac{1.89}{0.941} = 2.01 \text{ mol Na}$$

$$(11.3 \text{ g C})\left(\frac{1 \text{ mol}}{12.01 \text{ g}}\right) = 0.941 \text{ mol C} \qquad \frac{0.941}{0.941} = 1.00 \text{ mol C}$$

$$(45.3 \text{ g O})\left(\frac{1 \text{ mol}}{16.00 \text{ g}}\right) = 2.83 \text{ mol O} \qquad \frac{2.83}{0.941} = 3.00 \text{ mol O}$$

The empirical formula is $Na_2CO_3$.

(e) 18.8% Na, 29.0% Cl, 52.3% O

$$(18.8\text{ g Na})\left(\frac{1\text{ mol}}{22.99\text{ g}}\right) = 0.818\text{ mol Na} \qquad \frac{0.818}{0.818} = 1.00\text{ mol Na}$$

$$(29.0\text{ g Cl})\left(\frac{1\text{ mol}}{35.45\text{ g}}\right) = 0.818\text{ mol Cl} \qquad \frac{0.818}{0.818} = 1.00\text{ mol Cl}$$

$$(52.3\text{ g O})\left(\frac{1\text{ mol}}{16.00\text{ g}}\right) = 3.27\text{ mol O} \qquad \frac{3.27}{0.818} = 4.00\text{ mol O}$$

The empirical formula is $NaClO_4$.

(f) 72.02% Mn, 27.98% O

$$(72.02\text{ g Mn})\left(\frac{1\text{ mol}}{54.94\text{ g}}\right) = 1.311\text{ mol Mn} \qquad \frac{1.311}{1.311} = 1.000\text{ mol Mn}$$

$$(27.98\text{ g O})\left(\frac{1\text{ mol}}{16.00\text{ g}}\right) = 1.749\text{ mol O} \qquad \frac{1.749}{1.311} = 1.334\text{ mol O}$$

Multiply both values by 3 to give whole numbers.

(1.000 mol Mn)(3) = 3.000 mol Mn; (1.334 mol O)(3) = 4.002 mol O

The empirical formula is $Mn_3O_4$.

38. Empirical formulas from percent composition.

(a) 64.1% Cu, 35.9% Cl

$$(64.1\text{ g Cu})\left(\frac{1\text{ mol}}{63.55\text{ g}}\right) = 1.01\text{ mol Cu} \qquad \frac{1.01}{1.01} = 1.00\text{ mol Cu}$$

$$(35.9\text{ g Cl})\left(\frac{1\text{ mol}}{35.45\text{ g}}\right) = 1.01\text{ mol Cl} \qquad \frac{1.01}{1.01} = 1.00\text{ mol Cl}$$

The empirical formula is CuCl.

(b) 47.2% Cu, 52.8% Cl

$$(47.2\text{ g Cu})\left(\frac{1\text{ mol}}{63.55\text{ g}}\right) = 0.743\text{ mol Cu} \qquad \frac{0.743}{0.743} = 1.00\text{ mol Cu}$$

$$(52.8\text{ g Cl})\left(\frac{1\text{ mol}}{35.45\text{ g}}\right) = 1.49\text{ mol Cl} \qquad \frac{1.49}{0.743} = 2.01\text{ mol Cl}$$

The empirical formula is $CuCl_2$.

(c) 51.9% Cr, 48.1% S

$$(51.9\text{ g Cr})\left(\frac{1\text{ mol}}{52.00\text{ g}}\right) = 0.998\text{ mol Cr} \qquad \frac{0.998}{0.998} = 1.00\text{ mol Cr}$$

$$(48.1\text{ g S})\left(\frac{1\text{ mol}}{32.07\text{ g}}\right) = 1.50\text{ mol S} \qquad \frac{1.50}{0.998} = 1.50\text{ mol S}$$

Multiply both values by 2 to give whole numbers.

(1.00 mol Cr)(2) = 2.00 mol Cr; (1.50 mol S)(2) = 3.00 mol S

The empirical formula is $Cr_2S_3$.

(d) 55.3% K, 14.6% P, 30.1% O

$$(55.3\text{ g K})\left(\frac{1\text{ mol}}{39.10\text{ g}}\right) = 1.41\text{ mol K} \qquad \frac{1.41}{0.471} = 2.99\text{ mol K}$$

$$(14.6\text{ g P})\left(\frac{1\text{ mol}}{30.97\text{ g}}\right) = 0.471\text{ mol P} \qquad \frac{0.471}{0.741} = 1.00\text{ mol P}$$

$$(30.1\text{ g O})\left(\frac{1\text{ mol}}{16.00\text{ g}}\right) = 1.88\text{ mol O} \qquad \frac{1.88}{0.741} = 3.99\text{ mol O}$$

The empirical formula is $K_3PO_4$.

(e) 38.9% Ba, 29.4% Cr, 31.7% O

$$(38.9\text{ g Ba})\left(\frac{1\text{ mol}}{137.3\text{ g}}\right) = 0.283\text{ mol Ba} \qquad \frac{0.283}{0.283} = 1.00\text{ mol Ba}$$

$$(29.4\text{ g Cr})\left(\frac{1\text{ mol}}{52.00\text{ g}}\right) = 0.565\text{ mol Cr} \qquad \frac{0.565}{0.283} = 2.00\text{ mol Cr}$$

$$(31.7\text{ g O})\left(\frac{1\text{ mol}}{16.00\text{ g}}\right) = 1.98\text{ mol O} \qquad \frac{1.98}{0.283} = 7.00\text{ mol O}$$

The empirical formula is $BaCr_2O_7$.

(f) 3.99% P, 82.3% Br, 13.7% Cl

$$(3.99 \text{ g P})\left(\frac{1 \text{ mol}}{30.97 \text{ g}}\right) = 0.129 \text{ mol P} \qquad \frac{0.129}{0.129} = 1.00 \text{ mol P}$$

$$(82.3 \text{ g Br})\left(\frac{1 \text{ mol}}{79.90 \text{ g}}\right) = 1.03 \text{ mol Br} \qquad \frac{1.03}{0.129} = 7.98 \text{ mol Br}$$

$$(13.7 \text{ g Cl})\left(\frac{1 \text{ mol}}{35.45 \text{ g}}\right) = 0.386 \text{ mol Cl} \qquad \frac{0.386}{0.129} = 2.99 \text{ mol Cl}$$

The empirical formula is $PBr_8Cl_3$.

39. Empirical formula

$$(3.996 \text{ g Sn})\left(\frac{1 \text{ mol}}{118.7 \text{ g}}\right) = 0.0337 \text{ mol Sn} \qquad \frac{0.0337}{0.0337} = 1.00 \text{ mol Sn}$$

$$(1.077 \text{ g O})\left(\frac{1 \text{ mol}}{16.00 \text{ g}}\right) = 0.0673 \text{ mol O} \qquad \frac{0.0673}{0.0337} = 2.00 \text{ mol O}$$

The empirical formula is $SnO_2$.

40. Empirical formula

5.454 g product − 3.054 g V = 2.400 g O

$$(3.054 \text{ g V})\left(\frac{1 \text{ mol}}{50.94 \text{ g}}\right) = 0.0600 \text{ mol V} \qquad \frac{0.0600}{0.0600} = 1.00 \text{ mol V}$$

$$(2.400 \text{ g O})\left(\frac{1 \text{ mol}}{16.00 \text{ g}}\right) = 0.1500 \text{ mol O} \qquad \frac{0.1500}{0.0600} = 2.50 \text{ mol O}$$

Multiplying both by 2 gives the empirical formula $V_2O_5$.

41. Molecular formula of hydroquinone

65.45% C, 5.45% H, 29.09% O; molar mass = 110.1

$$(65.45 \text{ g C})\left(\frac{1 \text{ mol}}{12.01 \text{ g}}\right) = 5.450 \text{ mol C} \qquad \frac{5.450}{1.818} = 2.998 \text{ mol C}$$

$$(5.45 \text{ g H})\left(\frac{1 \text{ mol}}{1.008 \text{ g}}\right) = 5.41 \text{ mol H} \qquad \frac{5.41}{1.818} = 2.98 \text{ mol H}$$

$$(29.09 \text{ g O})\left(\frac{1 \text{ mol}}{16.00 \text{ g}}\right) = 1.818 \text{ mol O} \qquad \frac{1.818}{1.818} = 1.000 \text{ mol O}$$

The empirical formula is $C_3H_3O$ making the empirical mass 55.05.

$$\frac{\text{molar mass}}{\text{empirical mass}} = \frac{110.1}{55.05} = 2$$

The molecular formula is twice that of the empirical formula.
Molecular formula = $(C_3H_3O)_2 = C_6H_6O_2$

42. Molecular formula of fructose
40.0% C, 6.7% H, 53.3% O; molar mass = 180.1

$$(40.0 \text{ g C})\left(\frac{1 \text{ mol}}{12.01 \text{ g}}\right) = 3.33 \text{ mol C} \qquad \frac{3.33}{3.33} = 1.00 \text{ mol C}$$

$$(6.7 \text{ g H})\left(\frac{1 \text{ mol}}{1.008 \text{ g}}\right) = 6.6 \text{ mol H} \qquad \frac{6.6}{3.33} = 2.0 \text{ mol H}$$

$$(53.3 \text{ g O})\left(\frac{1 \text{ mol}}{16.00 \text{ g}}\right) = 3.33 \text{ mol O} \qquad \frac{3.33}{3.33} = 1.00 \text{ mol O}$$

The empirical formula is $CH_2O$ making the empirical mass 33.03.

$$\frac{\text{molar mass}}{\text{empirical mass}} = \frac{180.1}{33.03} = 5.994$$

The molecular formula is six times that of the empirical formula.
Molecular formula = $(CH_2O)_6 = C_6H_{12}O_6$

43. $$(0.350 \text{ mol } P_4)\left(\frac{6.022 \times 10^{23} \text{ molecules}}{\text{mol}}\right)\left(\frac{4 \text{ atoms P}}{\text{molecule } P_4}\right) = 8.43 \times 10^{23} \text{ atoms P}$$

44. $$(10.0 \text{ g K})\left(\frac{1 \text{ mol}}{39.10 \text{ g}}\right)\left(\frac{1 \text{ mol Na}}{1 \text{ mol K}}\right)\left(\frac{22.99 \text{ g}}{\text{mol}}\right) = 5.88 \text{ g Na}$$

45. $(1.79 \times 10^{-23}$ g/atom$)(6.022 \times 10^{23}$ atoms/molar mass$) = 10.8$ g/molar mass

46. $$(6.022 \times 10^{23} \text{ sheets})\left(\frac{4.60 \text{ cm}}{500 \text{ sheets}}\right)\left(\frac{1 \text{ m}}{100 \text{ cm}}\right) = 5.54 \times 10^{19} \text{ m}$$

47. $$\left(\frac{6.022 \times 10^{23} \text{ dollars}}{5.0 \times 10^{9} \text{ people}}\right) = 1.2 \times 10^{14} \text{ dollars/person}$$

48. The conversion is: $mi^3 \rightarrow ft^3 \rightarrow in.^3 \rightarrow cm^3 \rightarrow$ drops

(a) $(1\ mi^3)\left(\frac{5280\ ft}{mile}\right)^3\left(\frac{12.0\ in.}{ft}\right)^3\left(\frac{2.54\ cm}{inch}\right)^3\left(\frac{20\ drops}{1.0\ cm^3}\right) = 8 \times 10^{16}$ drops

(b) $(6.022 \times 10^{23}\ drops)\left(\frac{1\ mi^3}{8 \times 10^{16}\ drops}\right) = 8 \times 10^{16}\ mi^3$

49. 1 mol Ag = 107.9 g Ag

(a) $(107.9\ g\ Ag)\left(\frac{1\ cm^3}{10.5\ g}\right) = 10.3\ cm^3$

(b) $10.3\ cm^3$ = volume of cube = $(\text{one side})^3$

side = $\sqrt[3]{10.3\ cm^3} = 2.18$ cm

50. The conversion is: L sol. → mL sol. → g sol. → g $H_2SO_4$ → mol $H_2SO_4$

$(1.00L)\left(\frac{1000\ mL}{1\ L}\right)\left(\frac{1.55\ g}{1.00\ mL}\right)\left(\frac{0.650\ g\ H_2SO_4}{1.00\ g}\right)\left(\frac{1\ mol}{98.09\ g}\right) = 10.3\ mol\ H_2SO_4$

51. The conversion is: mL sol. → g sol. → g $HNO_3$ → mol $HNO_3$

$(100.\ mL)\left(\frac{1.42\ g}{1.00\ mL}\right)\left(\frac{0.720\ g\ HNO_3}{1.000\ g}\right)\left(\frac{1\ mol}{63.02\ g}\right) = 1.62\ mol\ HNO_3$

52. (a) Determine the molar mass of each compound.

$CO_2$, 44.01 g; $O_2$, 32.00 g; $H_2O$, 18.02 g; $CH_3OH$, 32.04 g. The 1.00 gram sample with the lowest molar mass will contain the most molecules. Thus, $H_2O$ will contain the most molecules.

(b) $(1.00\ g\ H_2O)\left(\frac{1\ mol}{18.02\ g}\right)\left(\frac{(3)\ (6.022 \times 10^{23}\ atoms)}{mol}\right) = 1.00 \times 10^{23}$ atoms

$(1.00\ g\ CH_3OH)\left(\frac{1\ mol}{32.04\ g}\right)\left(\frac{(6)\ (6.022 \times 10^{23}\ atoms)}{mol}\right) = 1.13 \times 10^{23}$ atoms

$(1.00\ g\ CO_2)\left(\frac{1\ mol}{44.01\ g}\right)\left(\frac{(3)\ (6.022 \times 10^{23}\ atoms)}{mol}\right) = 4.24 \times 10^{22}$ atoms

$(1.00\ g\ O_2)\left(\frac{1\ mol}{32.00\ g}\right)\left(\frac{(2)\ (6.022 \times 10^{23}\ atoms)}{mol}\right) = 3.76 \times 10^{22}$ atoms

The 1.00 g sample of $CH_3OH$ contains the most atoms

53. 1 mol $Fe_2S_3$ = 207.9 g $Fe_2S_3$ = $6.022 \times 10^{23}$ formula units

$$(6.022 \times 10^{23} \text{ atoms})\left(\frac{1 \text{ formula unit}}{5 \text{ atoms}}\right)\left(\frac{207.9 \text{ g Fe}_2\text{S}_3}{6.022 \times 10^{23} \text{ formula units}}\right) = 41.58 \text{ g Fe}_2\text{S}_3$$

54. From the formula, 2 Li (13.88 g) combine with 1 S (32.07 g).

$$\left(\frac{13.88 \text{ g Li}}{32.07 \text{ g S}}\right)(20.0 \text{ g S}) = 8.66 \text{ g Li}$$

55. (a) $HgCO_3$

| | | |
|---|---|---|
| | Hg | 200.6 g |
| | C | 12.01 g |
| 3 | O | 48.00 g |
| | | 260.6 g |

$$\left(\frac{200.6 \text{ g}}{260.6 \text{ g}}\right)(100) = 76.98\% \text{ Hg}$$

(b) $Ca(ClO_3)_2$

| | | |
|---|---|---|
| 6 | O | 96.00 g |
| 2 | Cl | 70.90 g |
| | Ca | 40.08 g |
| | | 207.0 g |

$$\left(\frac{96.00 \text{ g}}{207.0 \text{ g}}\right)(100) = 46.38\% \text{ O}$$

(c) $C_{10}H_{14}N_2$

| | | |
|---|---|---|
| 2 | N | 28.02 g |
| 10 | C | 120.1 g |
| 14 | H | 14.11 g |
| | | 162.2 g |

$$\left(\frac{28.02 \text{ g}}{162.2 \text{ g}}\right)(100) = 17.28\% \text{ N}$$

(d) $C_{55}H_{72}MgN_4O_5$

| | | |
|---|---|---|
| | Mg | 24.31 g |
| 55 | C | 660.55 g |
| 72 | H | 72.58 g |
| 4 | N | 56.04 g |
| 5 | O | 80.00 g |
| | | 893.5 g |

$$\left(\frac{24.31 \text{ g}}{893.5 \text{ g}}\right)(100) = 2.721\% \text{ Mg}$$

56. According to the formula, 1 mol (65.39 g) Zn combines with 1 mol (32.07 g) S.

$$(19.5 \text{ g Zn})\left(\frac{32.07 \text{ g S}}{65.39 \text{ g Zn}}\right) = 9.56 \text{ g S}$$

19.5 g Zn require 9.56 g S for complete reaction. Therefore, there is not sufficient S present (9.40 g) to react with the Zn.

57. Molecular formula of aspirin

60.0% C, 4.48% H, 35.5% O; molar mass of aspirin = 180.2

$$(60.0 \text{ g C})\left(\frac{1 \text{ mol}}{12.01 \text{ g}}\right) = 5.00 \text{ mol C} \qquad \frac{5.00}{2.22} = 2.25 \text{ mol C}$$

$$(4.48 \text{ g H})\left(\frac{1 \text{ mol}}{1.008 \text{ g}}\right) = 4.44 \text{ mol H} \qquad \frac{4.44}{2.22} = 2.00 \text{ mol H}$$

$$(35.5 \text{ g O})\left(\frac{1 \text{ mol}}{16.00 \text{ g}}\right) = 2.22 \text{ mol O} \qquad \frac{2.22}{2.22} = 1.00 \text{ mol O}$$

Multiplying each by 4 give the empirical formula $C_9H_8O_4$. The empirical mass is 180.2. Since the empirical mass equals the molar mass, the molecular formula is the same as the empirical formula, $C_9H_8O_4$.

58. Calculate the percent oxygen in $Al_2(SO_4)_3$.

| | |
|---|---|
| 2 Al | 53.96 |
| 3 S | 96.21 |
| 12 O | 192.0 |
| | 342.2 |

$$\left(\frac{192.0}{342.2}\right)(100) = 56.11\% \text{ O}$$

Now take 56.11% of 8.50 g.

$(8.50 \text{ g O})(0.5611) = 4.77 \text{ g O}$

59. Empirical formula of gallium arsenide; 48.2% Ga, 51.8% As

$$(48.2 \text{ g Ga})\left(\frac{1 \text{ mol}}{69.72 \text{ g}}\right) = 0.691 \text{ mol Ga} \qquad \frac{0.691}{0.691} = 1.00 \text{ mol Ga}$$

$$(51.8 \text{ g As})\left(\frac{1 \text{ mol}}{74.92 \text{ g}}\right) = 0.691 \text{ mol As} \qquad \frac{0.691}{0.691} = 1.00 \text{ mol As}$$

The empirical formula is GaAs.

60. (a) 7.79% C, 92.21% Cl

$$(7.79 \text{ g C})\left(\frac{1 \text{ mol}}{12.01 \text{ g}}\right) = 0.649 \text{ mol C} \qquad \frac{0.649}{0.649} = 1.00 \text{ mol C}$$

$$(92.21 \text{ g Cl})\left(\frac{1 \text{ mol}}{35.45 \text{ g}}\right) = 2.601 \text{ mol Cl} \qquad \frac{2.601}{0.649} = 4.01 \text{ mol Cl}$$

The empirical formula is $CCl_4$. The empirical mass is 153.8 which equals the molar mass, therefore the molecular formula is $CCl_4$.

(b) 10.13% C, 89.87% Cl

$$(10.13 \text{ g C})\left(\frac{1 \text{ mol}}{12.01 \text{ g}}\right) = 0.8435 \text{ mol C} \qquad \frac{0.8435}{0.8435} = 1.000 \text{ mol C}$$

$$(89.87 \text{ g Cl})\left(\frac{1 \text{ mol}}{35.45 \text{ g}}\right) = 2.535 \text{ mol Cl} \qquad \frac{2.535}{0.8435} = 3.005 \text{ mol Cl}$$

The empirical formula is $CCl_3$. The empirical mass is 118.4.

$$\frac{\text{molar mass}}{\text{empirical mass}} = \frac{236.7}{118.4} = 1.999$$

The molecular formula is twice that of the empirical formula.
Molecular formula = $C_2Cl_6$.

(c) 25.26% C, 74.74% Cl

$$(25.26 \text{ g C})\left(\frac{1 \text{ mol}}{12.01 \text{ g}}\right) = 2.103 \text{ mol C} \qquad \frac{2.103}{2.103} = 1.000 \text{ mol C}$$

$$(74.74 \text{ g Cl})\left(\frac{1 \text{ mol}}{35.45 \text{ g}}\right) = 2.108 \text{ mol Cl} \qquad \frac{2.103}{2.108} = 1.002 \text{ mol Cl}$$

The empirical formula is CCl. The empirical mass is 47.46.

$$\frac{\text{molar mass}}{\text{empirical mass}} = \frac{284.8}{47.46} = 6.000$$

The molecular formula is six times that of the empirical formula.
Molecular formula = $C_6Cl_6$.

(d) 11.25% C, 88.75% Cl

$$(11.25 \text{ g C})\left(\frac{1 \text{ mol}}{12.01 \text{ g}}\right) = 0.9367 \text{ mol C} \qquad \frac{0.9367}{0.9367} = 1.000 \text{ mol C}$$

$$(88.75 \text{ g Cl})\left(\frac{1 \text{ mol}}{35.45 \text{ g}}\right) = 2.504 \text{ mol Cl} \qquad \frac{2.504}{0.9367} = 2.673 \text{ mol Cl}$$

Multiplying each by 3 give the empirical formula $C_3Cl_8$. The empirical mass is 319.6.
Since the molar mass is also 319.6 the molecular formula is $C_3Cl_8$.

The conversion is: s $\rightarrow$ min $\rightarrow$ hr $\rightarrow$ day $\rightarrow$ yr

$$(6.022 \times 10^{23} \text{ s})\left(\frac{1 \text{ min}}{60 \text{ s}}\right)\left(\frac{1 \text{ hr}}{60 \text{ min}}\right)\left(\frac{1 \text{ day}}{24 \text{ hr}}\right)\left(\frac{1 \text{ year}}{365 \text{ days}}\right) = 1.910 \times 10^{16} \text{ years}$$

62. The conversion is: g Cu → mol → atom

$$(2.5 \text{ g Cu})\left(\frac{1 \text{ mol}}{63.55 \text{ g}}\right)\left(\frac{6.022 \times 10^{23} \text{ atoms}}{\text{mol}}\right) = 2.4 \times 10^{23} \text{ atoms Cu}$$

63. The conversion is: molecules → mol → g     1 trillion = $10^{12}$

$$(1000. \times 10^{12} \text{ molecules } C_3H_8O_3)\left(\frac{1 \text{ mol}}{6.022 \times 10^{23} \text{ molecules}}\right)\left(\frac{92.09 \text{ g}}{\text{mol}}\right) = 1.529 \times 10^{-7} \text{ g } C_3H_8O_3$$

64. $$(5.0 \times 10^{9} \text{ people})\left(\frac{1 \text{ mol people}}{6.022 \times 10^{23} \text{ people}}\right) = 8.3 \times 10^{-15} \text{ mol people}$$

65. Empirical formula

23.3% Co, 25.3% Mo, 51.4% Cl

$$(23.3 \text{ g Co})\left(\frac{1 \text{ mol}}{58.93 \text{ g}}\right) = 0.395 \text{ mol Co} \qquad \frac{0.395}{0.264} = 1.50$$

$$(25.3 \text{ g Mo})\left(\frac{1 \text{ mol}}{95.94 \text{ g}}\right) = 0.264 \text{ mol Mo} \qquad \frac{0.264}{0.264} = 1.00$$

$$(51.4 \text{ g Cl})\left(\frac{1 \text{ mol}}{35.45 \text{ g}}\right) = 1.45 \text{ mol Cl} \qquad \frac{1.45}{0.264} = 5.49$$

Multiplying by 2 gives the empirical formula $Co_3Mo_2Cl_{11}$.

66. The conversion is: g Al → mol Al → mol Mg → g Mg

$$(18 \text{ g Al})\left(\frac{1 \text{ mol}}{26.98 \text{ g}}\right)\left(\frac{2 \text{ mol Mg}}{1 \text{ mol Al}}\right)\left(\frac{24.31 \text{ g}}{\text{mol}}\right) = 32 \text{ g Mg}$$

67. (10.0 g N)(0.177) = 1.77 g N

$$(1.77 \text{ g N})\left(\frac{1 \text{ mol}}{14.01 \text{ g}}\right) = 0.126 \text{ mol N}$$

$$(3.8 \times 10^{23} \text{ atoms H})\left(\frac{1 \text{ mol}}{6.022 \times 10^{23} \text{ atoms}}\right) = 0.63 \text{ mol H}$$

To determine mol C first find grams H and subtract the grams of H and N from the grams of the sample.

$$(0.63 \text{ mol H})\left(\frac{1.008 \text{ g}}{\text{mol}}\right) = 0.64 \text{ g H}$$

| | |
|---|---|
| 10.0 g | sample |
| - 1.77 g | N |
| -0.64 g | H |
| 7.6 g | C |

$$(7.6 \text{ g C})\left(\frac{1 \text{ mol}}{12.01 \text{ g}}\right) = 0.63 \text{ mol C}$$

$$\text{N} \quad \frac{0.126}{0.126} = 1.00$$

$$\text{H} \quad \frac{0.63}{0.126} = 5.0$$

$$\text{C} \quad \frac{0.63}{0.126} = 5.0$$

The formula is $C_5H_5N$

68. Let $x$ = molar mass of $A_2O$

$$0.400x = 16.00 \text{ g O (Since } A_2O \text{ has only one mol of O atoms)}$$

$$x = 40.0 \text{ g O/mol } A_2O$$

$$40.0 = 16.00 + 2y \qquad y = \text{molar mass of A}$$

$$40.0 - 16.00 = 2y$$

$$12.0\frac{\text{g}}{\text{mol}} = y$$

Look in the periodic table for the element that has 12.0 g/mol.
The element is carbon. The mystery element is carbon.

69. (a) $CH_2O$ (divide by 6)
(b) $C_4H_9$ (divide by 2)
(c) $CH_2O$ (divide by 3)
(d) $C_{25}H_{52}$ (divide by 1)
(e) $C_6H_2Cl_2O$ (divide by 2)

CHAPTER 8

# CHEMICAL EQUATIONS

1. The purpose of balancing chemical equations is to conform to the Law of Conservation of Mass. Ratios of reactants and products can then be easily determined.

2. The coefficients in a balanced chemical equation represent the number of moles (or molecules or formula units) of each of the chemical species in the reaction.

3. (a) Yes. It is necessary to conserve atoms to follow the Law of Conservation of Mass.

   (b) No. Molecules can be taken apart and rearranged to form different molecules in reactions.

   (c) Moles of molecules are not conserved (b). Moles of atoms are conserved (a).

4. A chemical changed that absorbs heat energy is said to be an *endothermic* reaction. The products are at a higher energy level than the reactants. A chemical change that liberates heat energy is said to be an *exothermic* reaction. The products are at a lower energy level than the reactants.

5. (a) $2\,H_2 + O_2 \rightarrow 2\,H_2O$
   (b) $3\,C + Fe_2O_3 \rightarrow 2\,Fe + 3\,CO$
   (c) $H_2SO_4 + 2\,NaOH \rightarrow 2\,H_2O + Na_2SO_4$
   (c) $Al_2(CO_3)_3 \xrightarrow{\Delta} Al_2O_3 + 3\,CO_2$
   (d) $2\,NH_4I + Cl_2 \rightarrow 2\,NH_4Cl + I_2$

6. (a) $H_2 + Br_2 \rightarrow 2\,HBr$
   (b) $4\,Al + 3\,C \xrightarrow{\Delta} Al_4C_3$
   (c) $Ba(ClO_3)_2 \xrightarrow{\Delta} BaCl_2 + 3\,O_2$
   (d) $CrCl_3 + 3\,AgNO_3 \rightarrow Cr(NO_3)_3 + 3\,AgCl$
   (e) $2\,H_2O_2 \rightarrow 2\,H_2O + O_2$

7. (a) combination
   (b) single displacement
   (c) double displacement
   (d) decomposition
   (e) single displacement

8. (a) combination
   (b) combination
   (c) decomposition
   (d) double displacement
   (e) decomposition

9. (a) $2\,MnO_2 + CO \rightarrow Mn_2O_3 + CO_2$
   (b) $Mg_3N_2 + 6\,H_2O \rightarrow 3\,Mg(OH)_2 + 2\,NH_3$
   (c) $4\,C_3H_5(NO_3)_3 \rightarrow 12\,CO_2 + 10\,H_2O + 6\,N_2 + O_2$
   (d) $4\,FeS + 7\,O_2 \rightarrow 2\,Fe_2O_3 + 4\,SO_2$
   (e) $2\,Cu(NO_3)_2 \rightarrow 2\,CuO + 4\,NO_2 + O_2$
   (f) $3\,NO_2 + H_2O \rightarrow 2\,HNO_3 + NO$
   (g) $2\,Al + 3\,H_2SO_4 \rightarrow Al_2(SO_4)_3 + 3\,H_2$
   (h) $4\,HCN + 5\,O_2 \rightarrow 2\,N_2 + 4\,CO_2 + 2\,H_2O$
   (i) $2\,B_5H_9 + 12\,O_2 \rightarrow 5\,B_2O_3 + 9\,H_2O$

10. (a) $2\,SO_2 + O_2 \rightarrow 2\,SO_3$
    (b) $4\,Al + 3\,MnO_2 \xrightarrow{\Delta} 3\,Mn + 2\,Al_2O_3$
    (c) $2\,Na + 2\,H_2O \rightarrow 2\,NaOH + H_2$
    (d) $2\,AgNO_3 + Ni \rightarrow Ni(NO_3)_2 + 2\,Ag$
    (e) $Bi_2S_3 + 6\,HCl \rightarrow 2\,BiCl_3 + 3\,H_2S$
    (f) $2\,PbO_2 \xrightarrow{\Delta} 2\,PbO + O_2$
    (g) $2\,LiAlH_4 \xrightarrow{\Delta} 2\,LiH + 2\,Al + 3\,H_2$
    (h) $2\,KI + Br_2 \rightarrow 2\,KBr + I_2$
    (i) $2\,K_3PO_4 + 3\,BaCl_2 \rightarrow 6\,KCl + Ba_3(PO_4)_2$

11. (a) $2\,H_2O \rightarrow 2\,H_2 + O_2$
    (b) $HC_2H_3O_2 + KOH \rightarrow KC_2H_3O_2 + H_2O$
    (c) $2\,P + 3\,I_2 \rightarrow 2\,PI_3$
    (d) $2\,Al + 3\,CuSO_4 \rightarrow 3\,Cu + Al_2(SO_4)_3$
    (e) $(NH_4)_2SO_4 + BaCl_2 \rightarrow 2\,NH_4Cl + BaSO_4$
    (f) $SF_4 + 2\,H_2O \rightarrow SO_2 + 4\,HF$
    (g) $Cr_2(CO_3)_3 \xrightarrow{\Delta} Cr_2O_3 + 3\,CO_2$

12. (a) $2\,Cu + S \rightarrow Cu_2S$
    (b) $2\,H_3PO_4 + 3\,Ca(OH)_2 \xrightarrow{\Delta} Ca_3(PO_4)_2 + 6\,H_2O$

(c) $2\ Ag_2O \xrightarrow{\Delta} 4\ Ag + O_2$
(d) $FeCl_3 + 3\ NaOH \rightarrow Fe(OH)_3 + 3\ NaCl$
(e) $Ni_3(PO_4)_2 + 3\ H_2SO_4 \rightarrow 3\ NiSO_4 + 2\ H_3PO_4$
(f) $ZnCO_3 + 2\ HCl \rightarrow ZnCl_2 + H_2O + CO_2$
(g) $3\ AgNO_3 + AlCl_3 \rightarrow 3\ AgCl + Al(NO_3)_3$

13. (a) $Ag(s) + H_2SO_4(aq) \rightarrow$ no reaction
(b) $Cl_2(g) + 2\ NaBr(aq) \rightarrow Br_2(l) + 2\ NaCl(aq)$
(c) $Mg(s) + ZnCl_2(aq) \rightarrow Zn(s) + MgCl_2(aq)$
(d) $Pb(s) + 2\ AgNO_3(aq) \rightarrow 2\ Ag(s) + Pb(NO_3)_2(aq)$

14. (a) $Cu(s) + FeCl_3(aq) \rightarrow$ no reaction
(b) $H_2(g) + Al_2O_3(s) \xrightarrow{\Delta}$ no reaction
(c) $2\ Al(s) + 6\ HBr(aq) \rightarrow 3\ H_2(g) + 2\ AlBr_3(aq)$
(d) $I_2(s) + HCl(aq) \rightarrow$ no reaction

15. (a) $H_2 + I_2 \rightarrow 2\ HI$
(b) $CaCO_3 \xrightarrow{\Delta} CaO + CO_2$
(c) $Mg + H_2SO_4 \rightarrow H_2 + MgSO_4$
(d) $FeCl_2 + 2\ NaOH \rightarrow Fe(OH)_2 + 2\ NaCl$

16. (a) $SO_2 + H_2O \rightarrow H_2SO_3$
(b) $SO_3 + H_2O \rightarrow H_2SO_4$
(c) $Ca + 2\ H_2O \rightarrow Ca(OH)_2 + H_2$
(d) $2\ Bi(NO_3)_3 + 3\ H_2S \rightarrow Bi_2S_3 + 6\ HNO_3$

17. (a) $2\ Ba + O_2 \rightarrow 2\ BaO$
(b) $2\ NaHCO_3 \xrightarrow{\Delta} Na_2CO_3 + H_2O + CO_2$
(c) $Ni + CuSO_4 \rightarrow NiSO_4 + Cu$
(d) $MgO + 2\ HCl \rightarrow MgCl_2 + H_2O$
(e) $H_3PO_4 + 3\ KOH \rightarrow K_3PO_4 + 3\ H_2O$

18. (a) $C + O_2 \xrightarrow{\Delta} CO_2$
(b) $2\ Al(ClO_3)_3 \xrightarrow{\Delta} 9\ O_2 + 2\ AlCl_3$

(c) $CuBr_2 + Cl_2 \rightarrow CuCl_2 + Br_2$

(d) $2\ SbCl_3 + 3\ (NH_4)_2S \rightarrow Sb_2S_3 + 6\ NH_4Cl$

(e) $2\ NaNO_3 \xrightarrow{\Delta} 2\ NaNO_2 + O_2$

19. (a) One mole of $MgBr_2$ reacts with two moles of $AgNO_3$ to yield one mole of $Mg(NO_3)_2$ and two moles of AgBr.
    (b) One mole of $N_2$ reacts with three moles of $H_2$ to produce two moles of $NH_3$.
    (c) Two moles of $C_3H_7OH$ react with nine moles of $O_2$ to form six moles of $CO_2$ and eight moles of $H_2O$.

20. (a) Two moles of Na react with one mole of $Cl_2$ to produce two moles of NaCl and release 822 kJ of energy. The reaction is exothermic.
    (b) One mole of $PCl_5$ absorbs 92.9 kJ of energy to produce one mole of $PCl_3$ and one mole of $Cl_2$. The reaction is endothermic.

21. (a) $CaO + H_2O \rightarrow Ca(OH)_2 + 65.3\ kJ$
    (b) $2\ Al_2O_3 + 3260\ kJ \rightarrow 4\ Al + 3\ O_2$

22. (a) $2\ Al + 3\ I_2 \rightarrow 2\ AlI_3 + heat$
    (b) $4\ CuO + CH_4 + heat \rightarrow 4\ Cu + CO_2 + 2\ H_2O$
    (c) $Fe_2O_3 + 2\ Al \rightarrow 2\ Fe + Al_2O_3 + heat$

23. (a) change in color and texture of the bread
    (b) change in texture of the white and the yoke
    (c) the flame (combustion), change in matchhead, odor

24. $P_4O_{10} + 12\ HClO_4 \rightarrow 6\ Cl_2O_7 + 4\ H_3PO_4$

    10 O + 12 (4 O)      6 (7 O) + 4 (4 O)

    10 O + 48 O      42 O + 16 O

    58 O      58 O

25. In 7 $Al_2(SO_4)_3$ there are:
    (a) 14 atoms of Al
    (b) 21 atoms of S
    (c) 84 atoms of O
    (d) 119 total atoms

26. A balanced equation tells us:
   (1) the types of atoms/molecules involved in the reaction
   (2) the relationship between quantities of the substances in the reaction

   A balanced equations gives no information about
   (1) the time required for the reaction
   (2) odors or colors which may result

27.

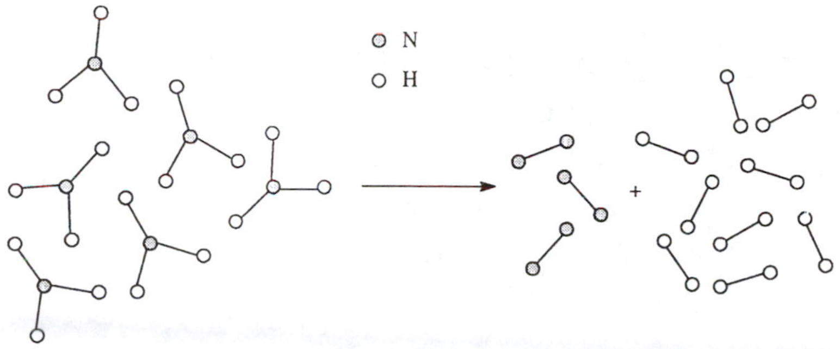

$$6\ NH_3 \xrightarrow{\Delta} 3\ N_2 + 9\ H_2$$

28. Zn metal is below Mg on the activity series.

29. $Ti + Ni(NO_3)_2 \rightarrow$ yes
   $Ti + Pb(NO_3)_2 \rightarrow$ yes
   $Ti + Mg(NO_3)_2 \rightarrow$ no

   Ti is above Ni and Pb in the activity series since both react. Ti is below Mg in the series since it will not replace Mg. From the printed activity series in the chapter Ni lies above Pb so the order is:

   Mg
   Ti
   Ni
   Pb

30. (a) $4\ K + O_2 \rightarrow 2\ K_2O$
   (b) $2\ Al + 3\ Cl_2 \rightarrow 2\ AlCl_3$
   (c) $CO_2 + H_2O \rightarrow H_2CO_3$
   (d) $CaO + H_2O \rightarrow Ca(OH)_2$

31. (a) $2\ HgO \xrightarrow{\Delta} 2\ Hg + O_2$
   (b) $2\ NaClO_3 \xrightarrow{\Delta} 2\ NaCl + 3\ O_2$

(c) $MgCO_3 \xrightarrow{\Delta} MgO + CO_2$
(d) $2\ PbO_2 \xrightarrow{\Delta} 2\ PbO + O_2$

32. (a) $Zn + H_2SO_4 \rightarrow H_2 + ZnSO_4$
(b) $2\ AlI_3 + 3\ Cl_2 \rightarrow 2\ AlCl_3 + 3\ I_2$
(c) $Mg + 2\ AgNO_3 \rightarrow Mg(NO_3)_2 + 2\ Ag$
(d) $2\ Al + 3\ CoSO_4 \rightarrow Al_2(SO_4)_3 + 3\ Co$

33. (a) $ZnCl_2 + 2\ KOH \rightarrow Zn(OH)_2 + 2\ KCl$
(b) $CuSO_4 + H_2S \rightarrow H_2SO_4 + CuS$
(c) $3\ Ca(OH)_2 + 2\ H_3PO_4 \rightarrow 6\ H_2O + Ca_3(PO_4)_2$
(d) $2\ (NH_4)_3PO_4 + 3\ Ni(NO_3)_2 \rightarrow 6\ NH_4NO_3 + Ni_3(PO_4)_2$
(e) $Ba(OH)_2 + 2\ HNO_3 \rightarrow 2\ H_2O + Ba(NO_3)_2$
(f) $(NH_4)_2S + 2\ HCl \rightarrow H_2S + 2\ NH_4Cl$

34. (a) $AgNO_3(aq) + KCl(aq) \rightarrow AgCl(s) + KNO_3(aq)$
(b) $Ba(NO_3)_2(aq) + MgSO_4(aq) \rightarrow Mg(NO_3)_2(aq) + BaSO_4(s)$
(c) $H_2SO_4(aq) + Mg(OH)_2(aq) \rightarrow 2\ H_2O(l) + MgSO_4(aq)$
(d) $MgO(s) + H_2SO_4(aq) \rightarrow H_2O(l) + MgSO_4(aq)$
(e) $Na_2CO_3(aq) + NH_4Cl(aq) \rightarrow$ no reaction

35. (a) $2\ C_2H_6 + 7\ O_2 \rightarrow 4\ CO_2 + 6\ H_2O$
(b) $2\ C_6H_6 + 15\ O_2 \rightarrow 12\ CO_2 + 6\ H_2O$
(c) $C_7H_{16} + 11\ O_2 \rightarrow 7\ CO_2 + 8\ H_2O$

36. 1. combustion of fossil fuels
2. destruction of the rain forests by burning
3. increased population

37. Carbon dioxide, methane, and water are all considered to be greenhouse gases. They each act to trap the heat near the surface of the earth in the same manner in which a greenhouse is warmed.

38. The effects of global warming can be reduced by:
    1. developing new energy sources (not dependant on fossil fuels)
    2. conservation of energy resources
    3. recycling
    4. decreased destruction of the rain forests and other forests

39. About half the carbon dioxide released into the atmosphere remains in the air. The rest is absorbed by plants and used in photosynthesis or is dissolved in the oceans.

CHAPTER 9

# CALCULATIONS FROM CHEMICAL EQUATIONS

1. The balanced equation is

$Ca_3P_2 + 6\,H_2O \rightarrow 3\,Ca(OH)_2 + 2\,PH_3$

(a) Correct: $(1\text{ mol } Ca_3P_2)\left(\dfrac{2\text{ mol } PH_3}{1\text{ mol } Ca_3P_2}\right) = 2\text{ mol } PH_3$

(b) Incorrect: 1 g $Ca_3P_2$ would produce 0.4 g $PH_3$

$$(1\text{ g } Ca_3P_2)\left(\frac{1\text{ mol}}{182.2\text{ g}}\right)\left(\frac{2\text{ mol } PH_3}{1\text{ mol } Ca_3P_2}\right)\left(\frac{33.99\text{ g}}{\text{mol}}\right) = 0.4\text{ g } PH_3$$

(c) Correct: see equation

(d) Correct: see equation

(e) Incorrect: 2 mol $Ca_3P_2$ requires 12 mol $H_2O$ to produce 4.0 mol $PH_3$.

$$(2\text{ mol } Ca_3P_2)\left(\frac{6\text{ mol } H_2O}{1\text{ mol } Ca_3P_2}\right) = 12\text{ mol } H_2O$$

(f) Correct: 2 mol $Ca_3P_2$ will react with 12 mol $H_2O$ (3 mol $H_2O$ are present in excess) and 6 mol $Ca(OH)_2$ will be formed.

$$(2\text{ mol } Ca_3P_2)\left(\frac{3\text{ mol } Ca(OH)_2}{1\text{ mol } Ca_3P_2}\right) = 6\text{ mol } Ca(OH)_2$$

(g) Incorrect: $(200.\text{ g } Ca_3P_2)\left(\dfrac{1\text{ mol}}{182.2\text{ g}}\right)\left(\dfrac{6\text{ mol } H_2O}{1\text{ mol } Ca_3P_2}\right)\left(\dfrac{18.02\text{ g}}{\text{mol}}\right) = 119\text{ g } H_2O$

The amount of water present (100. g) is less than needed to react with 200. g $Ca_3P_2$. $H_2O$ is the limiting reactant.

(h) Incorrect: $H_2O$ is the limiting reactant.

$$(100.\text{ g } H_2O)\left(\frac{1\text{ mol}}{18.02\text{ g}}\right)\left(\frac{2\text{ mol } PH_3}{6\text{ mol } H_2O}\right)\left(\frac{33.99\text{ g}}{\text{mol}}\right) = 62.9\text{ g } PH_3$$

2. The balanced equation is

$2\,CH_4 + 3\,O_2 + 2\,NH_3 \rightarrow 2\,HCN + 6\,H_2O$

(a) Correct

(b) Incorrect: $(16 \text{ mol } O_2)\left(\dfrac{2 \text{ mol HCN}}{3 \text{ mol } O_2}\right) = 10.7 \text{ mol HCN}$ (not 12 mol HCN)

(c) Correct

(d) Incorrect: $(12 \text{ mol HCN})\left(\dfrac{6 \text{ mol } H_2O}{2 \text{ mol HCN}}\right) = 36 \text{ mol } H_2O$ (not 4 mol $H_2O$)

(e) Correct

(f) Incorrect: $O_2$ is the limiting reactant

$(3 \text{ mol } O_2)\left(\dfrac{2 \text{ mol HCN}}{3 \text{ mol } O_2}\right) = 2 \text{ mol HCN}$ (not 3 mol HCN)

3. (a) $(25.0 \text{ g } KNO_3)\left(\dfrac{1 \text{ mol}}{101.1 \text{ g}}\right) = 0.247 \text{ mol } KNO_3$

(b) $(56 \text{ mmol NaOH})\left(\dfrac{1 \text{ mol}}{1000 \text{ mmol}}\right) = 0.056 \text{ mol NaOH}$

(c) $(5.4 \times 10^2 \text{ g } (NH_4)_2C_2O_4)\left(\dfrac{1 \text{ mol}}{124.1 \text{ g}}\right) = 4.4 \text{ mol } (NH_4)_2C_2O_4$

(d) The conversion is: mL sol $\rightarrow$ g sol $\rightarrow$ g $H_2SO_4$ $\rightarrow$ mol $H_2SO_4$

$(16.8 \text{ mL solution})\left(\dfrac{1.727 \text{ g}}{\text{mL}}\right)\left(\dfrac{0.800 \text{ g } H_2SO_4}{\text{g solution}}\right)\left(\dfrac{1 \text{ mol}}{98.09 \text{ g}}\right) = 0.237 \text{ mol } H_2SO_4$

4. (a) $(2.10 \text{ kg } NaHCO_3)\left(\dfrac{1000 \text{ g}}{\text{kg}}\right)\left(\dfrac{1 \text{ mol}}{84.01 \text{ g}}\right) = 25.0 \text{ mol } NaHCO_3$

(b) $(525 \text{ mg } ZnCl_2)\left(\dfrac{1 \text{ g}}{1000 \text{ mg}}\right)\left(\dfrac{1 \text{ mol}}{136.3 \text{ g}}\right) = 3.85 \times 10^{-3} \text{ mol } ZnCl_2$

(c) $(9.8 \times 10^{24} \text{ molecules } CO_2)\left(\dfrac{1 \text{ mol}}{6.022 \times 10^{23} \text{ molecules}}\right) = 16 \text{ mol } CO_2$

(d) $(250 \text{ mL } C_2H_5OH)\left(\dfrac{0.789 \text{ g}}{\text{mL}}\right)\left(\dfrac{1 \text{ mol}}{46.07 \text{ g}}\right) = 4.3 \text{ mol } C_2H_5OH$

5. (a) $(2.55 \text{ mol Fe(OH)}_3)\left(\frac{106.9 \text{ g}}{\text{mol}}\right) = 273 \text{ g Fe(OH)}_3$

(b) $(125 \text{ kg CaCO}_3)\left(\frac{1000 \text{ g}}{\text{kg}}\right) = 1.25 \times 10^5 \text{ g CaCO}_3$

(c) $(10.5 \text{ mol NH}_3)\left(\frac{17.03 \text{ g}}{\text{mol}}\right) = 179 \text{ g NH}_3$

(d) $(72 \text{ mmol HCl})\left(\frac{1 \text{ mol}}{1000 \text{ mmol}}\right)\left(\frac{36.46 \text{ g}}{\text{mol}}\right) = 2.6 \text{ g HCl}$

(e) $(500.0 \text{ mL Br}_2)\left(\frac{3.119 \text{ g}}{\text{mL}}\right) = 1559.5 \text{ g Br}_2 = 1.560 \times 10^3 \text{ g Br}_2$

6. (a) $(0.00844 \text{ mol NiSO}_4)\left(\frac{154.8 \text{ g}}{\text{mol}}\right) = 1.31 \text{ g NiSO}_4$

(b) $(0.0600 \text{ mol HC}_2\text{H}_3\text{O}_2)\left(\frac{60.05 \text{ g}}{\text{mol}}\right) = 3.60 \text{ g HC}_2\text{H}_3\text{O}_2$

(c) $(0.725 \text{ mol Bi}_2\text{S}_3)\left(\frac{514.2 \text{ g}}{\text{mol}}\right) = 373 \text{ g Bi}_2\text{S}_3$

(d) $(4.50 \times 10^{21} \text{ molecules C}_6\text{H}_{12}\text{O}_6)\left(\frac{1 \text{ mol}}{6.022\times10^{23} \text{ molecules}}\right)\left(\frac{180.2 \text{ g}}{\text{mol}}\right) = 1.35 \text{ g C}_6\text{H}_{12}\text{O}_6$

(e) $(75 \text{ mL solution})\left(\frac{1.175 \text{ g}}{\text{mL}}\right)\left(\frac{0.200 \text{ g K}_2\text{CrO}_4}{\text{g solution}}\right) = 18 \text{ g K}_2\text{CrO}_4$

7. 10.0 g $H_2O$ or 10.0 g $H_2O_2$

Water has a lower molar mass than hydrogen peroxide. 10.0 grams of water contain more moles, and therefore more molecules than 10.0 g of $H_2O_2$.

8. Larger number of molecules: 25.0 g HCl or 85 g $C_6H_{12}O_6$

$$(25.0 \text{ g HCl})\left(\frac{1 \text{ mol}}{36.46 \text{ g}}\right)\left(\frac{6.022\times10^{23} \text{ molecules}}{\text{mol}}\right) = 4.13 \times 10^{23} \text{ molecules HCl}$$

$$(85.0 \text{ g C}_6\text{H}_{12}\text{O}_6)\left(\frac{1 \text{ mol}}{180.2 \text{ g}}\right)\left(\frac{6.022\times10^{23} \text{ molecules}}{\text{mol}}\right) = 2.84 \times 10^{23} \text{ molecules C}_6\text{H}_{12}\text{O}_6$$

HCl contains more molecules

9. Mole ratios

$$2\ C_3H_7OH + 9\ O_2 \rightarrow 6\ CO_2 + 8\ H_2O$$

(a) $\dfrac{6\ \text{mol}\ CO_2}{2\ \text{mol}\ C_3H_7OH}$

(b) $\dfrac{2\ \text{mol}\ C_3H_7OH}{9\ \text{mol}\ O_2}$

(c) $\dfrac{9\ \text{mol}\ O_2}{6\ \text{mol}\ CO_2}$

(d) $\dfrac{8\ \text{mol}\ H_2O}{2\ \text{mol}\ C_3H_7OH}$

(e) $\dfrac{6\ \text{mol}\ CO_2}{8\ \text{mol}\ H_2O}$

(f) $\dfrac{8\ \text{mol}\ H_2O}{9\ \text{mol}\ O_2}$

10. Mole ratios

$$3\ CaCl_2 + 2\ H_3PO_4 \rightarrow Ca_3(PO_4)_2 + 6\ HCl$$

(a) $\dfrac{3\ \text{mol}\ CaCl_2}{1\ \text{mol}\ Ca_3(PO_4)_2}$

(b) $\dfrac{6\ \text{mol}\ HCl}{2\ \text{mol}\ H_3PO_4}$

(c) $\dfrac{3\ \text{mol}\ CaCl_2}{2\ \text{mol}\ H_3PO_4}$

(d) $\dfrac{1\ \text{mol}\ Ca_3(PO_4)_2}{2\ \text{mol}\ H_3PO_4}$

(e) $\dfrac{6\ \text{mol}\ HCl}{1\ \text{mol}\ Ca_3(PO_4)_2}$

(f) $\dfrac{2\ \text{mol}\ H_3PO_4}{6\ \text{mol}\ HCl}$

11. $C_2H_5OH + 3\ O_2 \rightarrow 2\ CO_2 + 3\ H_2O$

$$(7.75\ \text{mol}\ C_2H_5OH)\left(\frac{2\ \text{mol}\ CO_2}{1\ \text{mol}\ C_2H_5OH}\right) = 15.5\ \text{mol}\ CO_2$$

12. Moles of $Cl_2$

$$4\ HCl + O_2 \rightarrow 2\ Cl_2 + 2\ H_2O$$

$$(5.60\ \text{mol}\ HCl)\left(\frac{2\ \text{mol}\ Cl_2}{4\ \text{mol}\ HCl}\right) = 2.80\ \text{mol}\ Cl_2$$

13. $MnO_2(s) + 4\ HCl(aq) \rightarrow Cl_2(g) + MnCl_2(aq) + 2\ H_2O(l)$

$$(1.05\ \text{mol}\ MnO_2)\left(\frac{4\ \text{mol}\ HCl}{1\ \text{mol}\ MnO_2}\right) = 4.20\ \text{mol}\ HCl$$

14. $Al_4C_3 + 12\ H_2O \rightarrow 4\ Al(OH)_3 + 3\ CH_4$

(a) $(100.\ g\ Al_4C_3)\left(\frac{1\ mol}{144.0\ g}\right)\left(\frac{12\ mol\ H_2O}{1\ mol\ Al_4C_3}\right) = 8.33\ mol\ H_2O$

(b) $(0.600\ mol\ CH_4)\left(\frac{4\ mol\ Al(OH)_3}{3\ mol\ CH_4}\right) = 0.800\ mol\ Al(OH)_3$

15. Grams of NaOH

$Ca(OH)_2 + Na_2CO_3 \rightarrow 2\ NaOH + CaCO_3$

The conversion is: $g\ Ca(OH)_2 \rightarrow mol\ Ca(OH)_2 \rightarrow mol\ NaOH \rightarrow g\ NaOH$

$(500.\ g\ Ca(OH)_2)\left(\frac{1\ mol}{74.10\ g}\right)\left(\frac{2\ mol\ NaOH}{1\ mol\ Ca(OH)_2}\right)\left(\frac{40.00\ g}{mol}\right) = 5 \times 10^2\ g\ NaOH$

16. Grams of $Zn_3(PO_4)_2$

$3\ Zn + 2\ H_3PO_4 \rightarrow Zn_3(PO_4)_2 + 3\ H_2$

The conversion is: $g\ Zn \rightarrow mol\ Zn \rightarrow mol\ Zn_3(PO_4)_2 \rightarrow g\ Zn_3(PO_4)_2$

$(10.0\ g\ Zn)\left(\frac{1\ mol}{65.39\ g}\right)\left(\frac{1\ mol\ Zn_3(PO_4)_2}{3\ mol\ Zn}\right)\left(\frac{386.1\ g}{mol}\right) = 19.7\ g\ Zn_3(PO_4)_2$

17. The balanced equation is $Fe_2O_3 + 3\ C \rightarrow 2\ Fe + 3\ CO$

The conversion is: $kg\ Fe_2O_3 \rightarrow kmol\ Fe_2O_3 \rightarrow kmol\ Fe \rightarrow kg\ Fe$

$(125\ kg\ Fe_2O_3)\left(\frac{1\ kmol}{159.7\ kg}\right)\left(\frac{2\ kmol\ Fe}{1\ kmol\ Fe_2O_3}\right)\left(\frac{55.85\ kg}{kmol}\right) = 87.4\ kg\ Fe$

18. The balanced equation is $3\ Fe + 4\ H_2O \rightarrow Fe_3O_4 + 4\ H_2$

Calculate the grams of both $H_2O$ and Fe to produce 375 g $Fe_3O_4$

$(375\ g\ Fe_3O_4)\left(\frac{1\ mol}{231.6\ g}\right)\left(\frac{4\ mol\ H_2O}{1\ mol\ Fe_3O_4}\right)\left(\frac{18.02\ g}{mol}\right) = 117\ g\ H_2O$

$(375\ g\ Fe_3O_4)\left(\frac{1\ mol}{231.6\ g}\right)\left(\frac{3\ mol\ Fe}{1\ mol\ Fe_3O_4}\right)\left(\frac{55.85\ g}{mol}\right) = 271\ g\ Fe$

19. The balanced equation is $2\ C_2H_6 + 7\ O_2 \rightarrow 4\ CO_2 + 6\ H_2O$

(a) $$(15.0 \text{ mol } C_2H_6)\left(\frac{7 \text{ mol } O_2}{2 \text{ mol } C_2H_6}\right) = 52.5 \text{ mol } O_2$$

(b) $$(8.00 \text{ g } H_2O)\left(\frac{1 \text{ mol}}{18.02 \text{ g}}\right)\left(\frac{4 \text{ mol } CO_2}{6 \text{ mol } H_2O}\right)\left(\frac{44.01 \text{ g}}{\text{mol}}\right) = 13.0 \text{ g } CO_2$$

(c) $$(75.0 \text{ g } C_2H_6)\left(\frac{1 \text{ mol}}{30.07 \text{ g}}\right)\left(\frac{4 \text{ mol } CO_2}{2 \text{ mol } C_2H_6}\right)\left(\frac{44.01 \text{ g}}{\text{mol}}\right) = 2.20 \times 10^2 \text{ g } CO_2$$

20. $4\ FeS_2 + 11\ O_2 \rightarrow 2\ Fe_2O_3 + 8\ SO_2$

(a) $$(1.00 \text{ mol } FeS_2)\left(\frac{2 \text{ mol } Fe_2O_3}{4 \text{ mol } FeS_2}\right) = 0.500 \text{ mol } Fe_2O_3$$

(b) $$(4.50 \text{ mol } FeS_2)\left(\frac{11 \text{ mol } O_2}{4 \text{ mol } FeS_2}\right) = 12.4 \text{ mol } O_2$$

(c) $$(1.55 \text{ mol } Fe_2O_3)\left(\frac{8 \text{ mol } SO_2}{2 \text{ mol } Fe_2O_3}\right) = 6.20 \text{ mol } SO_2$$

(d) $$(0.512 \text{ mol } FeS_2)\left(\frac{8 \text{ mol } SO_2}{4 \text{ mol } FeS_2}\right)\left(\frac{64.07 \text{ g}}{\text{mol}}\right) = 65.6 \text{ g } SO_2$$

(e) $$(40.6 \text{ g } SO_2)\left(\frac{1 \text{ mol}}{64.07 \text{ g}}\right)\left(\frac{11 \text{ mol } O_2}{8 \text{ mol } SO_2}\right) = 0.871 \text{ mol } O_2$$

(f) $$(221 \text{ g } Fe_2O_3)\left(\frac{1 \text{ mol}}{159.7 \text{ g}}\right)\left(\frac{4 \text{ mol } FeS_2}{2 \text{ mol } Fe_2O_3}\right)\left(\frac{120.0 \text{ g}}{\text{mol}}\right) = 332 \text{ g } FeS_2$$

21. (a)

| $KOH$ | + | $HNO_3$ | $\rightarrow$ | $KNO_3$ | + | $H_2O$ |
|---|---|---|---|---|---|---|
| 16.0 g | | 12.0 g | | | | |

Choose one of the products and calculate its mass that would be produced from each given reactant. Using $KNO_3$ as the product:

$$(16.0 \text{ g KOH})\left(\frac{1 \text{ mol}}{56.10 \text{ g}}\right)\left(\frac{1 \text{ mol } KNO_3}{1 \text{ mol KOH}}\right)\left(\frac{101.1 \text{ g}}{\text{mol}}\right) = 28.8 \text{ g } KNO_3$$

$$(12.0 \text{ g } HNO_3)\left(\frac{1 \text{ mol}}{63.02 \text{ g}}\right)\left(\frac{1 \text{ mol } KNO_3}{1 \text{ mol KOH}}\right)\left(\frac{101.1 \text{ g}}{\text{mol}}\right) = 19.3 \text{ g } KNO_3$$

Since $HNO_3$ produces less $KNO_3$, it is the limiting reactant and KOH is in excess.

(b) $2\,NaOH + H_2SO_4 \rightarrow Na_2SO_4 + 2\,H_2O$
10.0 g 10.0 g

Choose one of the products and calculate its mass that would be produced from each given reactant. Using $H_2O$ as the product:

$$(10.0\text{ g NaOH})\left(\frac{1\text{ mol}}{40.00\text{ g}}\right)\left(\frac{2\text{ mol H}_2\text{O}}{2\text{ mol NaOH}}\right)\left(\frac{18.02\text{ g}}{\text{mol}}\right) = 4.51\text{ g H}_2\text{O}$$

$$(10.0\text{ g H}_2\text{SO}_4)\left(\frac{1\text{ mol}}{98.09\text{ g}}\right)\left(\frac{2\text{ mol H}_2\text{O}}{1\text{ mol H}_2\text{SO}_4}\right)\left(\frac{18.02\text{ g}}{\text{mol}}\right) = 3.67\text{ g H}_2\text{O}$$

Since $H_2SO_4$ produces less $H_2O$, it is the limiting reactant and NaOH is in excess.

22. (a) $2\,Bi(NO_3)_3 + H_2S \rightarrow Bi_2S_3 + 6\,HNO_3$
50.0 g 6.00 g

Choose one of the products and calculate its mass that would be produced from each given reactant. Using $Bi(NO_3)_3$ as the product:

$$(50.0\text{ g Bi(NO}_3)_3)\left(\frac{1\text{ mol}}{395.0\text{ g}}\right)\left(\frac{1\text{ mol Bi}_2\text{S}_3}{2\text{ mol Bi(NO}_3)_3}\right)\left(\frac{514.2\text{ g}}{\text{mol}}\right) = 32.5\text{ g Bi}_2\text{S}_3$$

$$(6.00\text{ g H}_2\text{S})\left(\frac{1\text{ mol}}{34.09\text{ g}}\right)\left(\frac{1\text{ mol Bi}_2\text{S}_3}{3\text{ mol H}_2\text{S}}\right)\left(\frac{514.2\text{ g}}{\text{mol}}\right) = 30.2\text{ g Bi}_2\text{S}_3$$

Since $H_2S$ produces less $Bi_2S_3$, it is the limiting reactant and $Bi(NO_3)_3$ is in excess.

(b) $3\,Fe + 4\,H_2O \rightarrow Fe_3O_4 + 4\,H_2$
40.0 g 16.0 g

Choose one of the products and calculate its mass that would be produced from each given reactant. Using $H_2$ as the product:

$$(40.0\text{ g Fe})\left(\frac{1\text{ mol}}{55.85\text{ g}}\right)\left(\frac{4\text{ mol H}_2}{3\text{ mol Fe}}\right)\left(\frac{2.016\text{ g}}{\text{mol}}\right) = 1.93\text{ g H}_2$$

$$(16.0\text{ g H}_2\text{O})\left(\frac{1\text{ mol}}{18.02\text{ g}}\right)\left(\frac{4\text{ mol H}_2}{4\text{ mol H}_2\text{O}}\right)\left(\frac{2.016\text{ g}}{\text{mol}}\right) = 1.79\text{ g H}_2$$

Since $H_2O$ produces less $H_2$, it is the limiting reactant and Fe is in excess.

23. Limiting reactant calculations

$C_3H_8 + 5\ O_2 \rightarrow 3\ CO_2 + 4\ H_2O$

(a) Reaction between 20.0 g $C_3H_8$ and 20.0 g $O_2$
Convert each amount to grams of $CO_2$

$$(20.0\text{ g } C_3H_8)\left(\frac{1\text{ mol}}{44.09\text{ g}}\right)\left(\frac{3\text{ mol } CO_2}{1\text{ mol } C_3H_8}\right)\left(\frac{44.01\text{ g}}{\text{mol}}\right) = 59.9\text{ g } CO_2$$

$$(20.0\text{ g } O_2)\left(\frac{1\text{ mol}}{32.00\text{ g}}\right)\left(\frac{3\text{ mol } CO_2}{5\text{ mol } O_2}\right)\left(\frac{44.01\text{ g}}{\text{mol}}\right) = 16.5\text{ g } CO_2$$

$O_2$ is the limiting reactant. The yield is 16.5 g $CO_2$.

(b) Reaction between 20.0 g $C_3H_8$ and 80.0 g $O_2$
Convert each amount to grams of $CO_2$

$$(20.0\text{ g } C_3H_8)\left(\frac{1\text{ mol}}{44.09\text{ g}}\right)\left(\frac{3\text{ mol } CO_2}{1\text{ mol } C_3H_8}\right)\left(\frac{44.01\text{ g}}{\text{mol}}\right) = 59.9\text{ g } CO_2$$

$$(80.0\text{ g } O_2)\left(\frac{1\text{ mol}}{32.00\text{ g}}\right)\left(\frac{3\text{ mol } CO_2}{5\text{ mol } O_2}\right)\left(\frac{44.01\text{ g}}{\text{mol}}\right) = 66.0\text{ g } CO_2$$

$C_3H_8$ is the limiting reactant. The yield is 59.9 g $CO_2$.

(c) Reaction between 2.0 mol $C_3H_8$ and 14.0 mol $O_2$
According to the equation, 2 mol $C_3H_8$ will react with 10 mol $O_2$. Therefore, $C_3H_8$ is the limiting reactant and 4.0 mol $O_2$ will remain unreacted.

$$(2.0\text{ mol } C_3H_8)\left(\frac{3\text{ mol } CO_2}{1\text{ mol } C_3H_8}\right) = 6.0\text{ mol } CO_2\text{ produced}$$

$$(2.0\text{ mol } C_3H_8)\left(\frac{4\text{ mol } H_2O}{1\text{ mol } C_3H_8}\right) = 8.0\text{ mol } H_2O\text{ produced}$$

When the reaction is completed, 6.0 mol $CO_2$, 8.0 $H_2O$, and 4.0 mol $O_2$ will be in the container.

24. Limiting reactant calculations

$C_3H_8 + 5\ O_2 \rightarrow 3\ CO_2 + 4\ H_2O$

(a) Reaction between 5.0 mol $C_3H_8$ and 5 mol $O_2$

$$(5.0\text{ mol } C_3H_8)\left(\frac{3\text{ mol } CO_2}{1\text{ mol } C_3H_8}\right) = 15.0\text{ mol } CO_2$$

$$(5.0 \text{ mol } O_2)\left(\frac{3 \text{ mol } CO_2}{5 \text{ mol } O_2}\right) = 3.0 \text{ mol } CO_2$$

The $O_2$ is the limiting reactant; 3.0 mol $CO_2$ produced.

(b) Reaction between 3.0 mol $C_3H_8$ and 20.0 mol $O_2$

$$(3.0 \text{ mol } C_3H_8)\left(\frac{3 \text{ mol } CO_2}{1 \text{ mol } C_3H_8}\right) = 9.0 \text{ mol } CO_2$$

$$(20.0 \text{ mol } O_2)\left(\frac{3 \text{ mol } CO_2}{5 \text{ mol } O_2}\right) = 12.0 \text{ mol } CO_2$$

The $C_3H_8$ is the limiting reactant; 9.0 mol $CO_2$ produced.

(c) Reaction between 20.0 mol $C_3H_8$ and 3.0 mol $O_2$
According to the equation, 1 mol $C_3H_8$ will react with 5 mol $O_2$, $O_2$ is clearly the limiting reactant.

$$(3.0 \text{ mol } O_2)\left(\frac{3 \text{ mol } CO_2}{5 \text{ mol } O_2}\right) = 1.8 \text{ mol } CO_2 \text{ produced}$$

25. $X_8 + 12\, O_2 \rightarrow 8\, XO_3$

$$(120.0 \text{ g } O_2)\left(\frac{1 \text{ mol}}{32.00 \text{ g}}\right)\left(\frac{1 \text{ mol } X_8}{12 \text{ mol } O_2}\right) = 0.3125 \text{ mol } X_8 \qquad 80.0 \text{ g } X_8 = 0.3125 \text{ mol } X_8$$

$$\frac{80.0 \text{ g}}{0.3125 \text{ mol}} = 256 \text{ g/mol } X_8$$

$$\text{molar mass } X = \frac{256 \frac{\text{g}}{\text{mol}}}{8} = 32.0 \frac{\text{g}}{\text{mol}}$$

Using the periodic table we find that the element with 32.0 g/mol is sulfur.

26. $X + 2\, HCl \rightarrow XCl_2 + H_2$

$$(2.42 \text{ g } H_2)\left(\frac{1 \text{ mol}}{2.016 \text{ g}}\right)\left(\frac{1 \text{ mol } X}{1 \text{ mol } H_2}\right) = 1.20 \text{ mol } X \qquad 78.5 \text{ g } X = 1.20 \text{ mol } X$$

$$\frac{78.5 \text{ g}}{1.20 \text{ mol}} = 65.4 \text{ g/mol}$$

Using the periodic table we find that the element with atomic mass 65.4 is zinc.

27. Limiting reactant calculation and percentage yield

$2\ Al + 3\ Br_2 \rightarrow 2\ AlBr_3$

Reaction between 25.0 g Al and 100. g $Br_2$
Calculate the grams of $AlBr_3$ from each reactant.

$$(25.0\ g\ Al)\left(\frac{1\ mol}{26.98\ g}\right)\left(\frac{2\ mol\ AlBr_3}{2\ mol\ Al}\right)\left(\frac{266.7\ g}{mol}\right) = 247\ g\ AlBr_3$$

$$(100.\ g\ Br_2)\left(\frac{1\ mol}{159.8\ g}\right)\left(\frac{2\ mol\ AlBr_3}{3\ mol\ Br_2}\right)\left(\frac{266.7\ g}{mol}\right) = 111\ g\ AlBr_3$$

$Br_2$ is limiting; 111 g $AlBr_3$ is the theoretical yield of product.

$$\text{Percent yield} = \left(\frac{\text{actual yield}}{\text{theoretical yield}}\right)(100) = \left(\frac{64.2\ g}{111\ g}\right)(100) = 57.8\%$$

28. Percent yield calculation

$Fe(s) + CuSO_4(aq) \rightarrow Cu(s) + FeSO_4(aq)$

$$(400.\ g\ CuSO_4)\left(\frac{1\ mol}{159.6\ g}\right)\left(\frac{1\ mol\ Cu}{1\ mol\ CuSO_4}\right)\left(\frac{63.55\ g}{mol}\right) = 159\ g\ Cu\ \text{(theoretical yield)}$$

$$\%\ \text{yield} = \left(\frac{\text{actual yield}}{\text{theoretical yield}}\right)(100) = \left(\frac{151\ g}{159\ g}\right)(100) = 95.0\%\ \text{yield of Cu}$$

29. The balanced equation is $3\ C + 2\ SO_2 \rightarrow CS_2 + 2\ CO_2$

Calculate the g C needed to produce 950 g $CS_2$ taking into account that the yield of $CS_2$ is 86.0%. First calculate the theoretical yield of $CS_2$.

$$\frac{950\ g\ CS_2}{0.860} = 1.1 \times 10^3\ g\ CS_2\ \text{(theoretical yield)}$$

Now calculate the grams of coke needed to produce $1.1 \times 10^3$ g $CS_2$.

$$(1.1 \times 10^3\ g\ CS_2)\left(\frac{1\ mol}{76.15\ g}\right)\left(\frac{3\ mol\ C}{1\ mol\ CS_2}\right)\left(\frac{12.01\ g}{mol}\right) = 5.2 \times 10^2\ g\ C$$

30. The balanced equation is $CaC_2 + 2\ H_2O \rightarrow C_2H_2 + Ca(OH)_2$

First calculate the grams of pure $CaC_2$ in the sample from the amount of $C_2H_2$ produced.

$$(0.540\ mol\ C_2H_2)\left(\frac{1\ mol\ CaC_2}{1\ mol\ C_2H_2}\right)\left(\frac{64.10\ g}{mol}\right) = 34.6\ g\ CaC_2\ \text{in the impure sample}$$

Now calculate the percent $CaC_2$ in the impure sample.

$$\left(\frac{34.6\text{ g CaC}_2}{44.5\text{ g sample}}\right)(100) = 77.8\%\text{ CaC}_2 \text{ in the impure sample}$$

31. No. There are not enough screwdrivers, wrenches or pliers. 2400 screwdrivers, 3600 wrenches and 1200 pliers are needed for 600 tool sets.

32. A subscript is used to indicate the number of atoms in a formula. It cannot be changed without changing the identity of the substance. Coefficients are used only to balance atoms in chemical equations. They may be changed as needed to achieve a balanced equation.

33. $4\text{ KO}_2 + 2\text{ H}_2\text{O} + 4\text{ CO}_2 \rightarrow 4\text{ KHCO}_3 + 3\text{ O}_2$

(a) $$\left(\frac{0.85\text{ g CO}_2}{\text{min}}\right)\left(\frac{1\text{ mol}}{44.01\text{ g}}\right)\left(\frac{4\text{ mol KO}_2}{4\text{ mol CO}_2}\right) = \frac{0.019\text{ mol KO}_2}{\text{min}}$$

$$\left(\frac{0.019\text{ mol KO}_2}{\text{min}}\right)(10.0\text{ min}) = 0.19\text{ mol KO}_2$$

(b) The conversion is: $\frac{\text{g CO}_2}{\text{min}} \rightarrow \frac{\text{mol CO}_2}{\text{min}} \rightarrow \frac{\text{mol O}_2}{\text{min}} \rightarrow \frac{\text{g O}_2}{\text{min}} \rightarrow \frac{\text{g O}_2}{\text{hr}}$

$$\left(\frac{0.85\text{ g CO}_2}{\text{min}}\right)\left(\frac{1\text{ mol}}{44.01\text{ g}}\right)\left(\frac{3\text{ mol O}_2}{4\text{ mol CO}_2}\right)\left(\frac{32.00\text{ g}}{\text{mol}}\right)\left(\frac{60.0\text{ min}}{1.0\text{ hr}}\right) = \frac{28\text{ g O}_2}{\text{hr}}$$

34. (a) $$(750\text{ g C}_6\text{H}_{12}\text{O}_6)\left(\frac{1\text{ mol}}{180.2\text{ g}}\right)\left(\frac{2\text{ mol C}_2\text{H}_5\text{OH}}{1\text{ mol C}_6\text{H}_{12}\text{O}_6}\right)\left(\frac{46.07\text{ g}}{\text{mol}}\right) = 380\text{ g C}_2\text{H}_5\text{OH}$$

$$(750\text{ g C}_6\text{H}_{12}\text{O}_6)\left(\frac{1\text{ mol}}{180.2\text{ g}}\right)\left(\frac{2\text{ mol CO}_2}{1\text{ mol C}_6\text{H}_{12}\text{O}_6}\right)\left(\frac{44.01\text{ g}}{\text{mol}}\right) = 370\text{ g CO}_2$$

(b) $$(380\text{ g C}_2\text{H}_5\text{OH})\left(\frac{1\text{ mL}}{0.79\text{ g}}\right) = 480\text{ mL C}_2\text{H}_5\text{OH}$$

35. $4\text{ P} + 5\text{ O}_2 \rightarrow \text{P}_4\text{O}_{10}$

$\text{P}_4\text{O}_{10} + 6\text{ H}_2\text{O} \rightarrow 4\text{ H}_3\text{PO}_4$

In the first reaction:

$$(20.0\text{ g P})\left(\frac{1\text{ mol}}{30.97\text{ g}}\right) = 0.646\text{ mol P}$$

$$(30.0 \text{ g } O_2)\left(\frac{1 \text{ mol}}{32.00 \text{ g}}\right) = 0.938 \text{ mol } O_2$$

This is a ratio of $\dfrac{0.646 \text{ mol P}}{0.938 \text{ mol } O_2} = \dfrac{3.44 \text{ mol P}}{5.00 \text{ mol } O_2}$

Therefore, P is the limiting reactant and the $P_4O_{10}$ produced is:

$$(0.646 \text{ mol P})\left(\frac{1 \text{ mol } P_4O_{10}}{4 \text{ mol P}}\right) = 0.162 \text{ mol } P_4O_{10}$$

In the second reaction:

$$(15.0 \text{ g } H_2O)\left(\frac{1 \text{ mol}}{18.02 \text{ g}}\right) = 0.832 \text{ mol } H_2O$$

and we have 0.162 mol $P_4O_{10}$. The ratio of $\dfrac{H_2O}{P_4O_{10}}$ is $\dfrac{0.832 \text{ mol}}{0.162 \text{ mol}} = \dfrac{5.14 \text{ mol}}{1.00 \text{ mol}}$

Therefore, $H_2O$ is the limiting reactant and the $H_3PO_4$ produced is:

$$(0.832 \text{ mol } H_2O)\left(\frac{4 \text{ mol } H_3PO_4}{6 \text{ mol } H_2O}\right)\left(\frac{97.99 \text{ g}}{\text{mol}}\right) = 54.4 \text{ g } H_3PO_4$$

36. $2\ CH_3OH + 3\ O_2 \rightarrow 2\ CO_2 + 4\ H_2O$

The conversion is: mL $CH_3OH$ → g $CH_3OH$ → mol $CH_3OH$ → mol $O_2$ → g $O_2$

$$(60.0 \text{ mL } CH_3OH)\left(\frac{0.72 \text{ g}}{\text{mL}}\right)\left(\frac{1 \text{ mol}}{32.04 \text{ g}}\right)\left(\frac{3 \text{ mol } O_2}{2 \text{ mol } CH_3OH}\right)\left(\frac{32.00 \text{ g}}{\text{mol}}\right) = 65 \text{ g } O_2$$

37. $7\ H_2O_2 + N_2H_4 \rightarrow 2\ HNO_3 + 8\ H_2O$

(a) $(0.33 \text{ mol } N_2H_4)\left(\dfrac{2 \text{ mol } HNO_3}{1 \text{ mol } N_2H_4}\right) = 0.66 \text{ mol } HNO_3$

(b) $(2.75 \text{ mol } H_2O)\left(\dfrac{7 \text{ mol } H_2O_2}{8 \text{ mol } H_2O}\right) = 2.41 \text{ mol } H_2O_2$

(c) $(8.72 \text{ mol } HNO_3)\left(\dfrac{8 \text{ mol } H_2O}{2 \text{ mol } HNO_3}\right) = 34.9 \text{ mol } H_2O$

(d) $(120 \text{ g } N_2H_4)\left(\dfrac{1 \text{ mol}}{32.05 \text{ g}}\right)\left(\dfrac{7 \text{ mol } H_2O_2}{1 \text{ mol } N_2H_4}\right)\left(\dfrac{34.02 \text{ g}}{\text{mol}}\right) = 8.9 \times 10^2 \text{ g } H_2O_2$

38. $4\ Ag + 2\ H_2S + O_2 \rightarrow 2\ Ag_2S + 2\ H_2O$

$$(1.1\ g\ Ag)\left(\frac{1\ mol}{107.9\ g}\right)\left(\frac{2\ mol\ Ag_2S}{4\ mol\ Ag}\right)\left(\frac{247.9\ g}{mol}\right) = 1.3\ g\ Ag_2S$$

$$(0.14\ g\ H_2S)\left(\frac{1\ mol}{34.09\ g}\right)\left(\frac{2\ mol\ Ag_2S}{2\ mol\ H_2S}\right)\left(\frac{247.9\ g}{mol}\right) = 1.0\ g\ Ag_2S$$

$$(0.080\ g\ O_2)\left(\frac{1\ mol}{32.00\ g}\right)\left(\frac{2\ mol\ Ag_2S}{1\ mol\ O_2}\right)\left(\frac{247.9\ g}{mol}\right) = 1.2\ g\ Ag_2S$$

The $H_2S$ is the limiting reactant so 1.0 g $Ag_2S$ is formed.

39. The balanced equation is $Zn + 2\ HCl \rightarrow ZnCl_2 + H_2$

180.0 g Zn - 35 g Zn = 145 g Zn reacted with HCl

(a) $$(145\ g\ Zn)\left(\frac{1\ mol}{65.39\ g}\right)\left(\frac{1\ mol\ H_2}{1\ mol\ Zn}\right) = 2.22\ mol\ H_2\ produced$$

(b) $$(145\ g\ Zn)\left(\frac{1\ mol}{65.39\ g}\right)\left(\frac{2\ mol\ HCl}{1\ mol\ Zn}\right)\left(\frac{36.46\ g}{mol}\right) = 162\ g\ HCl\ reacted$$

40. $Fe + CuSO_4 \rightarrow Cu + FeSO_4$
2.0 mol   3.0 mol

(a) 2.0 mol Fe react with 2.0 mol $CuSO_4$ to yield 2.0 mol Cu and 2.0 mol $FeSO_4$. 1.0 mol $CuSO_4$ is unreacted. At the completion of the reaction, there will be 2.0 mol Cu, 2.0 mol $FeSO_4$, and 1.0 mol $CuSO_4$.

(b) Determine which reactant is limiting and then calculate the g $FeSO_4$ produced from that reactant.

$$(20.0\ g\ Fe)\left(\frac{1\ mol}{55.85\ g}\right)\left(\frac{1\ mol\ Cu}{1\ mol\ Fe}\right)\left(\frac{63.55\ g}{mol}\right) = 22.8\ g\ Cu$$

$$(40.0\ g\ CuSO_4)\left(\frac{1\ mol}{159.6\ g}\right)\left(\frac{1\ mol\ Cu}{1\ mol\ CuSO_4}\right)\left(\frac{63.55\ g}{mol}\right) = 15.9\ g\ Cu$$

Since $CuSO_4$ produces less Cu, it is the limiting reactant. Determine the mass of $FeSO_4$ produced from 40.0 g $CuSO_4$.

$$(40.0\ g\ CuSO_4)\left(\frac{1\ mol}{159.6\ g}\right)\left(\frac{1\ mol\ FeSO_4}{1\ mol\ CuSO_4}\right)\left(\frac{151.9\ g}{mol}\right) = 38.1\ g\ FeSO_4\ produced$$

Calculate the mass of unreacted Fe.

$$(40.0\ g\ CuSO_4)\left(\frac{1\ mol}{159.6\ g}\right)\left(\frac{1\ mol\ Fe}{1\ mol\ CuSO_4}\right)\left(\frac{55.85\ g}{mol}\right) = 14.0\ g\ Fe\ will\ react$$

Unreacted Fe = 20.0 g − 14.0 g = 6.0 g. Therefore, at the completion of the reaction, 15.9 g Cu, 38.1 g $FeSO_4$, 6.0 g Fe, and no $CuSO_4$ remain.

41. Limiting reactant calculation

$CO(g) + 2\,H_2(g) \rightarrow CH_3OH(l)$

Reaction between 40.0 g CO and 10.0 g $H_2$: determine the limiting reactant by calculating the amount of $CH_3OH$ that would be formed from each reactant.

$$(40.0\text{ g CO})\left(\frac{1\text{ mol}}{28.01\text{ g}}\right)\left(\frac{1\text{ mol }CH_3OH}{1\text{ mol CO}}\right)\left(\frac{32.04\text{ g}}{\text{mol}}\right) = 45.8\text{ g }CH_3OH$$

$$(10.0\text{ g }H_2)\left(\frac{1\text{ mol}}{2.016\text{ g}}\right)\left(\frac{1\text{ mol }CH_3OH}{2\text{ mol }H_2}\right)\left(\frac{32.04\text{ g}}{\text{mol}}\right) = 79.5\text{ g }CH_3OH$$

CO is limiting; $H_2$ is in excess; 45.8 g $CH_3OH$ will be produced.
Calculate the mass of unreacted $H_2$:

$$(40.0\text{ g CO})\left(\frac{1\text{ mol}}{28.01\text{ g}}\right)\left(\frac{2\text{ mol }H_2}{1\text{ mol CO}}\right)\left(\frac{2.016\text{ g}}{\text{mol}}\right) = 5.76\text{ g }H_2\text{ react}$$

10.0 g $H_2$ − 5.76 g $H_2$ = 4.2 g $H_2$ remain unreacted

42. The balanced equation is $C_6H_{12}O_6 \rightarrow 2\,C_2H_5OH + 2\,CO_2$

(a) First calculate the theoretical yield.

$$(750\text{ g }C_6H_{12}O_6)\left(\frac{1\text{ mol}}{180.2\text{ g}}\right)\left(\frac{2\text{ mol }C_2H_5OH}{1\text{ mol }C_6H_{12}O_6}\right)\left(\frac{46.07\text{ g}}{\text{mol}}\right) = 3.8 \times 10^2\text{ g }C_2H_5OH \text{ (theoretical yield)}$$

Then take 84.6% of the theoretical yield to obtain the actual yield.

$$\text{actual yield} = \frac{(\text{theoretical yield})(84.6)}{100} = \frac{(3.8\times10^2\text{ g }C_2H_5OH)(84.6)}{100}$$

$$= 3.2 \times 10^2\text{ g }C_2H_5OH$$

(b) 475 g $C_2H_5OH$ represents 84.6% of the theoretical yield. Calculate the theoretical yield.

$$\text{theoretical yield} = \frac{475\text{ g}}{0.846} = 561\text{ g }C_2H_5OH$$

Now calculate the g $C_6H_{12}O_6$ needed to produce 561 g $C_2H_5OH$.

$$(561\text{ g }C_2H_5OH)\left(\frac{1\text{ mol}}{46.07\text{ g}}\right)\left(\frac{1\text{ mol }C_6H_{12}O_6}{2\text{ mol }C_2H_5OH}\right)\left(\frac{180.2\text{ g}}{\text{mol}}\right) = 1.10 \times 10^3\text{ g }C_6H_{12}O_6$$

43. The balanced equations are:

$CaCl_2 + 2\ AgNO_3 \rightarrow Ca(NO_3)_2 + 2\ AgCl$

$MgCl_2 + 2AgNO_3 \rightarrow Mg(NO_3)_2 + 2\ AgCl$

1 mol of each salt will produce the same amount (2 mol) of AgCl. $MgCl_2$ has a higher percentage of Cl than $CaCl_2$ because Mg has a lower atomic mass than Ca. Therefore, on an equal mass basis, $MgCl_2$ will produce more AgCl than will $CaCl_2$.

Calculations show that 1.00 g $MgCl_2$ produces 3.01 g AgCl, and 1.00 g $CaCl_2$ produces 2.56 g AgCl.

44. The balanced equation is $Li_2O + H_2O \rightarrow 2\ LiOH$

The conversion is: g $H_2O$ → mol $H_2O$ → mol $Li_2O$ → g $Li_2O$ → kg $Li_2O$

$$\left(\frac{2500\ g\ H_2O}{\text{astronaut day}}\right)\left(\frac{1\ mol}{18.02\ g}\right)\left(\frac{1\ mol\ Li_2O}{1\ mol\ H_2O}\right)\left(\frac{29.88\ g}{mol}\right)\left(\frac{1\ kg}{1000\ g}\right) = \frac{4.1\ kg\ Li_2O}{\text{astronaut day}}$$

$$\left(\frac{4.1\ kg\ Li_2O}{\text{astronaut day}}\right)(30\ \text{days})(3\ \text{astronauts}) = 3.7 \times 10^2\ kg\ Li_2O$$

45. The balanced equation is

$H_2SO_4 + 2\ NaCl \rightarrow Na_2SO_4 + 2\ HCl$

First calculate the g HCl to be produced

$$(20.0\ L\ HCl\ \text{solution})\left(\frac{1000\ mL}{1\ L}\right)\left(\frac{1.20\ g}{1.00\ mL}\right)(0.420) = 1.01 \times 10^4\ g\ HCl$$

Then calculate the g $H_2SO_4$ required to produce the HCl

$$(1.01 \times 10^4\ g\ HCl)\left(\frac{1\ mol}{36.46\ g}\right)\left(\frac{1\ mol\ H_2SO_4}{2\ mol\ HCl}\right)\left(\frac{98.09\ g}{1\ mol}\right) = 1.36 \times 10^4\ g\ H_2SO_4$$

Finally, calculate the kg $H_2SO_4$ (96%)

$$(1.36 \times 10^4\ g\ H_2SO_4)\left(\frac{1.00\ g\ H_2SO_4\ \text{solution}}{0.96\ g\ H_2SO_4}\right)\left(\frac{1\ kg}{1000\ g}\right) = 14\ kg\ \text{concentrated}\ H_2SO_4$$

46. The balanced equation is

$Al(OH)_3(s) + 3\ HCl(aq) \rightarrow AlCl_3(aq) + 3\ H_2O(l)$

The conversion is: L HCl → g HCl → mol HCl → mol $Al(OH)_3$ → g $Al(OH)_3$

$$(2.5\ L)\left(\frac{3.0\ g\ HCl}{L}\right)\left(\frac{1\ mol}{36.46\ g}\right)\left(\frac{1\ mol\ Al(OH)_3}{3\ mol\ HCl}\right)\left(\frac{78.00\ g}{mol}\right) = 5.3\ g\ Al(OH)_3$$

Now calculate the number of tablets that contain 5.3 g $Al(OH)_3$

$$(5.3\ g\ Al(OH)_3)\left(\frac{1000\ mg}{g}\right)\left(\frac{1\ tablet}{400.\ mg}\right) = 13\ tablets$$

47. The balanced equation is $2\ KClO_3 \rightarrow 2\ KCl + 3\ O_2$

12.82 g mixture − 9.45 g residue = 3.37 g $O_2$ lost by heating

Because the $O_2$ lost came only from $KClO_3$, we can use it to calculate the amount of $KClO_3$ in the mixture.

$$(3.37\ g\ O_2)\left(\frac{1\ mol}{32.00\ g}\right)\left(\frac{2\ mol\ KClO_3}{3\ mol\ O_2}\right)\left(\frac{122.6\ g}{mol}\right) = 8.61\ g\ KClO_3 \text{ in the mixture}$$

$$\left(\frac{8.61\ g\ KClO_3}{12.82\ g\ sample}\right)(100) = 67\%\ KClO_3$$

CHAPTER 10

# MODERN ATOMIC THEORY

1. An electron orbital is a region in space around the nucleus of an atom where an electron is most probably found.

2. A second electron may enter an orbital already occupied by an electron if its spin is opposite that of the electron already in the orbital and all other orbitals of the same sublevel contain an electron.

3. All the electrons in the atom are located in the orbitals closest to the nucleus.

4. Both 1s and 2s orbitals are spherical in shape and located symmetrically around the nucleus. The sizes of the spheres are different – the radius of the 2s orbital is larger than the 1s. The electrons in 2s orbitals are located further from the nucleus.

5. The energy sublevels are s, p, d, and f.

6. 1s, 2s, 2p, 3s, 3p, 4s, 3d, 4p

7. s – 2 electrons per shell
   p – 6 electrons per shell after the first energy level
   d – 10 electrons per shell after the second energy level

8. The main difference is that the Bohr orbit has an electron traveling a specific path while an orbital is a region in space where the electron is most probably found.

9. Bohr's model was inadequate since it could not account for atoms more complex than hydrogen. It was modified by Schrodinger into the modern concept of the atom in which electrons exhibit wave and particle properties. The motion of electrons is determined only by probability functions as a region in space, or a cloud surrounding the nucleus.

10. s orbital

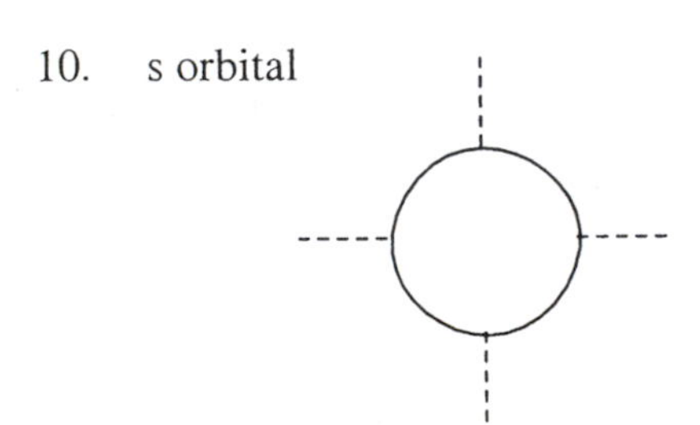

p orbitals

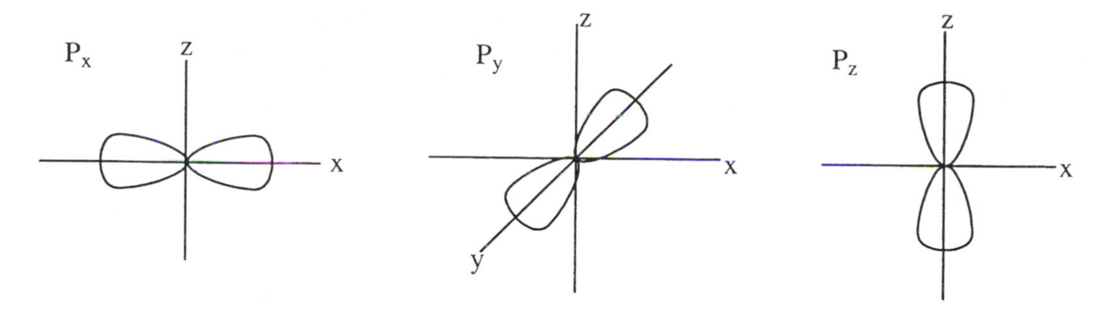


11. 3 is the third energy level
    d indicates an energy sublevel
    7 indicates the number of electrons in the d sublevel

12. Transition elements are found in the center of the periodic table. The last electrons for these elements are found in the d or f orbitals.
    Representative elements are located on either side of the periodic table (Group IA – VIIA). The valence electrons for these elements are found in s and/or p orbitals.

13. Elements in the s-block all have one or two electrons in the outermost shell. These valence electrons are located in an s-orbital.

14.

| Atomic # | Symbol |
|---|---|
| 8 | O |
| 16 | S |
| 34 | Se |
| 52 | Te |
| 84 | Po |

All of these elements have an outermost electron structure of $s^2p^4$

15. F, Cl, Br, I, At (Halogens)

16. The greatest number of elements in any period is 32. The 6$^{th}$ period has this number of electrons.

17. The elements in Group A always have their last electrons in the outermost energy level, while the last electrons in Group B lie in an inner level.

18. Pairs of elements which are out of sequence with respect to atomic masses are: Ar and K; Co and Ni; Te and I; Th and Pa; U and Np; La and Rf; Pu and Am.

19. (a) H 1 proton (c) Sc 21 protons
    (b) B 5 protons (d) U 92 protons

20. (a) F 9 protons (c) Br 35 protons
    (b) Ag 47 protons (d) Sb 51 protons

21. (a) B $1s^22s^22p^1$
    (b) Ti $1s^22s^22p^63s^23p^64s^23d^2$
    (c) Zn $1s^22s^22p^63s^23p^64s^23d^{10}$
    (d) Sr $1s^22s^22p^63s^23p^64s^23d^{10}4p^65s^2$

22. (a) Cl $1s^22s^22p^63s^23p^5$
    (b) Ag $1s^22s^22p^63s^23p^64s^23d^{10}4p^65s^14d^{10}$
    (c) Li $1s^22s^1$
    (d) Fe $1s^22s^22p^63s^23p^64s^23d^6$
    (e) I $1s^22s^22p^63s^23p^64s^23d^{10}4p^65s^24d^{10}5p^5$

23. The spectral lines of hydrogen are produced by energy emitted when an electron falls from a higher energy level to a lower energy level (closer to the nucleus).

24. Bohr said that a number of orbits were available for electrons, each corresponding to an energy level. When an electron falls from a higher energy orbit to a lower energy orbit, energy is given off as a specific wavelength of light. Only those energies in the visible range are seen in the hydrogen spectrum. Each line corresponds to a change from one orbit to another.

25. 9 orbitals in the third energy level: 3s, $3p_x$, $3p_y$, $3p_z$ plus five d orbitals

26. 32 electrons in the fourth energy level

27. (a) (7p 7n) $2e^-$ $5e^-$ $^{14}_{7}N$

    (b) (17p 18n) $2e^-$ $8e^-$ $7e^-$ $^{35}_{17}Cl$

    (c) (30p 35n) $2e^-$ $8e^-$ $18e^-$ $2e^-$ $^{65}_{30}Zn$

| | | Nucleus | Electrons | Symbol |
|---|---|---|---|---|
| | (d) | 40p 51n | $2e^-$ $8e^-$ $18e^-$ $10e^-$ $2e^-$ | $^{91}_{40}Zr$ |
| | (e) | 53p 74n | $2e^-$ $8e^-$ $18e^-$ $18e^-$ $7e^-$ | $^{127}_{53}I$ |
| 28. | (a) | 14p 14n | $2e^-$ $8e^-$ $4e^-$ | $^{28}_{14}Si$ |
| | (b) | 16p 16n | $2e^-$ $8e^-$ $6e^-$ | $^{32}_{16}S$ |
| | (c) | 18p 22n | $2e^-$ $8e^-$ $8e^-$ | $^{40}_{18}Ar$ |
| | (d) | 23p 28n | $2e^-$ $8e^-$ $11e^-$ $2e^-$ | $^{51}_{23}V$ |
| | (e) | 15p 16n | $2e^-$ $8e^-$ $5e^-$ | $^{31}_{15}P$ |

29. (a) Mg (c) Ni
    (b) Al (d) Mn

30. (a) Sc (c) Sn
    (b) Zn (d) Cs

31.

| | atomic No. | electron structure |
|---|---|---|
| (a) | 8 | $1s^22s^22p^4$ |
| (b) | 11 | $1s^22s^22p^63s^1$ |
| (c) | 17 | $1s^22s^22p^63s^23p^5$ |
| (d) | 23 | $1s^22s^22p^63s^23p^64s^23d^3$ |
| (e) | 28 | $1s^22s^22p^63s^23p^64s^23d^8$ |
| (f) | 34 | $1s^22s^22p^63s^23p^64s^23d^{10}4p^4$ |

32.

| | atomic No. | electron structure |
|---|---|---|
| (a) | 9 | $[He]2s^22p^5$ |
| (b) | 26 | $[Ar]4s^23d^6$ |
| (c) | 31 | $[Ar]4s^23d^{10}4p^1$ |
| (d) | 39 | $[Kr]5s^24d^1$ |
| (e) | 52 | $[Kr]5s^24d^{10}5p^4$ |
| (f) | 10 | $[He]2s^22p^6$ |

33. (a) $^{32}_{16}S$ (b) $^{60}_{28}Ni$

34. (a) (13p 14n) $2e^-$ $8e^-$ $3e^-$ $^{27}_{13}Al$

(b) (22p 26n) $2e^-$ $8e^-$ $10e^-$ $2e^-$ $^{48}_{22}Ti$

35. The eleventh electron of sodium is located in the third energy level because the first and second levels are filled.

36. The last electron in potassium is located in the fourth energy level because the 4s orbital is at a lower energy level than the 3d orbital

37. Noble gases all have filled s and p orbitals in the outermost energy level.

38. Noble gases each have filled s and p orbitals in the outermost energy level.

39. Moving from left to right in any period of elements, the atomic number increases by one from one element to the next and the atomic radius generally decreases. Each period (except period 1) begins with an alkali metal and ends with a noble gas. There is a trend in properties of the elements changing from metallic to nonmetallic from the beginning to the end of the period.

40. All the elements in a Group have the same number of outer shell electrons.

41. The outermost energy level contains one electron in an s orbital.

42. All of these elements have a $s^2d^{10}$ electron configuration in their outermost energy levels.

43. (a) and (g)
(b) and (d)
(c) and (f)

44. (a) and (f)
    (e) and (h)

45. 12, 38 since they are in the same periodic group

46. 7, 33 since they are in the same periodic group

47. (a) K, metal (c) S, nonmetal
    (b) Pu, metal (d) Sb, metalloid

48. (a) I, nonmetal (c) Mo, metal
    (b) W, metal (d) Ge, metalloid

49. Period 6, lanthanide series, contains the first element with an electron in an f orbital.

50. Period 4 group IIIB contains the first element with an electron in a d orbital.

51. Group VIIA contain 7 valence electrons.
    Group VIIB contain 2 electrons in the outermost level and 5 electrons in an inner d orbital.
    Group A elements are representative while Group B elements are transition elements.

52. Group IIIA contain 3 valence electrons
    Group IIIB contain 2 electrons in the outermost level and one electron in an inner d orbital.
    Group A elements are representative while Group B elements are transition elements.

53. $1s^3$ Li Atomic No. 3
    $1s^3 2s^3 2p^9$ P Atomic No. 15
    $1s^3 2s^3 2p^9 3s^3 3p^9$ Co Atomic No. 27

54. Nitrogen has more valence electrons on more energy levels. More varied electron transitions are possible.

55. (a) all 100
    (b) all but H, He, Li, Be (96)
    (c) 80 (all but those from H to Ca)
    (d) the 44 elements beginning after Ba

56. (a) $\frac{2}{2} \times 100 = 100\%$

(b) $\frac{4}{4} \times 100 = 100\%$

(c) $\frac{10}{54} \times 100 = 19\%$

(d) $\frac{8}{34} \times 100 = 24\%$

(e) $\frac{11}{55} \times 100 = 20\%$

57. (a) :Ȯ: 2 pairs of valence electrons

(b) ·P̈· 1 pair of valence electrons

(c) :Ï: 3 pairs of valence electrons

(d) :Xe: 4 pairs of valence electrons

(e) Rb· 0 pairs of valence electrons

58. $\dfrac{1.5}{1.0 \times 10^{-8}} = 150{,}000{,}000 = \dfrac{1.5 \times 10^{8}}{1}$

59. (a) Ne (c) F

(b) Ge (d) N

60. The outermost electron structure for both sulfur and oxygen is $s^2p^4$.

61. Transition elements are found in Groups IB – VIIB and VIII.

62. In transition elements the last electron added is in a d or f orbital. The last electron added in a representative element is in an s or p orbital.

63. Elements number 8, 16, 34, 52, 84 all have 6 electrons in their outer shell.

64. The outermost energy level is the $7^{th}$. Element 87 contains one electron in the 7s orbital.

65. If 36 is a noble gas, 35 would be in periodic group VIIA and 37 would be in periodic group IA.

66. Answers will vary but should at least include a statement about: (1) Numbering of the elements and their relationship to atomic structure; (2) division of the elements into periods and groups; (3) division of the elements into metals, nonmetals, and metalloids; (4) identification and location of the representative and transition elements.

67. (a) $[Rn]7s^2 5f^{14} 6d^{10} 7p^5$

 (b) 7 valence electrons, $7s^2 7p^5$

 (c) F, Cl, Br, I, At

 (d) halogen family, Period 7

68. (a) The two elements are isotopes.

 (b) The two elements are adjacent to each other in the same period.

69. Most gases are located in the upper right part of the periodic table (H is an exception). They are nonmetals. Liquids show no pattern. Neither do solids, except the vast majority of solids are metals.

CHAPTER 11

# CHEMICAL BONDS: THE FORMATION OF COMPOUNDS FROM ATOMS

1. smallest Cl, Mg, Na, K, Rb largest.

2. More energy is required for neon because it has a very stable outer shell electron structure consisting of an octet of electrons in filled orbitals (noble gas electron structure). Sodium, an alkali metal, has a relatively unstable outer shell electron structure with a single electron in an unfilled orbital. The sodium electron is also farther away from the nucleus and is shielded by more inner electron shells than are neon outer shell electrons.

3. When a third electron is removed from beryllium, it must come from a very stable electron shell structure corresponding to that of the noble gas, helium. In addition, the third electron must be removed from a +2 beryllium ion, which increases the difficulty of removing it.

4. The first ionization energy decreases from top to bottom because in the successive alkali metals, the outermost electron is farther away from the nucleus and is more shielded from the positive nucleus by additional electron shells.

5. The first ionization energy decreases from top to bottom because the outermost electrons in the successive noble gases are farther away from the nucleus and are more shielded by additional inner electron shells.

6. Barium and beryllium are in the same family. The electron to be removed from barium is, however, located in an energy level farther away from the nucleus than is the energy level holding the electron in beryllium. Hence, it requires less energy to remove the electron from barium than to remove the electron from beryllium. Barium, therefore, has a lower ionization energy than beryllium.

7. The first electron removed from a sodium atom is the one outer-shell electron, which is shielded from most of its nuclear charge by the electrons of lower levels. To remove a second electron from the sodium ion requires breaking into the noble gas structure. This requires much more energy than that necessary to remove the first electron, because the $Na^+$ is already positive.

8. (a) Ka > Na
   (b) Na > Mg
   (c) O > F
   (d) I > Br
   (e) Zr > Ti

9. The first element in each group has the smallest radius.

10. Atomic size increases down a column since each successive element has an additional energy level which contains electrons located farther from the nucleus.

11.

| Group | IA | IIA | IIIA | IVA | VA | VIA | VIIA |
|---|---|---|---|---|---|---|---|
| | E• | E: | E: | •E: | •E: | •E: | •E: |

12. Lewis structures:

Cs• Ba: Tl: •Pb: •Po: •At: :Rn:

Each of these is a representative element and has the same number of electrons in its outer shell as its periodic group.

13. (a) Elements with the highest electronegativities are found in the upper right hand corner of the periodic table.

(b) Elements with the lowest electronegativities are found in the lower left of the periodic table.

14. Valence electrons are the electrons found in the outermost energy level of an atom.

15. By losing one electron, a potassium atom acquires a noble gas structure and becomes a $K^+$ ion. To become a $K^{2+}$ ion requires the loss of a second electron and breaking into the noble gas structure of the $K^+$ ion. This requires too much energy to generally occur.

16. An aluminum ion has a +3 charge because it has lost 3 electrons in acquiring a noble gas electron structure.

17. Magnesium atom is larger because it has electrons in the 3rd shell, while a magnesium ion does not. Also, the ion has 12 protons and 10 electrons, creating a charge imbalance and drawing the electrons in towards the nucleus more closely.

18. A bromine atom is smaller because it has one less electron than the bromine ion in its outer shell. Also, the ion has 35 protons and 36 electrons, creating a charge imbalance, resulting in a lessening of the attraction of the electrons towards the nucleus.

19.

| | + | − |
|---|---|---|
| (a) | H | O |
| (b) | Na | F |
| (c) | H | N |
| (d) | Pb | S |
| (e) | N | O |
| (f) | H | C |

20.

| | + | − |
|---|---|---|
| (a) | H | Cl |
| (b) | Li | H |
| (c) | C | Cl |
| (d) | I | Br |
| (e) | Mg | H |
| (f) | O | F |

21. (a) ionic (b) covalent (c) covalent (d) ionic

22. (a) covalent (b) ionic (c) covalent (d) ionic

23. Magnesium has an electron structure $1s^2 2s^2 2p^6 3s^2$, while the structure for chlorine is $1s^2 2s^2 2p^6 3s^2 3p^5$. When these two elements react with each other, each magnesium atom loses its $3s^2$ electrons, one to each of two chlorine atoms. The resulting structures for both magnesium and chlorine are noble gas configurations.

24. (a) $F + 1\ e^- \rightarrow F^-$ (b) $Ca \rightarrow Ca^{2+} + 2\ e^-$

25. (a) Mg: + ·F: + ·F: ⟶ $MgF_2$

(b) 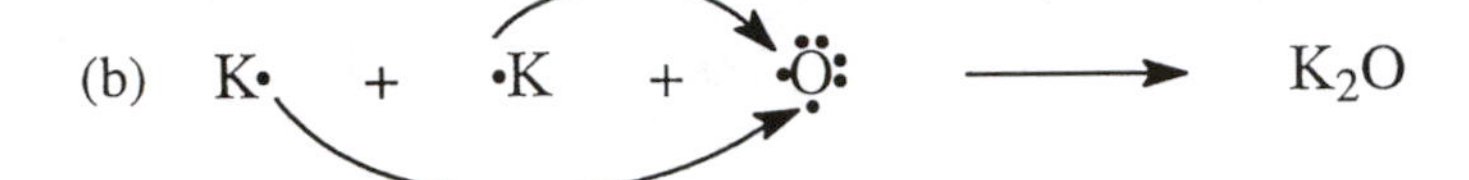


26. (a) Ca: + O: ⟶ CaO

(b) Na• + •Br: ⟶ NaBr

27. Valence electrons: H (1) K (1) Mg (2) He (2) Al (3)

28. Valence electrons: Si (4) N (5) P (5) O (6) Cl (7)

29. Noble gas structures:

(a) Calcium atom, lose $2\ e^-$

(b) Sulfur atom, gain $2\ e^-$

(c) Helium, none

30. (a) Chloride ion, none
    (b) Nitrogen atom, gain 3 $e^-$ or lose 5 $e^-$
    (c) Potassium atom, lose 1 $e^-$

31. (a) A magnesium ion, $Mg^{2+}$, is larger than an aluminum ion, $Al^{3+}$. Both ions have the same number of electrons (isoelectronic), with identical distribution of these electrons in their energy levels. The increased nuclear charge of the $Al^{3+}$ ion pulls the electrons in the $Al^{3+}$ ion closer to its nucleus, making it smaller than the $Mg^{2+}$ ion.

    (b) The $Fe^{2+}$ ion is larger than the $Fe^{3+}$ ion because the $Fe^{2+}$ ion has one more electron than the $Fe^{3+}$ ion. Both ions have the same nuclear charge.

32. (a) A potassium atom is larger than a potassium ion because the atom has one more energy level containing an electron than the ion.

    (b) A bromide ion is larger than a bromine atom because it has, within the same principle energy level, one more electron than a bromine atom.

33. (a) NaH, $Na_2O$ (c) $AlH_3$, $Al_2O_3$
    (b) $CaH_2$, CaO (d) $SnH_4$, $SnO_2$

34. (a) $SbH_3$, $Sb_2O_3$ (c) HCl, $Cl_2O_7$
    (b) $H_2Se$, $SeO_3$ (d) $CCl_4$, $CO_2$

35.
$Li_2SO_4$ lithium sulfate
$K_2SO_4$ potassium sulfate
$Rb_2SO_4$ rubidium sulfate
$Cs_2SO_4$ cesium sulfate
$Fr_2SO_4$ francium sulfate

36.
$BeBr_2$ beryllium bromide
$BaBr_2$ barium bromide
$MgBr_2$ magnesium bromide
$RaBr_2$ radium bromide
$SrBr_2$ strontium bromide

37. Lewis structures:

(a) Na• (b) $[:\ddot{\underset{..}{Br}}:]^-$ (c) $[:\ddot{\underset{..}{O}}:]^{2-}$

38. (a) $\underset{\cdot}{Ga}:$ (b) $[Ga]^{3+}$ (c) $[Ca]^{2+}$

39. (a) covalent (c) ionic
    (b) ionic (d) covalent

40. (a) covalent (c) covalent
(b) ionic (d) covalent

41. (a) ionic (b) covalent (c) covalent

42. (a) covalent (b) covalent (c) ionic

43. (a) H:H (b) :N:::N: (c) :Cl:Cl:

44. (a) :O::O: (b) :Br:Br: (c) :I:I:

45. (a) :Cl:N:Cl:
:Cl:

(c) H H
H:C:C:H
H H

(b) H:O:C::O:
:O:
H

(d) [Na]$^+$ [:O:N::O:
:O:]$^-$

46. (a) :S:H
H

(c) H:N:H
H

(b) :S::C::S:

(d) [H
H:N:H
H]$^+$ [:Cl:]$^-$

47. (a) $[Ba]^{2+}$

(d) [:C:::N:]$^-$

(b) $[Al]^{3+}$

(e) [:O::C:O:
:O:
H]$^-$

(c) [:O:S:O:
:O:]$^{2-}$

48. (a) [:I:]$^-$

(d) [:O:Cl:O:
:O:]$^-$

(b) [:S:]$^{2-}$

(e) [:O:N::O:
:O:]$^-$

(c) [:O:C::O:
:O:]$^{2-}$

49. (a) $H_2O$, polar (b) HBr, polar (c) $CF_4$, nonpolar

50. (a) $F_2$, nonpolar (b) $CO_2$, nonpolar (c) $NH_3$, polar

51. (a) 4 electron pairs, tetrahedral
    (b) 4 electron pairs, tetrahedral
    (c) 3 electron pairs, trigonal planar

52. (a) 2 electron pairs, linear
    (b) 4 electron pairs, tetrahedral
    (c) 4 electron pairs, tetrahedral

53. (a) tetrahedral (b) pyramidal (c) tetrahedral

54. (a) tetrahedral (b) pyramidal (c) tetrahedral

55. (a) tetrahedral (b) pyramidal (c) bent

56. (a) tetrahedral (b) bent (c) bent

57. Oxygen

58. Potassium

59. (a) Hg (b) Be (c) N (d) Fr (e) Au

60. (a) Zn (b) Be (c) N

61. (1) Fluorine is missing one electron from its 2p level while neon has a full energy level.
    (2) Fluorine's valence electrons are closer to the nucleus. The attraction between the electrons and the positive nucleus is greater.

62. Lithium has a +1 charge after the first electron is removed. It takes more energy to overcome that charge and to remove another electron than to remove a single electron from an uncharged He atom.

63. Yes. Each of these elements have an $ns^1$ electron and they could lose that electron in the same way elements in group IA do. They would then form +1 ions and ionic compounds such as CuCl, AgCl, and AuCl.

64. $SnBr_2$, $GeBr_2$

65. The bond between sodium and chlorine is ionic. An electron has been transferred from a sodium atom to a chlorine atom. The substance is composed of ions not molecules. Use of the word molecule implies covalent bonding.

66. A covalent bond results from the sharing of a pair of electrons between two atoms, while an ionic bond involves the transfer of one or more electrons from one atom to another.

67. This structure shown is incorrect since the bond is ionic. It should be represented as:

$$[Na]^{+}\,[:\ddot{\underset{..}{O}}:]^{2-}\,[Na]^{+}$$

68. The four most electronegative elements are F, O, N, Cl

69. highest F, O, S, H, Mg, Cs lowest

70. It is possible for a molecule to be nonpolar even though it contains polar bonds. If the molecule is symmetrical, the polarities of the bonds will cancel (in a manner similar to a positive and negative number of the same size) resulting in a nonpolar molecule. An example is $CO_2$ which is linear and nonpolar.

71. Both molecules contain polar bonds. $CO_2$ is symmetrical about the C atom, so the polarities cancel. In CO, there is only one polar bond, therefore the molecule is polar.

72. (a) 105° (b) 107° (c) 109.5° (d) 109.5°

73. (a) Both use the p orbitals for bonding. B uses one s and two p orbitals while N uses three p orbitals for bonding.

    (b) $BF_3$ is trigonal planar while $NF_3$ is pyramidal.

    (c) $BF_3$ no lone pairs
    $NF_3$ one lone pair

    (d) $BF_3$ has 3 very polar covalent bonds. $NF_3$ has 3 covalent bonds

74. Fluorine's electronegativity is greater than any other element. Ionic bonds form between atoms of widely different electronegativities. Therefore, Cs–F, Rb–F, K–F or Fr–F would be ionic substances with the greatest electronegativity difference.

75. Each element in a particular column has the same number of valence electrons and therefore the same Lewis structure.

76. S $\dfrac{1.40\ g}{32.07\ \frac{g}{mol}} = 0.0437\ mol$ $\dfrac{0.0437}{0.0437} = 1.00$

O $\dfrac{2.10\ g}{16.00\ \frac{g}{mol}} = 0.131\ mol$ $\dfrac{0.131}{0.0437} = 3.00$

Empirical formula is $SO_3$

:Ö:
|
:Ö—S=Ö:

77. We need to know the molecular formula before we can draw the Lewis structure. From the data, determine the empirical and then the molecular formula.

C $\dfrac{14.5\ g}{12.01\ \frac{g}{mol}} = 1.21\ mol$ $\dfrac{1.21}{1.21} = 1.00$

Cl $\dfrac{85.5\ g}{35.45\ \frac{g}{mol}} = 2.41\ mol$ $\dfrac{2.41}{1.21} = 2.01$

$CCl_2$ is the empirical formula

empirical mass = 1(12.01) + 2(35.45) = 82.91

$\dfrac{166}{82.91} = 2.00$

Therefore, the molecular formula is $(CCl_2)_2$ or $C_2Cl_4$

:Cl:  :Cl:
C=C
:Cl:  :Cl:

CHAPTER 12

# THE GASEOUS STATE OF MATTER

1. In Figure 12.1, color is the evidence of diffusion; bromine is colored and air is colorless. If hydrogen and oxygen had been the two gases, this would not work because both gases are colorless. Two ways could be used to show the diffusion. The change of density would be one method. Before diffusion the gas in the flask containing hydrogen would be much less dense. After diffusion, the gas densities in both flasks would be equal. A second method would require the introduction of spark gaps into both flasks. Before diffusion, neither gas would show a reaction when sparked. After diffusion, the gases in both flasks would explode because of the mixture of hydrogen and oxygen.

2. The air pressure inside the balloon is greater than the air pressure outside the balloon. The pressure inside must equal the sum of the outside air pressure plus the pressure exerted by the stretched rubber of the balloon.

3. The major components of dry air are nitrogen and oxygen.

4. 1 torr = 1 mm Hg

5. The molecules of $H_2$ at 100°C are moving faster. Temperature is a measure of average kinetic energy. At higher temperatures, the molecules will have more kinetic energy.

6. 1 atm corresponds to 4 L.

7. The pressure times the volume at any point on the curve is equal to the same value. This is an inverse relationship as is Boyle's law. (PV = k)

8. If $T_2 < T_1$, the volume of the cylinder would decrease (the piston would move downward).

9. The pressure inside the bottle is less than atmospheric pressure. We come to this conclusion because the water inside the bottle is higher than the water in the trough (outside the bottle).

10. The density of air is given as 1.29 g/L. Any gas with a greater density is listed below air on the table. Any five of these gases would be correct. ($O_2$, $H_2S$, HCl, $F_2$, $CO_2$).

11. Basic assumptions of Kinetic Molecular Theory include:

    (a) Gases consist of tiny particles.

(b) The distance between particles is great compared to the size of the particles.

(c) Gas particles move in straight lines. They collide with one another and with the walls of the container with no loss of energy.

(d) Gas particles have no attraction for each other.

(e) The average kinetic energy of all gases is the same at any given temperature. It varies directly with temperature.

12. The order of increasing molecular velocities is the order of decreasing molar masses.

    molecular velocity increases →

    Rn, $F_2$, $N_2$ $CH_4$, He, $H_2$

    ← molar mass increases

    At the same temperature the kinetic energies of the gases are the same and equal to $\frac{1}{2}mv^2$. For the kinetic energies to be the same, the velocities must increase as the molar masses decrease.

13. Average kinetic energies of all these gases are the same, as the gases are all at the same temperature.

14. Gases are described by the following parameters:

    (a) pressure (c) temperature
    (b) volume (d) number of moles

15. An ideal gas is one which follows the described gas laws at all P, V and T and whose behavior is described exactly by the Kinetic Molecular Theory.

16. A gas is least likely to behave ideally at low temperatures. Under this condition, the velocities of the molecules decrease and attractive forces between the molecules begin to play a significant role.

17. A gas is least likely to behave ideally at high pressures. Under this condition, the molecules are forced close enough to each other so that their volume is no longer small compared to the volume of the container. Attractive forces may also occur here and sooner or later, the gas will liquefy.

18. Equal volumes of $H_2$ and $O_2$ at the same T and P:

    (a) have equal number of molecules (Avogadro's law)

(b) mass $O_2$ = 16 times mass of $H_2$

(c) moles $O_2$ = moles $H_2$

(d) average kinetic energies are the same (T same)

(e) rate $H_2$ = 4 times the rate of $O_2$ (Graham's Law of Effusion)

(f) density $O_2$ = 16 times the density of $H_2$

$$\text{density } O_2 = \left(\frac{\text{mass } O_2}{\text{volume } O_2}\right) \qquad \text{density } H_2 = \left(\frac{\text{mass } H_2}{\text{volume } H_2}\right)$$

volume $O_2$ = volume $H_2$

$$\left(\frac{\text{mass } O_2}{\text{den } O_2}\right) = \left(\frac{\text{mass } H_2}{\text{den } H_2}\right) \qquad \text{density } O_2 = \left(\frac{\text{mass } O_2}{\text{mass } H_2}\right)(\text{den } H_2)$$

$$\text{density } O_2 = \left(\frac{32}{2}\right)(\text{density } H_2) = 16(\text{density } H_2)$$

19. Behavior of gases as described by the Kinetic Molecular Theory.

(a) Boyle's law. Boyle's law states that the volume of a fixed mass of gas is inversely proportional to the pressure, at constant temperature. The Kinetic Molecular Theory assumes the volume occupied by gases is mostly empty space. Decreasing the volume of a gas by compressing it, increases the concentration of gas molecules, resulting in more collisions of the molecules and thus increased pressure upon the walls of the container.

(b) Charles' law. Charles' law states that the volume of a fixed mass of gas is directly proportional to the absolute temperature, at constant pressure. According to Kinetic Molecular Theory, the kinetic energies of gas molecules are proportional to the absolute temperature. Increasing the temperature of a gas causes the molecules to move faster, and in order for the pressure not to increase, the volume of the gas must increase.

(c) Dalton's law. Dalton's law states that the pressure of a mixture of gases is the sum of the pressures exerted by the individual gases. According to the Kinetic Molecular Theory, there are no attractive forces between gas molecules; therefore, in a mixture of gases, each gas acts independently and the total pressure exerted will be the sum of the pressures exerted by the individual gases.

20. $N_2(g) + O_2(g) \rightarrow 2\ NO(g)$

    1 vol + 1 vol → 2 vol

    According to Avogadro's Law, equal volumes of nitrogen and oxygen at the same temperature and pressure contain the same number of molecules. In the reaction, nitrogen and oxygen molecules react in a 1:1 ratio. Since two volumes of nitrogen monoxide are produced, one molecule of nitrogen and one molecule of oxygen must produce two molecules of nitrogen monoxide. Therefore each nitrogen and oxygen molecule must be made up on two atoms.

21. We refer gases to STP because some reference point is needed to relate volume to moles. A temperature and pressure must be specified to determine the moles of gas in a given volume, and 0°C and 760 torr are convenient reference points.

22. Conversion of oxygen to ozone is an endothermic reaction. Evidence for this statement is that energy (286 kJ/3mol $O_2$) is required to convert $O_2$ to $O_3$.

23. Heating a mole of $N_2$ gas at constant pressure has the following effects:

    (a) Density will decrease. Heating the gas at constant pressure will increase its volume. The mass does not change, so the increased volume results in a lower density.

    (b) Mass does not change. Heating a substance does not change its mass.

    (c) Average kinetic energy of the molecules increases. This is a basic assumption of the Kinetic Molecular Theory.

    (d) Average velocity of the molecules will increase. Increasing the temperature increases the average kinetic energies of the molecules; hence, the average velocity of the molecules will increase also.

    (e) Number of $N_2$ molecules remains unchanged. Heating does not alter the number of molecules present, except if extremely high temperatures were attained. Then, the $N_2$ molecules might dissociate into N atoms resulting in fewer $N_2$ molecules.

24. Oxygen atom = O  Oxygen molecule = $O_2$  Ozone molecule = $O_3$
    An oxygen molecule contains 16 electrons.

25. (a) $(715\text{ mm Hg})\left(\dfrac{1\text{ atm}}{760\text{ mm Hg}}\right) = 0.941\text{ atm}$

    (b) $(715\text{ mm Hg})\left(\dfrac{1\text{ in. Hg}}{25.4\text{ mm Hg}}\right) = 28.1\text{ in. Hg}$

(c) $(715\ \text{mm Hg})\left(\dfrac{14.7\ \text{lb/in.}^2}{760\ \text{mm Hg}}\right) = 13.8\ \text{lb/in.}^2$

26. (a) $(715\ \text{mm Hg})\left(\dfrac{1\ \text{torr}}{1\ \text{mm Hg}}\right) = 715\ \text{torr}$

(b) $(715\ \text{mm Hg})\left(\dfrac{1013\ \text{mbar}}{760\ \text{mm Hg}}\right) = 953\ \text{mbar}$

(c) $(715\ \text{mm Hg})\left(\dfrac{101.325\ \text{kPa}}{760\ \text{mm Hg}}\right) = 95.3\ \text{kPa}$

27. (a) $(28\ \text{mm Hg})\left(\dfrac{1\ \text{atm}}{760\ \text{mm Hg}}\right) = 0.037\ \text{atm}$

(b) $(6000.\ \text{cm Hg})\left(\dfrac{1\ \text{atm}}{76\ \text{cm Hg}}\right) = 78.95\ \text{atm}$

(c) $(795\ \text{torr})\left(\dfrac{1\ \text{atm}}{760\ \text{torr}}\right) = 1.05\ \text{atm}$

(d) $(5.00\ \text{kPa})\left(\dfrac{1\ \text{atm}}{101.325\ \text{kPa}}\right) = 0.0493\ \text{atm}$

28. (a) $(62\ \text{mm Hg})\left(\dfrac{1\ \text{atm}}{760\ \text{mm Hg}}\right) = 0.082\ \text{atm}$

(b) $(4250.\ \text{cm Hg})\left(\dfrac{1\ \text{atm}}{76\ \text{cm Hg}}\right) = 55.92\ \text{atm}$

(c) $(225\ \text{torr})\left(\dfrac{1\ \text{atm}}{760\ \text{torr}}\right) = 0.296\ \text{atm}$

(d) $(0.67\ \text{kPa})\left(\dfrac{1\ \text{atm}}{101.325\ \text{kPa}}\right) = 0.0066\ \text{atm}$

29. $P_1V_1 = P_2V_2$ or $V_2 = \dfrac{P_1V_1}{P_2}$.

(a) $\dfrac{(500.\ \text{mm Hg})(400.\ \text{mL})}{760\ \text{mm Hg}} = 2.6 \times 10^2\ \text{mL}$

(b) $\dfrac{(500.\ \text{torr})(400.\ \text{mL})}{250\ \text{torr}} = 8.0 \times 10^2\ \text{mL}$

30. $P_1V_1 = P_2V_2$ or $V_2 = \dfrac{P_1V_1}{P_2}$.

(a) $\dfrac{(500.\ \text{mm Hg})(1\ \text{atm}/760\ \text{mm Hg})(400.\ \text{mL})}{2.00\ \text{atm}} = 132\ \text{mL}$

(b) $\dfrac{(500.\ \text{mm Hg})(1\ \text{torr}/1\ \text{mm Hg})(400.\ \text{mL})}{325\ \text{torr}} = 615\ \text{mL}$

31. $P_2 = \dfrac{P_1V_1}{V_2}$ $\quad \dfrac{(640.\ \text{mm Hg})(500.\ \text{mL})}{855\ \text{mL}} = 374\ \text{mm Hg}$

32. $P_2 = \dfrac{P_1V_1}{V_2}$ $\quad \dfrac{(640.\ \text{mm Hg})(500.\ \text{mL})}{450.\ \text{mL}} = 711\ \text{mm Hg}$

33. $\dfrac{V_1}{T_1} = \dfrac{V_2}{T_2}$ or $V_2 = \dfrac{V_1T_2}{T_1}$; Temperatures must be in Kelvin (°C + 273)

(a) $\dfrac{(6.00\ \text{L})(273\ \text{K})}{248\ \text{K}} = 6.60\ \text{L}$

(b) $\dfrac{(6.00\ \text{L})(100.\ \text{K})}{248\ \text{K}} = 2.42\ \text{L}$

34. $\dfrac{V_1}{T_1} = \dfrac{V_2}{T_2}$ or $V_2 = \dfrac{V_1T_2}{T_1}$; Temperatures must be in Kelvin (°C + 273)

0.0°F = −18°C

(a) $\dfrac{(6.00\ \text{L})(255\ \text{K})}{248\ \text{K}} = 6.17\ \text{L}$

(b) $\dfrac{(6.00\ \text{L})(345\ \text{K})}{248\ \text{K}} = 8.35\ \text{L}$

35. Use the combined gas law $\dfrac{P_1V_1}{T_1} = \dfrac{P_2V_2}{T_2}$ or $V_2 = \dfrac{P_1V_1T_2}{P_2T_1}$

$V_2 = \dfrac{(740\ \text{mm Hg})(410\ \text{mL})(273\ \text{K})}{(760\ \text{mm Hg})(300.\ \text{K})} = 3.6 \times 10^2\ \text{mL}$

36. Use the combined gas law $\frac{P_1V_1}{T_1} = \frac{P_2V_2}{T_2}$ or $V_2 = \frac{P_1V_1T_2}{P_2T_1}$

$$V_2 = \frac{(740 \text{ mm Hg})(410 \text{ mL})(523 \text{ K})}{(680 \text{ mm Hg})(300. \text{ K})} = 7.8 \times 10^2 \text{ mL}$$

37. Use the combined gas law $\frac{P_1V_1}{T_1} = \frac{P_2V_2}{T_2}$ or $V_2 = \frac{P_1V_1T_2}{P_2T_1}$

$$V_2 = \frac{(0.950 \text{ atm})(1400. \text{ mL})(275 \text{ K})}{(4.0 \text{ torr})(1 \text{ atm}/760 \text{ torr})(291 \text{ K})} = 2.4 \times 10^5 \text{ L}$$

38. Use the combined gas law $\frac{P_1V_1}{T_1} = \frac{P_2V_2}{T_2}$ or $V_2 = \frac{P_1V_1T_2}{P_2T_1}$

$$V_2 = \frac{(2.50 \text{ atm})(22.4 \text{ L})(268 \text{ K})}{(1.50 \text{ atm})(300. \text{ K})} = 33.4 \text{ L}$$

39. $P_{total} = P_{N_2} + P_{H_2O\ vapor} = 720.$ torr

$P_{H_2O\ vapor} = 17.5$ torr

$P_{N_2} = 720.$ torr $-$ 17.5 torr $=$ 703 torr

40. $P_{total} = P_{N_2} + P_{H_2O\ vapor} = 705$ torr

$P_{H_2O\ vapor} = 23.8$ torr

$P_{N_2} = 705$ torr $-$ 23.8 torr $=$ 681 torr

41. $P_{total} = P_{N_2} + P_{H_2} + P_{O_2}$

$= 200.$ torr $+$ 600. torr $+$ 300. torr $=$ 1100. torr $= 1.100 \times 10^3$ torr

42. $P_{total} = P_{H_2} + P_{N_2} + P_{O_2}$

$= 325$ torr $+$ 475 torr $+$ 650. torr $=$ 1450. torr $= 1.450 \times 10^3$ torr

43. $P_{total} = P_{CH_4} + P_{H_2O\ vapor}$

$P_{H_2O\ vapor} = 23.8$ torr

$P_{CH_4} = 720.$ torr $-$ 23.8 torr $=$ 696 torr

To calculate the volume of dry methane, note that the temperature is constant, so $P_1V_1 = P_2V_2$

$$V_2 = \frac{P_1V_1}{P_2} = \frac{(696 \text{ torr})(2.50 \text{ L})}{(760. \text{ torr})} = 2.29 \text{ L}$$

44. $P_{total} = P_{C_3H_8} + P_{H_2O\ vapor}$

$P_{H_2O\ vapor} = 20.5$ torr

$P_{C_3H_8} = 745\ \text{torr} - 20.5\ \text{torr} = 725\ \text{torr}$

To calculate the volume of dry propane, note that the temperature is constant, so $P_1V_1 = P_2V_2$

$$V_2 = \frac{P_1V_1}{P_2} = \frac{(725\ \text{torr})(1.25\ \text{L})}{(760.\ \text{torr})} = 1.19\ \text{L}\ C_3H_8$$

45. 1 mol of a gas occupies 22.4 L at STP

$$(2.5\ \text{mol})\left(\frac{22.4\ \text{L}}{\text{mol}}\right) = 56\ \text{L}\ Cl_2$$

46. $(1.25\ \text{mol})\left(\frac{22.4\ \text{L}}{\text{mol}}\right) = 28.0\ \text{L}\ N_2$

47. $(2500\ \text{mL})\left(\frac{1\ \text{L}}{1000\ \text{mL}}\right)\left(\frac{1\ \text{mol}}{22.4\ \text{L}}\right)\left(\frac{44.01\ \text{g}\ CO_2}{\text{mol}}\right) = 4.9\ \text{g}\ CO_2$

48. $(1.75\ \text{L})\left(\frac{1\ \text{mol}}{22.4\ \text{L}}\right)\left(\frac{17.03\ \text{g}\ NH_3}{\text{mol}}\right) = 1.33\ \text{g}\ NH_3$

49. (a) $(1.0\ \text{mol}\ NO_2)\left(\frac{22.4\ \text{L}}{\text{mol}}\right) = 22.4\ \text{L}\ NO_2$

(b) $(17.05\ \text{g}\ NO_2)\left(\frac{1\ \text{mol}}{46.01\ \text{g}}\right)\left(\frac{22.4\ \text{L}}{\text{mol}}\right) = 8.30\ \text{L}\ NO_2$

(c) $(1.20 \times 10^{24}\ \text{molecules}\ NO_2)\left(\frac{1\ \text{mol}}{6.022 \times 10^{23}\ \text{molecules}}\right)\left(\frac{22.4\ \text{L}}{\text{mol}}\right) = 44.6\ \text{L}\ NO_2$

50. (a) $(0.5\ \text{mol}\ H_2S)\left(\frac{22.4\ \text{L}}{\text{mol}}\right) = 11\ \text{L}\ H_2S$

(b) $(22.41\ \text{g}\ H_2S)\left(\frac{1\ \text{mol}}{34.09\ \text{g}}\right)\left(\frac{22.4\ \text{L}}{\text{mol}}\right) = 14.7\ \text{L}\ H_2S$

(c) $(8.55 \times 10^{23}\ \text{molecules}\ H_2S)\left(\frac{1\ \text{mol}}{6.022 \times 10^{23}\ \text{molecules}}\right)\left(\frac{22.4\ \text{L}}{\text{mol}}\right) = 31.8\ \text{L}\ H_2S$

51. $(1.00\ \text{L NH}_3)\left(\dfrac{1\ \text{mol}}{22.4\ \text{L}}\right)\left(\dfrac{6.022\times10^{23}\ \text{molecules}}{\text{mol}}\right) = 2.69 \times 10^{22}$ molecules $NH_3$

52. $(1.00\ \text{L CH}_4)\left(\dfrac{1\ \text{mol}}{22.4\ \text{L}}\right)\left(\dfrac{6.022\times10^{23}\ \text{molecules}}{\text{mol}}\right) = 2.69 \times 10^{22}$ molecules $CH_4$

53. (a) $d = \left(\dfrac{83.80\ \text{g Kr}}{\text{mol}}\right)\left(\dfrac{1\ \text{mol}}{22.4\ \text{L}}\right) = 3.74\ \text{g/L Kr}$

(b) $d = \left(\dfrac{80.07\ \text{g SO}_3}{\text{mol}}\right)\left(\dfrac{1\ \text{mol}}{22.4\ \text{L}}\right) = 3.57\ \text{g/L SO}_3$

54. (a) $d = \left(\dfrac{4.003\ \text{g He}}{\text{mol}}\right)\left(\dfrac{1\ \text{mol}}{22.4\ \text{L}}\right) = 0.179\ \text{g/L He}$

(b) $d = \left(\dfrac{56.10\ \text{g C}_4\text{H}_8}{\text{mol}}\right)\left(\dfrac{1\ \text{mol}}{22.4\ \text{L}}\right) = 2.50\ \text{g/L C}_4\text{H}_8$

55. (a) $d = \left(\dfrac{38.00\ \text{g F}_2}{\text{mol}}\right)\left(\dfrac{1\ \text{mol}}{22.4\ \text{L}}\right) = 1.70\ \text{g/L F}_2$

(b) Assume 1.00 mol of $F_2$ and determine the volume using the ideal gas equation, PV = nRT.

$$V = \frac{nRT}{P} = \frac{(1.00\ \text{mol})(0.0821\ \text{L atm/mol K})(300.\ \text{K})}{1.00\ \text{atm}} = 24.6\ \text{L at } 27^\circ\text{C and } 1.00\ \text{atm}$$

$$d = \frac{38.00\ \text{g}}{24.6\ \text{L}} = 1.54\ \text{g/L F}_2$$

56. (a) $d = \left(\dfrac{70.90\ \text{g Cl}_2}{\text{mol}}\right)\left(\dfrac{1\ \text{mol}}{22.4\ \text{L}}\right) = 3.17\ \text{g/L Cl}_2$

(b) Assume 1.00 mol of $Cl_2$ and determine the volume using the ideal gas equation, PV = nRT.

$$V = \frac{nRT}{P} = \frac{(1.00\ \text{mol})(0.0821\ \text{L atm/mol K})(295\ \text{K})}{0.500\ \text{atm}} = 48.4\ \text{L at } 22^\circ\text{C and } 0.500\ \text{atm}$$

$$d = \frac{70.90\ \text{g}}{48.4\ \text{L}} = 1.46\ \text{g/L Cl}_2$$

57. $PV = nRT \qquad V = \frac{nRT}{P} \qquad V = \frac{(2.3 \text{ mol})(0.0821 \text{ L atm/mol K})(300. \text{ K})}{\frac{750 \text{ torr}}{760 \frac{\text{torr}}{\text{atm}}}} = 57 \text{ L Ne}$

58. $PV = nRT \qquad V = \frac{nRT}{P} \qquad V = \frac{(0.75 \text{ mol})(0.0821 \text{ L atm/mol K})(298 \text{ K})}{\frac{725 \text{ torr}}{760 \frac{\text{torr}}{\text{atm}}}} = 19 \text{ L Kr}$

59. When working with gases, the identity of the gas does not affect the volume, as long as the number of moles are known.

Total moles = mol $H_2$ + mol $CO_2$ = 5.00 mol + 0.500 mol = 5.50 mol

$$V = (5.50 \text{ mol})\left(\frac{22.4 \text{ L}}{\text{mol}}\right) = 123 \text{ L}$$

60. When working with gases, the identity of the gas does not affect the volume, as long as the number of moles are known.

Total moles = mol $N_2$ + mol HCl = 2.50 mol + 0.750 mol = 3.25 mol

$$V = (3.25 \text{ mol})\left(\frac{22.4 \text{ L}}{\text{mol}}\right) = 72.8 \text{ L}$$

61. (a) $4\,NH_3(g) + 5\,O_2(g) \rightarrow 4\,NO(g) + 6\,H_2O(g)$

$$(5.5 \text{ mol NO})\left(\frac{4 \text{ mol } NH_3}{4 \text{ mol NO}}\right) = 5.5 \text{ mol } NH_3$$

(b) Limiting reactant problem. Remember, volume-volume relationships are the same as mole-mole relationships when dealing with gases at constant T and P.

$$(12 \text{ L } O_2)\left(\frac{4 \text{ mol NO}}{5 \text{ mol } O_2}\right) = 9.6 \text{ L NO (from } O_2)$$

$$(10. \text{ L } NH_3)\left(\frac{4 \text{ mol NO}}{4 \text{ mol } NH_3}\right) = 10. \text{ L NO (from } NH_3)$$

Oxygen is the limiting reactant, 9.6 L NO is formed.

(c) Limiting reactant problem. Remember, volume-volume relationships are the same as mole-mole relationships when dealing with gases at constant T and P.

$$(3.0 \text{ L } NH_3)\left(\frac{4 \text{ mol NO}}{4 \text{ mol } NH_3}\right) = 3.0 \text{ L NO (from } NH_3)$$

$$(3.0\text{ L }O_2)\left(\frac{4\text{ mol NO}}{5\text{ mol }O_2}\right) = 2.4\text{ L NO (from }O_2)$$

Oxygen is the limiting reactant, 2.4 L NO is formed.

62. $4\,NH_3(g) + 5\,O_2(g) \rightarrow 4\,NO(g) + 6\,H_2O(g)$

(a) $(7.0\text{ mol }O_2)\left(\frac{4\text{ mol }NH_3}{5\text{ mol }O_2}\right) = 5.6\text{ mol }NH_3$

(b) $(800.\text{ mL }O_2)\left(\frac{4\text{ mol NO}}{5\text{ mol }O_2}\right) = 640.\text{ mL NO} = 0.640\text{ L NO}$

(c) $(60.\text{ L NO})\left(\frac{1\text{ mol}}{22.4\text{ L}}\right)\left(\frac{5\text{ mol }O_2}{4\text{ mol NO}}\right)\left(\frac{32.00\text{ g}}{\text{mol}}\right) = 1.1\times10^2\text{ g }O_2$

63. The balanced equation is

$$4\,FeS + 7\,O_2 \xrightarrow{\Delta} 2\,Fe_2O_3 + 4\,SO_2$$

$$(600.\text{ g FeS})\left(\frac{1\text{ mol}}{87.92\text{ g}}\right)\left(\frac{7\text{ mol }O_2}{4\text{ mol FeS}}\right)\left(\frac{22.4\text{ L}}{\text{mol}}\right) = 268\text{ L }O_2$$

64. $4\,FeS + 7\,O_2 \xrightarrow{\Delta} 2\,Fe_2O_3 + 4\,SO_2$

$$(600.\text{ g FeS})\left(\frac{1\text{ mol}}{87.92\text{ g}}\right)\left(\frac{4\text{ mol }SO_2}{4\text{ mol FeS}}\right)\left(\frac{22.4\text{ L}}{\text{mol}}\right) = 153\text{ L }SO_2$$

65. (a)

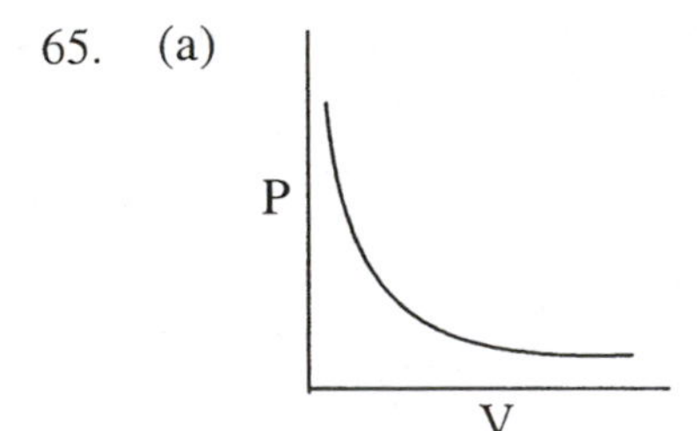


(c)

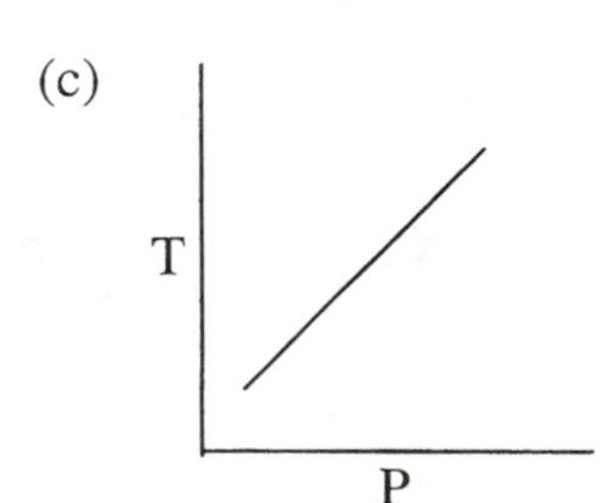


(b)

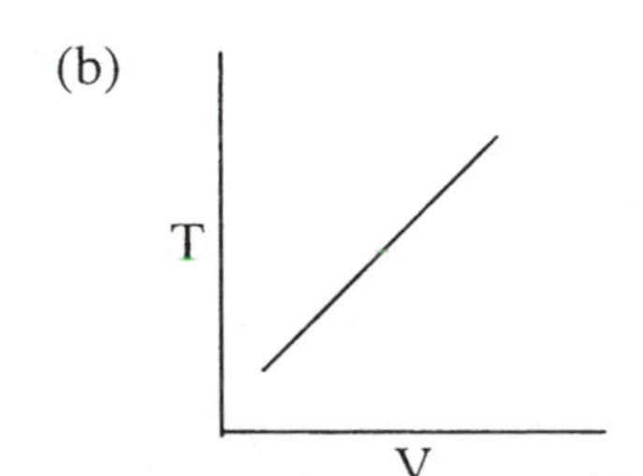


(d)

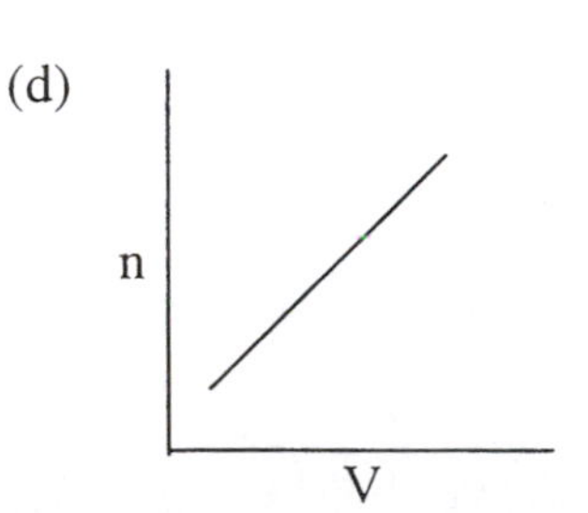

66. The can is a sealed unit and very likely still contains some of the aerosol. As the can is heated, pressure builds up in it eventually causing the can to explode and rupture with possible harm from flying debris.

67. One mole of an ideal gas occupies 22.4 liters at standard conditions. (0°C and 1 atm pressure)

$PV = nRT$

$(1.00\ \text{atm})(V) = (1.00\ \text{mol})(0.0821\ \text{L atm/mol K})(273\ \text{K})$

$V = 22.4\ \text{L}$

68. Solve for volume using $PV = nRT$

(a) $$V = \frac{(0.2\ \text{mol}\ Cl_2)(0.0821\ \text{L atm/mol K})(321\ \text{K})}{(80\ \text{cm}/76\ \text{cm})\ \text{atm}} = 5\ \text{L}\ Cl_2$$

(b) $$V = \frac{(4.2\ \text{mol}\ NH_3)\left(\frac{1\ \text{mol}}{17.03\ \text{g}}\right)\left(\frac{0.0821\ \text{L atm}}{\text{mol K}}\right)(161\ \text{K})}{0.65\ \text{atm}} = 5.0\ \text{L}\ NH_3$$

(c) Assume 25°C for room temperature

$$V = \frac{(21\ \text{g}\ SO_3)\left(\frac{1\ \text{mol}}{80.07\ \text{g}}\right)\left(\frac{0.0821\ \text{L atm}}{\text{mol K}}\right)(298\ \text{K})}{\frac{110\ \text{kPa}}{101.3\ \frac{\text{kPa}}{\text{atm}}}} = 5.9\ \text{L}\ SO_3$$

21 g $SO_3$ has the greatest volume

69. Assume 1 mol of each gas

(a) $SF_6 = 146.1\ \text{g/mol}$

$$d = \left(\frac{146.1\ \text{g}}{\text{mol}}\right)\left(\frac{1\ \text{mol}}{22.4\ \text{L}}\right) = 6.52\ \text{g/L}\ SF_6$$

(b) Assume 25°C and 1 atm pressure

$$V(\text{at }25°\text{C}) = (22.4\ \text{L})\left(\frac{298\ \text{K}}{273\ \text{K}}\right) = 24.5\ \text{L}$$

$C_2H_6 = 30.07\ \text{g/mol}$

$$d = \left(\frac{30.07\ \text{g}}{\text{mol}}\right)\left(\frac{1\ \text{mol}}{24.5\ \text{L}}\right) = 1.23\ \text{g/L}\ C_2H_6$$

(c) He at $-80°C$ and 2.15 atm

$$V = \frac{(1\text{ mol})(0.0821\text{ L atm/mol K})(193\text{ K})}{2.15\text{ atm}} = 7.37\text{ L}$$

$$d = \left(\frac{4.003\text{ g}}{\text{mol}}\right)\left(\frac{1\text{ mol}}{7.37\text{ L}}\right) = 0.543\text{ g/L He}$$

$SF_6$ has the greatest density

70. (a) Empirical formula. Assume 100. g starting material

$$\frac{80.0\text{ g C}}{12.01\text{ g/mol}} = 6.66\text{ mol C} \qquad \frac{6.66}{6.66} = 1$$

$$\frac{20.0\text{ g H}}{1.008\text{ g/mol}} = 19.8\text{ mol H} \qquad \frac{19.8}{6.66} = 2.97$$

Empirical formula = $CH_3$

Empirical mass = 12.01 g + 3.024 g = 15.03 g/mol

(b) Molecular formula. $\left(\frac{2.01\text{ g}}{1.5\text{ L}}\right)\left(\frac{22.4\text{ L}}{\text{mol}}\right) = 30.\text{ g/mol}$ (molar mass)

$\frac{30.\text{ g/mol}}{15.03\text{ g/mol}} = 2$; Molecular formula is $C_2H_6$

(c) Valence electrons = 2(4) + 6 = 14

```
    H   H
    |   |
H — C — C — H
    |   |
    H   H
```

Lewis structure

71. PV = nRT

(a) $\left(\frac{(790\text{ torr})(1\text{ atm})}{760\text{ torr}}\right)(2.0\text{ L}) = (n)(0.0821\text{ L atm/mol K})(298\text{ K})$

n = 0.085 mol (total moles)

(b) mol $N_2$ = total moles − mol $O_2$ − mol $CO_2$

$$= 0.085\text{ mol} - \frac{0.65\text{ g }O_2}{32.00\text{ g/mol}} - \frac{0.58\text{ g }CO_2}{44.01\text{ g/mol}}$$

mol $N_2$ = 0.085 mol − 0.020 mol $O_2$ − 0.013 mol $CO_2$ = 0.052 mol

$$(0.052\text{ mol }N_2)\left(\frac{28.02\text{ g }N_2}{\text{mol}}\right) = 1.5\text{ g }N_2$$

(c) $P_{O_2} = (790 \text{ torr})\left(\dfrac{0.020 \text{ mol } O_2}{0.085 \text{ mol}}\right) = 1.9 \times 10^2 \text{ torr}$

$P_{CO_2} = (790 \text{ torr})\left(\dfrac{0.013 \text{ mol } CO_2}{0.085 \text{ mol}}\right) = 1.2 \times 10^2 \text{ torr}$

$P_{N_2} = (790 \text{ torr})\left(\dfrac{0.051 \text{ mol } N_2}{0.085 \text{ mol}}\right) = 4.7 \times 10^2 \text{ torr}$

72. $2\ CO + O_2 \rightarrow 2\ CO_2$

Calculate the moles of $O_2$ and $CO_2$ to find the limiting reactant.

$PV = nRT$

$O_2$: $(1.8 \text{ atm})(0.500 \text{ L } O_2) = (n)(0.0821 \text{ L atm/mol K})(288 \text{ K})$

mol $O_2$ = 0.038 mol

CO: $\left(\dfrac{800 \text{ mm Hg} \times 1 \text{ atm}}{760 \text{ mm Hg}}\right)(0.500 \text{ L}) = (n)(0.0821 \text{ L atm/mol K})(333 \text{ K})$

mol CO = 0.019 mol *Limiting Reactant*

0.038 mol $O_2$ will react with 0.076 mol CO.

$(0.019 \text{ mol CO})\left(\dfrac{2 \text{ mol } CO_2}{2 \text{ mol CO}}\right)\left(\dfrac{22.4 \text{ L}}{\text{mol}}\right) = 0.43 \text{ L } CO_2 = 430 \text{ mL } CO_2$

73. $PV = nRT$ or $PV = \left(\dfrac{g}{\text{molar mass}}\right)RT$

$1.4 \text{ g/cm}^3 = 1.4 \times 10^3 \text{ g/L}$

$(1.3 \times 10^9 \text{ atm})(1.0 \text{ L}) = \left(\dfrac{1.4 \times 10^3 \text{ g}}{2.0 \text{ g/mol}}\right)(0.0821 \text{ L atm/mol K})(T)$

$T = \dfrac{(1.3 \times 10^9 \text{ atm})(1.0 \text{ L})(2.0 \text{ g/mol})}{(0.0821 \text{ L atm/mol K})(1.4 \times 10^3 \text{ g})} = 2.3 \times 10^7 \text{ K}$

74. (a) Assume atmospheric pressure of 14.7 lb/in.$^2$ to begin with.
Total pressure in the ball = 14.7 lb/in.$^2$ + 13 lb/in.$^2$ = 28 lb/in.$^2$

$PV = nRT$

$(28 \text{ lb/in.}^2)\left(\dfrac{1 \text{ atm}}{14.7 \text{ lb/in.}^2}\right)(2.24 \text{ L}) = (n)(0.0821 \text{ L atm/mol K})(293 \text{ K})$

n = 0.18 mol air

(b) mass of air in the ball

$$m = (0.18\ \text{mol})\left(\frac{29\ \text{g}}{\text{mol}}\right) = 5.2\ \text{g air}$$

(c) Actually the pressure changes when the temperature changes. Since pressure is directly proportional to moles we can calculate the change in moles required to keep the pressure the same at 30°C as it was at 20°C.

$$PV = nRT$$

$$(28\ \text{lb/in.}^2)\left(\frac{1\ \text{atm}}{14.7\ \text{lb/in.}^2}\right)(2.24\ \text{L}) = (n)(0.0821\ \text{L atm/mol K})(303\ \text{K})$$

$n = 0.17$ mol of air required to keep the pressure the same at 30°C.

0.01 mol air (0.18 – 0.17) must be allowed to escape from the ball.

$$(0.01\ \text{mol air})\left(\frac{29\ \text{g}}{\text{mol}}\right) = 0.29\ \text{g} \text{ or } 0.3\ \text{g air}$$

75. Use the combined gas laws to calculate the bursting temperature ($T_2$).

$$\frac{P_1V_1}{T_1} = \frac{P_2V_2}{T_2}$$

$P_1 = 65$ cm　　$P_2 = 1.00$ atm (76 cm)
$V_1 = 1.75$ L　　$V_2 = 2.00$ L
$T_1 = 20°C$ (293 K)　　$T_2 = T_2$

$$T_2 = \frac{P_2V_2T_1}{P_1V_1} = \frac{(76\ \text{cm})(2.00\ \text{L})(293\ \text{K})}{(65\ \text{cm})(1.75\ \text{K})} = 392\ \text{K}\quad (119°\text{C})$$

76. $P_1V_1 = P_2V_2$ or $P_2 = \dfrac{P_1V_1}{V_2}$

$$P_2 = \frac{(1.0\ \text{atm})(2500\ \text{L})}{25\ \text{L}} = 1.0 \times 10^2\ \text{atm}$$

77. To double the volume of a gas, at constant pressure, the temperature (K) must be doubled.

$$\frac{V_1}{T_1} = \frac{V_2}{T_2} \qquad V_2 = 2\,V_1$$

$$\frac{V_1}{T_1} = \frac{2\,V_1}{T_2} \qquad T_2 = \frac{2\,V_1T_1}{V_1} \qquad T_2 = 2\,T_1$$

$$T_2 = 2(300.\ \text{K}) = 600.\ \text{K} = 327°\text{C}$$

78. $V$ = volume at 22°C and 740 torr

$2\ V$ = volume after change in temperature (P constant)

$V$ = volume after change in pressure (T constant)

Since temperature is constant, $P_1V_1 = P_2V_2$ or $P_2 = \dfrac{P_1V_1}{V_2}$

$$P_2 = (740\ \text{torr})\left(\frac{2\ V}{V}\right) = 1.5 \times 10^3\ \text{torr (pressure to change 2 V to V)}$$

79. Volume is constant, so $\dfrac{P_1}{T_1} = \dfrac{P_2}{T_2}$ or $T_2 = \dfrac{T_1P_2}{P_1}$;

$$T_2 = \frac{(500.\ \text{torr})(295\ \text{K})}{700.\ \text{torr}} = 211\ \text{K} = -62°\text{C}$$

80. The volume of the cylinder remains constant, so
$-196°\text{C} + 273 = 77\ \text{K}$

$$\frac{P_1}{T_1} = \frac{P_2}{T_2} \text{ or } P_2 = \frac{P_1T_2}{T_1};$$

$$P_2 = \frac{(252\ \text{atm})(77\ \text{K})}{298\ \text{K}} = 65\ \text{atm}$$

81. The volume of the tires remains constant (until they burst), so

$$\frac{P_1}{T_1} = \frac{P_2}{T_2} \text{ or } T_2 = \frac{T_1P_2}{P_1};$$

$71.0°\text{F} = 21.7°\text{C} = 295\ \text{K}$

$$T_2 = \frac{(44\ \text{psi})(295\ \text{K})}{30.\ \text{psi}} = 433\ \text{K} = 160°\text{C} = 320°\text{F}$$

82. Use the combined gas laws.

$$\frac{P_1V_1}{T_1} = \frac{P_2V_2}{T_2} \text{ or } V_2 = \frac{P_1V_1T_2}{P_2T_1}$$

$$V_2 = \frac{(760\ \text{torr})(5.30\ \text{L})(343\ \text{K})}{(830\ \text{torr})(273\ \text{K})} = 6.1\ \text{L}$$

83. Use the combined gas laws.

$$\frac{P_1V_1}{T_1} = \frac{P_2V_2}{T_2} \quad \text{or} \quad P_2 = \frac{P_1V_1T_2}{V_2T_1}$$

$$P_2 = \frac{(1.00\ \text{atm})(800.\ \text{mL})(303\ \text{K})}{(250.\ \text{mL})(273\ \text{K})} = 3.55\ \text{atm}$$

84. Use the combined gas law $\frac{P_1V_1}{T_1} = \frac{P_2V_2}{T_2}$ or $V_2 = \frac{P_1V_1T_2}{P_2T_1}$

First calculate the volume at STP.

$$V_2 = \frac{(400.\ \text{torr})(600.\ \text{mL})(273\ \text{K})}{(760.\ \text{torr})(313\ \text{K})} = 275\ \text{mL}$$

At STP, a mole of any gas has a volume of 22.4 L

$$(0.275\ \text{L})\left(\frac{1\ \text{mol}}{22.4\ \text{L}}\right)\left(\frac{6.022 \times 10^{23}\ \text{molecules}}{1\ \text{mol}}\right) = 7.39 \times 10^{21}\ \text{molecules}$$

Each molecule of $N_2O$ contains 3 atoms, so:

$$(7.39 \times 10^{21}\ \text{molecules})\left(\frac{3\ \text{atoms}}{1\ \text{molecule}}\right) = 2.22 \times 10^{22}\ \text{atoms}$$

85. Each gas acts independently, so calculate the pressure of each gas in the 10 L container and add them together.

$$CO_2 \qquad \frac{(5.00\ \text{L})(500.\ \text{torr})}{10.0\ \text{L}} = 250.\ \text{torr}$$

$$CH_4 \qquad \frac{(3.00\ \text{L})(400.\ \text{torr})}{10.0\ \text{L}} = 120.\ \text{torr}$$

$$P_{total} = P_{CO_2} + P_{CH_4} = 250.\ \text{torr} + 120.\ \text{torr} = 370.\ \text{torr}$$

86. The number of moles of gas is proportional to pressure. (T and V constant)

$$\frac{P_1}{n_1} = \frac{P_2}{n_2} \quad \text{or} \quad n_2 = \frac{n_1P_2}{P_1}$$

(a) $(60.0\ \text{mol}\ H_2)\left(\frac{850\ \text{lb/in.}^2}{1500\ \text{lb/in.}^2}\right) = 34\ \text{mol}\ H_2$

(b) $(60.0\ \text{mol}\ H_2)\left(\frac{2.016\ \text{g}}{\text{mol}}\right) = 121\ \text{g}\ H_2$

87. The conversion is: $m^3 \rightarrow cm^3 \rightarrow mL \rightarrow L \rightarrow mol$

$$(1.00\ m^3)\left(\frac{100\ cm}{1\ m}\right)^3\left(\frac{1\ mL}{1\ cm^3}\right)\left(\frac{1\ L}{1000\ mL}\right)\left(\frac{1\ mol}{22.4\ L}\right) = 44.6\ mol\ Cl_2$$

88. First calculate the moles of gas and then convert moles to molar mass.

$$(0.560\ L)\left(\frac{1\ mol}{22.4\ L}\right) = 0.0250\ mol$$

$$\frac{1.08\ g}{0.0250\ mol} = 43.2\ g/mol \text{ (molar mass)}$$

89. At STP 22.4 L of $CH_4$ has a mass of 16.04 g. Using 1.00 g as the mass of the sample: $\frac{22.4\ L}{16.04\ g} = \frac{1.40\ L}{1.00\ g}$

$P_1 = 1$ atm $\quad P_2 = 1$ atm
$V_1 = 1.40$ L $\quad V_2 = 1.0$ L
$T_1 = 273$ K $\quad T_2 = T_2$

Since the pressure is constant, $\frac{V_1}{T_1} = \frac{V_2}{T_2}$

$$T_2 = \frac{V_2T_1}{V_1} = \frac{(1.00\ L)(273\ K)}{1.40\ L} = 195\ K = -78°C$$

90. The conversion is: $g/L \rightarrow g/mol$

$$\left(\frac{1.78\ g}{L}\right)\left(\frac{22.4\ L}{mol}\right) = 39.9\ g/mol \text{ (molar mass)}$$

91. $PV = nRT$

(a) $$V = \frac{nRT}{P} = \frac{(0.510\ mol)(0.0821\ L\ atm/mol\ K)(320.\ K)}{1.6\ atm} = 8.4\ L\ H_2$$

(b) $$n = \frac{PV}{RT} = \frac{(0.789\ atm)(16.0\ L)}{(0.0821\ L\ atm/mol\ K)(300.\ K)} = 0.513\ mol\ CH_4$$

The molar mass for $CH_4$ is 16.04 g/mol

$(16.04\ g/mol)(0.513\ mol) = 8.23\ g\ CH_4$

(c) $PV = nRT$, but $n = \frac{m}{M}$ where M is the molar mass and m is the mass of the gas.

Thus, $PV = \frac{mRT}{M}$. To determine density, $d = m/V$.

Solving $PV = \frac{mRT}{M}$ for $\frac{m}{V}$ produces $\frac{m}{V} = \frac{PM}{RT}$.

$$d = \frac{m}{V} = \frac{(4.00\ \text{atm})(44.01\ \text{g/mol})}{(0.0821\ \text{L atm/mol K})(253\ \text{K})} = 8.48\ \text{g/L}\ CO_2$$

(d) Since $d = \frac{m}{V} = \frac{PM}{RT}$ from part (c), solve for M (molar mass)

$$M = \frac{dRT}{P} = \frac{(2.58\ \text{g/L})(0.0821\ \text{L atm/mol K})(300.\ \text{K})}{1.00\ \text{atm}} = 63.5\ \text{g/mol (molar mass)}$$

92. $PV = nRT$

$$n = \frac{PV}{RT} = \frac{(0.813\ \text{atm})(0.215\ \text{L})}{(0.0821\ \text{L atm/mol K})(303\ \text{K})} = 7.03 \times 10^{-3}\ \text{mol}$$

$$\text{molar mass} = \left(\frac{1.15\ \text{g}}{7.03 \times 10^{-3}\ \text{mol}}\right) = 164\ \text{g/mol}$$

93. $PV = nRT$

$$T = \frac{PV}{nR} = \frac{(4.15\ \text{atm})(0.250\ \text{L})}{(4.50\ \text{mol})(0.0821\ \text{L atm/mol K})} = 2.81\ \text{K}$$

94. $PV = nRT$

$$n = \frac{PV}{RT} = \frac{(0.500\ \text{atm})(5.20\ \text{L})}{(0.0821\ \text{L atm/mol K})(250\ \text{K})} = 0.13\ \text{mol}\ N_2$$

95. $C_2H_2(g) + 2\ HF(g) \rightarrow C_2H_4F_2(g)$

1.0 mol $C_2H_2$ $\rightarrow$ 1.0 mol $C_2H_4F_2$

$$(5.0\ \text{mol HF})\left(\frac{1\ \text{mol}\ C_2H_4F_2}{2\ \text{mol HF}}\right) = 2.5\ \text{mol}\ C_2H_4F_2$$

$C_2H_2$ is the limiting reactant. 1.0 mol $C_2H_4F_2$ forms, no moles $C_2H_2$ remain.
According to the equation, 2.0 mol HF yields 1.0 mol $C_2H_4F_2$. Therefore,

5.0 mol HF − 2.0 mol HF = 3.0 mol HF unreacted

The flask contains 1.0 mol $C_2H_4F_2$ and 3.0 mol HF when the reaction is complete.

$$P = \frac{nRT}{V} = \frac{(4.0\text{ mol})(0.0821\text{ L atm/mol K})(273\text{ K})}{10.0\text{ L}} = 9.0\text{ atm}$$

96. $(8.30\text{ mol Al})\left(\frac{3\text{ mol }H_2}{2\text{ mol Al}}\right)\left(\frac{22.4\text{ L}}{\text{mol}}\right) = 279\text{ L }H_2\text{ at STP}$

97. According to Graham's Law of Effusion, the rates of effusion are inversely proportional to the molar mass.

$$\frac{\text{rate He}}{\text{rate }N_2} = \sqrt{\frac{\text{molar mass }N_2}{\text{molar mass He}}} = \sqrt{\frac{28.02}{4.003}} = \sqrt{7.000} = 2.646$$

Helium effuses 2.646 times faster than nitrogen.

98. (a) According to Graham's Law of Effusion, the rates of effusion are inversely proportional to the molar mass.

$$\frac{\text{rate He}}{\text{rate }CH_4} = \sqrt{\frac{16.04}{4.003}} = \sqrt{4.007} = 2.002$$

Helium effuses twice as fast as $CH_4$.

(b) $x$ = distance He travels

$100 - x$ = distance $CH_4$ travels

$D_{He} = 2\,D_{CH_4}$ D = distance traveled

$x = 2(100 - x)$

$3x = 200$

$x = 66.7$ cm

The gases meet 66.7 cm from the helium end.

99. Assume 100. g of material to start with. Calculate the empirical formula.

C $(85.7\text{ g})\left(\frac{1\text{ mol}}{12.01\text{ g}}\right) = 7.14\text{ mol}$ $\frac{7.14}{7.14} = 1.00\text{ mol}$

H $(14.3\text{ g})\left(\frac{1\text{ mol}}{1.008\text{ g}}\right) = 14.2\text{ mol}$ $\frac{14.2}{7.14} = 1.99\text{ mol}$

The empirical formula is $CH_2$. To determine the molecular formula, the molar mass must be known.

$$\left(\frac{2.50\text{ g}}{\text{L}}\right)\left(\frac{22.4\text{ L}}{\text{mol}}\right) = 56.0\text{ g/mol (molar mass)}$$

The empirical formula mass is 14.0 $\quad \frac{56.0}{14.0} = 4$

Therefore, the molecular formula is $(CH_2)_4 = C_4H_8$

100. $2\ CO(g) + O_2(g) \rightarrow 2\ CO_2(g)$ Determine the limiting reactant

$$(10.0 \text{ mol CO})\left(\frac{2 \text{ mol } CO_2}{2 \text{ mol CO}}\right) = 10.0 \text{ mol } CO_2 \text{ (from CO)}$$

$$(8.0 \text{ mol } O_2)\left(\frac{2 \text{ mol } CO_2}{1 \text{ mol } O_2}\right) = 16 \text{ mol } CO_2 \text{ (from } O_2)$$

CO: the limiting reactant, $O_2$: in excess, 3.0 mol $O_2$ unreacted.

(a) 10.0 mol CO react with 5.0 mol $O_2$

10.0 mol $CO_2$ and 3.0 mol $O_2$ are present, no CO will be present.

(b) $$P = \frac{nRT}{V} = \frac{(13 \text{ mol})(0.0821 \text{ L atm/mol K})(273 \text{ K})}{10. \text{ L}} = 29 \text{ atm}$$

101. $2\ KClO_3(s) \xrightarrow{\Delta} 2\ KCl(s) + 3\ O_2(g)$

First calculate the moles of $O_2$ produced. Then calculate the grams of $KClO_3$ required to produce the $O_2$. Then calculate the % $KClO_3$.

$$(0.25 \text{ L } O_2)\left(\frac{1 \text{ mol}}{22.4 \text{ L}}\right) = 0.011 \text{ mol } O_2$$

$$(0.011 \text{ mol } O_2)\left(\frac{2 \text{ mol } KClO_3}{3 \text{ mol } O_2}\right)\left(\frac{122.6 \text{ g}}{\text{mol}}\right) = 0.90 \text{ g } KClO_3 \text{ in the sample}$$

$$\left(\frac{0.90 \text{ g}}{1.20 \text{ g}}\right)(100) = 75\% \ KClO_3 \text{ in the mixture}$$

102. Some ammonia gas dissolves in the water squirted into the flask, lowering the pressure inside the flask. The atmospheric pressure outside is greater than the pressure inside the flask and pushes water from the beaker up the tube and into the flask, filling the flask.

103. (a) The pressure of the helium is simply the difference between levels of Hg; 250 mm Hg (250 torr).

(b) The pressure of the oxygen is the difference between the pressure of the atmosphere and the difference in the levels of Hg.

$$\begin{aligned} P_{O_2} &= P_{atm} + 300 \text{ mm Hg} \\ &= 760 \text{ mm Hg} + 300 \text{ mm Hg} \\ &= 1060 \text{ mm Hg} \ \ (1060 \text{ torr}) \end{aligned}$$

104. Assume 1.00 L of air. The mass of 1.00 L is 1.29 g.

$$\frac{P_1V_1}{T_1} = \frac{P_2V_2}{T_2}$$

$$V_2 = \frac{P_1V_1T_2}{P_2T_1} = \frac{(760 \text{ torr})(1.00 \text{ L})(290.\text{ K})}{(450 \text{ torr})(273 \text{ K})} = 1.8 \text{ L}$$

$$d = \frac{m}{V} = \frac{1.29 \text{ g}}{1.8 \text{ L}} = 0.72 \text{ g/L}$$

105. Each gas behaves as though it were alone in a 4.0 L system.

(a) After expansion: $P_1V_1 = P_2V_2$

For $CO_2$ $\quad P_2 = \dfrac{P_1V_1}{V_2} = \dfrac{(150.\text{ torr})(3.0 \text{ L})}{4.0 \text{ L}} = 1.1 \times 10^2 \text{ torr}$

For $H_2$ $\quad P_2 = \dfrac{P_1V_1}{V_2} = \dfrac{(50.\text{ torr})(1.0 \text{ L})}{4.0 \text{ L}} = 13 \text{ torr}$

(b) $P_{total} = P_{H_2} + P_{CO_2} = 110 \text{ torr} + 13 \text{ torr} = 120 \text{ torr}$ (2 sig. figures)

106. $P_1 = 40 \text{ atm}$ $\qquad P_2 = P_2$

$V_1 = 50.0 \text{ L}$ $\qquad V_2 = 50.0 \text{ L}$

$T_1 = 25°\text{C} = 298 \text{ K}$ $\qquad T_2 = 25°\text{C} + 125°\text{C} = 177°\text{C} = 450.\text{ K}$

Gas cylinders have constant volume, so pressure varies directly with temperature.

$$P_2 = \frac{P_1T_2}{T_1} = \frac{(40.0 \text{ atm})(450.\text{ K})}{298 \text{ K}} = 60.4 \text{ atm}$$

CHAPTER 13

# WATER AND THE PROPERTIES OF LIQUIDS

1. The potential energy is greater in the liquid water than in the ice. The heat necessary to melt the ice increases the potential energy of the liquid, thus allowing the molecules greater freedom of motion.

2. At 0°C, all three substances, $H_2S$, $H_2Se$, and $H_2Te$, are gases, because they all have boiling points below 0°C.

3. The pressure of the atmosphere must be 1.00 atmosphere, otherwise the water would be boiling at some other temperature.

4. 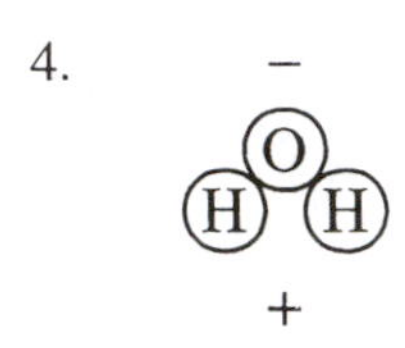


5. Melting point, boiling point, heat of fusion, heat of vaporization, density, and crystallization structure in the solid state are some of the physical properties of water that would be very different, if the molecules were linear and nonpolar instead of bent and highly polar. For example, the boiling point, melting point, heat of fusion and heat of vaporization would be lower because linear molecules have no dipole moment and the attraction among molecules would be much less.

6. Prefixes preceding the word hydrate are used in naming hydrates, indicating the number of molecules of water present in the formulas. The prefixes used are:

| | | | | |
|---|---|---|---|---|
| mono = 1 | di = 2 | tri = 3 | tetra = 4 | penta = 5 |
| hexa = 6 | hepta = 7 | octa = 8 | nona = 9 | deca = 10 |

7. The distillation setup in Figure 13.10 would be satisfactory for separating salt and water, but not for separating ethyl alcohol and water. In the first case, the water is easily vaporized, the salt is not, so the water boils off and condenses to a pure liquid. In the second case, both substances are easily vaporized, so both would vaporize (though not to and equal degree) and the condensed liquid would contain both substances.

8. The thermometer would be at about 70°C. The liquid is boiling, which means its vapor pressure equals the confining pressure. From Table 13.1, we find that ethyl alcohol has a vapor pressure of 543 torr at 70°C.

9. The water in both containers would have the same vapor pressure, for it is a function of the temperature of the liquid.

10. In Figure 13.1, it would be case (b) in which the atmosphere would reach saturation. The vapor pressure of water is the same in both (a) and (b), but since (a) is an open container the vapor escapes into the atmosphere and doesn't reach saturation.

11. If ethyl ether and ethyl alcohol were both placed in a closed container, (a) both substances would be present in the vapor, for both are volatile liquids; (b) ethyl ether would have more molecules in the vapor because it has a higher vapor pressure at a given temperature.

12. The vapor pressure observed in (c) would remain unchanged. The presence of more water in (b) does not change the magnitude of the vapor pressure of the water. The temperature of the water determines the magnitude of the vapor pressure.

13. At 30 torr, $H_2O$ would boil at approximately 29°C, ethyl alcohol at 14°C, and ethyl ether at some temperature below 0°C.

14. (a) At a pressure of 500 torr, water boils at 88°C.
    (b) The normal boiling point of ethyl alcohol is 78°C.
    (c) At a pressure of 0.50 atm (380 torr), ethyl ether boils at 16°C.

15. Based on Figure 13.5:

    (a) Line BC is horizontal because the temperature remains constant during the entire process of melting. The energy input is absorbed in changing from the solid to the liquid state.

    (b) During BC, both solid and liquid phases are present.

    (c) The line DE represents the change from liquid water to steam (vapor) at the boiling temperature of water.

16. Physical properties of water:
    (a) melting point, 0°C
    (b) boiling point, 100°C (at 1 atm pressure)
    (c) colorless
    (d) odorless
    (e) tasteless
    (f) heat of fusion, 335 J/g (80 cal/g)
    (g) heat of vaporization, 2.26 kJ/g (540 cal/g)
    (h) density = 1.0 g/mL (at 4°C)
    (i) specific heat = 4.184 J/g°C

17. For water, to have its maximum density, the temperature must be 4°C, and the pressure sufficient to keep it liquid. $d = 1.0$ g/mL

18. Apply heat to an ice-water mixture, the heat energy is absorbed to melt the ice, rather than warm the water, so the temperature remains constant until all the ice has melted.

19. Ice at 0°C contains less heat energy than water at 0°C. Heat must be added to convert ice to water, so the water will contain that much additional heat energy.

20. Ice floats in water because it is less dense than water. The density of ice at 0°C is 0.915 g/mL. Liquid water, however, has a density of 1.00 g/mL. Ice will sink in ethyl alcohol, which has a density of 0.789 g/mL.

21. The heat of vaporization of water would be lower if water molecules were linear instead of bent. If linear, the molecules of water would be nonpolar. The relatively high heat of vaporization of water is a result of the molecule being highly polar and having strong dipole-dipole hydrogen bonding attraction for other water molecules.

22. Ethyl alcohol exhibits hydrogen bonding; ethyl ether does not. This is indicated by the high heat of vaporization of ethyl alcohol, even though its molar mass is much less than the molar mass of ethyl ether.

23. A linear water molecule, being nonpolar, would exhibit less hydrogen bonding than the highly polar, bent, water molecule. The polar molecule has a greater intermolecular attractive force than a nonpolar molecule.

24. Ammonia exhibits hydrogen bonding; methane does not. The ammonia molecule is polar; the methane molecule is not.

25. Water, at 80°C, will have fewer hydrogen bonds than water at 40°C. At the higher temperature, the molecules of water are moving faster than at the lower temperature. This results in less hydrogen bonding at the higher temperature.

26. $H_2NCH_2CH_2NH_2$ has two polar $NH_2$ groups. It should, therefore, show more hydrogen bonding and a higher boiling point (117°C) versus 49°C for $CH_3CH_2CH_2NH_2$.

27. Rubbing alcohol feels cold when applied to the skin, because the evaporation of the alcohol absorbs heat from the skin. The alcohol has a fairly high vapor pressure (low boiling point) and evaporates quite rapidly. This produces the cooling effect.

28. (a) Order of increasing rate of evaporation: Mercury, acetic acid, water, toluene, benzene, carbon tetrachloride, methyl alcohol, bromine.

    (b) Highest boiling point is mercury. Lowest boiling point is bromine.

29. Water boils when its vapor pressure equals the prevailing atmospheric pressure over the water. In order for water to boil at 50°C, the pressure over the water would need to be reduced to a point equal to the vapor pressure of the water (92.5 torr).

30. In a pressure cooker, the temperature at which the water boils increases above its normal boiling point, because the water vapor (steam) formed by boiling cannot escape. This results in an increased pressure over water and, consequently, an increased boiling temperature.

31. Vapor pressure varies with temperature. The temperature at which the vapor pressure of a liquid equals the prevailing pressure is the boiling point of the liquid.

32. As temperature increases, molecular velocities increase. At higher molecular velocities, it becomes easier for molecules to break away from the attractive forces in the liquid.

33. Water has a relatively high boiling point because there is a high attraction between molecules due to hydrogen bonding.

34. Ammonia would have a higher vapor pressure than $SO_2$ at −40°C because it has a lower boiling point ($NH_3$ is more volatile than $SO_2$).

35. As the temperature of a liquid increases, the kinetic energy of the molecules as well as the vapor pressure of the liquid increases. When the vapor pressure of the liquid equals the external pressure, boiling begins with many of the molecules having enough energy to escape from the liquid. Bubbles of vapor are formed throughout the liquid and these bubbles rise to the surface, escaping as boiling continues.

36. HF has a higher boiling point than HCl because of the strong hydrogen bonding in HF (F is the most electronegative element). Neither $F_2$ nor $Cl_2$ will have hydrogen bonding, so the compound, $F_2$, with the lower molar mass, has the lower boiling point.

37. The boiling liquid remains at constant temperature because the added heat energy is being used to convert the liquid to a gas, i.e., to supply the heat of vaporization for the liquid at its boiling point.

38. 34.6°C, the boiling point of ether. (See Table 13.2)

39. The lake freezes from the top down because, as the temperature drops to freezing or below, the water on the surface tends to cool faster than the water that lies deeper. As the surface water freezes, the ice formed floats because the ice is less dense than the liquid water below it.

40. If the lake is in an area where the temperature is below freezing for part of the year, the expected temperature would be 4°C at the bottom of the lake. This is because the surface water would cool to 4°C (maximum density) and sink.

41. The formation of hydrogen and oxygen from water is an endothermic reaction, due to the following evidence:

    (a) Energy must continually be provided to the system for the reaction to proceed. The reaction will cease when the energy source is removed.

    (b) The reverse reaction, burning hydrogen in oxygen, releases energy as heat.

42. (a) The word anhydride originates from the Greek, anhydrous, meaning waterless. An anhydride is an oxide that reacts with water to form an acid or base.

    (b) An acid anhydride is an oxide of a nonmetal.

    (c) A basic anhydride is an oxide of a metal.

43. Acid anhydride: [$HClO_4$, $Cl_2O_7$] [$H_2CO_3$, $CO_2$] [$H_3PO_4$, $P_2O_5$]

44. Acid anhydride: [$H_2SO_3$, $SO_2$] [$H_2SO_4$, $SO_3$] [$HNO_3$, $N_2O_5$]

45. Basic anhydrides: [$LiOH$, $Li_2O$] [$NaOH$, $Na_2O$] [$Mg(OH)_2$, $MgO$]

46. Basic anhydrides: [$KOH$, $K_2O$] [$Ba(OH)_2$, $BaO$] [$Ca(OH)_2$, $CaO$]

47. (a) $Ba(OH)_2 \xrightarrow{\Delta} BaO + H_2O$
    (b) $2\,CH_3OH + 3\,O_2 \rightarrow 2\,CO_2 + 4\,H_2O$
    (c) $2\,Rb + 2\,H_2O \rightarrow 2\,RbOH + H_2$
    (d) $SnCl_2 \bullet 2\,H_2O \xrightarrow{\Delta} SnCl_2 + 2\,H_2O$
    (e) $HNO_3 + NaOH \rightarrow NaNO_3 + H_2O$
    (f) $CO_2 + H_2O \rightarrow H_2CO_3$

48. (a) $Li_2O + H_2O \rightarrow 2\,LiOH$
    (b) $2\,KOH \xrightarrow{\Delta} K_2O + H_2O$
    (c) $Ba + 2\,H_2O \rightarrow Ba(OH)_2 + H_2$
    (d) $Cl_2 + H_2O \rightarrow HCl + HClO$
    (e) $SO_3 + H_2O \rightarrow H_2SO_4$
    (f) $H_2SO_3 + 2\,KOH \rightarrow K_2SO_3 + 2\,H_2O$

49. (a) barium bromide dihydrate
    (b) aluminum chloride hexahydrate
    (c) iron(III) phosphate tetrahydrate

50. (a) magnesium ammonium phosphate hexahydrate
    (b) iron(II) sulfate heptahydrate
    (c) tin(IV) chloride pentahydrate

51. Deionized water is water from which the ions have been removed.
    (a) Hard water contains dissolved calcium and magnesium salts.
    (b) Soft water is free of ions that cause hardness ($Ca^{2+}$ and $Mg^{2+}$) but it may contain other ions such as $Na^+$ and $K^+$.

52. Deionized water is water from which the ions have been removed.
    (a) Distilled water has been vaporized by boiling and recondensed. It is free of nonvolatile impurities, but may still contain any volatile impurities that were initially present in the water.
    (b) Natural waters are generally not pure, but contain dissolved minerals and suspended matter, and can even contain harmful bacteria.

53. $$(100.\text{ g CoCl}_2 \bullet 6\text{ H}_2\text{O})\left(\frac{1\text{ mol}}{238.0\text{ g}}\right) = 0.420\text{ mol CoCl}_2 \bullet 6\text{ H}_2\text{O}$$

54. $$(100.\text{ g FeI}_2 \bullet 4\text{ H}_2\text{O})\left(\frac{1\text{ mol}}{381.7\text{ g}}\right) = 0.262\text{ mol FeI}_2 \bullet 4\text{ H}_2\text{O}$$

55. $$(100.\text{ g CoCl}_2 \bullet 6\text{ H}_2\text{O})\left(\frac{1\text{ mol}}{238.0\text{ g}}\right)\left(\frac{6\text{ mol H}_2\text{O}}{1\text{ mol CoCl}_2 \bullet 6\text{ H}_2\text{O}}\right) = 2.52\text{ mol H}_2\text{O}$$

56. $$(100.\text{ g FeI}_2 \bullet 4\text{ H}_2\text{O})\left(\frac{1\text{ mol}}{381.7\text{ g}}\right)\left(\frac{4\text{ mol H}_2\text{O}}{1\text{ mol FeI}_2 \bullet 4\text{ H}_2\text{O}}\right) = 1.05\text{ mol H}_2\text{O}$$

57. Assume 1 mol of the compound.
$$\left(\frac{\text{g H}_2\text{O}}{\text{g MgSO}_4 \bullet 7\text{ H}_2\text{O}}\right)(100) = \left(\frac{(7)(18.02\text{ g})}{246.5\text{ g}}\right)(100) = 51.17\%\text{ H}_2\text{O}$$

58. Assume 1 mol of hydrate.
$$\%\text{ H}_2\text{O} = \frac{\text{g H}_2\text{O}}{\text{g Al}_2(\text{SO}_4)_3 \bullet 18\text{ H}_2\text{O}} = \left(\frac{(18)(18.02\text{ g})}{666.5\text{ g}}\right)(100) = 48.67\%\text{ H}_2\text{O}$$

59. Assume 100. g of the compound.
$$(0.142)(100.\text{ g}) = 14.2\text{ g H}_2\text{O}$$
$$(0.858)(100.\text{ g}) = 85.8\text{ g Pb(C}_2\text{H}_3\text{O}_2)_2$$
$$(14.2\text{ g H}_2\text{O})\left(\frac{1\text{ mol}}{18.02\text{ g}}\right) = 0.788\text{ mol H}_2\text{O}$$

$$(85.8 \text{ g Pb}(C_2H_3O_2)_2)\left(\frac{1 \text{ mol}}{325.3 \text{ g}}\right) = 0.264 \text{ mol Pb}(C_2H_3O_2)_2$$

In the formula for the hydrate, there is one mole of $Pb(C_2H_3O_2)_2$, so divide each of the moles by 0.264.

$$\frac{0.264 \text{ mol Pb}(C_2H_3O_2)_2}{0.264 \text{ mol}} = 1 \text{ Pb}(C_2H_3O_2)_2$$

$$\frac{0.788 \text{ mol } H_2O}{0.264 \text{ mol}} = 2.98 \text{ } H_2O$$

Therefore, the formula is $Pb(C_2H_3O_2)_2 \bullet 3\ H_2O$.

60. 25.0 g hydrate − 16.9 g $FePO_4$ = 8.1 g $H_2O$ driven off

$$(8.1 \text{ g } H_2O)\left(\frac{1 \text{ mol}}{18.02 \text{ g}}\right) = 0.45 \text{ mol } H_2O \qquad \frac{0.45}{0.112} = 4.0$$

$$(16.9 \text{ g FePO}_4)\left(\frac{1 \text{ mol}}{150.8 \text{ g}}\right) = 0.112 \text{ mol FePO}_4 \qquad \frac{0.112}{0.112} = 1.00$$

The formula is $FePO_4 \bullet 4\ H_2O$.

61. (a) Heat water 20.°C → 100.°C

$$E_a = (m)(\text{specific heat})(\Delta T) = (120. \text{ g})\left(\frac{4.184 \text{ J}}{\text{g°C}}\right)(80\text{°C}) = 4.0 \times 10^4 \text{ J}$$

(b) Convert water to steam: heat of vaporization = $2.26 \times 10^3$ J/g

$$E_b = (m)(\text{heat of vaporization}) = (120. \text{ g})(2.26 \times 10^3 \text{ J/g}) = 2.71 \times 10^5 \text{ J}$$

$$E_{total} = E_a + E_b = (4.0 \times 10^4 \text{ J}) + (2.71 \times 10^5 \text{ J}) = 3.11 \times 10^5 \text{ J}$$

62. (a) Cool water 24°C → 0°C

$$E_a = (m)(\text{specific heat})(\Delta T) = (126 \text{ g})\left(\frac{4.184 \text{ J}}{\text{g°C}}\right)(24\text{°C}) = 1.3 \times 10^4 \text{ J}$$

(b) Convert water to ice

$$E_b = (m)(\text{heat of fusion}) = (126 \text{ g})(335 \text{ J/g}) = 4.22 \times 10^4 \text{ J}$$

$$E_{total} = E_a + E_b = (1.3 \times 10^4 \text{ J}) + (4.22 \times 10^4 \text{ J}) = 5.5 \times 10^4 \text{ J}$$

63. Energy released in cooling the water

$$E = (m)(\text{specific heat})(\Delta T) = (300. \text{ g})\left(\frac{1 \text{ cal}}{\text{g°C}}\right)(25\text{°C}) = 7.5 \times 10^3 \text{ cal}$$

Energy required to melt the ice

$$E = (m)(\text{heat of fusion}) = (100. \text{ g})(80. \text{ cal/g}) = 8.0 \times 10^3 \text{ cal}$$

Less energy is released in cooling the water than is required to melt the ice. Ice will remain and the water will be at 0°C.

64. Energy to heat the water = energy to condense the steam

$$(300.\ \text{g})\left(\frac{1\ \text{cal}}{\text{g}^\circ\text{C}}\right)(100.^\circ\text{C} - 25^\circ\text{C}) = (\text{m})(540\ \text{cal/g})$$

42 g = m (grams of steam required to heat the water to 100.°C)
42 g of steam are required to heat 300. g of water to 100.°C. Since only 35 g of steam are added to the system, the final temperature will be less than 100.°C. Not sufficient steam.

65. Energy lost by warm water = energy gained by the ice

$x$ = final temperature

$$\text{mass}(H_2O) = (1.5\ \text{L}\ H_2O)\left(\frac{1000\ \text{mL}}{\text{L}}\right)\left(\frac{1.0\ \text{g}}{\text{mL}}\right) = 1500\ \text{g}$$

$$(1500\ \text{g})\left(\frac{1\ \text{cal}}{\text{g}^\circ\text{C}}\right)(75^\circ\text{C} - x) = (75\ \text{g})(80.\ \frac{\text{cal}}{\text{g}}) + (75\ \text{g})\left(\frac{1\ \text{cal}}{\text{g}^\circ\text{C}}\right)(x - 0^\circ\text{C})$$

$$(112{,}500\ \text{cal}) - (1500x\ \text{cal/}^\circ\text{C}) = 6.0\times10^3\ \text{cal} + 75x\ \text{cal/}^\circ\text{C}$$
$$106{,}500\ \text{cal} = 1575x\ \text{cal/}^\circ\text{C}$$
$$68^\circ\text{C} = x$$

66. E = (m)(heat of fusion)
(500. g)(335 J/g) = 167,000 J needed to melt the ice
9560 J < 167,500 J
Since 167,500 J are required to melt all the ice, and only 9560 J are available, the system will be at 0°C. It will be a mixture of ice and water.

67. (a) $2\ Na + 2\ H_2O \rightarrow 2\ NaOH + H_2$

$$(1.00\ \text{g Na})\left(\frac{1\ \text{mol}}{22.99\ \text{g}}\right)\left(\frac{2\ \text{mol}\ H_2O}{2\ \text{mol Na}}\right)\left(\frac{18.02\ \text{g}}{\text{mol}}\right) = 0.784\ \text{g}\ H_2O$$

(b) $MgO + H_2O \rightarrow Mg(OH)_2$

$$(1.00\ \text{g MgO})\left(\frac{1\ \text{mol}}{40.31\ \text{g}}\right)\left(\frac{1\ \text{mol}\ H_2O}{1\ \text{mol MgO}}\right)\left(\frac{18.02\ \text{g}}{\text{mol}}\right) = 0.447\ \text{g}\ H_2O$$

(c) $N_2O_5 + H_2O \rightarrow 2\ HNO_3$

$$(1.00\ \text{g}\ N_2O_5)\left(\frac{1\ \text{mol}}{108.0\ \text{g}}\right)\left(\frac{1\ \text{mol}\ H_2O}{1\ \text{mol}\ N_2O_5}\right)\left(\frac{18.02\ \text{g}}{\text{mol}}\right) = 0.167\ \text{g}\ H_2O$$

68. (a) $2\,K + 2\,H_2O \rightarrow 2\,KOH + H_2$

$$(1.00\text{ mol K})\left(\frac{2\text{ mol }H_2O}{2\text{ mol K}}\right)\left(\frac{18.02\text{ g}}{\text{mol}}\right) = 18.0\text{ g }H_2O$$

(b) $Ca + 2\,H_2O \rightarrow Ca(OH)_2 + H_2$

$$(1.00\text{ mol Ca})\left(\frac{2\text{ mol }H_2O}{1\text{ mol Ca}}\right)\left(\frac{18.02\text{ g}}{\text{mol}}\right) = 36.0\text{ g }H_2O$$

(c) $SO_3 + H_2O \rightarrow H_2SO_4$

$$(1.00\text{ mol }SO_3)\left(\frac{1\text{ mol }H_2O}{1\text{ mol }SO_3}\right)\left(\frac{18.02\text{ g}}{\text{mol}}\right) = 18.0\text{ g }H_2O$$

69. Steam molecules will cause a more severe burn. Steam molecules contain more energy at 100°C than water molecules at 100°C due to the energy absorbed during the vaporization stage (heat of vaporization).

70. The alcohol has a higher vapor pressure than water and thus evaporates faster than water. When the alcohol evaporates it absorbs energy from the water, cooling the water. Eventually the water will lose enough energy to change from a liquid to a solid (freeze).

71. When one leaves the swimming pool, water starts to evaporate from the skin of the body. Part of the energy needed for evaporation is absorbed from the skin, resulting in the cool feeling.

72.

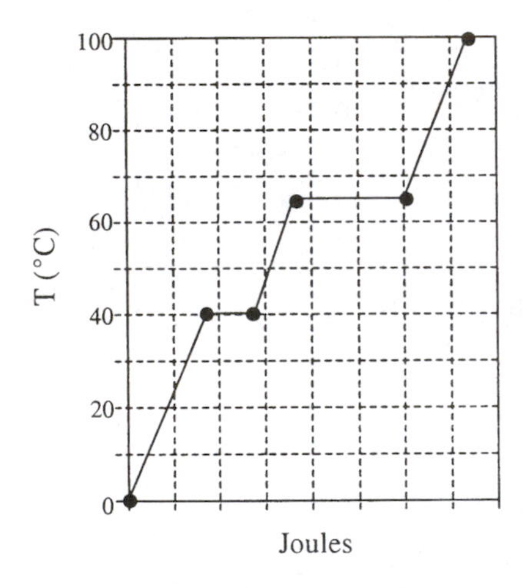


(a) From 0°C to 40.°C solid X warms until at 40.°C it begins to melt. The temperature remains at 40.0°C until all of X is melted. After that, liquid X will warm steadily to 65°C where it will boil and remain at 65°C until all of the liquid becomes vapor. Beyond 65°C, the vapor will warm steadily until 100°C.

(b)

| | | | |
|---|---|---|---|
| Joules needed (0°C to 40°C) | = (60. g)(3.5 J/g°C)(40.°C) | = | 8400 J |
| Joules needed at 40°C | = (60. g)(80. J/g) | = | 4800 J |
| Joules needed (40°C to 65°C) | = (60. g)(3.5 J/g°C )(25°C) | = | 5300 J |
| Joules needed at 65°C | = (60. g)(190 J/g) | = | 11,000 J |
| Joules needed (65°C to 100°C) | = (60. g)(3.5 J/g°C )(35°C) | = | 7400 J |
| Total Joules needed | | | 37,000 J |

73. As the temperature of a liquid increases, the molecules gain kinetic energy thereby increasing their escaping tendency (vapor pressure).

74. Since boiling occurs when vapor pressure equals atmospheric pressure, the graph in Figure 13.4 indicates that water will boil at about 75°C at 270 torr pressure.

75. $CuSO_4$ (anhydrous) is gray white. When exposed to moisture, it turns bright blue forming $CuSO_4 \bullet 5\ H_2O$. The color change is an indicator of moisture in the environment.

76. $MgSO_4 \bullet 7\ H_2O$ $\qquad$ $Na_2HPO_4 \bullet 12\ H_2O$

77. Soap can soften hard water by forming a precipitate with, and thus removing, the calcium and magnesium ions. This precipitate is a greasy scum and is very objectionable, so it is a poor way to soften water.

78. Chlorine is commonly used to destroy bacteria in water. Ozone and ultraviolet radiation are also used in some places.

79. Ozone, $O_3$

80. When organic pollutants in water are oxidized by dissolved oxygen, there may not be sufficient dissolved oxygen to sustain marine life, such as fish. Most marine life forms depend on dissolved oxygen for cellular respiration.

81. Liquids that are stored in ceramic containers should never be drunk, for they are likely to have dissolved some of the lead from the ceramic. If the ceramic is glazed, the liquid is less apt to dissolve lead from the ceramic.

82. $Na_2\ \text{zeolite}(s) + Mg^{2+}(aq) \rightarrow Mg\ \text{zeolite}(s) + 2\ Na^+(aq)$

83. Softening of hard water using sodium carbonate:
$CaCl_2(aq) + Na_2CO_3(aq) \rightarrow CaCO_3(s) + 2\ NaCl(aq)$

84. (a) Melt ice: $E_a$ = (m)(heat of fusion) = (225 g)(80. cal/g) = 18,000 cal

(b) Warm the water: $E_b$ = (m)(specific heat)($\Delta$T)

$$= (225\ g)\left(\frac{1\ cal}{g°C}\right)(100.°C) = 22{,}500\ cal$$

(c) Vaporize the water:

$E_c$ = (m)(heat of vaporization) = (225 g)(540 cal/g) = 121,500 cal

$E_{total} = E_a + E_b + E_c = 1.6 \times 10^5$ cal

85. The heat of vaporization of water is 2.26 kJ/g.

$$(2.26\ \text{kJ/g})\left(\frac{18.02\ \text{g}}{\text{mol}}\right) = 40.7\ \text{kJ/mol}$$

86. $$E = (\text{m})(\text{specific heat})(\Delta T) = (250.\ \text{g})\left(\frac{0.096\ \text{cal}}{\text{g}^\circ\text{C}}\right)(150. - 20.0^\circ\text{C}) = 3.1 \times 10^3\ \text{cal (3.1 kcal)}$$

87. Heat lost by warm water = heat gained by ice
m = grams of ice to lower temperature of water to 0.0°C.

$$(120.\ \text{g})\left(\frac{1\ \text{cal}}{\text{g}^\circ\text{C}}\right)(45^\circ\text{C} - 0.0^\circ\text{C}) = (\text{m})(80.\ \text{cal/g})$$

68 g = m (grams of ice melted)
68 g of ice melted. Therefore, 150. g − 68 g = 82 g ice remains.

88. Energy liberated when steam at 100.0°C condenses to water at 100.0°C

$$(50.0\ \text{mol steam})\left(\frac{18.02\ \text{g}}{\text{mol}}\right)\left(\frac{2.26\ \text{kJ}}{\text{g}}\right)\left(\frac{1000\ \text{J}}{\text{kJ}}\right) = 2.04 \times 10^6\ \text{J}$$

Energy liberated in cooling water from 100.0°C to 30.0°C

$$(50.0\ \text{mol}\ H_2O)\left(\frac{18.02\ \text{g}}{\text{mol}}\right)\left(\frac{4.184\ \text{J}}{\text{g}^\circ\text{C}}\right)(100.0^\circ\text{C} - 30.0^\circ\text{C}) = 2.64 \times 10^5\ \text{J}$$

Total energy liberated

$$2.04 \times 10^6\ \text{J} + 2.64 \times 10^5\ \text{J} = 2.30 \times 10^6\ \text{J}$$

89. Energy to warm the ice from −10.0°C to 0°C

$$(100.\ \text{g})\left(\frac{2.01\ \text{J}}{\text{g}^\circ\text{C}}\right)(10.0^\circ\text{C}) = 2010\ \text{J}$$

Energy to melt the ice at 0°C

$$(100.\ \text{g})(335\ \text{J/g}) = 33{,}500\ \text{J}$$

Energy to heat the water from 0°C to 20.0°C

$$(100.\ \text{g})\left(\frac{4.184\ \text{J}}{\text{g}^\circ\text{C}}\right)(20.0^\circ\text{C}) = 8370\ \text{J}$$

$E_{total}$ = 2010 J + 33,500 J + 8370 J = $4.39 \times 10^4$ J = 43.9 kJ

90. $2\,H_2O \rightarrow 2\,H_2 + O_2$

The conversion is $L\,O_2 \rightarrow mol\,O_2 \rightarrow mol\,H_2O \rightarrow g\,H_2O$

$$(25.0\text{ L }O_2)\left(\frac{1\text{ mol}}{22.4\text{ L}}\right)\left(\frac{2\text{ mol }H_2O}{1\text{ mol }O_2}\right)\left(\frac{18.02\text{ g}}{\text{mol}}\right) = 40.2\text{ g }H_2O$$

91. The conversion is $\frac{\text{mol}}{\text{day}} \rightarrow \frac{\text{molecules}}{\text{day}} \rightarrow \frac{\text{molecules}}{\text{hr}} \rightarrow \frac{\text{molecules}}{\text{min}} \rightarrow \frac{\text{molecules}}{\text{s}}$

$$\left(\frac{1.00\text{ mol }H_2O}{\text{day}}\right)\left(\frac{6.022 \times 10^{23}\text{ molecules}}{\text{mol}}\right)\left(\frac{1.00\text{ day}}{24\text{ hr}}\right)\left(\frac{1\text{ hr}}{60\text{ min}}\right)\left(\frac{1\text{ min}}{60\text{ s}}\right)$$

$= 6.97 \times 10^{18}$ molecules/s

92. Liquid water has a density of 1.00 g/mL.

$$d = \frac{m}{V} \qquad V = \frac{m}{d} = \frac{18.02\text{ g}}{1.00\text{ g/mL}} = 18.0\text{ mL} \quad \text{(volume of 1 mole)}$$

1.00 mole of water vapor at STP has a volume of 22.4 L (gas)

93. Mass solution − mass $H_2O$ = mass $H_2SO_4$

(122 mL)(1.26 g/mL) − (100. mL)(1.00 g/mL) = 54 g $H_2SO_4$

94. $2\,H_2 + O_2 \rightarrow 2\,H_2O$

(a) $(80.0\text{ mL }H_2)\left(\frac{1\text{ mL }O_2}{2\text{ mL }H_2}\right) = 40.0\text{ mL }O_2$ react with 80.0 mL of $H_2$

Since 60.0 mL of $O_2$ are available, some oxygen remains unreacted.

(b) 60.0 mL − 40.0 mL = 20.0 mL $O_2$ unreacted.

95. Energy absorbed by the student when steam at 100.°C changes to water at 100.°C

$$(1.5\text{ g steam})\left(\frac{2.26\text{ kJ}}{\text{g}}\right) = 3.4\text{ kJ} \quad (3.4 \times 10^3\text{ J})$$

Energy absorbed when water cools from 100.°C to 20.0°C

E = (m)(specific heat)($\Delta$t)

$$(1.5\text{ g})\left(\frac{4.184\text{ J}}{\text{g°C}}\right)(100.°\text{C} - 20.0°\text{C}) = 5.0 \times 10^2\text{ J}$$

Total = $3.4 \times 10^3$ J + $5.0 \times 10^2$ J = $3.9 \times 10^3$ J

CHAPTER 14

# SOLUTIONS

1. (a)

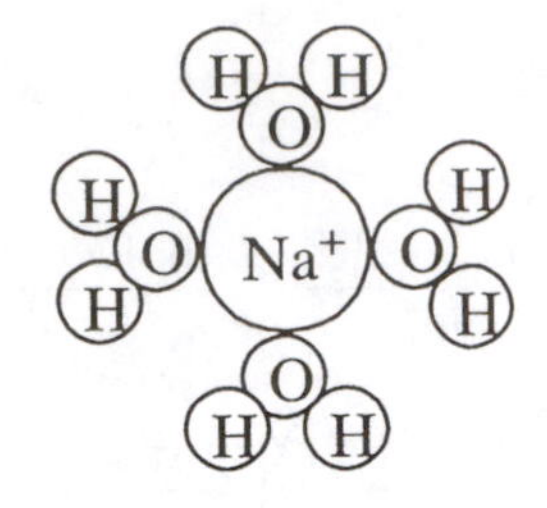


(b)

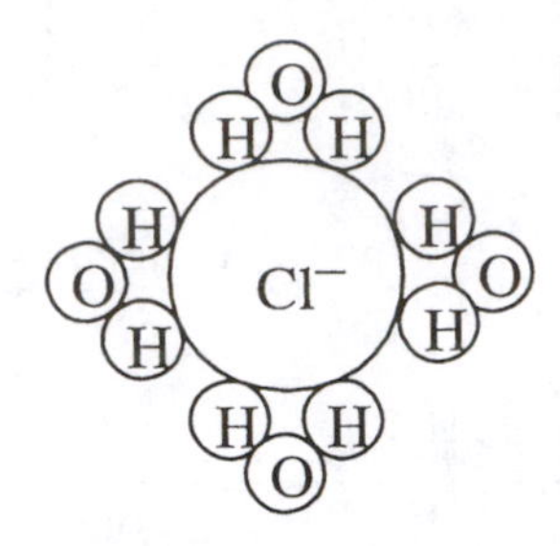


These diagrams are intended to illustrate the orientation of the water molecules about the ions, not the number of water molecules.

2. From Table 14.3, approximately 4.5 g of NaF would be soluble in 100 g of water at 50°C.

3. From Figure 14.3, solubilities in water at 25°C are:
   (a) KCl 35 g/100 g $H_2O$
   (b) $KClO_3$ 9 g/100 g $H_2O$
   (c) $KNO_3$ 39 g/100 g $H_2O$

4. Potassium fluoride has a relatively high solubility when compared to lithium or sodium fluoride. For lithium and sodium halides, the order of solubility (in order of increasing solubilities) is:

   $F^-$ $Cl^-$ $Br^-$ $I^-$

   For potassium halides, the order of increasing solubilities is:

   $Cl^-$ $Br^-$ $F^-$ $I^-$

5. (a) $KClO_3$ at 60°C, 25g
   (b) HCl at 20°C, 72 g
   (c) $Li_2SO_4$ at 80°C, 31 g
   (d) $KNO_3$ at 0°C, 14 g

6. $KNO_3$

7. A one molal solution in camphor will show a greater freezing point depression than a 2 molar solution in benzene.

$$\Delta t_f = \left(\frac{1 \text{ mol solute}}{\text{kg camphor}}\right)\left(\frac{40^\circ\text{C kg camphor}}{\text{mol solute}}\right) = 40^\circ\text{C (freezing point depression)}$$

$$\Delta t_f = \left(\frac{2 \text{ mol solute}}{\text{kg benzene}}\right)\left(\frac{5.1^\circ\text{C kg benzene}}{\text{mol solute}}\right) = 10.2^\circ\text{C (freezing point depression)}$$

8.

| Cube | 1 cm | 0.01 cm |
| --- | --- | --- |
| Volume | 1 $cm^3$ | $1 \times 10^{-6}$ $cm^3$ |
| Number/1 cm cube | 1 | $10^6$ [$(1\ cm^3)/(1 \times 10^{-6}\ cm^3) = 10^6$ cubes] |
| Area of face | 1 $cm^2$ | $1 \times 10^{-4}$ $cm^2$ |
| Total surface area | 6 $cm^2$ | $6 \times 10^2$ $cm^2$ |

$$(1 \times 10^6 \text{ cubes})(6 \text{ faces/cube})(1 \times 10^{-4} \text{ cm}^2\text{/face}) = 6 \times 10^2 \text{ cm}^2$$

9. $\dfrac{63 \text{ g } NH_4Cl}{150 \text{ g } H_2O} = \dfrac{42 \text{ g } NH_4Cl}{100 \text{ g } H_2O}$ From Figure 14.3, the solubility of $NH_4Cl$ in water is approximately 42 g/100 g $H_2O$ at 30°C, 46 g/100 g $H_2O$ at 40°C. Therefore, the solution of 63 g/150 g of water would be saturated at 10°C, 20°C, and 30°C. The solution would be unsaturated at 40°C and 50°C.

10. The dissolving process involves solvent molecules attaching to the solute ions or molecules. This rate decreases as more of the solvent molecules are already attached to solute molecules. As the solution becomes more saturated, the number of unused solvent molecules decreases. Also, the rate of recrystallization increases as the concentration of dissolved solute increases.

11. A supersaturated solution of $NaC_2H_3O_2$ may be prepared in the following sequence:

    (a) Determine the mass of $NaC_2H_3O_2$ necessary to saturate a specific amount of water at room temperature.

    (b) Place a bit more $NaC_2H_3O_2$ in the water than the amount needed to saturate the solution.

    (c) Heat the solution until all the solid dissolves.

    (d) Cover the container and allow it to cool undisturbed. The cooled solution, which should contain no solid $NaC_2H_3O_2$, is supersaturated.

    To test for supersaturation, add one small crystal of $NaC_2H_3O_2$ to the solution. Immediate crystallization is an indication that the solution was supersaturated.

12. Because the concentration of water is greater in the thistle tube, the water will flow through the membrane from the thistle tube to the urea solution in the beaker. The solution level in the thistle tube will fall.

13. The two components of a solution are the solute and the solvent. The solute is dissolved into the solvent or is the least abundant component. The solvent is the dissolving agent or the most abundant component.

14. It is not always apparent which component in a solution is the solute. For example, in a solution composed of equal volumes of two liquids, the designation of solute and solvent would be simply a matter of preference on the part of the person making the designation.

15. The ions or molecules of a dissolved solute do not settle out because the individual particles are so small that the force of molecular collisions is large compared to the force of gravity.

16. Yes. It is possible to have one solid dissolved in another solid. Metal alloys are of this type. Atoms of one metal are dissolved among atoms of another metal.

17. Orange. The three reference solutions are KCl, $KMnO_4$, and $K_2Cr_2O_7$. The all contain $K^+$ ions in solution. The different colors must result from the different anions dissolved in the solutions: $MnO_4^-$ (purple) and $Cr_2O_7^{2-}$ (orange). Therefore, it is predictable that the $Cr_2O_7^{2-}$ ion present in an aqueous solution of $Na_2Cr_2O_7$ will impart an orange color to the solution.

18. Hexane and benzene are both nonpolar molecules. There are no strong intermolecular forces between molecules of either substance or with each other, so they are miscible. Sodium chloride consists of ions strongly attracted to each other by electrical attractions. The hexane molecules, being nonpolar, have no strong forces to pull the ions apart, so sodium chloride is insoluble in hexane.

19. Coca Cola has two main characteristics, taste and fizz (carbonation). The carbonation is due to a dissolved gas, carbon dioxide. Since dissolved gases become less soluble as temperature increases, warm Coca Cola would be flat, with little to no carbonation. It is, therefore, unappealing to most people.

20. Air is considered to be a solution because it is a homogeneous mixture of several gaseous substances and does not have a fixed composition.

21. A teaspoon of sugar would definitely dissolve more rapidly in 200 mL of hot coffee than in 200 mL of iced tea. The much greater thermal agitation of the hot coffee will help break the sugar molecules away from the undissolved solid and disperse them throughout the solution. Other solutes in coffee and tea would have no significant effect. The temperature difference is the critical factor.

22. The solubility of gases in liquids is greatly affected by the pressure of a gas above the liquid. The greater the pressure, the more soluble the gas. There is very little effect of pressure regarding the dissolution of solids in liquids.

23. For a given mass of solute, the smaller the particles, the faster the dissolution of the solute. This is due to the smaller particles having a greater surface area exposed to the dissolving action of the solvent.

24. In a saturated solution, the net rate of dissolution is zero. There is no further increase in the amount of dissolved solute, even though undissolved solute is continuously dissolving, because dissolved solute is continuously coming out of solution, crystallizing at a rate equal to the rate of dissolving.

25. When crystals of $AgNO_3$ and NaCl are mixed, the contact between the individual ions is not intimate enough for the double displacement reaction to occur. When solutions of the two chemicals are mixed, the ions are free to move and come into intimate contact with each other, allowing the reaction to occur easily. The AgCl formed is insoluble.

26. A 16 molar solution of nitric acid is a solution that contains 16 moles $HNO_3$ per liter of solution.

27. The two solutions contain the same number of chloride ions. One liter of 1 M NaCl contains 1 mole of NaCl, therefore 1 mole of chloride ions. 0.5 liter of 1 M $MgCl_2$ contains 0.5 mol of $MgCl_2$ and 1 mole of chloride ions.

$$(0.5\ \text{L})\left(\frac{1\ \text{mol MgCl}_2}{\text{L}}\right)\left(\frac{2\ \text{mol Cl}^-}{1\ \text{mol MgCl}_2}\right) = 1\ \text{mol Cl}^-$$

28. The champagne would spray out of the bottle all over the place. The rise in temperature and the increase in kinetic energy of the molecules by shaking both act to decrease the solubility of gas within the liquid. The pressure inside the bottle would be great. As the cork is popped, much of the gas would escape from the liquid very rapidly, causing the champagne to spray.

29. The number of grams of NaCl in 750 mL of 5 molar solution is

$$(0.75\ \text{L})\left(\frac{5\ \text{mol NaCl}}{\text{L}}\right)\left(\frac{58.44\ \text{g}}{1\ \text{mol}}\right) = 200\ \text{g NaCl}$$

Dissolve the 200 g of NaCl in a minimum amount of water, then dilute the resulting solution to a final volume of 750 mL.

30. A semipermeable membrane will allow water molecules to pass through in both directions. If it has pure water on one side and 10% sugar solutions on the other side of the membrane, there is a higher concentration of water molecules on the pure water side. Therefore, there are more water molecule impacts per second on the pure water side of the membrane. The net result is more water molecules pass from the pure water to the sugar solution.

31. The urea solution will have the greater osmotic pressure because it has 1.67 mol solute/kg $H_2O$, while the glucose solution has only 0.83 mol solute/kg $H_2O$.

32. A lettuce leaf immersed in salad dressing containing salt and vinegar will become limp and wilted as a result of osmosis. As the water inside the leaf flows into the dressing where the solute concentration is higher the leaf becomes limp from fluid loss. In water, osmosis proceeds in the opposite direction flowing into the lettuce leaf maintaining a high fluid content and crisp leaf.

33. The concentration of solutes (such as salts) is higher in seawater than in body fluids. The survivors who drank seawater suffered further dehydration from the transfer of water by osmosis from body tissues to the intestinal tract.

34. Ranking of the specified bases in descending order of the volume of each required to react with 1 liter of 1 M HCl. The volume of each required to yield 1 mole of $OH^-$ ion is shown.

| | | |
|---|---|---|
| (a) | 1 M NaOH | 1 liter |
| (b) | 0.6 M $Ba(OH)_2$ | 0.8 liter |
| (c) | 2 M KOH | 0.5 liter |
| (d) | 1.5 M $Ca(OH)_2$ | 0.33 liter |

35. The boiling point of a liquid or solution is the temperature at which the vapor pressure of the liquid equals the pressure of the atmosphere. Since a solution containing a nonvolatile solute has a lower vapor pressure than the pure solvent, the boiling point of the solution must be at a higher temperature than for the pure solvent. This will result in the vapor pressure of the solution equaling the atmospheric pressure.

36. The freezing point is the temperature at which a liquid changes to a solid. The vapor pressure of a solution is lower than that of a pure solvent. Therefore, the vapor pressure curve of the solution intersects the vapor pressure curve of the pure solvent, at a temperature lower than the freezing point of the pure solvent. (See Figure 14.8a) At this point of intersection, the vapor pressure of the solution equals the vapor pressure of the pure solvent.

37. A glass filled with Seven-Up and crushed ice would be colder than a glass of water and crushed ice. The ice will keep the water at its freezing point. The Seven-Up will have a lower freezing point because it contains dissolved solutes.

38. Water and ice are different phases of the same substance in equilibrium at the freezing point of water, 0°C. The presence of the methanol lowers the vapor pressure and hence the freezing point of water. If the ratio of alcohol to water is high, the freezing point can be lowered as much as 10°C or more.

39. Effectiveness in lowering the freezing point of 500. g water:

(a) 100. g (2.17 mol) of ethyl alcohol is more effective than 100. g (0.292 mol) of sucrose.

(b) 20.0 g (0.435 mol) of ethyl alcohol is more effective than 100. g (0.292 mol) of sucrose.

(c) 20.0 g (0.625 mol) of methyl alcohol is more effective than 20.0 g (0.435 mol) of ethyl alcohol.

40. 5 molal NaCl = 5 mol NaCl/kg $H_2O$; 5 molar NaCl = 5 mol NaCl/L of solution. The volume of the 5 molal solution will be larger than 1 liter (1 L $H_2O$ + 5 mol NaCl). The volume of the 5 molar solution is exactly 1 L (5 mol NaCl + sufficient $H_2O$ to produce 1 L of solution). The molarity of a 5 molal solution is therefore, less than 5 molar.

41. Reasonably soluble: (a) KOH (b) $NiCl_2$ (d) $AgC_2H_3O_2$ (e) $Na_2CrO_4$
Insoluble: (c) ZnS

42. Reasonably soluble: (c) $CaCl_2$ (d) $Fe(NO_3)_3$
Insoluble: (a) $PbI_2$ (b) $MgCO_3$ (e) $BaSO_4$

43. Mass percent calculations.

(a) 25.0 g NaBr + 100. g $H_2O$ = 125 g solution

$$\left(\frac{25.0 \text{ g NaBr}}{125 \text{ g solution}}\right)(100) = 20.0\% \text{ NaBr}$$

(b) 1.20 g $K_2SO_4$ + 10.0 g $H_2O$ = 11.2 g solution

$$\left(\frac{1.20 \text{ g } K_2SO_4}{11.2 \text{ g solution}}\right)(100) = 10.7\% \text{ } K_2SO_4$$

44. (a) 40.0 g $Mg(NO_3)_2$ + 500. g $H_2O$ = 540. g solution

$$\left(\frac{40.0 \text{ g } Mg(NO_3)_2}{540. \text{ g solution}}\right)(100) = 7.41\% \text{ } Mg(NO_3)_2$$

(b) 17.5 g $NaNO_3$ + 250. g $H_2O$ = 268 g solution

$$\left(\frac{17.5 \text{ g } NaNO_3}{268 \text{ g solution}}\right)(100) = 6.53\% \text{ } NaNO_3$$

45. A 12.5% $AgNO_3$ solution contains 12.5 g $AgNO_3$ per 100. g solution

$$(30.0 \text{ g } AgNO_3)\left(\frac{100. \text{ g solution}}{12.5 \text{ g } AgNO_3}\right) = 240. \text{ g solution}$$

46. A 12.5% $AgNO_3$ solution contains 12.5 g $AgNO_3$ per 100. g solution

$$(0.400 \text{ mol } AgNO_3)\left(\frac{169.9 \text{ g}}{\text{mol}}\right)\left(\frac{100. \text{ g solution}}{12.5 \text{ g } AgNO_3}\right) = 544 \text{ g solution}$$

47. Mass percent calculations.

(a) 60.0 g NaCl + 200.0 g $H_2O$ = 260.0 g solution

$$\left(\frac{60.0 \text{ g NaCl}}{260.0 \text{ g solution}}\right)(100) = 23.1\% \text{ NaCl}$$

(b) $$(0.25 \text{ mol } HC_2H_3O_2)\left(\frac{60.03 \text{ g}}{\text{mol}}\right) = 15 \text{ g } HC_2H_3O_2$$

$$(3.0 \text{ mol } H_2O)\left(\frac{18.02 \text{ g}}{\text{mol}}\right) = 54 \text{ g } H_2O$$

$$\left(\frac{15 \text{ g } HC_2H_3O_2}{69 \text{ g solution}}\right)(100) = 22\% \text{ } HC_2H_3O_2$$

48. Mass percent calculation.

(a) 145.0 g NaOH + 1500 g $H_2O$ = 1645 g solution

$$\left(\frac{145.0 \text{ g NaOH}}{1645 \text{ g solution}}\right)(100) = 8.815\% \text{ NaOH}$$

(b) 1.0 molal solution of $C_6H_{12}O_6$ = $\left(\frac{1 \text{ mol } C_6H_{12}O_6}{1000. \text{ g } H_2O}\right)$

$$(1.0 \text{ mol } C_6H_{12}O_6)\left(\frac{180.2 \text{ g}}{\text{mol}}\right) = 180 \text{ g } C_6H_{12}O_6$$

1000. g $H_2O$ + 180 g $C_6H_{12}O_6$ = 1180 g solution

$$\left(\frac{180 \text{ g } C_6H_{12}O_6}{1180 \text{ g solution}}\right)(100) = 15\% \text{ } C_6H_{12}O_6$$

49. $$(65 \text{ g solution})\left(\frac{5.0 \text{ g KCl}}{100. \text{ g solution}}\right) = 3.3 \text{ g KCl}$$

50. $$(250. \text{ g solution})\left(\frac{15.0 \text{ g } K_2CrO_4}{100. \text{ g solution}}\right) = 37.5 \text{ g } K_2CrO_4$$

51. Mass/volume percent.

$$\left(\frac{22.0 \text{ g } CH_3OH}{100. \text{ mL solution}}\right)(100) = 22.0\% \; CH_3OH$$

52. Mass/volume percent.

$$\left(\frac{4.20 \text{ g NaCl}}{12.5 \text{ mL solution}}\right)(100) = 33.6\% \text{ NaCl}$$

53. Volume percent.

$$\left(\frac{10.0 \text{ mL } CH_3OH}{40.0 \text{ mL solution}}\right)(100) = 25.0\% \; CH_3OH$$

54. Volume percent.

$$\left(\frac{2.0 \text{ mL } C_6H_{14}}{9.0 \text{ mL solution}}\right)(100) = 22\% \; C_6H_{14}$$

55. Molarity problems $\left(M = \frac{\text{mol}}{\text{L}}\right)$

(a) $\left(\frac{0.10 \text{ mol}}{250 \text{ mL}}\right)\left(\frac{1000 \text{ mL}}{\text{L}}\right) = 0.40 \text{ M}$

(b) $\left(\frac{2.5 \text{ mol NaCl}}{0.650 \text{ L}}\right) = 3.8 \text{ M NaCl}$

(c) $\left(\frac{53.0 \text{ g } Na_2CrO_4}{1.00 \text{ L}}\right)\left(\frac{1 \text{ mol}}{162.0 \text{ g}}\right) = 0.327 \text{ M } Na_2CrO_4$

(d) $\left(\frac{260 \text{ g } C_6H_{12}O_6}{800. \text{ mL}}\right)\left(\frac{1000 \text{ mL}}{\text{L}}\right)\left(\frac{1 \text{ mol}}{180.2 \text{ g}}\right) = 1.8 \text{ M } C_6H_{12}O_6$

56. (a) $\left(\frac{0.025 \text{ mol HCl}}{10. \text{ mL}}\right)\left(\frac{1000 \text{ mL}}{\text{L}}\right) = 2.5 \text{ M HCl}$

(b) $\left(\frac{0.35 \text{ mol } BaCl_2 \bullet 2H_2O}{593 \text{ mL}}\right)\left(\frac{1000 \text{ mL}}{\text{L}}\right) = 0.59 \text{ M } BaCl_2 \bullet 2\,H_2O$

(c) $\left(\frac{1.5 \text{ g } Al_2(SO_4)_3}{2.00 \text{ L}}\right)\left(\frac{1 \text{ mol}}{342.2 \text{ g}}\right) = 2.19 \times 10^{-3} \text{ M } Al_2(SO_4)_3$

(d) $$\left(\frac{0.0282\text{ g Ca(NO}_3)_2}{1.00\text{ mL}}\right)\left(\frac{1000\text{ mL}}{\text{L}}\right)\left(\frac{1\text{ mol}}{164.1\text{ g}}\right) = 0.172\text{ M Ca(NO}_3)_2$$

57. $$\text{Molarity} = \frac{\text{mol solute}}{\text{L solution}} \quad \text{or} \quad \text{mol solute} = (\text{L solution})(\text{Molarity})$$

(a) $$(40.0\text{ L})\left(\frac{1.0\text{ mol LiCl}}{\text{L}}\right) = 40.\text{ mol LiCl}$$

(b) $$(25.0\text{ mL})\left(\frac{1\text{ L}}{1000\text{ mL}}\right)\left(\frac{3.0\text{ mol H}_2\text{SO}_4}{\text{L}}\right) = 0.0750\text{ mol H}_2\text{SO}_4$$

58. $$\text{Molarity} = \frac{\text{mol solute}}{\text{L solution}} \quad \text{or} \quad \text{mol solute} = (\text{L solution})(\text{Molarity})$$

(a) $$(349\text{ mL})\left(\frac{1\text{ L}}{1000\text{ mL}}\right)\left(\frac{0.0010\text{ mol NaOH}}{\text{L}}\right) = 3.5 \times 10^{-4}\text{ mol NaOH}$$

(b) $$(5000.\text{ mL})\left(\frac{1\text{ L}}{1000\text{ mL}}\right)\left(\frac{3.1\text{ mol CoCl}_2}{\text{L}}\right) = 16\text{ mol CoCl}_2$$

59. (a) $$(150\text{ L})\left(\frac{1.0\text{ mol NaCl}}{\text{L}}\right)\left(\frac{58.44\text{ g}}{\text{mol}}\right) = 8.8 \times 10^3\text{ g NaCl}$$

(b) $$(260\text{ mL})\left(\frac{18\text{ mol H}_2\text{SO}_4}{1000\text{ mL}}\right)\left(\frac{98.09\text{ g}}{\text{mol}}\right) = 4.6 \times 10^2\text{ g H}_2\text{SO}_4$$

60. (a) $$(0.035\text{ L})\left(\frac{10.0\text{ mol HCl}}{\text{L}}\right)\left(\frac{36.46\text{ g}}{\text{mol}}\right) = 13\text{ g HCl}$$

(b) $$(8.00\text{ mL})\left(\frac{1\text{ L}}{1000\text{ mL}}\right)\left(\frac{8.00\text{ mol Na}_2\text{C}_2\text{O}_4}{\text{L}}\right)\left(\frac{134.0\text{ g}}{\text{mol}}\right) = 8.58\text{ g Na}_2\text{C}_2\text{O}_4$$

61. (a) $$(0.430\text{ mol})\left(\frac{1\text{ L}}{0.256\text{ mol}}\right)\left(\frac{1000\text{ mL}}{\text{L}}\right) = 1.68 \times 10^3\text{ mL}$$

(b) $$(20.0\text{ g KCl})\left(\frac{1\text{ mol}}{74.55\text{ g}}\right)\left(\frac{1\text{ L}}{0.256\text{ mol}}\right)\left(\frac{1000\text{ mL}}{\text{L}}\right) = 1.05 \times 10^3\text{ mL}$$

62. (a) $$(10.0\text{ mol})\left(\frac{1\text{ L}}{0.256\text{ mol}}\right)\left(\frac{1000\text{ mL}}{\text{L}}\right) = 3.91 \times 10^4\text{ mL}$$

(b) The conversion is: g $Cl^-$ → mol $Cl^-$ → mol KCl → L → mL

$$(71.0 \text{ g Cl}^-)\left(\frac{1 \text{ mol}}{35.45 \text{ g}}\right)\left(\frac{1 \text{ mol KCl}}{1 \text{ mol Cl}^-}\right)\left(\frac{1 \text{ L}}{0.256 \text{ mol KCl}}\right)\left(\frac{1000 \text{ mL}}{\text{L}}\right) = 7.82 \times 10^3 \text{ mL}$$

63. (a) First calculate the moles of HCl in each solution. Then calculate the molarity.

$$(100. \text{ mL})\left(\frac{1 \text{ L}}{1000 \text{ mL}}\right)\left(\frac{1.0 \text{ mol}}{\text{L}}\right) = 0.10 \text{ mol HCl}$$

$$(150. \text{ mL})\left(\frac{1 \text{ L}}{1000 \text{ mL}}\right)\left(\frac{2.0 \text{ mol}}{\text{L}}\right) = 0.30 \text{ mol HCl}$$

Total mol = 0.40 mol HCl

Total volume = 100. mL + 150. mL = 250. mL (0.250 L)

$$\frac{0.40 \text{ mol HCl}}{0.250 \text{ L}} = 1.6 \text{ M HCl}$$

(b) First calculate the moles of NaCl in each solution. Then calculate the molarity.

$$(25.0 \text{ mL})\left(\frac{1 \text{ L}}{1000 \text{ mL}}\right)\left(\frac{1.25 \text{ mol}}{\text{L}}\right) = 0.0313 \text{ mol NaCl}$$

$$(75.0 \text{ mL})\left(\frac{1 \text{ L}}{1000 \text{ mL}}\right)\left(\frac{2.0 \text{ mol}}{\text{L}}\right) = 0.150 \text{ mol NaCl}$$

Total mol = 0.181 mol NaCl

Total volume = 25.0 mL + 75.0 mL = 100. mL = 0.100 L

$$\frac{0.181 \text{ mol NaCl}}{0.100 \text{ L}} = 1.81 \text{ M NaCl}$$

64. Dilution problem

$V_1M_1 = V_2M_2$

(a) $V_1 = 200.$ mL $\qquad V_2 = 400.$ mL

$M_1 = 12$ M $\qquad M_2 = M_2$

$(200. \text{ mL})(12 \text{ M}) = (400. \text{ mL})(M_2)$

$$M_2 = \frac{(200. \text{ mL})(12 \text{ M})}{400. \text{ mL}} = 6.0 \text{ M HCl}$$

(b) $V_1 = 60.0 \text{ mL}$ $\quad V_2 = 560. \text{ mL}$
$M_1 = 0.60 \text{ M}$ $\quad M_2 = M_2$
$(60.0 \text{ mL})(0.60 \text{ M}) = (560. \text{ mL})(M_2)$

$$M_2 = \frac{(60.0 \text{ mL})(0.60 \text{ M})}{560. \text{ mL}} = 0.064 \text{ M } ZnSO_4$$

65. $V_1M_1 = V_2M_2$

(a) $(V_1)(15 \text{ M}) = (50. \text{ mL})(6.0 \text{ M})$

$$V_1 = \frac{(50.0 \text{ mL})(6.0 \text{ M})}{15 \text{ M}} = 20. \text{ mL } 15 \text{ M } NH_3$$

(b) $(V_1)(18 \text{ M}) = (250 \text{ mL})(10.00 \text{ M})$

$$V_1 = \frac{(250 \text{ mL})(10.00 \text{ M})}{18 \text{ M}} = 140 \text{ mL } 18 \text{ M } H_2SO_4$$

66. $V_1M_1 = V_2M_2$

(a) $(V_1)(12 \text{ M}) = (400. \text{ mL})(6.0 \text{ M})$

$$V_1 = \frac{(400.0 \text{ mL})(6.0 \text{ M})}{12 \text{ M}} = 2.0 \times 10^2 \text{ mL } 12 \text{ M HCl}$$

(b) $(V_1)(16 \text{ M}) = (100. \text{ mL})(2.5 \text{ M})$

$$V_1 = \frac{(100. \text{ mL})(2.5 \text{ M})}{16 \text{ M}} = 16 \text{ mL } 16 \text{ M } HNO_3$$

67. $(0.250 \text{ L})\left(\frac{0.750 \text{ mol}}{\text{L}}\right) = 0.19 \text{ mol } H_2SO_4$

(a) Final volume after mixing

$250. \text{ mL} + 150. \text{ mL} = 400. \text{ mL} = 0.400 \text{ L}$

$$\frac{0.19 \text{ mol } H_2SO_4}{0.400 \text{ L}} = 0.48 \text{ M } H_2SO_4$$

(b) $(250. \text{ mL})\left(\frac{1 \text{ L}}{1000 \text{ mL}}\right)\left(\frac{0.70 \text{ mol } H_2SO_4}{\text{L}}\right) = 0.18 \text{ mol } H_2SO_4$

Total moles $= 0.19 \text{ mol} + 0.18 \text{ mol} = 0.37 \text{ mol } H_2SO_4$

Final volume $= 250. \text{ mL} + 250. \text{ mL} = 500. \text{ mL} = 0.500 \text{ L}$

$$\frac{0.37 \text{ mol } H_2SO_4}{0.500 \text{ L}} = 0.74 \text{ M } H_2SO_4$$

68. $(0.250\ L)\left(\frac{0.750\ mol}{L}\right) = 0.19\ mol\ H_2SO_4$

(a) $(400.\ mL)\left(\frac{1\ L}{1000\ mL}\right)\left(\frac{2.50\ mol\ H_2SO_4}{L}\right) = 1.00\ mol\ H_2SO_4$

Total moles = 0.19 mol + 1.00 mol = 1.19 mol $H_2SO_4$

Final volume = 250. mL + 400. mL = 650. mL = 0.650 L

$$\frac{1.19\ mol\ H_2SO_4}{0.650\ L} = 1.83\ M\ H_2SO_4$$

(b) Final volume after mixing

250. mL + 375 mL = 625 mL = 0.625 L

$$\frac{0.19\ mol\ H_2SO_4}{0.625\ L} = 0.30\ M\ H_2SO_4$$

69. $BaCl_2 + K_2CrO_4 \rightarrow BaCrO_4 + 2\ KCl$

(a) mL $BaCl_2$ → mol $BaCl_2$ → mol $BaCrO_4$ → g $BaCrO_4$

$$(100.0\ mL\ BaCl_2)\left(\frac{0.300\ mol}{1000\ mL}\right)\left(\frac{1\ mol\ BaCrO_4}{1\ mol\ BaCl_2}\right)\left(\frac{253.3\ g}{mol}\right) = 7.60\ g\ BaCrO_4$$

(b) mL $K_2CrO_4$ → mol $K_2CrO_4$ → mol $BaCl_2$ → mL $BaCl_2$

$$(50.0\ mL\ K_2CrO_4)\left(\frac{0.300\ mol}{1000\ mL}\right)\left(\frac{1\ mol\ BaCl_2}{1\ mol\ K_2CrO_4}\right)\left(\frac{1000\ mL}{1.0\ mol}\right) = 15\ mL\ of\ 1.0\ M\ BaCl_2$$

70. $3\ MgCl_2 + 2\ Na_3PO_4 \rightarrow Mg_3(PO_4)_2 + 6\ NaCl$

(a) mL $MgCl_2$ → mol $MgCl_2$ → mol $Na_3PO_4$ → mL $Na_3PO_4$

$$(50.0\ mL\ MgCl_2)\left(\frac{0.250\ mol}{1000\ mL}\right)\left(\frac{2\ mol\ Na_3PO_4}{3\ mol\ MgCl_2}\right)\left(\frac{1000\ mL}{0.250\ mol}\right) = 33.3\ mL\ of\ 0.250\ M\ Na_3PO_4$$

(b) mL $MgCl_2$ → mol $MgCl_2$ → mol $Mg_3(PO_4)_2$ → g $Mg_3(PO_4)_2$

$$(50.0\ mL\ MgCl_2)\left(\frac{0.250\ mol}{1000\ mL}\right)\left(\frac{1\ mol\ Mg_3(PO_4)_2}{3\ mol\ MgCl_2}\right)\left(\frac{262.9\ g}{mol}\right) = 1.10\ g\ Mg_3(PO_4)_2$$

71. The balanced equation is

$$6\,FeCl_2 + K_2Cr_2O_7 + 14\,HCl \rightarrow 6\,FeCl_3 + 2\,CrCl_3 + 2\,KCl + 7\,H_2O$$

(a) $(2.0 \text{ mol } FeCl_2)\left(\dfrac{2 \text{ mol KCl}}{6 \text{ mol } FeCl_2}\right) = 0.67 \text{ mol KCl}$

(b) $(1.0 \text{ mol } FeCl_2)\left(\dfrac{2 \text{ mol } CrCl_3}{6 \text{ mol } FeCl_2}\right) = 0.33 \text{ mol } CrCl_3$

(c) $(0.050 \text{ mol } K_2Cr_2O_7)\left(\dfrac{6 \text{ mol } FeCl_2}{1 \text{ mol } K_2Cr_2O_7}\right) = 0.30 \text{ mol } FeCl_2$

(d) $(0.025 \text{ mol } FeCl_2)\left(\dfrac{1 \text{ mol } K_2Cr_2O_7}{6 \text{ mol } FeCl_2}\right)\left(\dfrac{1000 \text{ mL}}{0.060 \text{ mol}}\right) = 69 \text{ mL of } 0.060\text{ M } K_2Cr_2O_7$

(e) $(15.0 \text{ mL } FeCl_2)\left(\dfrac{6.0 \text{ mol}}{1000 \text{ mL}}\right)\left(\dfrac{14 \text{ mol HCl}}{6 \text{ mol } FeCl_2}\right)\left(\dfrac{1000 \text{ mL}}{6.0 \text{ mol}}\right) = 35 \text{ mL of } 6.0\text{ M HCl}$

72. $2\,KMnO_4 + 16\,HCl \rightarrow 2\,MnCl_2 + 5\,Cl_2 + 8\,H_2O + 2\,KCl$

(a) $(0.050 \text{ mol } KMnO_4)\left(\dfrac{5 \text{ mol } Cl_2}{2 \text{ mol } KMnO_4}\right) = 0.13 \text{ mol } Cl_2$

(b) $(1.0 \text{ L } KMnO_4)\left(\dfrac{2.0 \text{ mol}}{\text{L}}\right)\left(\dfrac{16 \text{ mol HCl}}{2 \text{ mol } KMnO_4}\right) = 16 \text{ mol HCl}$

(c) $(200. \text{ mL } KMnO_4)\left(\dfrac{0.50 \text{ mol}}{1000 \text{ mL}}\right)\left(\dfrac{16 \text{ mol HCl}}{2 \text{ mol } KMnO_4}\right)\left(\dfrac{1000 \text{ mL}}{6.0 \text{ mol}}\right) = 1.3 \times 10^2 \text{ mL of } 6\text{ M HCl}$

(d) $(75.0 \text{ mL HCl})\left(\dfrac{6.0 \text{ mol}}{1000 \text{ mL}}\right)\left(\dfrac{5 \text{ mol } Cl_2}{16 \text{ mol HCl}}\right)\left(\dfrac{22.4 \text{ L}}{\text{mol}}\right) = 3.2 \text{ L } Cl_2$

73. $\text{Molality} = m = \dfrac{\text{mol solute}}{\text{kg solvent}}$

(a) $\left(\dfrac{14.0 \text{ g } CH_3OH}{100. \text{ g } H_2O}\right)\left(\dfrac{1000 \text{ g}}{\text{kg}}\right)\left(\dfrac{1 \text{ mol}}{32.04 \text{ g}}\right) = \left(\dfrac{4.37 \text{ mol } CH_3OH}{\text{kg } H_2O}\right) = 4.37\, m\ CH_3OH$

(b) $\left(\dfrac{2.50 \text{ mol } C_6H_6}{250 \text{ g } C_6H_{14}}\right)\left(\dfrac{1000 \text{ g}}{\text{kg}}\right) = \left(\dfrac{10. \text{ mol } C_6H_6}{\text{kg } C_6H_{14}}\right) = 10.\, m\ C_6H_6$

74. Molality $= m = \dfrac{\text{mol solute}}{\text{kg solvent}}$

(a) $\left(\dfrac{1.0 \text{ g } C_6H_{12}O_6}{1.0 \text{ g } H_2O}\right)\left(\dfrac{1000 \text{ g}}{\text{kg}}\right)\left(\dfrac{1 \text{ mol}}{180.2 \text{ g}}\right) = \left(\dfrac{5.5 \text{ mol } C_6H_{12}O_6}{\text{kg } H_2O}\right) = 5.5\ m\ C_6H_{12}O_6$

(b) $\left(\dfrac{0.250 \text{ mol } I_2}{1.0 \text{ kg } H_2O}\right) = 0.25\ m\ I_2$

75. (a) $\left(\dfrac{2.68 \text{ g } C_{10}H_8}{38.4 \text{ g } C_6H_6}\right)\left(\dfrac{1 \text{ mol}}{128.2 \text{ g } C_{10}H_8}\right)\left(\dfrac{1000 \text{ g } C_6H_6}{\text{kg}}\right) = 0.544\ m$

(b) $K_f$ (for benzene) $= \dfrac{5.1°C}{m}$ Freezing point of benzene $= 5.5°C$

$\Delta t_f = (0.544\ m)\left(\dfrac{5.1°C}{m}\right) = 2.8°C$

Freezing point of solution $= 5.5°C - 2.8°C = 2.7°C$

(c) $K_b$ (for benzene) $= \dfrac{2.53°C}{m}$ Boiling point of benzene $= 80.1°C$

$\Delta t_b = (0.544\ m)\left(\dfrac{2.53°C}{m}\right) = 1.38°C$

Boiling point of solution $= 80.1°C + 1.38°C = 81.5°C$

76. (a) $\left(\dfrac{100.0 \text{ g } C_2H_6O_2}{150.0 \text{ g } H_2O}\right)\left(\dfrac{1 \text{ mol}}{62.07 \text{ g}}\right)\left(\dfrac{1000 \text{ g}}{\text{kg}}\right) = 10.74\ m$

(b) $\Delta t_b = mK_b = (10.74\ m)\left(\dfrac{0.512°C}{m}\right) = 5.50°C$ (Increase in boiling point)

Boiling point $= 100.00°C + 5.50°C = 105.50°C$

(c) $\Delta t_f = mK_f = (10.74\ m)\left(\dfrac{1.86°C}{m}\right) = 20.0°C$ (Decrease in freezing point)

Freezing point $= 0.00°C - 20.0°C = -20.0°C$

77. Freezing point of acetic acid is 16.6°C $K_f$ acetic acid $= \dfrac{3.90°C}{m}$

$\Delta t_f = 16.6°C - 13.2°C = 3.4°C$

$\Delta t_f = mK_f$

$m = \dfrac{3.4°C}{3.90°C/m} = 0.87\ m$

Convert 8.00 g unknown/60.0 g $HC_2H_3O_2$ to g/mol (molar mass)

Conversion: $\dfrac{\text{g unknown}}{\text{g } HC_2H_3O_2} \rightarrow \dfrac{\text{g unknown}}{\text{kg } HC_2H_3O_2} \rightarrow \dfrac{\text{g}}{\text{mol}}$

$$\left(\frac{8.00 \text{ g unknown}}{60.0 \text{ g } HC_2H_3O_2}\right)\left(\frac{1000 \text{ g}}{\text{kg}}\right)\left(\frac{1 \text{ kg } HC_2H_3O_2}{0.87 \text{ mol unknown}}\right) = 153 \text{ g/mol}$$

78. $\Delta t_f = 2.50°C \qquad K_f \text{ (for } H_2O) = \dfrac{1.86°C}{m}$

$\Delta t_f = mK_f$

$$m = \frac{2.50°C}{1.86°C/m} = 1.34\ m$$

Covert 4.80 g unknown/22.0 g $H_2O$ to g/mol (molar mass)

$$\left(\frac{4.80 \text{ g unknown}}{22.0 \text{ g } H_2O}\right)\left(\frac{1000 \text{ g}}{\text{kg}}\right)\left(\frac{1 \text{ kg } H_2O}{1.34 \text{ mol unknown}}\right) = 163 \text{ g/mol}$$

79. First calculate the g NaOH to neutralize the HCl.

$NaOH + HCl \rightarrow NaCl + H_2O$

$$(0.15 \text{ L HCl})\left(\frac{1.0 \text{ mol}}{\text{L}}\right)\left(\frac{1 \text{ mol NaOH}}{1 \text{ mol HCl}}\right)\left(\frac{40.00 \text{ g}}{\text{mol}}\right) = 6.0 \text{ g NaOH required to neutralize the acid}$$

Now calculate the grams of 10% NaOH solution that contains 6.0 g NaOH

$$\frac{6.0 \text{ g NaOH}}{x} = \frac{10.0 \text{ g NaOH}}{100.0 \text{ g } 10.0\% \text{ NaOH solution}}$$

$x = 60.$ g 10% NaOH solution

80. $1.0\ m \text{ HCl} = \dfrac{1 \text{ mol HCl}}{1 \text{ kg } H_2O} = \dfrac{36.46 \text{ g HCl}}{1000 \text{ g } H_2O}$

Total mass of solution = 1000 g + 36.46 g = 1036.46 g

Therefore, $1.0\ m \text{ HCl} = \dfrac{1 \text{ mol HCl}}{1036.46 \text{ g HCl solution}}$

$NaOH + HCl \rightarrow NaCl + H_2O$

Calculate the g NaOH to neutralize HCl

$$(250. \text{ g solution})\left(\frac{1 \text{ mol HCl}}{1036.46 \text{ solution}}\right)\left(\frac{1 \text{ mol NaOH}}{1 \text{ mol HCl}}\right)\left(\frac{40.00 \text{ g}}{\text{mol}}\right) = 9.648 \text{ g NaOH}$$

Calculate the grams of 10.0% NaOH solution that contains 9.648 g NaOH.

$$\frac{9.648 \text{ g NaOH}}{x} = \frac{10.0 \text{ g NaOH}}{100.0 \text{ g } 10.0\% \text{ NaOH solution}}$$

$x = 96.5$ g 10% NaOH solution

81. (a) $(1.0 \text{ L syrup})\left(\frac{1000 \text{ mL}}{\text{L}}\right)\left(\frac{1.06 \text{ g}}{\text{mL}}\right)\left(\frac{15.0 \text{ g sugar}}{100. \text{ g syrup}}\right) = 1.6 \times 10^2 \text{ g sugar}$

(b) $\left(\frac{1.6 \times 10^2 \text{ g } C_{12}H_{22}O_{11}}{\text{L}}\right)\left(\frac{1 \text{ mol}}{342.3 \text{ g}}\right) = 0.47 \text{ M}$

(c) $m = \frac{\text{mol sugar}}{\text{kg } H_2O}$ 15% sugar by mass $= 15.0 \text{ g } C_{12}H_{22}O_{11} + 85.0 \text{ g } H_2O$

$$\left(\frac{15.0 \text{ g } C_{12}H_{22}O_{11}}{85.0 \text{ g } H_2O}\right)\left(\frac{1000 \text{ g } H_2O}{1 \text{ kg } H_2O}\right)\left(\frac{1 \text{ mol}}{342.3 \text{ g } C_{12}H_{22}O_{11}}\right) = 0.516 \text{ m}$$

82. $K_f = \frac{5.1°C}{m}$ $\Delta t_f = 0.614°C$

$$\left(\frac{3.84 \text{ g } C_4H_2N}{250. \text{ g } C_6H_6}\right)\left(\frac{1000 \text{ g}}{\text{kg}}\right) = \frac{15.4 \text{ g } C_4H_2N}{\text{kg } C_6H_6}$$

$\Delta t_f = mK_f$

$$m = \frac{0.614°C}{5.1°C/m} = 0.12\ m = \frac{0.12 \text{ mol } C_4H_2N}{\text{kg } C_6H_6}$$

$$\left(\frac{15.4 \text{ g } C_4H_2N}{\text{kg } C_6H_6}\right)\left(\frac{1 \text{ kg } C_6H_6}{0.12 \text{ mol } C_4H_2N}\right) = 1.3 \times 10^2 \text{ g/mol}$$

Empirical mass ($C_4H_2N$) = 64.07 g

$\frac{130 \text{ g}}{64.07 \text{ g}} = 2.0$ (number of empirical formulas per molecular formula)

Therefore, the molecular formula is twice the empirical formula, or $C_8H_4N_2$.

83. $(12.0 \text{ mol HCl})\left(\frac{36.46 \text{ g}}{\text{mol}}\right) = 438 \text{ g HCl in } 1.00 \text{ L solution}$

$(1.00 \text{ L})\left(\frac{1.18 \text{ g solution}}{\text{mL}}\right)\left(\frac{1000 \text{ mL}}{\text{L}}\right) = 1180 \text{ g solution}$

1180 g solution − 438 g HCl = 742 g $H_2O$ (0.742 kg $H_2O$)

Since molality $= \frac{\text{mol HCl}}{\text{kg } H_2O} = \frac{12.0 \text{ mol HCl}}{0.742 \text{ kg } H_2O} = 16.2\ m \text{ HCl}$

84. First calculate the g $KNO_3$ in the solution.

The conversion is: $\frac{\text{mg } K^+}{\text{mL}} \rightarrow \frac{\text{g } K^+}{\text{mL}} \rightarrow \frac{\text{g } KNO_3}{\text{mL}} \rightarrow \text{g } KNO_3$

$$\left(\frac{5.5 \text{ mg } K^+}{\text{mL}}\right)\left(\frac{1 \text{ g}}{1000 \text{ mg}}\right)\left(\frac{101.1 \text{ g } KNO_3}{39.10 \text{ g } K^+}\right)(450 \text{ mL}) = 6.4 \text{ g } KNO_3$$

Now calculate the mol $KNO_3$ and the molarity.

$$(6.4 \text{ g } KNO_3)\left(\frac{1 \text{ mol}}{101.1 \text{ g}}\right) = 0.063 \text{ mol } KNO_3$$

$$\frac{0.063 \text{ mol } KNO_3}{0.450 \text{ L}} = 0.14 \text{ M}$$

85. $(25.0 \text{ g KCl})\left(\frac{100. \text{ g solution}}{5.50 \text{ g KCl}}\right) = 455 \text{ g solution}$

Alternate solution:

$$\left(\frac{25.0 \text{ g KCl}}{x}\right) = \left(\frac{5.50 \text{ g KCl}}{100. \text{ g solution}}\right)$$

$x = 455$ g solution

86. (a) $(500. \text{ mL solution})\left(\frac{0.90 \text{ g NaCl}}{100. \text{ mL solution}}\right) = 4.5 \text{ g NaCl}$

(b) $\left(\frac{4.5 \text{ g NaCl}}{x \text{ mL}}\right)(100) = 9.0\%$ $\qquad x$ = volume of 9.0% solution

$$x = \frac{4.5 \text{ g NaCl}}{9.0\%} = 50. \text{ mL (4.5 g NaCl in solution)}$$

500. mL − 50. mL = 450. mL $H_2O$ must evaporate

87. From Figure 14.4, the solubility of $KNO_3$ in $H_2O$ at 20°C is 32 g per 100 g $H_2O$.

$(50.0 \text{ g } KNO_3)\left(\frac{100. \text{ g } H_2O}{32.0 \text{ g } KNO_3}\right) = 156 \text{ g } H_2O$ to produce a saturated solution.

175 g $H_2O$ − 156 g $H_2O$ = 19 g $H_2O$ must be evaporated.

88. $(150 \text{ mL alcohol})\left(\frac{100. \text{ mL solution}}{70.0 \text{ mL alcohol}}\right) = 210 \text{ mL solution}$

89. (a) $(1.00 \text{ L solution})\left(\frac{1000 \text{ mL solution}}{\text{L solution}}\right)\left(\frac{1.21 \text{ g}}{\text{mL}}\right)\left(\frac{35.0 \text{ g } HNO_3}{100. \text{ g solution}}\right) = 424 \text{ g } HNO_3$

(b) $(500.\text{ g HNO}_3)\left(\frac{1000\text{ mL solution}}{424\text{ g HNO}_3}\right)\left(\frac{1.00\text{ L}}{1000\text{ mL}}\right) = 1.18\text{ L solution}$

90. Assume 1.000 L (1000. mL) of solution

$$\left(\frac{1000.\text{ mL}}{\text{L}}\right)\left(\frac{1.21\text{ g solution}}{\text{mL}}\right)\left(\frac{35.0\text{ g HNO}_3}{100.\text{ g solution}}\right)\left(\frac{1\text{ mol}}{63.02\text{ g}}\right) = 6.72\text{ M HNO}_3$$

91. First calculate the molarity of the solution

$$\left(\frac{80.0\text{ g H}_2\text{SO}_4}{500.\text{ mL}}\right)\left(\frac{1000\text{ mL}}{\text{L}}\right)\left(\frac{1\text{ mol}}{98.09\text{ g}}\right) = 1.63\text{ M H}_2\text{SO}_4$$

$M_1V_1 = M_2V_2$

$(1.63\text{ M})(500.\text{ mL}) = (0.10\text{ M})(V_2)$

$$V_2 = \frac{(1.63\text{ M})(500.\text{ mL})}{0.10\text{ M}} = 8.2 \times 10^3\text{ mL} = 8.2\text{ L}$$

92. Note that the problem asks for the volume of water to be added, not the final volume of the solution.

$M_1V_1 = M_2V_2$

$(1.40\text{ M})(300.\text{ mL}) = (0.500\text{ M})(V_2)$

$$V_2 = \frac{(1.40\text{ M})(300.\text{ mL})}{0.500\text{ M}} = 840.\text{ mL (final volume)}$$

840. mL − 300. mL = 540. mL water to be added

93. $M_1V_1 = M_2V_2$

$(16\text{ M})(10.0\text{ mL}) = (M_2)(500.\text{ mL})$

$$M_2 = \frac{(16\text{ M})(10.0\text{ mL})}{500.0\text{ M}} = 0.32\text{ M HNO}_3$$

94. $(V_1)(5.00\text{ M}) = (250\text{ mL})(0.625\text{ M})$

$$V_1 = \frac{(250\text{ mL})(0.625\text{ M})}{5.00\text{ M}} = 31\text{ mL 5.00 M KOH}$$

To make 250. mL of 0.625 M KOH, take 31.3 mL of 5.00 M KOH and dilute with water to a volume of 250. mL.

95. $Mg + 2\,HCl \rightarrow MgCl_2 + H_2(g)$

(a) mL HCl → mol HCl → mol $H_2$

$$(200.\ \text{mL HCl})\left(\frac{3.00\ \text{mol}}{1000\ \text{mL}}\right)\left(\frac{1\ \text{mol}\ H_2}{2\ \text{mol HCl}}\right) = 0.300\ \text{mol}\ H_2$$

(b) PV = nRT

$$P = (720\ \text{torr})\left(\frac{1\ \text{atm}}{760\ \text{torr}}\right) = 0.95\ \text{atm}$$

T = 27°C = 300. K
n = 0.300 mol

$$V = \frac{nRT}{P} = \frac{(0.300\ \text{mol})(0.0821\ \text{l atm/mol K})(300.\ \text{K})}{0.95\ \text{atm}} = 7.8\ \text{L}\ H_2$$

96. $Mg + 2\,HCl \rightarrow MgCl_2 + H_2(g)$
L $H_2$ → mol $H_2$ → mol HCl → M HCl

$$(3.50\ \text{L}\ H_2)\left(\frac{1\ \text{mol}}{22.4\ \text{L}}\right)\left(\frac{2\ \text{mol HCl}}{1\ \text{mol}\ H_2}\right)\left(\frac{1}{0.150\ \text{L}}\right) = 2.08\ \text{M HCl}$$

97. Use $M_1V_1 = M_2V_2$

$$1\ \text{drop} = \frac{1}{20.}\text{mL} = 0.050\ \text{mL}$$

100. mL + 0.050 mL = 100.050 mL

(17.8 M)(0.050 mL) = (M)(100.050 mL)

$$M = \frac{(17.8\ \text{M})(0.050\ \text{mL})}{100.050\ \text{mL}} = 8.9 \times 10^{-3}\ \text{M}$$

98. $Mg(OH)_2 + 2\,HCl \rightarrow MgCl_2 + 2\,H_2O$

$Al(OH)_3 + 3\,HCl \rightarrow AlCl_3 + 3\,H_2O$

Calculate the moles of HCl neutralized by each base.

$$(12.0\ \text{g}\ Mg(OH)_2)\left(\frac{1\ \text{mol}}{58.33\ \text{g}}\right)\left(\frac{2\ \text{mol HCl}}{1\ \text{mol}\ Mg(OH)_2}\right) = 0.400\ \text{mol HCl}$$

$$(10.0\ \text{g}\ Al(OH)_3)\left(\frac{1\ \text{mol}}{78.00\ \text{g}}\right)\left(\frac{3\ \text{mol HCl}}{1\ \text{mol}\ Al(OH)_3}\right) = 0.385\ \text{mol HCl}$$

12.0 g $Mg(OH)_2$ reacts with more HCl than 10.0 g $Al(OH)_3$. Therefore, $Mg(OH)_2$ is more effective in neutralizing stomach acid.

99. (a) With equal masses of $CH_3OH$ and $C_2H_5OH$, the substance with the lower molar mass will represent more moles of solute in solution. Therefore, the $CH_3OH$ will be more effective than $C_2H_5OH$ as an antifreeze.

(b) Equal molal solutions will lower the freezing point of the solution by the same amount.

100. Calculate molarity and molality. Assume 1000 mL of solution to calculate the amounts of $H_2SO_4$ and $H_2O$ in the solution.

$$(1000 \text{ mL solution})\left(\frac{1.29 \text{ g}}{\text{mL}}\right) = 1.29 \times 10^3 \text{ g solution}$$

$$(1.29 \times 10^3 \text{ g solution})\left(\frac{38 \text{ g } H_2SO_4}{100 \text{ g solution}}\right) = 4.9 \times 10^2 \text{ g } H_2SO_4$$

$$1.29 \times 10^3 \text{ g solution} - 4.9 \times 10^2 \text{ g } H_2SO_4 = 8.0 \times 10^2 \text{ g } H_2O \text{ in the solution}$$

$$m = \left(\frac{490 \text{ g } H_2SO_4}{8.0 \times 10^2 \text{ g } H_2O}\right)\left(\frac{1000 \text{ g}}{\text{kg}}\right)\left(\frac{1 \text{ mol}}{98.09 \text{ g}}\right) = 6.2\ m\ H_2SO_4$$

$$\text{M} = \left(\frac{4.9 \times 10^2 \text{ g } H_2SO_4}{\text{L}}\right)\left(\frac{1 \text{ mol}}{98.09 \text{ g}}\right) = 5.0 \text{ M } H_2SO_4$$

101. $1.00 \text{ lb} = 453.6 \text{ g sugar } (C_{12}H_{22}O_{11})$

$$(4.00 \text{ lb } H_2O)\left(\frac{453.6 \text{ g}}{\text{lb}}\right) = 1.81 \times 10^3 \text{ g } H_2O \ \ (1.81 \text{ kg } H_2O)$$

$$(453.6 \text{ g } C_{12}H_{22}O_{11})\left(\frac{1 \text{ mol}}{342.3 \text{ g}}\right) = 1.33 \text{ mol } C_{12}H_{22}O_{11}$$

$K_f$ (for $H_2O$) = 1.86°C kg solvent/mol solute

$$\Delta t_f = mK_f = \left(\frac{1.33 \text{ mol } C_{12}H_{22}O_{11}}{1.81 \text{ kg } H_2O}\right)\left(\frac{1.86\,°\text{C kg } H_2O}{\text{mol } C_{12}H_{22}O_{11}}\right) = 1.37\,°\text{C}$$

Freezing point of solution = 0°C − 1.37°C = −1.37°C = 29.5°F

If the sugar solution is placed outside, where the temperature is 20°F, the solution will freeze.

102. Freezing point depression is 5.4°C

(a) $\Delta t_f = mK_f$

$$m = \frac{\Delta t_f}{K_f} = \frac{5.4\,°\text{C}}{1.86\,°\text{C kg solvent/mol solute}} = 2.9\ m$$

(b) $K_b \text{ (for } H_2O) = \frac{0.512°\text{C kg solvent}}{\text{mol solute}} = \frac{0.512°\text{C}}{m}$

$$\Delta t_b = mK_b = (2.9\ m)\left(\frac{0.512°\text{C}}{m}\right) = 1.5°\text{C}$$

$$\text{Boiling point} = 100°\text{C} + 1.5°\text{C} = 101.5°\text{C}$$

103. Freezing point depression = 0.372°C $K_f = \frac{1.86°\text{C}}{m}$

$$\Delta t_f = mK_f$$

$$m = \frac{0.372°\text{C}}{1.86°\text{C}/m} = 0.200\ m$$

$$(6.20 \text{ g } C_2H_6O_2)\left(\frac{1 \text{ mol}}{62.07 \text{ g}}\right) = 0.100 \text{ mol } C_2H_6O_2$$

$$(0.100 \text{ mol } C_2H_6O_2)\left(\frac{1 \text{ kg } H_2O}{0.200 \text{ mol } C_2H_6O_2}\right)\left(\frac{1000 \text{ g } H_2O}{\text{kg } H_2O}\right) = 500.\text{ g } H_2O$$

104. (a) Freezing point depression = 20.0°C

$$12.0 \text{ L } H_2O\left(\frac{1000 \text{ mL}}{\text{L}}\right)\left(\frac{1.00 \text{ g}}{\text{mL}}\right) = 1.20 \times 10^4 \text{ g } H_2O$$

$$\Delta t_f = mK_f$$

$$m = \frac{20.0°\text{C}}{1.86°\text{C}/m} = 10.8\ m$$

$$(1.20 \times 10^4 \text{ g } H_2O)\left(\frac{10.8 \text{ mol } C_2H_6O_2}{1000 \text{ g } H_2O}\right)\left(\frac{62.07 \text{ g}}{\text{mol}}\right) = 8.04 \times 10^3 \text{ g } C_2H_6O_2$$

(b) $(8.04 \times 10^3 \text{ g } C_2H_6O_2)\left(\frac{1.00 \text{ mL}}{1.11 \text{ g}}\right) = 7.24 \times 10^3 \text{ mL } C_2H_6O_2$

(c) $1.8(-20.0) + 32 = -4.0°\text{F}$

105. Yes, a saturated solution can also be a dilute solution. For example, the solubility of AgCl in water at 25°C is $1.3 \times 10^{-5}$ mol/L. Thus, the solution formed by dissolving AgCl in water is both saturated and very dilute.

106. $HCl + NaOH \rightarrow NaCl + H_2O$
1 mol 1 mol
g NaOH → mol NaOH → mol HCl → L HCl

$$(12 \text{ g NaOH})\left(\frac{1 \text{ mol}}{40.00 \text{ g}}\right)\left(\frac{1 \text{ mol HCl}}{1 \text{ mol NaOH}}\right)\left(\frac{1 \text{ L HCl}}{0.65 \text{ mol HCl}}\right) = 0.46 \text{ L HCl (460 mL)}$$

107. $HNO_3 + NaHCO_3 \rightarrow NaNO_3 + H_2O + CO_2$

First calculate the grams of $NaHCO_3$ in the sample.

mL $HNO_3$ $\rightarrow$ L $HNO_3$ $\rightarrow$ mol $HNO_3$ $\rightarrow$ mol $NaHCO_3$ $\rightarrow$ g $NaHCO_3$

$$(150 \text{ mL } HNO_3)\left(\frac{1 \text{ L}}{1000 \text{ mL}}\right)\left(\frac{0.055 \text{ mol}}{\text{L}}\right)\left(\frac{1 \text{ mol } NaHCO_3}{1 \text{ mol } HNO_3}\right)\left(\frac{84.01 \text{ g}}{\text{mol}}\right)$$

$= 0.69$ g $NaHCO_3$ in the sample

$$\left(\frac{0.69 \text{ g}}{1.48 \text{ g}}\right)(100) = 47\% \ NaHCO_3$$

108. (a) Dilution problem: $M_1V_1 = M_2V_2$

$$(1.5 \text{ M})(8.4 \text{ L}) = (17.8 \text{ M})(V_2)$$

$$V_2 = \frac{(1.5 \text{ M})(8.4 \text{ L})}{17.8 \text{ M}} = 0.71 \text{ L}$$

0.71 L of 17.8 M $H_2SO_4$ is to be diluted to 8.4 L.

8.4 L − 0.71 L = 7.7 L $H_2O$ must be added

(b) $$\left(\frac{17.8 \text{ mol}}{1000. \text{ mL}}\right)(1.00 \text{ mL}) = 0.0178 \text{ mol}$$

(c) $$\left(\frac{1.5 \text{ mol}}{1000. \text{ mL}}\right)(1.00 \text{ mL}) = 0.0015 \text{ mol}$$

109. Freezing point depression is 3.6°C

$\Delta t_f = mK_f$

$$m = \frac{\Delta t_f}{K_f} = \frac{3.6°\text{C}}{1.86°\text{C}/m} = 1.9\ m \text{ solution}$$

$$\Delta t_b = mK_b = (1.9\ m)\left(\frac{0.512°\text{C}}{m}\right) = 0.97°\text{C}$$

Boiling point = 100.00°C + 0.97°C = 100.97°C

110. moles $HNO_3$ total = moles $HNO_3$ from 3.00 M + moles $HNO_3$ from 12.0 M

$$M_TV_T = M_{3.00\,M}V_{3.00\,M} + M_{12.0\,M}V_{12.0\,M}$$

Assume preparation of 1000. mL of 6 M solution

Let $y$ = volume of 3.00 M solution; volume of 12.0 M = 1000. mL − $y$

$(6.00\ M)(1000.\ mL) = (3.00\ M)(y) + (12.0\ M)(1000.\ mL - y)$

$6000.\ mL = 3.00\,y\ mL + 12{,}000\ mL - 12.0\,y$

$6000.\ mL = 9.00\,y \qquad y = \dfrac{6000.\ mL}{9.00} = 667\ mL\ 3\ M$

$1000.\ mL - 667\ mL = 333\ mL\ 12\ M$

Mix together 667 mL 3.00 M $HNO_3$ and 333 mL of 12.0 M $HNO_3$ to get 1000. mL of 6.00 M $HNO_3$.

111. $HBr + NaOH \rightarrow NaBr + H_2O$

First calculate the molarity of the diluted HBr solution.

The reaction is 1 mol HBr to 1 mol NaOH, so

$M_AV_A = M_BV_B$

$(M_A)(100.0\ mL) = (0.37\ M)(88.4\ mL)$

$M_A = \dfrac{(0.37\ M)(88.4\ mL)}{100.00\ mL} = 0.33\ M\ HBr$ (diluted solution)

Now calculate the molarity of the HBr before dilution.

$M_1V_1 = M_2V_2$

$(M_1)(20.0\ mL) = (0.33\ M)(240.\ mL)$

$M_1 = \dfrac{(0.33\ M)(240.\ mL)}{20.0\ mL} = 4.0\ M\ HBr$ (original solution)

112. $Ba(NO_3)_2 + 2\ KOH \rightarrow Ba(OH)_2 + 2\ KNO_3$

This is a limiting reactant problem. First calculate the moles of each reactant and determine the limiting reactant.

$M \times L = \left(\dfrac{moles}{L}\right)(L) = moles$

$\left(\dfrac{0.642\ mol}{L}\right)(0.0805\ L) = 0.0517\ mol\ Ba(NO_3)_2$

$\left(\dfrac{0.743\ mol}{L}\right)(0.0445\ L) = 0.0331\ mol\ KOH$

According to the equation, twice as many moles of KOH as $Ba(NO_3)_2$ are needed, so KOH is the limiting reactant.

$(0.0331\ mol\ KOH)\left(\dfrac{1\ mol\ Ba(OH)_2}{2\ mol\ KOH}\right)\left(\dfrac{171.3\ g}{mol}\right) = 2.84\ g\ Ba(OH)_2$ is formed

113. $(300.\ \text{g solution})\left(\dfrac{5.0\ \text{g sucrose}}{100.\ \text{g solution}}\right) = 15\ \text{g sucrose}$

$(y\ \text{g 2.0\% solution})\left(\dfrac{2.0\ \text{g sucrose}}{100.\ \text{g solution}}\right) = 15\ \text{g sucrose}$

$y = \left(\dfrac{100.\ \text{g solution}}{2.0\ \text{g sucrose}}\right)(15\ \text{g sucrose}) = 750\ \text{g 2\% solution}$

114. (a) $\left(\dfrac{0.25\ \text{mol}}{\text{L}}\right)(0.0458\ \text{L}) = 0.011\ \text{mol}\ Li_2CO_3$

(b) $\left(\dfrac{0.25\ \text{mol}}{\text{L}}\right)(0.75\ \text{L})\left(\dfrac{73.89\ \text{g}}{\text{mol}}\right) = 14\ \text{g}\ Li_2CO_3$

(c) $(6.0\ \text{g}\ Li_2CO_3)\left(\dfrac{1\ \text{mol}}{73.89\ \text{g}}\right)\left(\dfrac{1000.\ \text{mL}}{0.25\ \text{mol}}\right) = 3.2 \times 10^2\ \text{mL solution}$

(d) Assume 1000. mL solution

$\left(\dfrac{1.22\ \text{g}}{\text{mL}}\right)(1000.\ \text{mL})\ \ 1220\ \text{g solution}$

$\left(\dfrac{0.25\ \text{mol}}{\text{L}}\right)\left(\dfrac{73.89\ \text{g}\ Li_2CO_3}{\text{mol}}\right) = 18\ \text{g}\ Li_2CO_3\ \text{per L solution}$

$\% = \left(\dfrac{\text{g solute}}{\text{g solvent}}\right)(100) = \left(\dfrac{18\ \text{g}}{1220\ \text{g}}\right)(100) = 1.5\%$

115. Calculate the total moles of HCl and divide by the total volume.

$\left(\dfrac{0.35\ \text{mol}}{\text{L}}\right)(0.4000\ \text{L}) = 0.14\ \text{mol HCl}$

$\left(\dfrac{0.65\ \text{mol}}{\text{L}}\right)(1.1\ \text{L}) = 0.72\ \text{mol HCl}$

0.86 mol HCl (total mol HCl)

Total volume = 1.1 L + 0.40 L = 1.5 L

$\text{Molarity} = \dfrac{0.86\ \text{mol}}{1.5\ \text{L}} = 0.57\ \text{M HCl}$

CHAPTER 15

# ACIDS, BASES, AND SALTS

1. The Arrhenius definition is restricted to aqueous solutions, while the Bronsted-Lowry definition is not.

2. An electrolyte must be present in the solution for the bulb to glow.

3. Electrolytes include acids, bases, and salts.

4. First, the orientation of the polar water molecules about the $Na^+$ and $Cl^-$ is different. The positive end (hydrogen) of the water molecule is directed towards $Cl^-$, while the negative end (oxygen) of the water molecule is directed towards the $Na^+$. Second, more water molecules will fit around $Cl^-$, since it is larger than the $Na^+$ ion.

5. The pH for a solution with a hydrogen ion concentration of 0.003 M will be between 2 and 3.

6. Tomato juice is more acidic than blood, since its pH is lower.

7. By the Arrhenius theory, an acid is a substance that produces hydrogen ions in aqueous solution. A base is a substance that produces hydroxide ions in aqueous solution. By the Bronsted-Lowry theory, an acid is a proton donor, while a base accepts protons. Since a proton is a hydrogen ion, then the two theories are very similar for acids, but not bases. A chloride ion can accept a proton (producing HCl), so it is a Bronsted-Lowry base, but would not be a base by the Arrhenius theory, since it does not produce hydroxide ions.

   By the Lewis theory, an acid is an electron pair acceptor, and a base is an electron pair donor. Many individual substances would be similarly classified as bases by Bronsted-Lowry or Lewis theories, since a substance with an electron pair to donate, can accept a proton. But, the Lewis definition is almost exclusively applied to reactions where the acid and base combine into a single molecule. The Bronsted-Lowry definition is usually applied to reactions that involve a transfer of a proton from the acid to the base. The Arrhenius definition is most often applied to individual substances, not to reactions. According to the Arrhenius theory, neutralization involves the reaction between a hydrogen ion and a hydroxide ion to form water.

   Neutralization, according to the Bronsted-Lowry theory, involves the transfer of a proton to a negative ion. The formation of a coordinate-covalent bond constitutes a Lewis neutralization.

8. Neutralization reactions:

Arrhenius: $HCl + NaOH \rightarrow NaCl + H_2O$ $(H^+ + OH^- \rightarrow H_2O)$
Bronsted-Lowry: $HCl + KCN \rightarrow HCN + KCl$ $(H^+ + CN^- \rightarrow HCN)$
Lewis: $AlCl_3 + NaCl \rightarrow AlCl_4^- + Na^+$

:Cl: Al :Cl: :Cl: + [:Cl:]$^-$ ⟶ [:Cl:Al:Cl: with :Cl: above and below]$^-$

9. (a) [:Br:]$^-$ (b) [:O:H]$^-$ (c) [:C:::N:]$^-$

These ions are considered to be bases according to the Bronsted-Lowry theory, because they can accept a proton at any of their unshared pairs of electrons. They are considered to be bases according to the Lewis acid-base theory, because they can donate an electron pair.

10. The classes of compounds containing electrolytes are acids, bases, and salts.

11. Names of the compounds in Table 15.3

| | | | |
|---|---|---|---|
| $H_2SO_4$ | sulfuric acid | $HC_2H_3O_2$ | acetic acid |
| $HNO_3$ | nitric acid | $H_2CO_3$ | carbonic acid |
| $HCl$ | hydrochloric acid | $HNO_2$ | nitrous acid |
| $HBr$ | hydrobromic acid | $H_2SO_3$ | sulfurous acid |
| $HClO_4$ | perchloric acid | $H_2S$ | hydrosulfuric acid |
| $NaOH$ | sodium hydroxide | $H_2C_2O_4$ | oxalic acid |
| $KOH$ | potassium hydroxide | $H_3BO_3$ | boric acid |
| $Ca(OH)_2$ | calcium hydroxide | $HClO$ | hypochlorous acid |
| $Ba(OH)_2$ | barium hydroxide | $NH_3$ | ammonia |
| | | $HF$ | hydrofluoric acid |

12. Hydrogen chloride dissolved in water conducts an electric current. HCl reacts with polar water molecules to produce $H_3O^+$ and $Cl^-$ ions, which conduct electric current. Hexane is a nonpolar solvent, so it cannot pull the HCl molecules apart. Since there are no ions in the hexane solution, it does not conduct an electric current. HCl does not ionize in hexane.

13. In their crystalline structure, salts exist as positive and negative ions in definite geometric arrangement to each other, held together by the attraction of the opposite charges. When dissolved in water, the salt dissociates as the ions are pulled away from each other by the polar water molecules.

14. Testing the electrical conductivity of the solutions shows that $CH_3OH$ is a nonelectrolyte, while NaOH is an electrolyte. This indicates that the OH group in $CH_3OH$ must be covalently bonded to the $CH_3$ group.

15. Molten NaCl conducts electricity because the ions are free to move. In the solid state, however, the ions are immobile and do not conduct electricity.

16. Dissociation is the separation of already existing ions in an ionic compound. Ionization is the formation of ions from molecules. The dissolving of NaCl is a dissociation, since the ions already exist in the crystalline compound. The dissolving of HCl in water is an ionization process, because ions are formed from HCl molecules and $H_2O$.

17. Strong electrolytes are those which are essentially 100% ionized or dissociated in water. Weak electrolytes are those which are only slightly ionized in water.

18. Ions are hydrated in solution because there is an electrical attraction between the charged ions and the polar water molecules.

19. The main distinction between water solutions of strong and weak electrolytes is the degree of ionization of the electrolyte. A solution of an electrolyte contains many more ions than a solution of a nonelectrolyte. Strong electrolytes are essentially 100% ionized. Weak electrolytes are only slightly ionized in water.

20. (a) In a neutral solution, the concentration of $H^+$ and $OH^-$ are equal.

    (b) In an acid solution, the concentration of $H^+$ is greater than the concentration of $OH^-$.

    (c) In a basic solution, the concentration of $OH^-$ is greater than the concentration of $H^+$.

21. The net ionic equation for an acid-base reaction in aqueous solutions is:
    $H^+ + OH^- \rightarrow H_2O$.

22. The HCl molecule is polar and, consequently, is much more soluble in the polar solvent, water, than in the nonpolar solvent, hexane. There is also a chemical reaction between HCl and $H_2O$ molecules. $HCl + H_2O \rightarrow H_3O^+ + Cl^-$

23. Pure water is neutral because when it ionizes it produces equal molar concentrations of acid $[H^+]$ and base $[OH^-]$ ions.

24. The fundamental difference between a colloidal dispersion and a true solution lies in the size of the particles. In a true solution particles are usually ions or hydrated molecules and are less than 1 nm in size. In colloidals the particles are aggregates of ions or molecules, ranging in size from 1-1000 nm.

25. The Tyndall effect is observed when a narrow beam of light is passed through a colloidal suspension. The light is reflected from the colloidal particles effectively illuminating the path of the light through the liquid. In a true solution the light path cannot be seen because the dissolved particles are too small to reflect light.

26. Dialysis is the process of removing dissolved solutes from a colloidal dispersion by use of a dialyzing membrane. The dissolved solutes pass through the membrane leaving the colloidal dispersion behind. Dialysis is used in artificial kidneys to remove soluble waste products from the blood.

27.

(a) $H_2SO_4 - HSO_4^-$; $H_2C_2H_3O_2^+ - HC_2H_3O_2$

(b) step 1: $H_2SO_4 - HSO_4^-$; $H_3O^+ - H_2O$

step 2: $HSO_4^- - SO_4^{2-}$; $H_3O^+ - H_2O$

(c) $HClO_4 - ClO_4^-$; $H_3O^+ - H_2O$

(d) $H_3O^+ - H_2O$; $CH_3OH - CH_3O^-$

28. Conjugate acid-base pairs:

(a) $HCl - Cl^-$; $NH_4^+ - NH_3$

(b) $HCO_3^- - CO_3^{2-}$; $H_2O - OH^-$

(c) $H_3O^+ - H_2O$; $H_2CO_3 - HCO_3^-$

(d) $HC_2H_3O_2 - C_2H_3O_2^-$; $H_3O^+ - H_2O$

29. Balancing equations

(a) $Mg(s) + 2\ HCl(aq) \rightarrow MgCl_2(aq) + H_2(g)$

(b) $BaO(s) + 2\ HBr(aq) \rightarrow BaBr_2(aq) + H_2O(l)$

(c) $2\ Al(s) + 3\ H_2SO_4(aq) \rightarrow Al_2(SO_4)_3(aq) + 3\ H_2(g)$

(d) $Na_2CO_3(aq) + 2\ HCl(aq) \rightarrow 2\ NaCl(aq) + H_2O(l) + CO_2(g)$

(e) $Fe_2O_3(s) + 6\ HBr(aq) \rightarrow 2\ FeBr_3(aq) + 3\ H_2O(l)$

(f) $Ca(OH)_2(aq) + H_2CO_3(aq) \rightarrow CaCO_3(s) + 2\ H_2O(l)$

30. (a) $NaOH(aq) + HBr(aq) \rightarrow NaBr(aq) + H_2O(l)$

(b) $KOH(aq) + HCl(aq) \rightarrow KCl(aq) + H_2O(l)$

(c) $Ca(OH)_2(aq) + 2\,HI(aq) \rightarrow CaI_2(aq) + 2\,H_2O(l)$

(d) $Al(OH)_3(s) + 3\,HBr(aq) \rightarrow AlBr_3(aq) + 3\,H_2O(l)$

(e) $Na_2O(s) + 2\,HClO_4(aq) \rightarrow 2\,NaClO_4(aq) + H_2O(l)$

(f) $3\,LiOH(aq) + FeCl_3(aq) \rightarrow Fe(OH)_3(s) + 3\,LiCl(aq)$

31. The following compounds are electrolytes:
(a) HCl acid in water (b) $CO_2$ acid in water (c) $CaCl_2$ salt

32. The following compounds are electrolytes:
(a) $NaHCO_3$ salt (e) RbOH base
(b) $AgNO_3$ salt (f) $K_2CrO_4$ salt
(c) HCOOH acid

33. Calculation of molarity of ions.

(a) $$(0.015\text{ M NaCl})\left(\frac{1\text{ mol Na}^+}{1\text{ mol NaCl}}\right) = 0.015\text{ M Na}^+$$

$$(0.015\text{ M NaCl})\left(\frac{1\text{ mol Cl}^-}{1\text{ mol NaCl}}\right) = 0.015\text{ M Cl}^-$$

(b) $$(4.25\text{ M NaKSO}_4)\left(\frac{1\text{ mol Na}^+}{1\text{ mol NaKSO}_4}\right) = 4.25\text{ M Na}^+$$

$$(4.25\text{ M NaKSO}_4)\left(\frac{1\text{ mol K}^+}{1\text{ mol NaKSO}_4}\right) = 4.25\text{ M K}^+$$

$$(4.25\text{ M NaKSO}_4)\left(\frac{1\text{ mol SO}_4^{2-}}{1\text{ mol NaKSO}_4}\right) = 4.25\text{ M SO}_4^{2-}$$

(c) $$(0.20\text{ M CaCl}_2)\left(\frac{1\text{ mol Ca}^{2+}}{1\text{ mol CaCl}_2}\right) = 0.20\text{ M Ca}^{2+}$$

$$(0.20\text{ M CaCl}_2)\left(\frac{2\text{ mol Cl}^-}{1\text{ mol CaCl}_2}\right) = 0.40\text{ M Cl}^-$$

(d) $$\left(\frac{22.0\text{ g KI}}{500.\text{ mL}}\right)\left(\frac{1\text{ mol}}{166.0\text{ g}}\right)\left(\frac{1000\text{ mL}}{\text{L}}\right) = 0.265\text{ M KI}$$

$$(0.265\text{ M KI})\left(\frac{1\text{ mol K}^+}{1\text{ mol KI}}\right) = 0.265\text{ M K}^+$$

$$(0.265\text{ M KI})\left(\frac{1\text{ mol I}^-}{1\text{ mol KI}}\right) = 0.265\text{ M I}^-$$

34. (a) $$(0.75\text{ M ZnBr}_2)\left(\frac{1\text{ mol Zn}^{2+}}{1\text{ mol ZnBr}_2}\right) = 0.75\text{ M Zn}^{2+}$$

$$(0.75\text{ M ZnBr}_2)\left(\frac{2\text{ mol Br}^-}{1\text{ mol ZnBr}_2}\right) = 1.5\text{ M Br}^-$$

(b) $$(1.65\text{ M Al}_2(\text{SO}_4)_3)\left(\frac{3\text{ mol SO}_4^{2-}}{1\text{ mol Al}_2(\text{SO}_4)_3}\right) = 4.95\text{ M SO}_4^{2-}$$

$$(1.65\text{ M Al}_2(\text{SO}_4)_3)\left(\frac{2\text{ mol Al}^{3+}}{1\text{ mol Al}_2(\text{SO}_4)_3}\right) = 3.30\text{ M Al}^{3+}$$

(c) $$\left(\frac{900.\text{ g (NH}_4)_2\text{SO}_4}{20.0\text{ L}}\right)\left(\frac{1\text{ mol}}{132.2\text{ g}}\right) = 0.340\text{ M (NH}_4)_2\text{SO}_4$$

$$(0.340\text{ M (NH}_4)_2\text{SO}_4)\left(\frac{2\text{ mol NH}_4^+}{1\text{ mol (NH}_4)_2\text{SO}_4}\right) = 0.680\text{ M NH}_4^+$$

$$(0.340\text{ M (NH}_4)_2\text{SO}_4)\left(\frac{1\text{ mol SO}_4^{2-}}{1\text{ mol (NH}_4)_2\text{SO}_4}\right) = 0.340\text{ M SO}_4^{2-}$$

(d) $$\left(\frac{0.0120\text{ g Mg(ClO}_3)_2}{0.00100\text{ L}}\right)\left(\frac{1\text{ mol}}{191.2\text{ g}}\right) = 0.0628\text{ M Mg(ClO}_3)_2$$

$$(0.0628\text{ M Mg(ClO}_3)_2)\left(\frac{1\text{ mol Mg}^{2+}}{1\text{ mol Mg(ClO}_3)_2}\right) = 0.0628\text{ M Mg}^{2+}$$

$$(0.0628\text{ M Mg(ClO}_3)_2)\left(\frac{2\text{ mol ClO}_3^-}{1\text{ mol Mg(ClO}_3)_2}\right) = 0.126\text{ M ClO}_3^-$$

35. The molarity of each ion, as calculated in Exercise 33 will be used to calculate the mass of each ion present in 100. mL of solution.

(a) $$(0.100\ \text{L})\left(\frac{0.015\ \text{mol Na}^+}{\text{L}}\right)\left(\frac{22.99\ \text{g}}{\text{mol}}\right) = 0.034\ \text{g Na}^+$$

$$(0.100\ \text{L})\left(\frac{0.015\ \text{mol Cl}^-}{\text{L}}\right)\left(\frac{35.45\ \text{g}}{\text{mol}}\right) = 0.053\ \text{g Cl}^-$$

(b) $$(0.100\ \text{L})\left(\frac{4.25\ \text{mol Na}^+}{\text{L}}\right)\left(\frac{22.99\ \text{g}}{\text{mol}}\right) = 9.77\ \text{g Na}^+$$

$$(0.100\ \text{L})\left(\frac{4.25\ \text{mol K}^+}{\text{L}}\right)\left(\frac{39.10\ \text{g}}{\text{mol}}\right) = 16.6\ \text{g K}^+$$

$$(0.100\ \text{L})\left(\frac{4.25\ \text{mol SO}_4^{2-}}{\text{L}}\right)\left(\frac{96.07\ \text{g}}{\text{mol}}\right) = 40.8\ \text{g SO}_4^{2-}$$

(c) $$(0.100\ \text{L})\left(\frac{0.20\ \text{mol Ca}^{2+}}{\text{L}}\right)\left(\frac{40.08\ \text{g}}{\text{mol}}\right) = 0.80\ \text{g Ca}^{2+}$$

$$(0.100\ \text{L})\left(\frac{0.40\ \text{mol Cl}^-}{\text{L}}\right)\left(\frac{35.45\ \text{g}}{\text{mol}}\right) = 1.4\ \text{g Cl}^-$$

(d) $$(0.100\ \text{L})\left(\frac{0.265\ \text{mol K}^+}{\text{L}}\right)\left(\frac{39.10\ \text{g}}{\text{mol}}\right) = 1.04\ \text{g K}^+$$

$$(0.100\ \text{L})\left(\frac{0.265\ \text{mol I}^-}{\text{L}}\right)\left(\frac{126.9\ \text{g}}{\text{mol}}\right) = 3.36\ \text{g I}^-$$

36. The molarity of each ion, as calculated in Exercise 34, will be used to calculate the mass of each ion present in 100 mL of solution.

(a) $$(0.100\ \text{L})\left(\frac{0.75\ \text{mol Zn}^{2+}}{\text{L}}\right)\left(\frac{65.39\ \text{g}}{\text{mol}}\right) = 4.9\ \text{g Zn}^{2+}$$

$$(0.100\ \text{L})\left(\frac{1.5\ \text{mol Br}^-}{\text{L}}\right)\left(\frac{79.90\ \text{g}}{\text{mol}}\right) = 12\ \text{g Br}^-$$

(b) $$(0.100\ \text{L})\left(\frac{3.30\ \text{mol Al}^{3+}}{\text{L}}\right)\left(\frac{26.98\ \text{g}}{\text{mol}}\right) = 8.90\ \text{g Al}^{3+}$$

$$(0.100\ \text{L})\left(\frac{4.95\ \text{mol SO}_4^{2-}}{\text{L}}\right)\left(\frac{96.07\ \text{g}}{\text{mol}}\right) = 47.6\ \text{g SO}_4^{2-}$$

(c) $$(0.100\ \text{L})\left(\frac{0.680\ \text{mol NH}_4^{+}}{\text{L}}\right)\left(\frac{18.04\ \text{g}}{\text{mol}}\right) = 1.23\ \text{g NH}_4^{+}$$

$$(0.100\ \text{L})\left(\frac{0.340\ \text{mol SO}_4^{2-}}{\text{L}}\right)\left(\frac{96.07\ \text{g}}{\text{mol}}\right) = 3.27\ \text{g SO}_4^{2-}$$

(d) $$(0.100\ \text{L})\left(\frac{0.0628\ \text{mol Mg}^{2+}}{\text{L}}\right)\left(\frac{24.31\ \text{g}}{\text{mol}}\right) = 0.153\ \text{g Mg}^{2+}$$

$$(0.100\ \text{L})\left(\frac{0.126\ \text{mol ClO}_3^{-}}{\text{L}}\right)\left(\frac{83.45\ \text{g}}{\text{mol}}\right) = 1.05\ \text{g ClO}_3^{-}$$

37. (a) $$(30.0\ \text{mL})\left(\frac{1.0\ \text{mol NaCl}}{1000\ \text{mL}}\right) = 0.030\ \text{mol NaCl}$$

$$(40.0\ \text{mL})\left(\frac{1.0\ \text{mol NaCl}}{1000\ \text{mL}}\right) = 0.040\ \text{mol NaCl}$$

Total mol NaCl = 0.030 mol + 0.040 mol = 0.070 mol NaCl

$$\frac{0.070\ \text{mol NaCl}}{0.070\ \text{L}} = 1.0\ \text{M NaCl}$$

$$(1.0\ \text{M NaCl})\left(\frac{1.0\ \text{mol Na}^{+}}{1.0\ \text{mol NaCl}}\right) = 1.0\ \text{M Na}^{+}$$

$$(1.0\ \text{M NaCl})\left(\frac{1.0\ \text{mol Cl}^{-}}{1.0\ \text{mol NaCl}}\right) = 1.0\ \text{M Cl}^{-}$$

(b) $HCl + NaOH \rightarrow NaCl + H_2O$

$$(30.0\ \text{mL HCl})\left(\frac{1\ \text{L}}{1000\ \text{mL}}\right)\left(\frac{1.0\ \text{mol}}{\text{L}}\right) = 0.030\ \text{mol HCl}$$

$$(30.0\ \text{mL NaOH})\left(\frac{1\ \text{L}}{1000\ \text{mL}}\right)\left(\frac{1.0\ \text{mol}}{\text{L}}\right) = 0.030\ \text{mol NaOH}$$

0.030 mol HCl reacts with 0.030 mol NaOH and produces 0.030 mol NaCl. The final volume is 0.060 L. 0.030 mol NaCl/0.060 L = 0.50 M NaCl. Since there is one mole each of sodium and chloride ions per mole of NaCl, the molar concentration of $Na^+$ and $Cl^-$ will be 0.50 M $Na^+$ and 0.50 M $Cl^-$.

(c) $KOH + HCl \rightarrow KCl + H_2O$

$$(100.0 \text{ mL})\left(\frac{1 \text{ L}}{1000 \text{ mL}}\right)\left(\frac{0.40 \text{ mol KOH}}{\text{L}}\right) = 0.040 \text{ mol KOH}$$

$$(100.0 \text{ mL})\left(\frac{1 \text{ L}}{1000 \text{ mL}}\right)\left(\frac{0.80 \text{ mol HCl}}{\text{L}}\right) = 0.080 \text{ mol HCl}$$

0.040 mol KOH reacts with 0.040 mol HCl. 0.040 mol HCl remains and 0.040 mol KCl is produced. The final volume is 200.0 mL and contains 0.040 mol HCl and 0.040 mol KCl. Moles of ions are: 0.040 mol $H^+$, 0.040 mol $K^+$, and 0.080 mol $Cl^-$. Concentrations of ions are:

$$\frac{0.040 \text{ mol H}^+}{0.200 \text{ L}} = 0.20 \text{ M H}^+ \qquad \text{molarity K}^+ = \text{molarity H}^+$$

$$\frac{0.080 \text{ mol Cl}^-}{0.200 \text{ L}} = 0.40 \text{ M Cl}^-$$

38. (a) 100.0 mL of 2.0 M KCl and 100.0 mL of 1.0 M $CaCl_2$ are mixed, giving a final volume of 200.0 mL and concentrations of 1.0 M KCl and 0.5 M $CaCl_2$. The concentration of $K^+$ will be 1.0 M and the concentration of $Ca^{2+}$ will be 0.5 M. The chloride ion concentration will be 2.0 M (1.0 M from the KCl and 2(0.5 M) from the $CaCl_2$).

(b) $$(35.0 \text{ mL})\left(\frac{1 \text{ L}}{1000 \text{ mL}}\right)\left(\frac{0.20 \text{ mol Ba(OH)}_2}{\text{L}}\right) = 0.0070 \text{ mol Ba(OH)}_2$$

$$(35.0 \text{ mL})\left(\frac{1 \text{ L}}{1000 \text{ mL}}\right)\left(\frac{0.20 \text{ mol H}_2\text{SO}_4}{\text{L}}\right) = 0.0070 \text{ mol H}_2\text{SO}_4$$

Final volume = 35.0 mL + 35.0 mL = 70.0 mL

| $H_2SO_4$ | + | $Ba(OH)_2$ | $\rightarrow$ | $BaSO_{4}(s)$ | + | $2\ H_2O$ |
|---|---|---|---|---|---|---|
| 0.0070 mol | | 0.0070 mol | | 0.0070 mol | | 0.014 mol |

The $H_2SO_4$ and the $Ba(OH)_2$ react completely producing insoluble $BaSO_4$ and $H_2O$. No ions are present in solution.

(c) $(0.500\ \text{L NaCl})\left(\dfrac{2.0\ \text{mol}}{\text{L}}\right) = 1.0\ \text{mol NaCl}$

$(1.00\ \text{L AgNO}_3)\left(\dfrac{1.00\ \text{mol}}{\text{L}}\right) = 1.0\ \text{mol AgNO}_3$

| $NaCl(aq)$ | + | $AgNO_3(aq)$ | → | $AgCl(s)$ | + | $NaNO_3(aq)$ |
|---|---|---|---|---|---|---|
| 1.0 mol | | 1.0 mol | | 1.0 mol | | 1.0 mol |

The AgCl is insoluble and produces no ions. The 1.0 mol $NaNO_3$ will produce 1.0 mol $Na^+$ ions and 1.0 mol $NO_3^-$ ions. The final volume of the solution is 1.5 L. The concentration of the ions are:

$\dfrac{1.0\ \text{mol Na}^+}{1.5\ \text{L}} = 0.67\ \text{M Na}^+$ $\qquad$ $\dfrac{1.0\ \text{mol NO}_3^-}{1.5\ \text{L}} = 0.67\ \text{M NO}_3^-$

39. The reaction of HCl and NaOH occurs on a 1:1 mole ratio.

$$HCl + NaOH \rightarrow NaCl + H_2O$$

At the endpoint in these titration reactions, equal moles of HCl and NaOH will have reacted. Moles = (molarity)(volume). At the endpoint, mol HCl = mol NaOH. Therefore, at the endpoint,
$M_AV_A = M_BV_B$
(a) $(37.70\ \text{mL})(0.728\ \text{M}) = (40.13\ \text{mL})(\text{M HCl})$

$$\text{M HCl} = \frac{(37.70\ \text{mL})(0.728\ \text{M})}{40.13\ \text{mL}} = 0.684\ \text{M HCl}$$

(b) $\dfrac{(33.66\ \text{mL})(0.306\ \text{M})}{19.00\ \text{mL}} = 0.542\ \text{M HCl}$

(c) $\dfrac{(18.00\ \text{mL})(0.555\ \text{M})}{27.25\ \text{mL}} = 0.367\ \text{M HCl}$

40. The reaction of HCl and NaOH occurs on a 1:1 mole ratio.

$$HCl + NaOH \rightarrow NaCl + H_2O$$

At the endpoint in these titration reactions, equal moles of HCl and NaOH will have reacted. Moles = (molarity)(volume). At the endpoint, mol HCl = mol NaOH. Therefore, at the endpoint,
$M_AV_A = M_BV_B$

(a) $\frac{(37.19 \text{ mL})(0.126 \text{ M})}{31.91 \text{ mL}} = 0.147 \text{ M NaOH}$

(b) $\frac{(48.04 \text{ mL})(0.482 \text{ M})}{24.02 \text{ mL}} = 0.964 \text{ M NaOH}$

(c) $\frac{(13.13 \text{ mL})(1.425 \text{ M})}{39.39 \text{ mL}} = 0.4750 \text{ M NaOH}$

41. (a) $SO_4^{2-}(aq) + Ba^{2+}(aq) \rightarrow BaSO_4(s)$

(b) $CaCO_3(s) + 2\,H^+(aq) \rightarrow Ca^{2+}(aq) + CO_2(g) + H_2O(l)$

(c) $Mg(s) + 2\,HC_2H_3O_2\,(aq) \rightarrow Mg^{2+}(aq) + H_2(g) + 2\,C_2H_3O_2^-(aq)$

42. (a) $H_2S(g) + Cd^{2+}(aq) \rightarrow CdS(s) + 2\,H^+(aq)$

(b) $Zn(s) + 2\,H^+(aq) \rightarrow Zn^{2+}(aq) + H_2(g)$

(c) $Al^{3+}(aq) + PO_4^{3-}(aq) \rightarrow AlPO_4(s)$

43. The more acidic solution is listed followed by an explanation.

(a) 1 molar $H_2SO_4$. The concentration of $H^+$ in 1 M $H_2SO_4$ is greater than 1 M, since there are two ionizable hydrogens per mole of $H_2SO_4$. In HCl the concentration of $H^+$ will be 1 M, since there is only one ionizable hydrogen per mole HCl.

(b) 1 molar HCl. HCl is a strong electrolyte, producing more $H^+$ than $HC_2H_3O_2$ which is a weak electrolyte.

44. The more acidic solution is listed followed by an explanation.

(a) 2 molar HCl. 2 M HCl will yield 2 M $H^+$ concentration. 1 M HCl will yield 1 M $H^+$ concentration.

(b) 1 molar $H_2SO_4$. Both are strong acids. The concentration of $H^+$ in 1 M $H_2SO_4$ is greater than in 1 M $HNO_3$ because $H_2SO_4$ has two ionizable hydrogens per mole whereas $HNO_3$ has only one ionizable hydrogen per mole.

45. $3\,HCl + Al(OH)_3 \rightarrow AlCl_3 + 3\,H_2O$

g $Al(OH)_3$ → mol $Al(OH)_3$ → mol HCl → mL HCl

0.245 M HCl contains 0.245 mol HCl/1000 mL

$$(10.0 \text{ g Al(OH)}_3)\left(\frac{1 \text{ mol}}{78.00 \text{ g}}\right)\left(\frac{3 \text{ mol HCl}}{1 \text{ mol Al(OH)}_3}\right)\left(\frac{1000 \text{ mL}}{0.245 \text{ mol}}\right)$$

$= 1.57 \times 10^3$ mL of 0.245 M HCl

46. $2\,HCl + Ca(OH)_2 \rightarrow CaCl_2 + 2\,H_2O$

M $Ca(OH)_2$ → mol $Ca(OH)_2$ → mol HCl → mL HCl

0.245 M HCl contains 0.245 mol HCl/1000 mL

$$(0.0500 \text{ L Ca(OH)}_2)\left(\frac{0.100 \text{ mol}}{\text{L}}\right)\left(\frac{2 \text{ mol HCl}}{1 \text{ mol Ca(OH)}_2}\right)\left(\frac{1000 \text{ mL}}{0.245 \text{ mol}}\right)$$

= 40.8 mL of 0.245 M HCl

47. $NaOH + HCl \rightarrow NaCl + H_2O$

First calculate the grams of NaOH in the sample.

L HCl → mol HCl → mol NaOH → g NaOH

$$(0.01825 \text{ L HCl})\left(\frac{0.2406 \text{ mol}}{\text{L}}\right)\left(\frac{1 \text{ mol NaOH}}{1 \text{ mol HCl}}\right)\left(\frac{40.00 \text{ g}}{\text{mol}}\right) = 0.1756 \text{ g NaOH in the sample}$$

$$\left(\frac{0.1756 \text{ g NaOH}}{0.200 \text{ g sample}}\right)(100) = 87.8\% \text{ NaOH}$$

48. $NaOH + HCl \rightarrow NaCl\ H_2O$

L HCl → mol HCl → mol NaOH → g NaOH

$$(0.04990 \text{ L HCl})\left(\frac{0.466 \text{ mol}}{\text{L}}\right)\left(\frac{1 \text{ mol NaOH}}{1 \text{ mol HCl}}\right)\left(\frac{40.00 \text{ g}}{\text{mol}}\right) = 0.930 \text{ g NaOH in the sample}$$

1.00 g sample − 0.930 g NaOH = 0.070 g NaCl in sample

$$\left(\frac{0.070 \text{ g NaCl}}{1.00 \text{ g sample}}\right)(100) = 7.0\% \text{ NaCl in the sample}$$

49. $Zn + 2\,HCl \rightarrow ZnCl_2 + H_2$

This is a limiting reactant problem. First find the moles of Zn and HCl from the given data and then identify the limiting reactant.

g Zn → mol Zn

$$(5.00 \text{ g Zn})\left(\frac{1 \text{ mol}}{65.39 \text{ g}}\right) = 0.0765 \text{ mol Zn}$$

$$(0.100 \text{ L HCl})\left(\frac{0.350 \text{ mol}}{\text{L}}\right) = 0.0350 \text{ mol HCl}$$

Therefore Zn is in excess and HCl is the limiting reactant.

$$(0.0350 \text{ mol HCl})\left(\frac{1 \text{ mol } H_2}{2 \text{ mol HCl}}\right) = 0.0175 \text{ mol } H_2 \text{ produced in the reaction}$$

$$T = 27°C = 300.\text{ K}$$

$$P = (700.\text{ torr})\left(\frac{1 \text{ atm}}{760 \text{ torr}}\right) = 0.921 \text{ atm}$$

$$PV = nRT$$

$$V = \frac{nRT}{P} = \frac{(0.0175 \text{ mol})(0.0821 \text{ L atm/mol K})(300.\text{ K})}{0.921 \text{ atm}} = 0.468 \text{ L } H_2$$

50. $Zn + 2\,HCl \rightarrow ZnCl_2 + H_2$

This is a limiting reactant problem. First find moles of Zn and HCl from the given data and then identify the limiting reactant.

g Zn $\rightarrow$ mol Zn

$$(5.00 \text{ g Zn})\left(\frac{1 \text{ mol}}{65.39 \text{ g}}\right) = 0.0765 \text{ mol Zn}$$

$$(0.200 \text{ L HCl})\left(\frac{0.350 \text{ mol}}{\text{L}}\right) = 0.0700 \text{ mol HCl}$$

Zn is in excess and HCl is the limiting reactant.

$$(0.0700 \text{ mol HCl})\left(\frac{1 \text{ mol } H_2}{2 \text{ mol HCl}}\right) = 0.0350 \text{ mol } H_2$$

$$T = 27°C = 300.\text{ K}$$

$$P = (700.\text{ torr})\left(\frac{1 \text{ atm}}{760 \text{ torr}}\right) = 0.921 \text{ atm}$$

$$PV = nRT$$

$$V = \frac{nRT}{P} = \frac{(0.0350 \text{ mol})(0.0821 \text{ L atm/mol K})(300.\text{ K})}{0.921 \text{ atm}} = 0.936 \text{ L } H_2$$

51. Calculation of the pH solutions:

(a) $H^+ = 0.01\text{ M} = 1 \times 10^{-2}\text{ M}$; $pH = -\log(1 \times 10^{-2}) = 2.0$

(b) $H^+ = 1.0\text{ M}$; $pH = -\log 1.0 = 0$

(c) $H^+ = 6.5 \times 10^{-9}\text{ M}$; $pH = -\log(6.5 \times 10^{-9}) = 8.19$

52. (a) $H^+ = 1 \times 10^{-7}$ M; $pH = -\log(1 \times 10^{-7}) = 7.0$

(b) $H^+ = 0.50$ M; $pH = -\log(5.0 \times 10^{-1}) = 0.30$

(c) $H^+ = 0.00010$ M $= 1.0 \times 10^{-4}$ M; $pH = -\log(1.0 \times 10^{-4}) = 4.00$

53. (a) Orange juice $= 3.7 \times 10^{-4}$ M $H^+$
$pH = -\log(3.7 \times 10^{-4}) = 3.43$

(b) Vinegar $= 2.8 \times 10^{-3}$ M $H^+$
$pH = -\log(2.8 \times 10^{-3}) = 2.55$

54. (a) Black coffee $= 5.0 \times 10^{-5}$ M $H^+$
$pH = -\log(5.0 \times 10^{-5}) = 4.30$

(b) Limewater $= 3.4 \times 10^{-11}$ M $H^+$
$pH = -\log(3.4 \times 10^{-11}) = 10.47$

55. $CaI_2 \rightarrow Ca^{2+} + 2\,I^-$

$$\left(\frac{0.520 \text{ mol } I^-}{L}\right)\left(\frac{1 \text{ mol } Ca^{2+}}{2 \text{ mol } I}\right) = \left(\frac{0.260 \text{ mol } Ca^{2+}}{L}\right) = 0.260 \text{ M } Ca^{2+}$$

56. Dilution problem

$V_1M_1 = V_2M_2$

$(100.\text{ mL})(12\text{ M}) = (V_2)(0.40\text{ M})$

$$V_2 = \frac{(100.\text{ mL})(12\text{ M})}{0.40\text{ M}} = 3.0 \times 10^3 \text{ mL}$$

57. $Ba(OH)_2 + 2\,HCl \rightarrow BaCl_2 + 2\,H_2O$

M HCl $\rightarrow$ mol HCl $\rightarrow$ mol $Ba(OH)_2$ $\rightarrow$ M $Ba(OH)_2$

$$\left(\frac{0.430 \text{ mol HCl}}{L}\right)\left(\frac{1 \text{ L}}{1000 \text{ mL}}\right)(29.26 \text{ mL}) = 0.0126 \text{ mol HCl}$$

$$(0.0126 \text{ mol HCl})\left(\frac{1 \text{ mol } Ba(OH)_2}{2 \text{ mol HCl}}\right) = 0.00630 \text{ mol } Ba(OH)_2$$

$$\frac{0.00630 \text{ mol } Ba(OH)_2}{0.02040 \text{ L}} = 0.309 \text{ M } Ba(OH)_2$$

58. The acetic acid solution freezes at a lower temperature than the alcohol solution. The acetic acid ionizes slightly while the alcohol does not. The ionization of the acetic acid increases its particle concentration in solution above that of the alcohol solution, resulting in a lower freezing point for the acetic acid solution.

59. It is more economical to purchase $CH_3OH$ at the same cost per pound as $C_2H_5OH$. Because $CH_3OH$ has a lower molar mass than $C_2H_5OH$, the $CH_3OH$ solution will contain more particles per pound in a given solution and therefore, have a greater effect on the freezing point of the radiator solution.

Assume 100. g of each compound.

$$CH_3OH: \quad \frac{100.\text{ g}}{34.04\text{ g/mol}} = 2.84\text{ mol}$$

$$CH_3CH_2OH: \quad \frac{100.\text{ g}}{46.07\text{ g/mol}} = 2.17\text{ mol}$$

60. A hydronium ion is a hydrated hydrogen ion.

$$\underset{\text{(hydrogen ion)}}{H^+} + H_2O \rightarrow \underset{\text{(hydronium ion)}}{H_3O^+}$$

61. Freezing point depression is directly related to the concentration of particles in the solution.

$C_{12}H_{22}O_{11}$ > $HC_2H_3O_2$ > $HCl$ > $CaCl_2$

1 mol > 1+ mol > 2 mol > 3 mol (particles in solution)

62. (a) 100°C $\quad pH = -\log(1 \times 10^{-6}) = 6.0 \quad$ pH of $H_2O$ is greater at 25°C

25°C $\quad pH = -\log(1 \times 10^{-7}) = 7.0$

(b) $1 \times 10^{-6} > 1 \times 10^{-7}$ so, $H^+$ concentration is higher at 100°C.

(c) The water is neutral at both temperatures, because the $H_2O$ ionizes into equal concentrations of $H^+$ and $OH^-$ at any temperature.

63. As the pH changes by 1 unit, the concentration of $H^+$ in solution changes by a factor of 10. For example, the pH of 0.10 M HCl is 1.00, while the pH of 0.0100 M HCl is 2.00.

64. $Na_2CO_3 + 2\,HCl \rightarrow 2\,NaCl + CO_2 + H_2O$

g $Na_2CO_3$ → mol $Na_2CO_3$ → mol HCl → M HCl

$$(0.452\text{ g }Na_2CO_3)\left(\frac{1\text{ mol}}{106.0\text{ g}}\right)\left(\frac{2\text{ mol HCl}}{1\text{ mol }Na_2CO_3}\right) = 0.00853\text{ mol HCl}$$

$$\frac{0.00853 \text{ mol}}{0.0424 \text{ L}} = 0.201 \text{ M HCl}$$

65. $2\ HCl + Ca(OH)_2 \rightarrow CaCl_2 + 2\ H_2O$

$g\ Ca(OH)_2 \rightarrow mol\ Ca(OH)_2 \rightarrow mol\ HCl \rightarrow mL\ HCl$

$$(2.00 \text{ g Ca(OH)}_2)\left(\frac{1 \text{ mol}}{74.10 \text{ g}}\right)\left(\frac{2 \text{ mol HCl}}{1 \text{ mol Ca(OH)}_2}\right)\left(\frac{1000 \text{ mL}}{0.1234 \text{ mol}}\right) = 437 \text{ mL of } 0.1234 \text{ M HCl}$$

66. $KOH + HNO_3 \rightarrow KNO_3 + H_2O$

$L\ HNO_3 \rightarrow mol\ HNO_3 \rightarrow mol\ KOH \rightarrow g\ KOH$

$$(0.05000 \text{ L HNO}_3)\left(\frac{0.240 \text{ mol}}{\text{L}}\right)\left(\frac{1 \text{ mol KOH}}{1 \text{ mol HNO}_3}\right)\left(\frac{56.11 \text{ g}}{\text{mol}}\right) = 0.673 \text{ g KOH}$$

67. pH of 1.0 L solution containing 0.1 mL of 1.0 M HCl

$$(0.1 \text{ mL})\left(\frac{1.0 \text{ L}}{1000 \text{ mL}}\right)\left(\frac{1 \text{ mol HCl}}{\text{L}}\right) = 1 \times 10^{-4} \text{ mol HCl}$$

$$\frac{1 \times 10^{-4} \text{ mol HCl}}{1.0 \text{ L}} = 1 \times 10^{-4} \text{ M HCl}$$

$1 \times 10^{-4}$ M HCl produces $1 \times 10^{-4}$ M $H^+$

$pH = -\log(1 \times 10^{-4}) = 4.0$

68. Dilution problem

$$V_1M_1 = V_2M_2 \qquad V_1 = \frac{V_2M_2}{M_1} = \frac{(50.0 \text{ L})(5.00 \text{ M})}{18.0 \text{ M}} = 13.9 \text{ L of } 18.0 \text{ M H}_2\text{SO}_4$$

69. $NaOH + HCl \rightarrow NaCl + H_2O$

$$(3.0 \text{ g NaOH})\left(\frac{1 \text{ mol}}{40.00 \text{ g}}\right) = 0.075 \text{ mol NaOH}$$

$$(500 \text{ mL HCl})\left(\frac{1 \text{ L}}{1000 \text{ mL}}\right)\left(\frac{0.10 \text{ mol}}{\text{L}}\right) = 0.050 \text{ mol HCl}$$

This solution is basic. The NaOH will neutralize the HCl with an excess of 0.025 mol of NaOH remaining.

70. $Ba(OH)_2(aq) + 2\ HCl(aq) \rightarrow BaCl_2(aq) + 2\ H_2O(l)$

$$(0.380 \text{ L Ba(OH)}_2)\left(\frac{0.35 \text{ mol}}{\text{L}}\right) = 0.13 \text{ mol Ba(OH)}_2$$

$0.13 \text{ mol Ba(OH)}_2 \rightarrow 0.26 \text{ mol OH}^-$

$$(0.5000 \text{ L HCl})\left(\frac{0.65 \text{ mol}}{\text{L}}\right) = 0.33 \text{ mol HCl}$$

$0.33 \text{ mol HCl} \rightarrow 0.33 \text{ mol H}^+$

0.33 mol $H^+$ will neutralize 0.26 mol $OH^-$ and leave 0.07 mol $H^+$ (0.33 - 0.26) remaining in solution.

Total volume = 500.0 mL + 380 mL = 880 mL (0.88 L)

$$[\text{H}^+] \text{ in solution} = \frac{0.07 \text{ mol H}^+}{0.88 \text{ L}} = 0.08 \text{ M H}^+$$

$$\text{pH} = -\log [\text{H}^+] = -\log (8 \times 10^{-2}) = 1.1$$

71. $$(0.05000 \text{ L HCl})\left(\frac{0.2000 \text{ mol}}{\text{L}}\right) = 0.01000 \text{ mol HCl} = 0.01000 \text{ mol H}^+ \text{ in } 50.00 \text{ mL HCl}$$

(a) no base added: $\text{pH} = -\log (0.2000) = 0.700$

(b) 10.00 mL base added: $(0.01000 \text{ L})\left(\frac{0.2000 \text{ mol}}{\text{L}}\right) = 0.002000 \text{ mol NaOH}$ $= 0.002000 \text{ mol OH}^-$

$(0.01000 \text{ mol H}^+) - (0.002000 \text{ mol OH}^-) = 0.00800 \text{ mol H}^+$ in 60.00 mL solution

$$[\text{H}^+] = \frac{0.00800 \text{ mol}}{0.06000 \text{ L}} \qquad \text{pH} = -\log\left(\frac{0.00800}{0.06000}\right) = 0.880$$

(c) 25.00 mL base added: $(0.02500 \text{ L})\left(\frac{0.2000 \text{ mol}}{\text{L}}\right) = 0.005000 \text{ mol NaOH} = \text{mol OH}^-$

$(0.01000 \text{ mol H}^+) - (0.005000 \text{ mol OH}^-) = 0.00500 \text{ mol H}^+$ in 75.00 mL solution

$$[\text{H}^+] = \frac{0.00500 \text{ mol}}{0.07500 \text{ L}} \qquad \text{pH} = -\log\left(\frac{0.00500}{0.07500}\right) = 1.2$$

(d) 49.00 mL base added: $(0.04900 \text{ L})\left(\frac{0.2000 \text{ mol}}{\text{L}}\right) = 0.009800 \text{ mol NaOH} = \text{mol OH}^-$

$(0.01000 \text{ mol H}^+) - (0.009800 \text{ mol OH}^-) = 0.00020 \text{ mol H}^+$ in 99.00 mL solution

$$[\text{H}^+] = \frac{0.00020 \text{ mol}}{0.09900 \text{ L}} \qquad \text{pH} = -\log\left(\frac{0.00020}{0.09900}\right) = 2.69$$

(e) 49.90 mL base added: $(0.04990 \text{ L})\left(\frac{0.2000 \text{ mol}}{\text{L}}\right) = 0.009980 \text{ mol NaOH} = \text{mol OH}^-$

$(0.01000 \text{ mol H}^+) - (0.009980 \text{ mol OH}^-) = 2 \times 10^{-5} \text{ mol H}^+$ in 99.9 mL solution

$$[\text{H}^+] = \frac{2 \times 10^{-5} \text{ mol}}{0.0999 \text{ L}} \qquad \text{pH} = -\log\left(\frac{2 \times 10^{-5}}{0.0999}\right) = 3.7$$

(f) 49.99 mL base added: $(0.04999\ \text{L})\left(\dfrac{0.2000\ \text{mol}}{\text{L}}\right) = 0.009998\ \text{mol NaOH} = \text{mol OH}^-$

$(0.01000\ \text{mol H}^+) - (0.009998\ \text{mol OH}^-) = 2 \times 10^{-6}\ \text{mol H}^+$ in 99.9 mL solution

$[\text{H}^+] = \dfrac{2\times10^{-6}\ \text{mol}}{0.09999\ \text{L}}$ $\quad \text{pH} = -\log\left(\dfrac{2\times10^{-6}}{9.999\times10^{-2}}\right) = 4.7$

(g) 50.00 mL of 0.2000 M NaOH neutralizes 50.00 mL of 0.2000 M HCl. No excess acid or base is in the solution. Therefore, the solution is neutral with a pH = 7.0

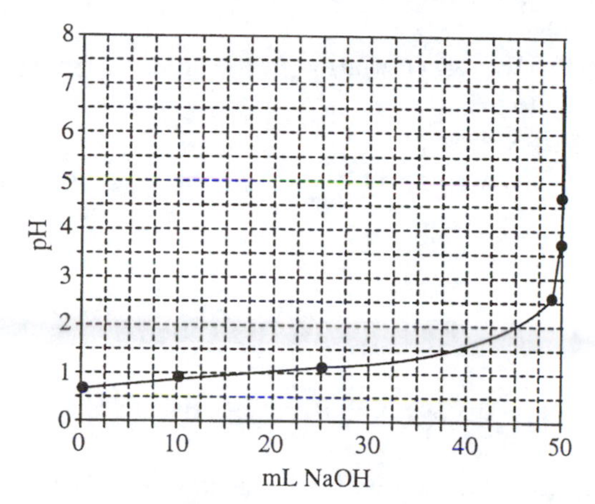


72. (a) $2\ \text{NaOH}(aq) + \text{H}_2\text{SO}_4(aq) \rightarrow \text{Na}_2\text{SO}_4(aq) + 2\ \text{H}_2\text{O}(l)$

(b) mol $H_2SO_4$ → mol NaOH → mL NaOH

$$(0.0050\ \text{mol H}_2\text{SO}_4)\left(\dfrac{2\ \text{mol NaOH}}{1\ \text{mol H}_2\text{SO}_4}\right)\left(\dfrac{1000\ \text{mL}}{0.10\ \text{mol}}\right) = 1.0 \times 10^2\ \text{mL NaOH}$$

(c) $$(0.0050\ \text{mol H}_2\text{SO}_4)\left(\dfrac{1\ \text{mol Na}_2\text{SO}_4}{1\ \text{mol H}_2\text{SO}_4}\right)\left(\dfrac{142.1\ \text{g}}{\text{mol}}\right) = 0.71\ \text{g Na}_2\text{SO}_4$$

73. mol acid = mol base (lactic acid has one acidic H)

$$\dfrac{1.0\ \text{g acid}}{\text{molar mass}} = (0.017\ \text{L})\left(\dfrac{0.65\ \text{mol}}{\text{L}}\right) = 90.\ \text{g/mol (molar mass)}$$

mass of empirical formula ($HC_3H_5O_3$) = 90.17 g/mol
molar mass = mass of empirical formula
Therefore the molecular formula is $HC_3H_5O_3$

74. $HNO_3 + KOH \rightarrow KNO_3 + H_2O$

$M_AV_A = M_BV_B$

$(M_A)(25\text{ mL}) = (0.60\text{ M})(50.0\text{ mL})$

$M_A = 1.2\text{ M}$ (diluted solution)

Dilution problem $M_1V_1 = M_2V_2$

$(M_1)(10.0\text{ mL}) = (1.2\text{ M})(100.00\text{ mL})$

$M_1 = 12\text{ M } HNO_3$ (original solution)

75. Yes, adding water changes the concentration of the acid, which changes the concentration of the $[H^+]$, and changes the pH.

No, the solution theoretically will never reach a pH of 7, but it will approach pH 7 as water is added.

76. $pH = -\log [H^+]$ $pH = 2,\ [H^+] = 1 \times 10^{-2}$ (solution X)
$pH = 4,\ [H^+] = 1 \times 10^{-4}$ (solution Y)

statement (c) is correct. The $[H^+]$ of X is 100 times that of Y.

77. (a) The solution is neutral, neither acidic nor basic ($pH = 7.0$).

(b) The solution is basic, $[OH^-] > 10^{-7}$; $pH = 12$.

(c) The solution is acidic.

(d) Cannot determine whether the solution is acidic or basic.

CHAPTER 16

# CHEMICAL EQUILIBRIUM

1. At 25°C both tubes would appear the same and contain more molecules in the gaseous state than the tube at 0°C, and less molecules in the gaseous state than the tube at 80°C.

2. The reaction is endothermic because the increased temperature increases the concentration of product ($NO_2$) present at equilibrium.

3. At equilibrium, the rate of the forward reaction equals the rate of the reverse reaction.

4. The sum of the pH and the pOH is 14. A solution whose pH is $-1$ would have a pOH of 15.

5. Acids stronger than acetic acid are: benzoic, cyanic, formic, hydrofluoric, and nitrous acids (all equilibrium constants are greater than the equilibrium constant for acetic acid). Acids weaker than acetic acid are: carbolic, hydrocyanic, and hypochlorous acids (all have equilibrium constants smaller than the equilibrium constant for acetic acid). All have one ionizable hydrogen atom.

6. The order of solubility will correspond to the order of the values of the solubility product constants of the salts being compared. This occurs because each salt in the comparison produces the same number of ions (two in this case) for each formula unit of salt that dissolves. This type of comparison would not necessarily be valid if the salts being compared gave different numbers of ions per formula unit of salt dissolving. The order is: $AgC_2H_3O_2$, $PbSO_4$, $BaSO_4$, $AgCl$, $BaCrO_4$, $AgBr$, $AgI$, $PbS$.

7. (a) $K_{sp}\ Mn(OH_2) = 2.0 \times 10^{-13}$; $K_{sp}\ Ag_2CrO_4 = 1.9 \times 10^{-12}$.

   Each salt gives 3 ions per formula units of salt dissolving. Therefore, the salt with the largest $K_{sp}$ (in this case $Ag_2CrO_4$) is more soluble.

   (b) $K_{sp}\ BaCrO_4 = 8.5 \times 10^{-11}$; $K_{sp}\ Ag_2CrO_4 = 1.9 \times 10^{-12}$. $Ag_2CrO_4$ has a greater molar solubility than $BaCrO_4$, even though its $K_{sp}$ is smaller, because the $Ag_2CrO_4$ produces more ions per formula unit of salt dissolving than $BaCrO_4$.

   $BaCrO_4(s) \leftrightarrows Ba^{2+} + CrO_4^{2-}$ $\qquad K_{sp} = [Ba^{2+}][CrO_4^{2-}]$

   Let $y$ = molar solubility of $BaCrO_4$

   $K_{sp} = [y][y] = 8.5 \times 10^{-11}$

   $y = \sqrt{8.5 \times 10^{-11}} = 9.2 \times 10^{-6}$ mol $BaCrO_4$/L

$Ag_2CrO_4(s) \leftrightarrows 2\ Ag^+ + CrO_4^{2-}$ $\qquad K_{sp} = [Ag^+]^2[CrO_4^{2-}]$

Let $y$ = molar solubility of $Ag_2CrO_4$

$K_{sp} = [2y]^2[y] = 1.9 \times 10^{-12}$

$$y = \sqrt[3]{\frac{1.9 \times 10^{-12}}{4}} = 7.8 \times 10^{-5} \text{ mol } Ag_2CrO_4/L$$

$Ag_2CrO_4$ has the greater solubility.

8. $HC_2H_3O_2 \leftrightarrows H^+ + C_2H_3O_2^-$

| Initial Concentrations | | Added | Concentration After Equilibrium Shifts |
|---|---|---|---|
| $HC_2H_3O_2$ | 1.00 M | -------- | 1.01 M |
| $H^+$ | $1.8 \times 10^{-5}$ M | 0.010 mol | $1.9 \times 10^{-5}$ |
| $C_2H_3O_2^-$ | 1.00 M | -------- | 0.99 M |

The initial concentration of $H^+$ in the buffer solution is very low ($1.8 \times 10^{-5}$ M) because of the large excess of acetate ions. 0.010 mole of HCl is added to one liter of the buffer solution. This will supply 0.010 M $H^+$. The added $H^+$ creates a stress on the right side of the equation. The equilibrium shifts to the left, using up almost all the added $H^+$, reducing the acetate ion by approximately 0.010 M, and increasing the acetic acid by approximately 0.010 M. The concentration of $H^+$ will not increase significantly and the pH is maintained relatively constant.

9. In a saturated sodium chloride solution, the equilibrium is

$Na^+(aq) + Cl^-(aq) \leftrightarrows NaCl(s)$

Bubbling in HCl gas increases the concentration of $Cl^-$, creating a stress, which will cause the equilibrium to shift to the right, precipitating solid NaCl.

10. The rate of a reaction increases when the concentration of one of the reactants increases. The increase in concentration causes the number of collisions between the reactants to increase. The rate of a reaction, being proportional to the frequency of such collisions, as a result will increase.

11. If pure HI is placed in a vessel at 700 K, some of it will decompose. Since the reaction is reversible ($H_2 + I_2 \leftrightarrows 2\ HI$) HI molecules will react to produce $H_2$ and $I_2$.

12. An increase in temperature causes the rate of reaction to increase, because it increases the velocity of the molecules. Faster moving molecules increase the number and effectiveness of the collisions between molecules resulting in an increase in the rate of the reaction.

13. $A + B \leftrightarrows C + D$

When A and B are initially mixed, the rate of the forward reaction to produce C and D is at its maximum. As the reaction proceeds, the rate of production of C and D decreases because the concentrations of A and B decrease. As C and D are produced, some of the collisions between C and D will result in the reverse reaction, forming A and B. Finally, an equilibrium is achieved in which the forward rate exactly equals the reverse rate.

14. $HC_2H_3O_2 + H_2O \leftrightarrows H_3O^+ + C_2H_3O_2^-$

As water is added (diluting the solution from 1.0 M to 0.10 M), the equilibrium shifts to the right, yielding a higher percent ionization.

15. The statement does not contradict Le Chatelier's Principle. The previous question deals with the case of dilution. If pure acetic acid is added to a dilute solution, the reaction will shift to the right, producing more ions in accordance with Le Chatelier's Principle. But, the concentration of the un-ionized acetic acid will increase faster than the concentration of the ions, thus yielding a smaller percent ionization.

16. At different temperatures, the degree of ionization of water varies, being higher at higher temperatures. Consequently, the pH of water can be different at different temperatures.

17. In pure water, $H^+$ and $OH^-$ are produced in equal quantities by the ionization of the water molecules, $H_2O \leftrightarrows H^+ + OH^-$. Since $pH = -\log[H^+]$, and $pOH = -\log[OH^-]$, they will always be identical for pure water. At 25 °C, they each have the value of 7, but at higher temperatures, the degree of ionization is greater, so the pH and pOH would both be less than 7, but still equal.

18. In water the silver acetate dissociates until the equilibrium concentration of ions is reached. In nitric acid solution, the acetate ions will react with hydrogen ions to form acetic acid molecules. The $HNO_3$ removes acetate ions from the silver acetate equilibrium allowing more silver acetate to dissolve. If HCl is used, a precipitate of silver chloride would be formed, since silver chloride is less soluble than silver acetate. Thus, more silver acetate would dissolve in HCl than in pure water.

$$AgC_2H_3O_2(s) \leftrightarrows Ag^+(aq) + C_2H_3O_2^-(aq)$$

19. When the salt, sodium acetate, is dissolved in water, the solution becomes basic. The dissolving reaction is

$$NaC_2H_3O_2(s) \xrightarrow{H_2O} Na^+(aq) + C_2H_3O_2^-(aq)$$

The acetate ion reacts with water (hydrolysis). The reaction does not go to completion, but some $OH^-$ ions are produced and at equilibrium the solution is basic.

$$C_2H_3O_2^-(aq) + H_2O(l) \leftrightarrows OH^-(aq) + HC_2H_3O_2(aq)$$

20. A buffer solution contains a weak acid or base plus a salt of that weak acid or base, such as dilute acetic acid and sodium acetate.

$$HC_2H_3O_2(aq) \leftrightharpoons H^+(aq) + C_2H_3O_2^-(aq)$$

$$NaC_2H_3O_2(aq) \leftrightharpoons Na^+(aq) + C_2H_3O_2^-(aq)$$

When a small amount of a strong acid ($H^+$) is added to this buffer solution, the $H^+$ reacts with the acetate ions to form un-ionized acetic acid, thus neutralizing the added acid. When a strong base, $OH^-$, is added it reacts with un-ionized acetic acid to neutralized the added base. As a result, in both cases, the approximate pH of the solution is maintained.

21. Reversible systems

(a) $H_2O(s) \overset{0°C}{\leftrightharpoons} H_2O(l)$

(b) $Na_2SO_4(s) \leftrightharpoons 2\ Na^+(aq) + SO_4^{2-}(aq)$

22. Reversible systems

(a) $H_2O(s) \overset{0°C}{\leftrightharpoons} H_2O(l)$

(b) $SO_2(l) \leftrightharpoons SO_2(g)$

23. Equilibrium system

$$4\ NH_3(g) + 3\ O_2(g) \leftrightharpoons 2\ N_2(g) + 6\ H_2O(g) + 1531\ kJ$$

(a) The reaction is exothermic with heat being evolved.

(b) The addition of $O_2$ will shift the reaction to the right until equilibrium is reestablished. The concentration of $N_2$, $H_2O$, and $O_2$ will be increased. The concentration of the $NH_3$ will be decreased.

24. Equilibrium system

$$4\ NH_3(g) + 3\ O_2(g) \leftrightharpoons 2\ N_2(g) + 6\ H_2O(g) + 1531\ kJ$$

(a) The addition of $N_2$ will shift the reaction to the left until equilibrium is reestablished. The concentration of $NH_3$, $O_2$ and $N_2$ will be increased. The concentration of $H_2O$ will be decreased.

(b) The addition of heat will shift the reaction to the left. This shift will use up the heat added.

25. $N_2(g) + 3\,H_2(g) \leftrightarrows 2\,NH_3(g) + 92.5$ kJ

| Change or stress imposed on the system at equilibrium | | Direction of reaction, left or right, to re-establish equilibrium | Changes in number of moles | | |
|---|---|---|---|---|---|
| | | | $N_2$ | $H_2$ | $NH_3$ |
| (a) | Add $N_2$ | right | I | D | I |
| (b) | Remove $H_2$ | left | I | D | D |
| (c) | Decrease volume of reaction vessel | right | D | D | I |
| (d) | Increase temperature | left | I | I | D |

I = Increase; D = Decrease; N = No Change;
? = insufficient information to determine

26. $N_2(g) + 3\,H_2(g) \leftrightarrows 2\,NH_3(g) + 92.5$ kJ

| Change or stress imposed on the system at equilibrium | | Direction of reaction, left or right, to re-establish equilibrium | Changes in number of moles | | |
|---|---|---|---|---|---|
| | | | $N_2$ | $H_2$ | $NH_3$ |
| (a) | Add $NH_3$ | left | I | I | D |
| (b) | Increase volume of reaction vessel | left | I | I | D |
| (c) | Add a catalyst | no change | N | N | N |
| (d) | Add $H_2$ and $NH_3$ | ? | ? | I | I |

I = Increase; D = Decrease; N = No Change;
? = insufficient information to determine

27. Direction of shift in equilibrium:

| Reaction | Increased Temperature | Increased Pressure (Volume Decreases) | Add Catalyst |
|---|---|---|---|
| (a) | right | right | no change |
| (b) | left | no change | no change |
| (c) | left | right | no change |

28. Direction of shift in equilibrium:

| Reaction | Increased Temperature | Increased Pressure (Volume Decreases) | Add Catalyst |
|---|---|---|---|
| (a) | right | left | no change |
| (b) | left | left | no change |
| (c) | left | left | no change |

29. Equilibrium shifts
    (a) right
    (b) left
    (c) none

30. Equilibrium shifts
    (a) left
    (b) right
    (c) right

31. (a) $K_{eq} = \dfrac{[Cl_2]^2[H_2O]^2}{[HCl]^4[O_2]}$    (c) $K_{eq} = \dfrac{[PCl_3][Cl_2]}{[PCl_5]}$

    (b) $K_{eq} = \dfrac{[NH_3]^2}{[N_2][H_2]^3}$

32. (a) $K_{eq} = \dfrac{[H^+][ClO_2^-]}{[HClO_2]}$    (c) $K_{eq} = \dfrac{[NO]^4[H_2O]^6}{[NH_3]^4[O_2]^5}$

    (b) $K_{eq} = \dfrac{[H^+][C_2H_3O_2^-]}{[HC_2H_3O_2]}$

33. (a) $K_{sp} = [Cu^{2+}][S^{2-}]$    (c) $K_{sp} = [Pb^{2+}][Br^-]^2$

    (b) $K_{sp} = [Ba^{2+}][SO_4^{2-}]$    (d) $K_{sp} = [Ag^+]^3[AsO_4^{3-}]$

34. (a) $K_{sp} = [Fe^{3+}][OH^-]^3$    (c) $K_{sp} = [Ca^{2+}][F^-]^2$

    (b) $K_{sp} = [Sb^{5+}]^2[S^{2-}]^5$    (d) $K_{sp} = [Ba^{2+}]^3[PO_4^{3-}]^2$

35. If the $H^+$ ion concentration is decreased:

    (a) pH is increased

    (b) pOH is decreased

(c) $[OH^-]$ is increased

(d) $K_w$ remains the same. $K_w$ is a constant at a given temperature.

36. If the $H^+$ ion concentration is increased:

(a) pH is decreased (pH of 1 is more acidic than that of 4)

(b) pOH is increased

(c) $[OH^-]$ is decreased

(d) $K_w$ remains unchanged. $K_w$ is a constant at a given temperature.

37. The basis for deciding if a salt dissolved in water produces an acidic, a basic, or a neutral solution, is whether or not the salt reacts with water (hydrolysis). Salts that contain an ion derived from a weak acid or base will hydrolyze to produce an acidic or a basic solution.

(a) KCl, neutral

(b) $Na_2CO_3$, basic

(c) $K_2SO_4$, neutral

(d) $(NH_4)_2SO_4$, acidic

38. The basis for deciding if a salt dissolved in water produces an acidic, a basic, or a neutral solution, is whether or not the salt reacts with water (hydrolysis). Salts that contain an ion derived from a weak acid or base will hydrolyze to produce an acidic or a basic solution.

(a) $Ca(CN)_2$, basic

(b) $BaBr_2$, neutral

(c) $NaNO_2$, basic

(d) NaF, basic

39. (a) $NO_2^-(aq) + H_2O(l) \leftrightarrows OH^-(aq) + HNO_2(aq)$

(b) $C_2H_3O_2^-(aq) + H_2O(l) \leftrightarrows OH^-(aq) + HC_2H_3O_2(aq)$

40. (a) $NH_4^+(aq) + H_2O(l) \leftrightarrows H_3O^+(aq) + NH_3(aq)$

(b) $SO_3^{2-}(aq) + H_2O(l) \leftrightarrows OH^-(aq) + HSO_3^-(aq)$

41. (a) $HCO_3^-(aq) + H_2O(l) \leftrightarrows OH^-(aq) + H_2CO_3(aq)$

(b) $NH_4^+(aq) + H_2O(l) \leftrightarrows H_3O^+(aq) + NH_3(aq)$

42. (a) $OCl^-(aq) + H_2O(l) \leftrightarrows OH^-(aq) + HOCl(aq)$

(b) $ClO_2^-(aq) + H_2O(l) \leftrightarrows OH^-(aq) + HClO_2(aq)$

43. When excess acid ($H^+$) gets into the blood stream it reacts with $HCO_3^-$ to form un-ionized $H_2CO_3$, thus neutralizing the acid and maintaining the approximate pH of the solution.

44. When excess base gets into the blood stream it reacts with $H^+$ to form water. Then $H_2CO_3$ ionizes to replace $H^+$, thus maintaining the approximate pH of the solution.

45. (a) $HC_2H_3O_2 \leftrightarrows H^+ + C_2H_3O_2^-$

Let $x$ = molarity of $HC_2H_3O_2$ ionizing to establish equilibrium. Equilibrium concentrations are:

$[H^+] = [C_2H_3O_2^-] = x$

$[HC_2H_3O_2] = 0.25 - x = 0.25$ (since $x$ is small)

$$K_a = \frac{[H^+][C_2H_3O_2^-]}{[HC_2H_3O_2]} = \frac{x^2}{0.25} = 1.8 \times 10^{-5}$$

$x^2 = (0.25)(1.8 \times 10^{-5})$

$x = \sqrt{(0.25)(1.8 \times 10^{-5})} = 2.1 \times 10^{-3}\text{ M} = [H^+]$

(b) $pH = -\log [H^+] = -\log (2.1 \times 10^{-3}) = 2.68$

(c) Percent ionization

$$\frac{[H^+]}{[HC_2H_3O_2]}(100) = \left(\frac{2.1 \times 10^{-3}}{0.25}\right)(100) = 0.84\%$$

46. (a) $HC_6H_5O(aq) \leftrightarrows H^+(aq) + C_6H_5O^-(aq)$

Let $x$ = molarity of $HC_6H_5O$ ionizing to establish equilibrium. Equilibrium concentrations are:
$[H^+] = [C_6H_5O^-] = x$
$[HC_6H_5O] = 0.25 - x = 0.25$ (since $x$ is small)

$$K_a = \frac{[H^+][C_6H_5O^-]}{[HC_6H_5O]} = \frac{x^2}{0.25} = 1.3 \times 10^{-10}$$

$x^2 = (0.25)(1.3 \times 10^{-10})$

$x = \sqrt{(0.25)(1.3 \times 10^{-10})} = 5.7 \times 10^{-6} = [H^+]$

(b) $pH = -\log [H^+] = -\log (5.7 \times 10^{-6}) = 5.24$

(c) Percent ionization

$$\frac{[H^+]}{[HC_6H_5O]}(100) = \left(\frac{5.7 \times 10^{-6}}{0.25}\right)(100) = 2.3 \times 10^{-3}\ \%$$

47. $HA \leftrightarrows H^+ + A^-$ $\quad 0.52\% = 0.0052$

$$[H^+] = [A^-] = (1.000\ M)(0.0052) = 5.2 \times 10^{-3}\ M$$

$$[HA] = 1.000\ M - 0.0052\ M = 0.9948\ M$$

$$K_a = \frac{[H^+][A^-]}{[HA]} = \frac{(5.2 \times 10^{-3})^2}{0.9948} = 2.7 \times 10^{-5}$$

48. $HA \leftrightarrows H^+ + A^-$ $\quad pH = 5 = -\log[H^+] \quad [H^+] = 1 \times 10^{-5}\ M$

$$[H^+] = [A^-] = 1 \times 10^{-5}\ M$$

$$[HA] = 0.15\ M - 1 \times 10^{-5}\ M = 0.15\ M$$

$$K_a = \frac{[H^+][A^-]}{[HA]} = \frac{(1 \times 10^{-5})^2}{0.15} = 7 \times 10^{-10}$$

49. $HC_2H_3O_2 \leftrightarrows H^+ + C_2H_3O_2^-$

$$K_a = \frac{[H^+][C_2H_3O_2^-]}{[HC_2H_3O_2]} = 1.8 \times 10^{-5}$$

Let $x$ = molarity of $HC_2H_3O_2$, which is ionized, to establish equilibrium. Equilibrium concentrations are:

$$[H^+] = [C_2H_3O_2^-] = x$$

$$[HC_2H_3O_2] = \text{initial concentration} - x = \text{initial concentration}$$

Since $K_a$ is small, the degree of ionization is small. Therefore, the approximation, initial concentration $- x =$ initial concentration, is valid.

(a) $[H^+] = [C_2H_3O_2^-] = x \qquad [HC_2H_3O_2] = 1.0\ M$

$$\frac{(x)(x)}{1.0} = 1.8 \times 10^{-5}$$

$$x^2 = (1.0)(1.8 \times 10^{-5})$$

$$x = \sqrt{1.8 \times 10^{-5}} = 4.2 \times 10^{-3}\ M$$

$$\left(\frac{4.2 \times 10^{-3}\ M}{1.0\ M}\right)(100) = 0.42\% \text{ ionized}$$

$$pH = -\log(4.2 \times 10^{-3}) = 2.38$$

(b) $[HC_2H_3O_2] = 0.10\ M$

$$\frac{(x)(x)}{0.10} = 1.8 \times 10^{-5}$$

$$x^2 = (0.10)(1.8 \times 10^{-5}) = 1.8 \times 10^{-6}$$

$$x = \sqrt{1.8 \times 10^{-6}} = 1.3 \times 10^{-3}\ M$$

$$\left(\frac{1.3 \times 10^{-3}\ M}{0.10\ M}\right)(100) = 1.3\%\text{ ionized}$$

$$pH = -\log(1.3 \times 10^{-3}) = 2.89$$

(c) $[HC_2H_3O_2] = 0.010\ M$

$$\frac{(x)(x)}{0.010} = 1.8 \times 10^{-5}$$

$$x^2 = (0.010)(1.8 \times 10^{-5}) = 1.8 \times 10^{-7}$$

$$x = \sqrt{1.8 \times 10^{-7}} = 4.2 \times 10^{-4}\ M$$

$$\left(\frac{4.2 \times 10^{-4}\ M}{0.010\ M}\right)(100) = 4.2\%\text{ ionized}$$

$$pH = -\log(4.2 \times 10^{-4}) = 3.38$$

50. $HClO \leftrightharpoons H^+ + ClO^-$

$$K_a = \frac{[H^+][ClO^-]}{[HClO]} = 3.5 \times 10^{-8}$$

Let $x$ = molarity of HClO, which is ionized, to establish equilibrium. Equilibrium concentrations are:

$$[H^+] = [ClO^-] = x$$

$$[HClO] = \text{initial concentration} - x = \text{initial concentration}$$

Since $K_a$ is small, the degree of ionization is small. Therefore, the approximation, initial concentration $- x$ = initial concentration, is valid.

(a) $[H^+] = [ClO^-] = x$ $\qquad$ $[HClO] = 1.0\ M$

$$\frac{(x)(x)}{1.0} = 3.5 \times 10^{-8} \qquad x^2 = (1.0)(3.5 \times 10^{-8})$$

$$x = \sqrt{3.5 \times 10^{-8}} = 1.9 \times 10^{-4}\ M$$

$$\left(\frac{1.9 \times 10^{-4}\text{ M}}{1.0\text{ M}}\right)(100) = 1.9 \times 10^{-2}\% \text{ ionized}$$

$$\text{pH} = -\log(1.9 \times 10^{-4}) = 3.72$$

(b) $[HClO] = 0.10$ M

$$\frac{(x)(x)}{0.10} = 3.5 \times 10^{-8} \qquad x^2 = (0.10)(3.5 \times 10^{-8})$$

$$x = \sqrt{3.5 \times 10^{-9}} = 5.9 \times 10^{-5}\text{ M}$$

$$\left(\frac{5.9 \times 10^{-5}\text{ M}}{0.10\text{ M}}\right)(100) = 0.059\% \text{ ionized}$$

$$\text{pH} = -\log(5.9 \times 10^{-5}) = 4.23$$

(c) $[HClO] = 0.010$ M

$$\frac{(x)(x)}{0.010} = 3.5 \times 10^{-8} \qquad x^2 = (0.010)(3.5 \times 10^{-8})$$

$$x = \sqrt{3.5 \times 10^{-10}} = 1.9 \times 10^{-5}\text{ M}$$

$$\left(\frac{1.9 \times 10^{-5}\text{ M}}{0.010\text{ M}}\right)(100) = 0.19\% \text{ ionized}$$

$$\text{pH} = -\log(1.9 \times 10^{-5}) = 4.72$$

51. $HA \leftrightarrows H^+ + A^-$ $\qquad K_a = \frac{[H^+][A^-]}{[HA]}$

First, find the $[H^+]$. This is calculated from the pH expression, $\text{pH} = -\log[H^+] = 3.7$. Enter $-3.7$ into the calculator and push the $10^x$ key.
This yields the $[H^+] = 2 \times 10^{-4}$.

$$[H^+] = [A^-] = 2 \times 10^{-4} \qquad [HA] = 0.37$$

$$K_a = \frac{[H^+][A^-]}{[HA]} = \frac{(2 \times 10^{-4})(2 \times 10^{-4})}{0.37} = 1 \times 10^{-7}$$

52. See problem 51 for a discussion of calculating $[H^+]$ from pH.

$$HA \leftrightarrows H^+ + A^- \qquad K_a = \frac{[H^+][A^-]}{[HA]} \qquad \text{pH} = 2.89$$

$$-\log[H^+] = 2.89 \qquad [H^+] = 1.3 \times 10^{-3}$$

$[H^+] = 1.3 \times 10^{-3} = [A^-]$ $[HA] = 0.23$

$$K_a = \frac{[H^+][A^-]}{[HA]} = \frac{(1.3 \times 10^{-3})(1.3 \times 10^{-3})}{0.23} = 7.3 \times 10^{-6}$$

53. 6.0 M HCl yields $[H^+] = 6.0$ M (100% ionized)

$pH = -\log 6.0 = -0.78$

$pOH = 14 - pH = 14 - (-0.78) = 14.78$

$$[OH^-] = \frac{K_w}{[H^+]} = \frac{1 \times 10^{-14}}{6.0} = 1.7 \times 10^{-15}$$

54. 1.0 M NaOH yields $[OH^-] = 1.0$ M (100% ionized)

$pOH = -\log 1.0 = 0.00$

$pH = 14 - pOH = 14.00$

$$[H^+] = \frac{K_w}{[OH^-]} = \frac{1.0 \times 10^{-14}}{1.0} = 1 \times 10^{-14}$$

55. $pH + pOH = 14.0$ $pOH = 14.0 - pH$

(a) 0.00010 M HCl $[H^+] = 0.00010\text{ M} = 1.0 \times 10^{-4}\text{ M}$

$pH = -\log (1.0 \times 10^{-4}) = 4.00$

$pOH = 14.0 - 4.00 = 10.0$

(b) 0.010 M NaOH $[OH^-] = 0.010\text{ M} = 1.0 \times 10^{-2}\text{ M}$

$pOH = -\log (1.0 \times 10^{-2}) = 2.00$

$pH = 14.0 - 2.00 = 12.0$

56. $pH + pOH = 14.0$

(a) 0.00250 M NaOH $[OH^-] = 2.5 \times 10^{-3}\text{ M}$

$pOH = -\log (2.5 \times 10^{-3}) = 2.60$

$pH = 14.0 - 2.60 = 11.4$

(b) $HClO \leftrightarrows H^+ + ClO^-$
$0.10\ M \quad x \quad x$

$$K_a = \frac{[H^+][ClO^-]}{[HClO]} = 3.5 \times 10^{-8}$$

$$\frac{(x)(x)}{0.10} = 3.5 \times 10^{-8}$$

$x^2 = (0.10)(3.5 \times 10^{-8}) \qquad x = \sqrt{3.5 \times 10^{-9}}$

$x = 5.9 \times 10^{-5} = [H^+]$

$pH = -\log(5.9 \times 10^{-5}) = 4.23$

$pOH = 14.0 - 4.23 = 9.8$

(c) $Fe(OH)_2(s) \leftrightarrows Fe^{2+} + 2\,OH^-$
$x \quad x \quad 2x$

$K_{sp} = [Fe^{2+}][OH^-]^2 = (x)(2x)^2 = 8.0 \times 10^{-16}$

$4x^3 = 8.0 \times 10^{-16}$

$$x = \sqrt[3]{\frac{8.0 \times 10^{-16}}{4}} = 5.8 \times 10^{-6}$$

$[OH^-] = 2x = 2(5.8 \times 10^{-6}) = 1.2 \times 10^{-5}$

$pOH = -\log(1.2 \times 10^{-5}) = 4.92$

$pH = 14.0 - 4.92 = 9.1$

57. Calculate the $[OH^-]$. $[OH^-] = \dfrac{K_w}{[H^+]}$

(a) $[H^+] = 1.0 \times 10^{-4} \qquad [OH^-] = \dfrac{1.0 \times 10^{-14}}{1.0 \times 10^{-4}} = 1.0 \times 10^{-10}$

(b) $[H^+] = 2.8 \times 10^{-6} \qquad [OH^-] = \dfrac{1.0 \times 10^{-14}}{2.8 \times 10^{-6}} = 3.6 \times 10^{-9}$

58. Calculate the $[OH^-]$. $[OH^-] = \dfrac{K_w}{[H^+]}$

(a) $[H^+] = 4.0 \times 10^{-9} \qquad [OH^-] = \dfrac{1.0 \times 10^{-14}}{4.0 \times 10^{-9}} = 2.5 \times 10^{-6}$

(b) $[H^+] = 8.9 \times 10^{-2} \qquad [OH^-] = \dfrac{1.0 \times 10^{-14}}{8.9 \times 10^{-2}} = 1.1 \times 10^{-13}$

59. Calculate the $[H^+]$. $[H^+] = \dfrac{K_w}{[OH^-]}$

(a) $[OH^-] = 6.0 \times 10^{-7}$ $\qquad [H^+] = \dfrac{1.0 \times 10^{-14}}{6.0 \times 10^{-7}} = 1.7 \times 10^{-8}$

(b) $[OH^-] = 1 \times 10^{-8}$ $\qquad [H^+] = \dfrac{1 \times 10^{-14}}{1 \times 10^{-8}} = 1 \times 10^{-6}$

60. Calculate the $[H^+]$. $[H^+] = \dfrac{K_w}{[OH^-]}$

(a) $[OH^-] = 4.5 \times 10^{-6}$ $\qquad [H^+] = \dfrac{1.0 \times 10^{-14}}{4.5 \times 10^{-6}} = 2.2 \times 10^{-9}$

(b) $[OH^-] = 7.3 \times 10^{-4}$ $\qquad [H^+] = \dfrac{1 \times 10^{-14}}{7.3 \times 10^{-4}} = 1.4 \times 10^{-11}$

61. The molar solubilities of the salts and their ions are indicated below the formulas in the equilibrium equations.

(a)
$$\underset{3.9 \times 10^{-5}}{BaSO_4(s)} \leftrightarrows \underset{3.9 \times 10^{-5}}{Ba^{2+}} + \underset{3.9 \times 10^{-5}}{SO_4^{2-}}$$

$K_{sp} = [Ba^{2+}][SO_4^{2-}] = (3.9 \times 10^{-5})^2 = 1.5 \times 10^{-9}$

(b)
$$\underset{7.8 \times 10^{-5}}{Ag_2CrO_4(s)} \leftrightarrows \underset{2(7.8 \times 10^{-5})}{2\,Ag^+} + \underset{7.8 \times 10^{-5}}{CrO_4^{2-}}$$

$K_{sp} = [Ag^+]^2[CrO_4^{2-}] = (15.6 \times 10^{-5})^2(7.8 \times 10^{-5}) = 1.9 \times 10^{-12}$

(c) First change g/L → mol/L

$$\left(\frac{0.67\text{ g }CaSO_4}{\text{L}}\right)\left(\frac{1\text{ mol}}{136.1\text{ g}}\right) = 4.9 \times 10^{-3}\text{ M }CaSO_4$$

$$\underset{4.9 \times 10^{-3}}{CaSO_4(s)} \leftrightarrows \underset{4.9 \times 10^{-3}}{Ca^{2+}} + \underset{4.9 \times 10^{-3}}{SO_4^{2-}}$$

$K_{sp} = [Ca^{2+}][SO_4^{2-}] = (4.9 \times 10^{-3})^2 = 2.4 \times 10^{-5}$

(d) First change g/L → mol/L

$$\left(\frac{0.0019\text{ g AgCl}}{\text{L}}\right)\left(\frac{1\text{ mol}}{143.4\text{ g}}\right) = 1.3 \times 10^{-5}\text{ M AgCl}$$

$$\underset{1.3 \times 10^{-5}}{AgCl(s)} \leftrightarrows \underset{1.3 \times 10^{-5}}{Ag^+} + \underset{1.3 \times 10^{-5}}{Cl^-}$$

$K_{sp} = [Ag^+][Cl^-] = (1.3 \times 10^{-5})^2 = 1.7 \times 10^{-10}$

62. The molar solubilities of the salts and their ions are indicated below the formulas in the equilibrium equations.

(a)
$$\begin{array}{llll} ZnS(s) & \leftrightarrows & Zn^{2+} \quad + & S^{2-} \\ 3.5 \times 10^{-12} & & 3.5 \times 10^{-12} & 3.5 \times 10^{-12} \end{array}$$

$$K_{sp} = [Zn^{2+}][S^{2-}] = (3.5 \times 10^{-12})^2 = 1.2 \times 10^{-23}$$

(b)
$$\begin{array}{llll} Pb(IO_3)_2(s) & \leftrightarrows & Pb^{2+} \quad + & 2\ IO_3^- \\ 4.0 \times 10^{-5} & & 4.0 \times 10^{-5} & 2(4.0 \times 10^{-5}) \end{array}$$

$$K_{sp} = [Pb^{2+}][IO_3^-]^2 = (4.0 \times 10^{-5})(8.0 \times 10^{-5})^2 = 2.6 \times 10^{-13}$$

(c) First change g/L → mol/L

$$\left(\frac{6.73 \times 10^{-3}\text{ g } Ag_3PO_4}{\text{L}}\right)\left(\frac{1\text{ mol}}{418.7\text{ g}}\right) = 1.61 \times 10^{-5}\text{ M } Ag_3PO_4$$

$$\begin{array}{llll} Ag_3PO_4(s) & \leftrightarrows & 3\ Ag^+ \quad + & PO_4^{3-} \\ 1.61 \times 10^{-5} & & 3(1.61 \times 10^{-5}) & 1.61 \times 10^{-5} \end{array}$$

$$K_{sp} = [Ag^+]^3[PO_4^{3-}] = (4.83 \times 10^{-5})^3(1.61 \times 10^{-5}) = 1.81 \times 10^{-18}$$

(d) First change g/L → mol/L

$$\left(\frac{2.33 \times 10^{-4}\text{ g } Zn(OH)_2}{\text{L}}\right)\left(\frac{1\text{ mol}}{99.41\text{ g}}\right) = 2.34 \times 10^{-6}\text{ M } Zn(OH)_2$$

$$\begin{array}{llll} Zn(OH)_2(s) & \leftrightarrows & Zn^{2+} \quad + & 2\ OH^- \\ 2.34 \times 10^{-6} & & 2.34 \times 10^{-6} & 2(2.34 \times 10^{-6}) \end{array}$$

$$K_{sp} = [Zn^{2+}][OH^-]^2 = (2.34 \times 10^{-6})(4.68 \times 10^{-6})^2 = 5.13 \times 10^{-17}$$

63. (a)
$$\begin{array}{llll} Ag_2SO_4(s) & \leftrightarrows & 2\ Ag^+ \quad + & SO_4^{2-} \\ x & & 2x & x \end{array}$$

$$K_{sp} = [Ag^+]^2[SO_4^{2-}] = (2x)^2(x) = 4x^3 = 1.5 \times 10^{-5}$$

$$x = \sqrt[3]{\frac{1.5 \times 10^{-5}}{4}} = 1.6 \times 10^{-2}\text{ M}$$

(b)
$$\begin{array}{llll} Mg(OH)_2(s) & \leftrightarrows & Mg^{2+} \quad + & 2\ OH^- \\ x & & x & 2x \end{array}$$

$$K_{sp} = [Mg^{2+}][OH^-]^2 = (x)(2x)^2 = 4x^3 = 7.1 \times 10^{-12}$$

$$x = \sqrt[3]{\frac{7.1 \times 10^{-12}}{4}} = 1.2 \times 10^{-4}\text{ M}$$

64. The molar solubilities of the salts and their ions will be represented in terms of $x$ below their formulas in the equilibrium equations.

(a)

$$\underset{x}{BaCO_3(s)} \leftrightarrows \underset{x}{Ba^{2+}} + \underset{x}{CO_3^{2-}}$$

$$K_{sp} = [Ba^{2+}][CO_3^{2-}] = x^2 = 2.0 \times 10^{-9}$$

$$x = \sqrt{2.0 \times 10^{-9}} = 4.5 \times 10^{-5}\ M$$

(b)

$$\underset{x}{AlPO_4(s)} \leftrightarrows \underset{x}{Al^{3+}} + \underset{x}{PO_4^{3-}}$$

$$K_{sp} = [Al^{3+}][PO_4^{3-}] = x^2 = 5.8 \times 10^{-19}$$

$$x = \sqrt{5.8 \times 10^{-19}} = 7.6 \times 10^{-10}\ M$$

65. (a)

$$\left(\frac{1.6 \times 10^{-2}\ \text{mol } Ag_2SO_4}{L}\right)(0.100\ L)\left(\frac{311.9\ g}{mol}\right) = 0.50\ g\ Ag_2SO_4$$

(b)

$$\left(\frac{1.2 \times 10^{-4}\ \text{mol } Mg(OH)_2}{L}\right)(0.100\ L)\left(\frac{58.33\ g}{mol}\right) = 7.0 \times 10^{-4}\ g\ Mg(OH)_2$$

66. (a)

$$\left(\frac{4.5 \times 10^{-5}\ \text{mol } BaCO_3}{L}\right)(0.100\ L)\left(\frac{197.3\ g}{mol}\right) = 8.9 \times 10^{-4}\ g\ BaCO_3$$

(b)

$$\left(\frac{7.6 \times 10^{-10}\ \text{mol } AlPO_4}{L}\right)(0.100\ L)\left(\frac{122.0\ g}{mol}\right) = 9.3 \times 10^{-9}\ g\ AlPO_4$$

67. The molar concentrations of ions, after mixing, are calculated and these concentrations are substituted into the equilibrium expression. The value obtained is compared to the $K_{sp}$ of the salt. If the value is greater than the $K_{sp}$, precipitation occurs. If the value is less than the $K_{sp}$, no precipitation occurs.

100. mL 0.010 M $Na_2SO_4$ $\rightarrow$ 100. mL 0.010 M $SO_4^{2-}$

100. mL 0.001 M $Pb(NO_3)_2$ $\rightarrow$ 100. mL 0.001 M $Pb^{2+}$

Volume after mixing = 200. mL

Concentrations after mixing: $SO_4^{2-}$ = 0.0050 M $Pb^{2+}$ = 0.0005 M

$$[Pb^{2+}][SO_4^{2-}] = (5.0 \times 10^{-3})(5 \times 10^{-4}) = 3 \times 10^{-6}$$

$K_{sp} = 1.3 \times 10^{-8}$ which is less than $3 \times 10^{-6}$, therefore, precipitation occurs.

68. The molar concentrations of ions, after mixing, are calculated and these concentrations are substituted into the equilibrium expression. The value obtained is compared to the $K_{sp}$ of the salt. If the value is greater than the $K_{sp}$, precipitation occurs. If the value is less than the $K_{sp}$, no precipitation occurs.

$50.0\text{ mL } 1.0 \times 10^{-4}\text{ M AgNO}_3 \rightarrow 50.0\text{ mL } 1.0 \times 10^{-4}\text{ M Ag}^+$

$100.\text{ mL } 1.0 \times 10^{-4}\text{ M NaCl} \rightarrow 100.\text{ mL } 1.0 \times 10^{-4}\text{ M Cl}^-$

Volume after mixing = 150. mL

Concentrations after mixing:

$$(1.0 \times 10^{-4}\text{ M Ag}^+)\left(\frac{50.0\text{ mL}}{150.\text{ mL}}\right) = 3.3 \times 10^{-5}\text{ M Ag}^+$$

$$(1.0 \times 10^{-4}\text{ M Cl}^-)\left(\frac{100.\text{ mL}}{150.\text{ mL}}\right) = 6.7 \times 10^{-5}\text{ M Cl}^-$$

$$[\text{Ag}^+][\text{Cl}^-] = (3.3 \times 10^{-5})(6.7 \times 10^{-5}) = 2.2 \times 10^{-9}$$

$K_{sp} = 1.7 \times 10^{-10}$ which is less than $2.2 \times 10^{-9}$, therefore, precipitation occurs.

69. The concentration of $Br^-$ = 0.10 M in 1.0 L of 0.10 M NaBr. Substitute this $Br^-$ concentration in the $K_{sp}$ expression and solve for the $[Ag^+]$ in equilibrium with 0.10 M $Br^-$.

$$K_{sp} = [\text{Ag}^+][\text{Br}^-] = 5.0 \times 10^{-13}$$

$$[\text{Ag}^+] = \frac{5.0 \times 10^{-13}}{[\text{Br}^-]} = \frac{5.0 \times 10^{-13}}{0.10} = 5.0 \times 10^{-12}\text{ M}$$

$$\left(\frac{5.0 \times 10^{-12}\text{ mol Ag}^+}{\text{L}}\right)\left(\frac{1\text{ mol AgBr}}{1\text{ mol Ag}^+}\right)(1.0\text{ L}) = 5.0 \times 10^{-12}\text{ mol AgBr will dissolve}$$

70. $$\left(\frac{0.10\text{ mol MgBr}_2}{\text{L}}\right)\left(\frac{2\text{ mol Br}^-}{1\text{ mol MgBr}_2}\right) = \left(\frac{0.20\text{ mol Br}^-}{\text{L}}\right) = 0.20\text{ M Br}^-\text{ in solution.}$$

Substitute the $Br^-$ concentration in the $K_{sp}$ expression and solve for $[Ag^+]$ in equilibrium with 0.20 M $Br^-$.

$$[\text{Ag}^+] = \frac{5.0 \times 10^{-13}}{[\text{Br}^-]} = \frac{5.0 \times 10^{-13}}{0.20} = 2.5 \times 10^{-12}\text{ M}$$

$$\left(\frac{2.5 \times 10^{-12}\text{ mol Ag}^+}{\text{L}}\right)\left(\frac{1\text{ mol AgBr}}{1\text{ mol Ag}^+}\right)(1.0\text{ L}) = 2.5 \times 10^{-12}\text{ mol AgBr will dissolve}$$

71. $HC_2H_3O_2 \leftrightarrows H^+ + C_2H_3O_2^-$

$$K_a = \frac{[H^+][C_2H_3O_2^-]}{[HC_2H_3O_2]} = 1.8 \times 10^{-5}$$

$$[H^+] = K_a\left(\frac{[HC_2H_3O_2]}{[C_2H_3O_2^-]}\right) \qquad [HC_2H_3O_2] = 0.20\ M \qquad [C_2H_3O_2^-] = 0.10\ M$$

$$[H^+] = (1.8 \times 10^{-5})\left(\frac{0.20}{0.10}\right) = 3.6 \times 10^{-5}\ M$$

$$pH = -\log(3.6 \times 10^{-5}) = 4.44$$

72. $HC_2H_3O_2 \leftrightarrows H^+ + C_2H_3O_2^-$

$$K_a = \frac{[H^+][C_2H_3O_2^-]}{[HC_2H_3O_2]} = 1.8 \times 10^{-5}$$

$$[H^+] = K_a\left(\frac{[HC_2H_3O_2]}{[C_2H_3O_2^-]}\right) \qquad [HC_2H_3O_2] = 0.20\ M \qquad [C_2H_3O_2^-] = 0.20\ M$$

$$[H^+] = (1.8 \times 10^{-5})\left(\frac{0.20}{0.20}\right) = 1.8 \times 10^{-5}\ M$$

$$pH = -\log(1.8 \times 10^{-5}) = 4.74$$

73. Initially, the solution of NaCl is neutral. $[H^+] = 1 \times 10^{-7}$

$pH = -\log(1 \times 10^{-7}) = 7.0$

Final $H^+ = 2.0 \times 10^{-2}$ M

$pH = -\log(2.0 \times 10^{-2}) = 1.70$

Change in pH $= 7.0 - 1.70 = 5.3$ units

74. Initially, $[H^+] = 1.8 \times 10^{-5}$

$pH = -\log(1.8 \times 10^{-5}) = 4.74$

Final $H^+ = 1.9 \times 10^{-5}$

$pH = -\log(1.9 \times 10^{-5}) = 4.72$

Change in pH $= 4.74 - 4.72 = 0.02$ units in the buffered solution

75. $H_2 + I_2 \leftrightarrows 2\,HI$
The reaction is a 1 to 1 mole ratio of hydrogen to iodine. The data given indicates that hydrogen is the limiting reactant.

$$(2.10\text{ mol } H_2)\left(\frac{2\text{ mol HI}}{1\text{ mol } H_2}\right) = 4.20\text{ mol HI}$$

76. $H_2 + I_2 \leftrightarrows 2\,HI$

(a) 2.00 mol $H_2$ and 2.00 mol $I_2$ will produce 4.00 mol HI assuming 100% yield. However, at 79% yield you get
$4.00\text{ mol HI} \times 0.79 = 3.16\text{ mol HI}$

(b) The addition of 0.27 mol $I_2$ makes the iodine present in excess and the 2.00 mol $H_2$ the limiting reactant. The yield increases to 85%.

$$(2.00\text{ mol } H_2)\left(\frac{2\text{ mol HI}}{1\text{ mol } H_2}\right)\left(\frac{0.85\text{ mol}}{1.00\text{ mol}}\right) = 3.4\text{ mol HI}$$

There will be 15% unreacted $H_2$ and $I_2$ plus the extra $I_2$ added.

$(0.15)(2.0\text{ mol } H_2) = 0.30\text{ mol } H_2$ present; also 0.30 mol $I_2$.

In addition to the 0.30 mol of unreacted $I_2$, will be the 0.27 mol $I_2$ added.

$0.27\text{ mol} + 0.30\text{ mol} = 0.57\text{ mol } I_2$ present.

(c) $$K = \frac{[HI]^2}{[H_2][I_2]}$$

The formation of 3.16 mol HI required the reaction of 1.58 mol $I_2$ and 1.58 mol $H_2$. At equilibrium, the concentrations are:

3.16 mol HI; $\quad 2.00 - 1.58 = 0.42\text{ mol } H_2 = 0.42\text{ mol } I_2$

$$K_{eq} = \frac{(3.16)^2}{(0.42)(0.42)} = 57$$

In the calculation of the equilibrium constant, the actual number of moles of reactants and products present at equilibrium can be used in the calculation in place of molar concentrations. This occurs because the reaction is gaseous and the liters of HI produced equals the sum of the liters of $H_2$ and $I_2$ reacting. In the equilibrium expression, the volumes will cancel.

77. $H_2 + I_2 \leftrightarrows 2\,HI$

$$(64.0\text{ g HI})\left(\frac{1\text{ mol}}{127.9\text{ g}}\right) = 0.500\text{ mol HI present}$$

$$(0.500 \text{ mol HI})\left(\frac{1 \text{ mol } I_2}{2 \text{ mol HI}}\right) = 0.250 \text{ mol } I_2 \text{ reacted}$$

$$(0.500 \text{ mol HI})\left(\frac{1 \text{ mol } H_2}{2 \text{ mol HI}}\right) = 0.250 \text{ mol } H_2 \text{ reacted}$$

$$(6.00 \text{ g } H_2)\left(\frac{1 \text{ mol}}{2.016 \text{ g}}\right) = 2.98 \text{ mol } H_2 \text{ initially present}$$

$$(200. \text{ g } I_2)\left(\frac{1 \text{ mol}}{253.8 \text{ g}}\right) = 0.788 \text{ mol } I_2 \text{ initially present}$$

At equilibrium, moles present are:

0.500 mol HI; $2.98 - 0.250 = 2.73$ mol $H_2$

$0.788 - 0.250 = 0.538$ mol $I_2$

78. $PCl_3(g) + Cl_2(g) \leftrightarrows PCl_5(g)$

$$K_{eq} = \frac{[PCl_5]}{[PCl_3][Cl_2]}$$

The concentrations are:

$$PCl_5 = \frac{0.22 \text{ mol}}{20. \text{ L}} = 0.011 \text{ M}$$

$$PCl_3 = \frac{0.10 \text{ mol}}{20. \text{ L}} = 0.0050 \text{ M}$$

$$Cl_2 = \frac{1.50 \text{ mol}}{20. \text{ L}} = 0.075 \text{ M}$$

$$K_{eq} = \frac{0.011}{(0.0050)(0.075)} = 29$$

79. $100°C - 30°C = 70°C$ temperature increase. This increase is equal to seven 10°C increments. The reaction rate will be increased by $2^7 = 128$ times.

80. **Hypochlorous acid** $HOCl \leftrightarrows H^+ + OCl^-$

Equilibrium concentrations:

$[H^+] = [OCl^-] = 5.9 \times 10^{-5}$ M

$[HOCl] = 0.1 - 5.9 \times 10^{-5} = 0.10$ M (neglecting $5.9 \times 10^{-5}$)

$$K_a = \frac{[H^+][OCl^-]}{[HOCl]} = \frac{(5.9 \times 10^{-5})(5.9 \times 10^{-5})}{0.10} = 3.5 \times 10^{-8}$$

**Propanoic acid** $HC_3H_5O_2 \leftrightarrows H^+ + C_3H_5O_2^-$

Equilibrium concentrations:

$[H^+] = [C_3H_5O_2^-] = 1.4 \times 10^{-3}$ M

$[HC_3H_5O_2] = 0.15 - 1.4 \times 10^{-3} = 0.15$ M (neglecting $1.4 \times 10^{-3}$)

$$K_a = \frac{[H^+][C_3H_5O_2^-]}{[HC_3H_5O_2]} = \frac{(1.4 \times 10^{-3})(1.4 \times 10^{-3})}{0.15} = 1.3 \times 10^{-5}$$

**Hydrocyanic acid** $HCN \leftrightarrows H^+ + CN^-$

Equilibrium concentrations:

$[H^+] = [CN^-] = 8.9 \times 10^{-6}$ M

$[HCN] = 0.20 - 8.9 \times 10^{-6} = 0.20$ M (neglecting $8.9 \times 10^{-6}$)

$$K_a = \frac{[H^+][CN^-]}{[HCN]} = \frac{(8.9 \times 10^{-6})^2}{0.20} = 4.0 \times 10^{-10}$$

81. Let $y$ = M $CaF_2$ dissolving

$$\begin{array}{ccccc} CaF_2(s) & \leftrightarrows & Ca^{2+} & + & 2\,F^- \\ y & & y & & 2y \end{array}$$

(a) $K_{sp} = [Ca^{2+}][F^-]^2 = (y)(2y)^2 = 4y^3 = 3.9 \times 10^{-11}$

$$y = \sqrt[3]{\frac{3.9 \times 10^{-11}}{4}} = 2.1 \times 10^{-4} \text{ M } (CaF_2 \text{ dissolved})$$

$$\left(\frac{2.1 \times 10^{-4} \text{ mol } CaF_2}{L}\right)\left(\frac{1 \text{ mol } Ca^{2+}}{1 \text{ mol } CaF_2}\right) = 2.1 \times 10^{-4} \text{ M } Ca^{2+}$$

$$\left(\frac{2.1 \times 10^{-4} \text{ mol } CaF_2}{L}\right)\left(\frac{2 \text{ mol } F^-}{1 \text{ mol } CaF_2}\right) = 4.2 \times 10^{-4} \text{ M } F^-$$

(b) $$\left(\frac{2.1 \times 10^{-4} \text{ mol } CaF_2}{L}\right)(0.500 \text{ L})\left(\frac{78.08 \text{ g}}{\text{mol}}\right) = 8.2 \times 10^{-3} \text{ g } CaF_2$$

82. The molar concentrations of ions, after mixing, are calculated and these concentrations are substituted into the equilibrium expression. The value obtained is compared to the $K_{sp}$ of the salt. If the value is greater than the $K_{sp}$, precipitation occurs. If the value is less than the $K_{sp}$, no precipitation occurs.

(a) $100.\text{ mL } 0.010\text{ M } Na_2SO_4 \rightarrow 100.\text{ mL } 0.010\text{ M } SO_4^{2-}$

$100.\text{ mL } 0.001\text{ M } Pb(NO_3)_2 \rightarrow 100.\text{ mL } 0.001\text{ M } Pb^{2+}$

Volume after mixing = 200. mL

Concentrations after mixing: $SO_4^{2-} = 0.0050\text{ M}$ $Pb^{2+} = 0.0005\text{ M}$

$[Pb^{2+}][SO_4^{2-}] = (5.0 \times 10^{-3})(5 \times 10^{-4}) = 3 \times 10^{-6}$

$K_{sp} = 1.3 \times 10^{-8}$ which is less than $3 \times 10^{-6}$, therefore, precipitation occurs.

(b) $50.0\text{ mL } 1.0 \times 10^{-4}\text{ M } AgNO_3 \rightarrow 50.0\text{ mL } 1.0 \times 10^{-4}\text{ M } Ag^+$

$100.\text{ mL } 1.0 \times 10^{-4}\text{ M NaCl} \rightarrow 100.\text{ mL } 1.0 \times 10^{-4}\text{ M } Cl^-$

Volume after mixing = 150. mL

Concentrations after mixing:

$$(1.0 \times 10^{-4})\left(\frac{50.0\text{ mL}}{150.\text{ mL}}\right) = 3.3 \times 10^{-5}\text{ M } Ag^+$$

$$(1.0 \times 10^{-4})\left(\frac{100.\text{ mL}}{150.\text{ mL}}\right) = 6.7 \times 10^{-5}\text{ M } Cl^-$$

$[Ag^+][Cl^-] = (3.3 \times 10^{-5})(6.7 \times 10^{-5}) = 2.2 \times 10^{-9}$

$K_{sp} = 1.7 \times 10^{-10}$ which is less than $2.2 \times 10^{-9}$, therefore, precipitation occurs.

(c) Convert g $Ca(NO_3)_2$ to g $Ca^{2+}$

$$\left(\frac{1.0\text{ g } Ca(NO_3)_2}{0.150\text{ L}}\right)\left(\frac{1\text{ mol}}{164.1\text{ g}}\right)\left(\frac{1\text{ mol } Ca^{2+}}{1\text{ mol } Ca(NO_2)_2}\right) = 0.041\text{ M } Ca^{2+}$$

$250\text{ mL } 0.01\text{ M NaOH} \rightarrow 250\text{ mL } 0.01\text{ M } OH^-$

Final volume = $4.0 \times 10^2$ mL

Concentration after mixing:

$$(0.041\text{ M } Ca^{2+})\left(\frac{150\text{ mL}}{4.0 \times 10^2\text{ mL}}\right) = 0.015\text{ M } Ca^{2+}$$

$$(0.01\text{ M } OH^-)\left(\frac{250\text{ mL}}{4.0 \times 10^2\text{ mL}}\right) = 0.0063\text{ M } OH^-$$

$[Ca^{2+}][OH^-]^2 = (0.015)(0.0063)^2 = 6.0 \times 10^{-7}$

$K_{sp} = 1.3 \times 10^{-6}$ which is greater than $6.0 \times 10^{-7}$, therefore, no precipitation occurs.

83. With a known $Ba^{2+}$ concentration, the $SO_4^{2-}$ concentration can be calculated using the $K_{sp}$ value. $BaSO_4(s) \leftrightarrows Ba^{2+} + SO_4^{2-}$

$K_{sp} = [Ba^{2+}][SO_4^{2-}] = 1.5 \times 10^{-9}$ $Ba^{2+} = 0.050$ M

(a) $$[SO_4^{2-}] = \frac{K_{sp}}{[Ba^{2+}]} = \frac{1.5 \times 10^{-9}}{0.050} = 3.0 \times 10^{-8} \text{ M } SO_4^{2-} \text{ in solution}$$

(b) M $SO_4^{2-}$ = M $BaSO_4$ in solution

$$\left(\frac{3.0 \times 10^{-8} \text{ mol } BaSO_4}{\text{L}}\right)(0.100 \text{ L})\left(\frac{233.4 \text{ g}}{\text{mol}}\right) = 7.0 \times 10^{-7} \text{ g } BaSO_4 \text{ remain in solution}$$

84. If $[Pb^{2+}][Cl^-]^2$ exceeds the $K_{sp}$, precipitation will occur.

$K_{sp} = [Pb^{2+}][Cl^-]^2 = 2.0 \times 10^{-5}$

0.050 M $Pb(NO_3)_2 \rightarrow$ 0.050 M $Pb^{2+}$

0.010 M NaCl $\rightarrow$ 0.010 M $Cl^-$

$(0.050)(0.010)^2 = 5.0 \times 10^{-6}$

$[Pb^{2+}][Cl^-]^2$ is smaller than the $K_{sp}$ value. Therefore, no precipitate of $PbCl_2$ will form.

85. $[Ba^{2+}][SO_4^{2-}] = 1.5 \times 10^{-9}$ $[Sr^{2+}][SO_4^{2-}] = 3.5 \times 10^{-7}$

Both cations are present in equal concentrations (0.10 M). Therefore, as $SO_4^{2-}$ is added, the $K_{sp}$ of $BaSO_4$ will be exceeded before that of $SrSO_4$. $BaSO_4$ precipitates first.

86. $2\,SO_2(g) + O_2(g) \leftrightarrows 2\,SO_3(g)$

$$K_{eq} = \frac{[SO_3]^2}{[SO_2]^2[O_2]} = \frac{(11.0)^2}{(4.20)^2(0.60 \times 10^{-3})} = 1.1 \times 10^4$$

87. $$(0.048 \text{ g } BaF_2)\left(\frac{1 \text{ mol}}{175.3 \text{ g}}\right) = 2.7 \times 10^{-4} \text{ mol } BaF_2$$

$$\left(\frac{2.7 \times 10^{-4} \text{ mol}}{0.015 \text{ L}}\right) = 1.8 \times 10^{-2} \text{ M } BaF_2 \text{ dissolved}$$

| $BaF_2(s)$ | $\leftrightarrows$ | $Ba^{2+}$ | + | $2\,F^-$ | |
|---|---|---|---|---|---|
| $1.8 \times 10^{-2}$ | | $1.8 \times 10^{-2}$ | | $2(1.8 \times 10^{-2})$ | (molar concentration) |

$K_{sp} = [Ba^{2+}][F^-]^2 = (1.8 \times 10^{-2})(3.6 \times 10^{-2})^2 = 2.3 \times 10^{-5}$

88. $N_2 + 3\,H_2 \leftrightarrows 2\,NH_3$

$$K_{eq} = \frac{[NH_3]^2}{[N_2][H_2]^3} = 4.0 \qquad \text{Let } y = [NH_3]$$

$$4.0 = \frac{y^2}{(2.0)(2.0)^3} \qquad y^2 = 64 \qquad y = \sqrt{64}$$

$$y = 8.0\text{ M} = [NH_3]$$

89. Total volume of mixture = 40.0 mL (0.0400 L)

$$K_{sp} = [Sr^{2+}][SO_4^{2-}] = 7.6 \times 10^{-7}$$

$$[Sr^{2+}] = \frac{(1.0 \times 10^{-3}\text{ M})(0.0250\text{ L})}{0.0400\text{ L}} = 6.3 \times 10^{-4}\text{ M}$$

$$[SO_4^{2-}] = \frac{(2.0 \times 10^{-3}\text{ M})(0.0150\text{ L})}{0.0400\text{ L}} = 7.5 \times 10^{-4}\text{ M}$$

$$[Sr^{2+}][SO_4^{2-}] = (6.3 \times 10^{-4})(7.5 \times 10^{-4}) = 4.7 \times 10^{-7}$$

$4.7 \times 10^{-7} < 7.6 \times 10^{-7}$ no precipitation should occur.

90. First change $g\ Hg_2I_2 \rightarrow mol\ Hg_2I_2$

$$\left(\frac{3.04 \times 10^{-7}\text{ g } Hg_2I_2}{\text{L}}\right)\left(\frac{1\text{ mol}}{655.0\text{ g}}\right) = 4.64 \times 10^{-10}\text{ M } Hg_2I_2 \quad \text{(molar solubility)}$$

| $Hg_2I_2$ | $\leftrightarrows$ | $Hg_2^{2+}$ | + | $2\,I^-$ |
|---|---|---|---|---|
| | | $4.64 \times 10^{-10}$ M | | $2(4.64 \times 10^{-10}$ M) |

$$K_{sp} = [Hg_2^{2+}][I^-]^2 = (4.64 \times 10^{-10})(9.28 \times 10^{-10})^2 = 4.00 \times 10^{-28}$$

91. $3\,O_2(g) + \text{heat} \leftrightarrows 2\,O_3(g)$

Three ways to increase ozone

(a) increase heat

(b) increase amount of $O_2$

(c) increase pressure

(d) remove $O_3$ as it is made

92. $H_2O(l) \leftrightarrows H_2O(g)$

Conditions on the second day
(a) the temperature could have been cooler
(b) the humidity in the air could have been higher
(c) the air pressure could have been greater

93. Treat this as an equilibrium where W = whole nuts, S = shell halves, and K = kernels

| W | $\leftrightarrows$ | 2 S | + | K | |
|---|---|---|---|---|---|
| 144 | | 0 | | 0 | amount before cracking |
| $144 - x$ | | $2x$ | | $x$ | $x$ = number of kernels after cracking |

$144 - x + 2x + x = 194$ total pieces

$144 + 2x = 194; \quad 2x = 50$

$x = 25$ kernels; 50 shell halves; 119 whole nuts left

$$K_{eq} = \frac{(2x)^2(x)}{144 - x} = \frac{(50)^2(25)}{119} = 5.3 \times 10^2$$

94. $CO(g) + H_2O(g) \leftrightarrows CO_2(g) + H_2(g)$

(c) is the correct answer

$$K_{eq} = \frac{[CO_2][H_2]}{[CO][H_2O]} = 1$$

With equal concentrations of products and reactants, the $K_{eq}$ value will equal 1.

95. (a) $K_{eq} = \dfrac{[O_3]^2}{[O_2]^3}$

(b) $K_{eq} = \dfrac{[H_2O(l)]}{[H_2O(g)]}$

(c) $K_{eq} = \dfrac{[MgO][CO_2]}{[MgCO_3]}$

(d) $K_{eq} = \dfrac{[Bi_2S_3][H^+]^6}{[Bi^{3+}]^2[H_2S]^3}$

96.

| 2A | + | B | $\leftrightarrows$ | C | |
|---|---|---|---|---|---|
| 1.0 M | | 1.0 M | | 0 | Initial conditions |
| 1.0 - 2(0.30) | | 1.0 - 0.30 | | 0.30 | Equilibrium concentrations |
| 0.4 M | | 0.7 M | | 0.30 M | |

$$K_{eq} = \frac{[C]}{[A]^2[B]} = \frac{0.30}{(0.4)^2(0.7)} = 3$$

97. Since the second reaction is the reverse of the first, the $K_{eq}$ value of the second reaction will be the reciprocal of the $K_{eq}$ value of the first reaction.

$$K_{eq} = \frac{[I_2][Cl_2]}{[ICl]^2} = 2.2 \times 10^{-3} \quad \text{(first reaction)}$$

$$K_{eq} = \frac{[ICl]^2}{[I_2][Cl_2]} \qquad K_{eq} = \frac{1}{2.2 \times 10^{-3}} = 450$$

98. $HNO_2(aq) \leftrightarrows H^+(aq) + NO_2^-(aq)$

$OH^-$ reacts with $H^+$ and equilibrium shifts to the right.

(a) After an initial increase, $[OH^-]$ will be neutralized and equilibrium shifts to the right.

(b) $[H^+]$ will be reduced (reacts with $OH^-$). Equilibrium shifts to the right.

(c) $[NO_2^-]$ increases as equilibrium shifts to the right.

(d) $[HNO_2]$ decreases and equilibrium shifts to the right

99.

| $SO_2(g)$ | + | $NO_2(g)$ | $\leftrightarrows$ | $SO_3(g)$ | + | $NO(g)$ | |
|---|---|---|---|---|---|---|---|
| 0.50 M | | 0.50 M | | 0 | | 0 | Initial conditions |
| $0.50 - x$ | | $0.50 - x$ | | $x$ | | $x$ | Equilibrium concentrations |

$$K_{eq} = \frac{[SO_3][NO]}{[SO_2][NO_2]} = \frac{x^2}{(0.50 - x)^2} = 90.$$

Take the square root of both sides

$$\frac{x}{0.50 - x} = 9.5 \qquad x = 0.45 \text{ M}$$

$[SO_3] = [NO] = 0.45$ M

$[SO_2] = [NO_2] = 0.05$ M

100. $CaSO_4(s) \leftrightarrows Ca^{2+}(aq) + SO_4^{2-}(aq)$

$K_{sp} = [Ca^{2+}][SO_4^{2-}] = 2.0 \times 10^{-4}$

Let $x$ = moles $CaSO_4$ that dissolve/L = $[Ca^{2+}] = [SO_4^{2-}]$

$(x)(x) = 2.0 \times 10^{-4} \qquad x = \sqrt{2.0 \times 10^{-4}}$

$x = 0.014$ M $CaSO_4$

M → moles → grams

$$\left(\frac{0.014 \text{ mol } CaSO_4}{\text{L}}\right)(0.600 \text{ L})\left(\frac{136.2 \text{ g}}{\text{mol}}\right) = 1.1 \text{ g } CaSO_4$$

101. $PbF_2(s) \leftrightarrows Pb^{2+} + 2\,F^-$

change g $PbF_2$ → mol $PbF_2$

$$\left(\frac{0.098 \text{ g } PbF_2}{0.400 \text{ L}}\right)\left(\frac{1 \text{ mol}}{245.2 \text{ g}}\right) = 1.0 \times 10^{-3} \text{ mol/L} = 1.0 \times 10^{-3} \text{ M } PbF_2$$

$K_{sp} = (Pb^{2+})(F^-)^2$

$[Pb^{2+}] = 1.0 \times 10^{-3}$; $[F^-] = 2(1.0 \times 10^{-3}) = 2.0 \times 10^{-3}$

$K_{sp} = (1.0 \times 10^{-3})(2.0 \times 10^{-3})^2 = 4.0 \times 10^{-9}$

CHAPTER 17

# OXIDATION–REDUCTION

1. (a) Iodine is oxidized. Its oxidation number increases from 0 to +5.

   (b) Chlorine is reduced. Its oxidation number decreases from 0 to – 1.

2. The higher metal on the list is more reactive.

   (a) Al (b) Ba (c) Ni

3. If the free element is higher on the list than the ion with which it is paired, the reaction occurs.

   (a) Yes. $Zn(s) + Cu^{2+}(aq) \rightarrow Zn^{2+}(aq) + Cu(s)$

   (b) No reaction

   (c) Yes. $Sn(s) + 2\ Ag^{+}(aq) \rightarrow Sn^{2+}(aq) + 2\ Ag(s)$

   (d) No reaction

   (e) Yes. $Ba(s) + FeCl_2(aq) \rightarrow BaCl_2(aq) + Fe(s)$

   (f) No reaction

   (g) Yes. $Ni(s) + Hg(NO_3)_2(aq) \rightarrow Ni(NO_3)_2(aq) + Hg(l)$

   (h) Yes. $2\ Al(s) + 3\ CuSO_4(aq) \rightarrow Al_2(SO_4)_3(aq) + 3\ Cu(s)$

4. (a) $2\ Al + Fe_2O_3 \rightarrow Al_2O_3 + 2\ Fe + \text{Heat}$

   (b) Al is above Fe in the activity series, which indicates Al is more active than Fe.

   (c) No. Iron is less active than aluminum and will not displace aluminum from its compounds.

   (d) Yes. Aluminum is above chromium in the activity series and will displace $Cr^{3+}$ from its compounds.

5. (a) $2\ Al(s) + 6\ HCl(aq) \rightarrow 2\ AlCl_3(aq) + 3\ H_2(g)$

   $2\ Al(s) + 3\ H_2SO_4(aq) \rightarrow Al_2(SO_4)_3(aq) + 3\ H_2(g)$

   (b) $2\ Cr(s) + 6\ HCl(aq) \rightarrow 2\ CrCl_3(aq) + 3\ H_2(g)$

   $2\ Cr(s) + 3\ H_2SO_4(aq) \rightarrow Cr_2(SO_4)_3(aq) + 3\ H_2(g)$

(c) $Au(s) + HCl(aq) \rightarrow$ no reaction

$Au(s) + H_2SO_4(aq) \rightarrow$ no reaction

(d) $Fe(s) + 2\,HCl(aq) \rightarrow FeCl_2(aq) + H_2(g)$

$Fe(s) + H_2SO_4(aq) \rightarrow FeSO_4(aq) + H_2(g)$

(e) $Cu(s) + HCl(aq) \rightarrow$ no reaction

$Cu(s) + H_2SO_4(aq) \rightarrow$ no reaction

(f) $Mg(s) + 2\,HCl(aq) \rightarrow MgCl_2(aq) + H_2(g)$

$Mg(s) + H_2SO_4(aq) \rightarrow MgSO_4(aq) + H_2(g)$

(g) $Hg(l) + HCl(aq) \rightarrow$ no reaction

$Hg(l) + H_2SO_4(aq) \rightarrow$ no reaction

(h) $Zn(s) + 2\,HCl(aq) \rightarrow ZnCl_2(aq) + H_2(g)$

$Zn(s) + H_2SO_4(aq) \rightarrow ZnSO_4(aq) + H_2(g)$

6. (a) Oxidation occurs at the anode. The reaction is

$2\,Cl^-(aq) \rightarrow Cl_2(g) + 2\,e^-$

(b) Reduction occurs at the cathode. The reaction is

$Ni^{2+}(aq) + 2\,e^- \rightarrow Ni(s)$

(c) The net chemical reaction is

$$Ni^{2+}(aq) + 2\,Cl^-(aq) \xrightarrow[\text{energy}]{\text{electrical}} Ni(s) + Cl_2(g)$$

7. In Figure 17.3, electrical energy is causing chemical reactions to occur. In Figure 17.4, chemical reactions are used to produce electrical energy.

8. (a) It would not be possible to monitor the voltage produced, but the reactions in the cell would still occur.

(b) If the salt bridge were removed, the reaction would stop. Ions must be mobile to maintain an electrical neutrality of ions in solution. The two solutions would be isolated with no complete electrical circuit.

9. Oxidation and reduction are complementary processes because one does not occur without the other. The loss of $e^-$ in oxidation is accompanied by a gain of $e^-$ in reduction.

10. $Ca^{2+} + 2\,e^- \rightarrow Ca$ cathode reaction, reduction

    $2\,Br^- \rightarrow Br_2 + 2\,e^-$ anode reaction, oxidation

11. During electroplating of metals, the metal is plated by reducing the positive ions of the metal in the solution. The plating will occur at the cathode, the source of the electrons. With an alternating current, the polarity of the electrode would be constantly changing, so at one instant the metal would be plating and the next instant the metal would be dissolving.

12. Since lead dioxide and lead(II) sulfate are insoluble, it is unnecessary to have salt bridges in the cells of a lead storage battery.

13. The electrolyte in a lead storage battery is dilute sulfuric acid. In the discharge cycle, $SO_4^{2-}$, is removed from solution as it reacts with $PbO_2$ and $H^+$ to form $PbSO_4(s)$ and $H_2O$. Therefore, the electrolyte solution contains less $H_2SO_4$ and becomes less dense.

14. If $Hg^{2+}$ ions are reduced to metallic mercury, this would occur at the cathode, because reduction takes place at the cathode.

15. In both electrolytic and voltaic cells, oxidation and reduction reactions occur. In an electrolytic cell an electric current is forced through the cell causing a chemical change to occur. In voltaic cells, spontaneous chemical changes occur, generating an electric current.

16. In some voltaic cells, the reactants at the electrodes are in solution. For the cell to function, these reactants must be kept separated. A salt bridge permits movement of ions in the cell. This keeps the solution neutral with respect to the charged particles (ions) in the solution.

17. The oxidation number of the underlined element is indicated by the number following the formula.

| | | | | | | | | |
|---|---|---|---|---|---|---|---|---|
| (a) | $\underline{Na}Cl$ | +1 | (c) | $\underline{Pb}O_2$ | +4 | (e) | $H_2\underline{S}O_3$ | +4 |
| (b) | $Fe\underline{Cl}_3$ | -1 | (d) | $Na\underline{N}O_3$ | +5 | (f) | $\underline{N}H_4Cl$ | -3 |

18. The oxidation number of the underlined element is indicated by the number following the formula.

| | | | | | | | | |
|---|---|---|---|---|---|---|---|---|
| (a) | $K\underline{Mn}O_4$ | +7 | (c) | $\underline{N}H_3$ | -3 | (e) | $K_2\underline{Cr}O_4$ | +6 |
| (b) | $\underline{I}_2$ | 0 | (d) | $K\underline{Cl}O_3$ | +5 | (f) | $K_2\underline{Cr}_2O_7$ | +6 |

19. The oxidation number of the underlined element is indicated by the number following the formula.

(a) $\underline{S}^{2-}$ −2 (c) $Na_2\underline{O}_2$ −1

(b) $\underline{N}O_2^-$ +3 (d) $\underline{Bi}^{3+}$ +3

20. The oxidation number of the underlined element is indicated by the number following the formula.

(a) $\underline{O}_2$ 0 (c) $Fe(\underline{O}H)_3$ −2

(b) $\underline{As}O_4^{3-}$ +5 (d) $\underline{I}O_3^-$ +5

21.

| | Balanced half-reaction | Changing Element | Type of reaction |
|---|---|---|---|
| (a) | $Zn^{2+} + 2\,e^- \rightarrow Zn$ | Zn | reduction |
| (b) | $2\,Br^- \rightarrow Br_2 + 2\,e^-$ | Br | oxidation |
| (c) | $MnO_4^- + 8\,H^+ + 5\,e^- \rightarrow Mn^{2+} + 4\,H_2O$ | Mn | reduction |
| (d) | $Ni \rightarrow Ni^{2+} + 2\,e^-$ | Ni | oxidation |

22.

| | Balanced reactions | Changing Element | Type of reaction |
|---|---|---|---|
| (a) | $SO_3^{2-} + H_2O \rightarrow SO_4^{2-} + 2\,H^+ + 2\,e^-$ | S | oxidation |
| (b) | $NO_3^- + 4\,H^+ + 3\,e^- \rightarrow NO + 2\,H_2O$ | N | reduction |
| (c) | $S_2O_4^{2-} + 2\,H_2O \rightarrow 2\,SO_3^{2-} + 4\,H^+ + 2\,e^-$ | S | oxidation |
| (d) | $Fe^{2+} \rightarrow Fe^{3+} + 1\,e^-$ | Fe | oxidation |

23. (1) $Cr + HCl \rightarrow CrCl_3 + H_2$

(a) Cr is oxidized, H is reduced

(b) HCl is the oxidizing agent, Cr the reducing agent

(2) $SO_4^{2-} + I^- + H^+ \rightarrow H_2S + I_2 + H_2O$

(a) I is oxidized, S is reduced

(b) $SO_4^{2-}$ is the oxidizing agent, $I^-$ the reducing agent

24. (1) $AsH_3 + Ag^+ + H_2O \rightarrow H_3AsO_4 + Ag + H^+$

(a) As is oxidized, Ag is reduced

(b) $Ag^+$ is the oxidizing agent, $AsH_3$ the reducing agent

(2) $Cl_2 + NaBr \rightarrow NaCl + Br_2$

(a) Br is oxidized, Cl is reduced

(b) $Cl_2$ is the oxidizing agent, NaBr the reducing agent

25. Balancing oxidation-reduction equations

(a) $Zn + S \rightarrow ZnS$

ox $Zn^0 \rightarrow Zn^{2+} + 2\,e^-$ Add half reactions

red $S^0 + 2\,e^- \rightarrow S^{2-}$ the $2\,e^-$ cancel

$Zn + S \rightarrow ZnS$

(b) $AgNO_3 + Pb \rightarrow Pb(NO_3)_2 + Ag$

ox $Pb^0 \rightarrow Pb^{2+} + 2\,e^-$

red $Ag^+ + 1\,e^- \rightarrow Ag^0$ Multiply by 2, add the half

$Pb + 2\,Ag^+ \rightarrow Pb^{2+} + 2\,Ag$ reactions, the $2\,e^-$ cancel

Transfer the coefficients to the original equation and complete the balancing by inspection.

$2\,AgNO_3 + Pb \rightarrow Pb(NO_3)_2 + 2\,Ag$

(c) $Fe_2O_3 + CO \rightarrow Fe + CO_2$

ox $C^{2+} \rightarrow C^{4+} + 2\,e^-$ Multiply by 3

red $Fe^{3+} + 3\,e^- \rightarrow Fe^0$ Multiply by 2, add, the $6\,e^-$ cancel

$3\,C^{2+} + 2\,Fe^{3+} \rightarrow 3\,C^{4+} + 2\,Fe$

Transfer the coefficients to the original equation (the coefficient 2 in front of the $Fe^{3+}$ becomes the subscript 2 in $Fe_2O_3$). Complete the balancing by inspection.

$Fe_2O_3 + 3\,CO \rightarrow 2\,Fe + 3\,CO_2$

(d) $H_2S + HNO_3 \rightarrow S + NO + H_2O$

$S^{2-} \rightarrow S^0 + 2\,e^-$ Multiply by 3

$N^{5+} + 3\,e^- \rightarrow N^{2+}$ Multiply by 2, add, the 6 $e^-$ cancel

$3\,S^{2-} + 2\,N^{5+} \rightarrow 3\,S + 2\,N^{2+}$

Transfer the coefficients to the original equations and complete the balancing by inspection.

$3\,H_2S + 2\,HNO_3 \rightarrow 3\,S + 2\,NO + 4\,H_2O$

(e) $MnO_2 + HBr \rightarrow MnBr_2 + Br_2 + H_2O$

$Br^- \rightarrow Br^0 + 1\,e^-$ Multiply by 2

$Mn^{4+} + 2\,e^- \rightarrow Mn^{2+}$ Add equations and the 2 $e^-$ cancel

$Mn^{4+} + 2\,Br^- \rightarrow Mn^{2+} + 2\,Br^0$

Transfer the coefficients to the original equation. The coefficient 2 in front of the $Br^-$ becomes the subscript 2 in the $Br_2$. Also, 2 more $Br^-$ ions are required to account for the 2 $Br^-$ ions that do not change oxidation numbers. These 2 are part of the compound $MnBr_2$.

$MnO_2 + 4\,HBr \rightarrow MnBr_2 + Br_2 + 2\,H_2O$

26. (a) Balancing oxidation-reduction equations

$Cl_2 + KOH \rightarrow KCl + KClO_3 + H_2O$

$Cl^0 \rightarrow Cl^{5+} + 5\,e^-$

$Cl^0 + e^- \rightarrow Cl^-$ Multiply by 5, add, the 5 $e^-$ cancel

$3\,Cl_2 \rightarrow Cl^{5+} + 5\,Cl^-$ 6 $Cl^0$ becomes 3 $Cl_2$

Transfer the coefficients to the original equations and complete the balancing by inspection.

$3\,Cl_2 + 6\,KOH \rightarrow KClO_3 + 5\,KCl + 3\,H_2O$

(b) $Ag + HNO_3 \rightarrow AgNO_3 + NO + H_2O$

$Ag^0 \rightarrow Ag^+ + e^-$ Multiply by 3, add, the

$N^{5+} + 3\,e^- \rightarrow N^{2+}$ 3 $e^-$ cancel

$3\,Ag + N^{5+} \rightarrow 3\,Ag^+ + N^{2+}$

Transfer the coefficients to the original equations and complete the balancing by inspection.

$3\,Ag + 4\,HNO_3 \rightarrow 3\,AgNO_3 + NO + 2\,H_2O$

(c) $CuO + NH_3 \rightarrow N_2 + Cu + H_2O$

$N^{3-} \rightarrow N^0 + 3\,e^-$ Multiply by 2

$Cu^{2+} + 2\,e^- \rightarrow Cu^0$ Multiply by 3, add, the 6 $e^-$ cancel

$2\,N^{3-} + 3\,Cu^{2+} \rightarrow N_2 + 3\,Cu$

Transfer the coefficients to the original equations and complete the balancing by inspection.

$3\,CuO + 2\,NH_3 \rightarrow N_2 + 3\,Cu + 3\,H_2O$

(d) $PbO_2 + Sb + NaOH \rightarrow PbO + NaSbO_2 + H_2O$

$Sb^0 \rightarrow Sb^{3+} + 3\,e^-$ Multiply by 2

$Pb^{4+} + 2\,e^- \rightarrow 2\,Pb^{2+}$ Multiply by 3, add, the 6 $e^-$ cancel

$2\,Sb + 3\,Pb^{4+} \rightarrow 2\,Sb^{3+} + 3\,Pb^{2+}$

Transfer the coefficients to the original equations and complete the balancing by inspection.

$3\,PbO_2 + 2\,Sb + 2\,NaOH \rightarrow 3\,PbO + 2\,NaSbO_2 + H_2O$

(e) $H_2O_2 + KMnO_4 + H_2SO_4 \rightarrow O_2 + MnSO_4 + K_2SO_4 + H_2O$

$O_2^{2-} \rightarrow O_2^0 + 2\,e^-$ Multiply by 5

$Mn^{7+} + 5\,e^- \rightarrow Mn^{2+}$ Multiply by 2, add, the 10 $e^-$ cancel

$5\,O_2^{2-} + 2\,Mn^{7+} \rightarrow 5\,O_2 + 2\,Mn^{2+}$

Transfer the coefficients to the original equations and complete the balancing by inspection.

$5\,H_2O_2 + 2\,KMnO_4 + 3\,H_2SO_4 \rightarrow 5\,O_2 + 2\,MnSO_4 + K_2SO_4 + 8\,H_2O$

27. (a) $Zn + NO_3^- \rightarrow Zn^{2+} + NH_4^+$ (acidic solution)

Step 1 Write half-reaction equations. Balance except H and O.

$Zn \rightarrow Zn^{2+}$

$NO_3^- \rightarrow NH_4^+$

Step 2 Balance H and O using $H_2O$ and $H^+$

$Zn \rightarrow Zn^{2+}$

$10\ H^+ + NO_3^- \rightarrow NH_4^+ + 3\ H_2O$

Step 3 Balance electrically with electrons

$Zn \rightarrow Zn^{2+} + 2\ e^-$

$10\ H^+ + NO_3^- + 8\ e^- \rightarrow NH_4^+ + 3\ H_2O$

Step 4 Equalize the loss and gain of electrons

$4\ (Zn \rightarrow Zn^{2+} + 2\ e^-)$

$10\ H^+ + NO_3^- + 8\ e^- \rightarrow NH_4^+ + 3\ H_2O$

Step 5 Add the half reactions – electrons cancel

$10\ H^+ + 4\ Zn + NO_3^- \rightarrow 4\ Zn^{2+} + NH_4^+ + 3\ H_2O$

(b) $NO_3^- + S \rightarrow NO_2 + SO_4^{2-}$ (acidic solution)

Step 1 Write half-reaction equations. Balance except H and O.

$S \rightarrow SO_4^{2-}$

$NO_3^- \rightarrow NO_2$

Step 2 Balance H and O using $H_2O$ and $H^+$

$4\ H_2O + S \rightarrow SO_4^{2-} + 8\ H^+$

$2\ H^+ + NO_3^- \rightarrow NO_2 + H_2O$

Step 3 Balance electrically with electrons

$4\ H_2O + S \rightarrow SO_4^{2-} + 8\ H^+ + 6\ e^-$

$2\ H^+ + NO_3^- + e^- \rightarrow NO_2 + H_2O$

Step 4 and 5 Equalize the loss and gain of electrons; add the half-reactions

$4\ H_2O + S \rightarrow SO_4^{2-} + 8\ H^+ + 6\ e^-$

$6\ (2\ H^+ + NO_3^- + e^- \rightarrow NO_2 + H_2O)$

$4\ H^+ + S + 6\ NO_3^- \rightarrow 6\ NO_2 + SO_4^{2-} + 2\ H_2O$

$4\ H_2O$ and $8\ H^+$ canceled from each side

(c) $PH_3 + I_2 \rightarrow H_3PO_2 + I^-$ (acidic solution)

Step 1 Write half-reaction equations. Balance except H and O.

$PH_3 \rightarrow H_3PO_2$

$I_2 \rightarrow 2\ I^-$

Step 2 Balance H and O using $H_2O$ and $H^+$

$2\ H_2O + PH_3 \rightarrow H_3PO_2 + 4\ H^+$

$I_2 \rightarrow 2\ I^-$

Step 3 Balance electrically with electrons

$2\ H_2O + PH_3 \rightarrow H_3PO_2 + 4\ H^+ + 4\ e^-$

$I_2 + 2\ e^- \rightarrow 2\ I^-$

Step 4 and 5 Equalize the loss and gain of electrons; add the half-reactions

$2\ H_2O + PH_3 \rightarrow H_3PO_2 + 4\ H^+ + 4\ e^-$

$2\ (I_2 + 2\ e^- \rightarrow 2\ I^-)$

$PH_3 + 2\ H_2O + 2\ I_2 \rightarrow H_3PO_2 + 4\ I^- + 4\ H^+$

(d) $Cu + NO_3^- \rightarrow Cu^{2+} + NO$ (acidic solution)

Step 1 Write half-reaction equations. Balance except H and O.

$Cu \rightarrow Cu^{2+}$

$NO_3^- \rightarrow NO$

Step 2 Balance H and O using $H_2O$ and $H^+$

$Cu \rightarrow Cu^{2+}$

$4H^+ + NO_3^- \rightarrow NO + 2\ H_2O$

Step 3 Balance electrically with electrons

$Cu \rightarrow Cu^{2+} + 2\,e^-$

$4H^+ + NO_3^- + 3\,e^- \rightarrow NO + 2\,H_2O$

Step 4 and 5 Equalize the loss and gain of electrons; add the half-reactions

$3\,(Cu \rightarrow Cu^{2+} + 2\,e^-)$

$\underline{2\,(4\,H^+ + NO_3^- + 3\,e^- \rightarrow NO + 2\,H_2O)}$

$3\,Cu + 8\,H^+ + 2\,NO_3^- \rightarrow 3\,Cu^{2+} + 2\,NO + 4\,H_2O$

(e) $ClO_3^- + Cl^- \rightarrow Cl_2$ (acidic solution)

Step 1 Write half-reaction equations. Balance except H and O.

$Cl^- \rightarrow Cl^0$

$ClO_3^- \rightarrow Cl^0$

Step 2 Balance H and O using $H_2O$ and $H^+$

$Cl^- \rightarrow Cl^0$

$6\,H^+ + ClO_3^- \rightarrow Cl^0 + 3\,H_2O$

Step 3 Balance electrically with electrons

$Cl^- \rightarrow Cl^0 + e^-$

$6\,H^+ + ClO_3^- + 5\,e^- \rightarrow Cl^0 + 3\,H_2O$

Step 4 and 5 Equalize the loss and gain of electrons; add the half-reactions

$5\,(Cl^- \rightarrow Cl^0 + e^-)$

$\underline{6\,H^+ + ClO_3^- + 5\,e^- \rightarrow Cl^0 + 3\,H_2O}$

$6\,H^+ + ClO_3^- + 5\,Cl^- \rightarrow 3\,Cl_2 + 3\,H_2O$

28. (a) $ClO_3^- + I^- \rightarrow I_2 + Cl^-$ (acidic solution)

Step 1 Write half-reaction equations. Balance except H and O.

$2\,I^- \rightarrow I_2$

$ClO_3^- \rightarrow Cl^-$

Step 2 Balance H and O using $H_2O$ and $H^+$

$2\,I^- \rightarrow I_2$

$6\,H^+ + ClO_3^- \rightarrow Cl^- + 3\,H_2O$

Step 3 Balance electrically with electrons

$2\,I^- \rightarrow I_2 + 2\,e^-$

$6\,H^+ + ClO_3^- + 6\,e^- \rightarrow Cl^- + 3\,H_2O$

Step 4 and 5 Equalize the loss and gain of electrons; add the half-reactions

$3\,(2\,I^- \rightarrow I_2 + 2\,e^-)$

$\underline{6\,H^+ + ClO_3^- + 6\,e^- \rightarrow Cl^- + 3\,H_2O}$

$6\,H^+ + ClO_3^- + 6\,I^- \rightarrow 3\,I_2 + Cl^- + 3\,H_2O$

(b) $Cr_2O_7^{2-} + Fe^{2+} \rightarrow Cr^{3+} + Fe^{3+}$ (acidic solution)

Step 1 Write half-reaction equations. Balance except H and O.

$Fe^{2+} \rightarrow Fe^{3+}$

$Cr_2O_7^{2-} \rightarrow 2\,Cr^{3+}$

Step 2 Balance H and O using $H_2O$ and $H^+$

$Fe^{2+} \rightarrow Fe^{3+}$

$14\,H^+ + Cr_2O_7^{2-} \rightarrow 2\,Cr^{3+} + 7\,H_2O$

Step 3 Balance electrically with electrons

$Fe^{2+} \rightarrow Fe^{3+} + e^-$

$14\,H^+ + Cr_2O_7^{2-} + 6\,e^- \rightarrow 2\,Cr^{3+} + 7\,H_2O$

Step 4 and 5 Equalize the loss and gain of electrons; add the half-reactions

$6\,(Fe^{2+} \rightarrow Fe^{3+} + e^-)$

$\underline{14\,H^+ + Cr_2O_7^{2-} + 6\,e^- \rightarrow 2\,Cr^{3+} + 7\,H_2O}$

$14\,H^+ + Cr_2O_7^{2-} + 6\,Fe^{2+} \rightarrow 2\,Cr^{3+} + 6\,Fe^{3+} + 7\,H_2O$

(c) $MnO_4^- + SO_2 \rightarrow Mn^{2+} + SO_4^{2-}$ (acidic solution)

Step 1 Write half-reaction equations. Balance except H and O.

$SO_2 \rightarrow SO_4^{2-}$

$MnO_4^- \rightarrow Mn^{2+}$

Step 2 Balance H and O using $H_2O$ and $H^+$

$2\,H_2O + SO_2 \rightarrow SO_4^{2-} + 4\,H^+$

$8\,H^+ + MnO_4^- \rightarrow Mn^{2+} + 4\,H_2O$

Step 3 Balance electrically with electrons

$2\ H_2O + SO_2 \rightarrow SO_4^{2-} + 4\ H^+ + 2\ e^-$

$8\ H^+ + MnO_4^- + 5\ e^- \rightarrow Mn^{2+} + 4\ H_2O$

Step 4 and 5 Equalize the loss and gain of electrons; add the half-reactions

$5\ (2\ H_2O + SO_2 \rightarrow SO_4^{2-} + 4\ H^+ + 2\ e^-)$

$2\ (8\ H^+ + MnO_4^- + 5\ e^- \rightarrow Mn^{2+} + 4\ H_2O)$

$2\ H_2O + 2\ MnO_4^- + 5\ SO_2 \rightarrow 4\ H^+ + 2\ Mn^{2+} + 5\ SO_4^{2-}$

$8\ H_2O$, $16\ H^+$, and $10\ e^-$ canceled from each side

(d) $H_3AsO_3 + MnO_4^- \rightarrow H_3AsO_4 + Mn^{2+}$ (acidic solution)

Step 1 Write half-reaction equations. Balance except H and O.

$H_3AsO_3 \rightarrow H_3AsO_4$

$MnO_4^- \rightarrow Mn^{2+}$

Step 2 Balance H and O using $H_2O$ and $H^+$

$H_2O + H_3AsO_3 \rightarrow 2\ H^+ + H_3AsO_4$

$8\ H^+ + MnO_4^- \rightarrow Mn^{2+} + 4\ H_2O$

Step 3 Balance electrically with electrons

$H_2O + H_3AsO_3 \rightarrow 2\ H^+ + H_3AsO_4 + 2\ e^-$

$8\ H^+ + MnO_4^- + 5\ e^- \rightarrow Mn^{2+} + 4\ H_2O$

Step 4 and 5 Equalize the loss and gain of electrons; add the half-reactions

$5\ (H_2O + H_3AsO_3 \rightarrow 2\ H^+ + H_3AsO_4 + 2\ e^-)$

$2\ (8\ H^+ + MnO_4^- + 5\ e^- \rightarrow Mn^{2+} + 4\ H_2O)$

$6\ H^+ + 5\ H_3AsO_3 + 2\ MnO_4^- \rightarrow 5\ H_3AsO_4 + 2\ Mn^{2+} + 3\ H_2O$

$5\ H_2O$, $10\ H^+$, and $10\ e^-$ canceled from each side

(e) $Cr_2O_7^{2-} + H_3AsO_3 \rightarrow Cr^{3+} + H_3AsO_4$ (acidic solution)

Step 1 Write half-reaction equations. Balance except H and O.

$H_3AsO_3 \rightarrow H_3AsO_4$

$Cr_2O_7^{2-} \rightarrow 2\ Cr^{3+}$

Step 2 Balance H and O using $H_2O$ and $H^+$

$H_2O + H_3AsO_3 \rightarrow 2\,H^+ + H_3AsO_4$

$14\,H^+ + Cr_2O_7^{2-} \rightarrow 2\,Cr^{3+} + 7\,H_2O$

Step 3 Balance electrically with electrons

$H_2O + H_3AsO_3 \rightarrow 2\,H^+ + H_3AsO_4 + 2\,e^-$

$14\,H^+ + Cr_2O_7^{2-} + 6\,e^- \rightarrow 2\,Cr^{3+} + 7\,H_2O$

Step 4 and 5 Equalize the loss and gain of electrons; add the half-reactions

$3\,(H_2O + H_3AsO_3 \rightarrow 2\,H^+ + H_3AsO_4 + 2\,e^-)$

$\underline{14\,H^+ + Cr_2O_7^{2-} + 6\,e^- \rightarrow 2\,Cr^{3+} + 7\,H_2O}$

$8\,H^+ + Cr_2O_7^{2-} + 3\,H_3AsO_3 \rightarrow 2\,Cr^{3+} + 3\,H_3AsO_4 + 4\,H_2O$

$3\,H_2O$, $6\,H^+$, and $6\,e^-$ canceled from each side

29. (a) $Cl_2 + IO_3^- \rightarrow Cl^- + IO_4^-$ (basic solution)

Step 1 Write half-reaction equations. Balance except H and O.

$IO_3^- \rightarrow IO_4^-$

$Cl_2 \rightarrow 2\,Cl^-$

Step 2 Balance H and O using $H_2O$ and $H^+$

$H_2O + IO_3^- \rightarrow IO_4^- + 2\,H^+$

$Cl_2 \rightarrow 2\,Cl^-$

Step 3 Add $OH^-$ ions to both sides (same number as $H^+$ ions)

$2\,OH^- + H_2O + IO_3^- \rightarrow IO_4^- + 2\,H^+ + 2\,OH^-$

$Cl_2 \rightarrow 2\,Cl^-$

Step 4 Combine $H^+$ and $OH^-$ to form $H_2O$; cancel $H_2O$ where possible

$2\,OH^- + H_2O + IO_3^- \rightarrow IO_4^- + 2\,H_2O$

$Cl_2 \rightarrow 2\,Cl^-$

$2\,OH^- + IO_3^- \rightarrow IO_4^- + H_2O$

$Cl_2 \rightarrow 2\,Cl^-$

Step 5 Balance electrically with electrons

$2\,OH^- + IO_3^- \rightarrow IO_4^- + H_2O + 2\,e^-$

$Cl_2 + 2\,e^- \rightarrow 2\,Cl^-$

Step 6 Electron loss and gain is balanced

Step 7 Add half-reactions

$2\ OH^- + IO_3^- + Cl_2 \rightarrow IO_4^- + 2\ Cl^- + H_2O$

(b) $MnO_4^- + ClO_2^- \rightarrow MnO_2 + ClO_4^-$ (basic solution)

Step 1 Write half-reaction equations. Balance except H and O.

$ClO_2^- \rightarrow ClO_4^-$

$MnO_4^- \rightarrow MnO_2$

Step 2 Balance H and O using $H_2O$ and $H^+$

$2\ H_2O + ClO_2^- \rightarrow ClO_4^- + 4\ H^+$

$MnO_4^- + 4\ H^+ \rightarrow MnO_2 + 2\ H_2O$

Step 3 Add $OH^-$ ions to both sides (same number as $H^+$ ions)

$4\ OH^- + 2\ H_2O + ClO_2^- \rightarrow ClO_4^- + 4\ H^+ + 4\ OH^-$

$4\ OH^- + MnO_4^- + 4\ H^+ \rightarrow MnO_2 + 2\ H_2O + 4\ OH^-$

Step 4 Combine $H^+$ and $OH^-$ to form $H_2O$; cancel $H_2O$ where possible

$4\ OH^- + 2\ H_2O + ClO_2^- \rightarrow ClO_4^- + 4\ H_2O$

$4\ H_2O + MnO_4^- \rightarrow MnO_2 + 2\ H_2O + 4\ OH^-$

$4\ OH^- + ClO_2^- \rightarrow ClO_4^- + 2\ H_2O$

$2\ H_2O + MnO_4^- \rightarrow MnO_2 + 4\ OH^-$

Step 5 Balance electrically with electrons

$4\ OH^- + ClO_2^- \rightarrow ClO_4^- + 2\ H_2O + 4\ e^-$

$2\ H_2O + MnO_4^- + 3\ e^- \rightarrow MnO_2 + 4\ OH^-$

Step 6 and 7 Equalize gain and loss of electrons; add half-reactions

$3\ (4\ OH^- + ClO_2^- \rightarrow ClO_4^- + 2\ H_2O + 4\ e^-)$

$\underline{4\ (2\ H_2O + MnO_4^- + 3\ e^- \rightarrow MnO_2 + 4\ OH^-)}$

$2\ H_2O + 4\ MnO_4^- + 3\ ClO_2^- \rightarrow 4\ MnO_2 + 3\ ClO_4^- + 4\ OH^-$

$6\ H_2O$, $12\ OH^-$, and $12\ e^-$ canceled from each side

(c) $Se \rightarrow SeO_3^{2-} + Se^{2-}$ (basic solution)

Step 1 Write half-reaction equations. Balance except H and O.

$Se \rightarrow SeO_3^{2-}$

$Se \rightarrow Se^{2-}$

Step 2 Balance H and O using $H_2O$ and $H^+$

$3\,H_2O + Se \rightarrow SeO_3^{2-} + 6\,H^+$

$Se \rightarrow Se^{2-}$

Step 3 Add $OH^-$ ions to both sides (same number as $H^+$ ions)

$6\,OH^- + 3\,H_2O + Se \rightarrow SeO_3^{2-} + 6\,H^+ + 6\,OH^-$

$Se \rightarrow Se^{2-}$

Step 4 Combine $H^+$ and $OH^-$ to form $H_2O$; cancel $H_2O$ where possible

$6\,OH^- + 3\,H_2O + Se \rightarrow SeO_3^{2-} + 6\,H_2O$

$Se \rightarrow Se^{2-}$

$6\,OH^- + Se \rightarrow SeO_3^{2-} + 3\,H_2O$

Step 5 Balance electrically with electrons

$6\,OH^- + Se \rightarrow SeO_3^{2-} + 3\,H_2O + 4\,e^-$

$Se + 2\,e^- \rightarrow Se^{2-}$

Step 6 and 7 Equalize gain and loss of electrons; add half-reactions

$6\,OH^- + Se \rightarrow SeO_3^{2-} + 3\,H_2O + 4\,e^-$

$\underline{2\,(Se + 2\,e^- \rightarrow Se^{2-})}$

$6\,OH^- + 3\,Se \rightarrow SeO_3^{2-} + 2\,Se^{2-} + 3\,H_2O$

(d) $Fe_3O_4 + MnO_4^- \rightarrow Fe_2O_3 + MnO_2$ (basic solution)

Step 1 Write half-reaction equations. Balance except H and O.

$2\,Fe_3O_4 \rightarrow 3\,Fe_2O_3$

$MnO_4^- \rightarrow MnO_2$

Step 2 Balance H and O using $H_2O$ and $H^+$

$H_2O + 2\,Fe_3O_4 \rightarrow 3\,Fe_2O_3 + 2\,H^+$

$4\,H^+ + MnO_4^- \rightarrow MnO_2 + 2\,H_2O$

Step 3 Add $OH^-$ ions to both sides (same number as $H^+$ ions)

$2\ OH^- + H_2O + 2\ Fe_3O_4 \rightarrow 3\ Fe_2O_3 + 2\ H^+ + 2\ OH^-$

$4\ OH^- + 4\ H^+ + MnO_4^- \rightarrow MnO_2 + 2\ H_2O + 4\ OH^-$

Step 4 Combine $H^+$ and $OH^-$ to form $H_2O$; cancel $H_2O$ where possible

$2\ OH^- + H_2O + 2\ Fe_3O_4 \rightarrow 3\ Fe_2O_3 + 2\ H_2O$

$4\ H_2O + MnO_4^- \rightarrow MnO_2 + 2\ H_2O + 4\ OH^-$

$2\ OH^- + 2\ Fe_3O_4 \rightarrow 3\ Fe_2O_3 + H_2O$

$2\ H_2O + MnO_4^- \rightarrow MnO_2 + 4\ OH^-$

Step 5 Balance electrically with electrons

$2\ OH^- + 2\ Fe_3O_4 \rightarrow 3\ Fe_2O_3 + H_2O + 2\ e^-$

$2\ H_2O + MnO_4^- + 3\ e^- \rightarrow MnO_2 + 4\ OH^-$

Step 6 and 7 Equalize gain and loss of electrons; add half-reactions

$3\ (2\ OH^- + 2\ Fe_3O_4 \rightarrow 3\ Fe_2O_3 + H_2O + 2\ e^-)$

$\underline{2\ (2\ H_2O + MnO_4^- + 3\ e^- \rightarrow MnO_2 + 4\ OH^-)}$

$H_2O + 6\ Fe_3O_4 + 2\ MnO_4^- \rightarrow 9\ Fe_2O_3 + 2\ MnO_2 + 2\ OH^-$

$3\ H_2O$, $6\ OH^-$, and $6\ e^-$ canceled from each side

(e) $BrO^- + Cr(OH)_4^- \rightarrow Br^- + CrO_4^{2-}$ (basic solution)

Step 1 Write half-reaction equations. Balance except H and O.

$Cr(OH)_4^- \rightarrow CrO_4^{2-}$

$BrO^- \rightarrow Br^-$

Step 2 Balance H and O using $H_2O$ and $H^+$

$Cr(OH)_4^- \rightarrow CrO_4^{2-} + 4\ H^+$

$2\ H^+ + BrO^- \rightarrow Br^- + H_2O$

Step 3 Add $OH^-$ ions to both sides (same number as $H^+$ ions)

$4\ OH^- + Cr(OH)_4^- \rightarrow CrO_4^{2-} + 4\ H^+ + 4\ OH^-$

$2\ OH^- + 2\ H^+ + BrO^- \rightarrow Br^- + H_2O + 2\ OH^-$

Step 4 Combine $H^+$ and $OH^-$ to form $H_2O$; cancel $H_2O$ where possible

$4\ OH^- + Cr(OH)_4^- \rightarrow CrO_4^{2-} + 4\ H_2O$

$2\ H_2O + BrO^- \rightarrow Br^- + H_2O + 2\ OH^-$

$H_2O + BrO^- \rightarrow Br^- + 2\ OH^-$

Step 5 Balance electrically with electrons

$4\ OH^- + Cr(OH)_4^- \rightarrow CrO_4^{2-} + 4\ H_2O + 3\ e^-$

$H_2O + BrO^- + 2\ e^- \rightarrow Br^- + 2\ OH^-$

Step 6 and 7 Equalize gain and loss of electrons; add half-reactions

$2\ (4\ OH^- + Cr(OH)_4^- \rightarrow CrO_4^{2-} + 4\ H_2O + 3\ e^-)$

$3\ (H_2O + BrO^- + 2\ e^- \rightarrow Br^- + 2\ OH^-)$

$2\ OH^- + 3\ BrO^- + 2\ Cr(OH)_4^- \rightarrow 3\ Br^- + 2\ CrO_4^{2-} + 5\ H_2O$

$3\ H_2O$, $6\ OH^-$ and $6\ e^-$ canceled from each side

30. (a) $MnO_4^- + SO_3^{2-} \rightarrow MnO_2 + SO_4^{2-}$ (basic solution)

Step 1 Write half-reaction equations. Balance except H and O.

$SO_3^{2-} \rightarrow SO_4^{2-}$

$MnO_4^- \rightarrow MnO_2$

Step 2 Balance H and O using $H_2O$ and $H^+$

$H_2O + SO_3^{2-} \rightarrow SO_4^{2-} + 2\ H^+$

$MnO_4^- + 4\ H^+ \rightarrow MnO_2 + 2\ H_2O$

Step 3 Add $OH^-$ ions to both sides (same number as $H^+$ ions)

$2\ OH^- + H_2O + SO_3^{2-} \rightarrow SO_4^{2-} + 2\ H^+ + 2\ OH^-$

$4\ OH^- + MnO_4^- + 4\ H^+ \rightarrow MnO_2 + 3\ H_2O + 4\ OH^-$

Step 4 Combine $H^+$ and $OH^-$ to form $H_2O$; cancel $H_2O$ where possible

$2\ OH^- + H_2O + SO_3^{2-} \rightarrow SO_4^{2-} + 2\ H_2O$

$MnO_4^- + 4\ H_2O \rightarrow MnO_2 + 2\ H_2O + 4\ OH^-$

$2\ OH^- + SO_3^{2-} \rightarrow SO_4^{2-} + H_2O$

$MnO_4^- + 2\ H_2O \rightarrow MnO_2 + 4\ OH^-$

Step 5 Balance electrically with electrons

$2\ OH^- + SO_3^{2-} \rightarrow SO_4^{2-} + H_2O + 2\ e^-$

$3\ e^- + MnO_4^- + 2\ H_2O \rightarrow MnO_2 + 4\ OH^-$

Step 6 and 7 Equalize gain and loss of electrons; add half-reactions

$3\ (2\ OH^- + SO_3^{2-} \rightarrow SO_4^{2-} + H_2O + 2\ e^-)$

$2\ (MnO_4^- + 2\ H_2O + 3\ e^- \rightarrow MnO_2 + 4\ OH^-)$

$H_2O + 2\ MnO_4^- + 3\ SO_3^{2-} \rightarrow 2\ MnO_2 + 3\ SO_4^{2-} + 2\ OH^-$

$3\ H_2O$, $4\ OH^-$, and $6\ e^-$ canceled from each side

(b) $ClO_2 + SbO_2^- \rightarrow ClO_2^- + Sb(OH)_6^-$ (basic solution)

Step 1 Write half-reaction equations. Balance except H and O.

$SbO_2^- \rightarrow Sb(OH)_6^-$

$ClO_2 \rightarrow ClO_2^-$

Step 2 Balance H and O using $H_2O$ and $H^+$

$4\ H_2O + SbO_2^- \rightarrow Sb(OH)_6^- + 2\ H^+$

$ClO_2 \rightarrow ClO_2^-$

Step 3 Add $OH^-$ ions to both sides (same number as $H^+$ ions)

$2\ OH^- + 4\ H_2O + SbO_2^- \rightarrow Sb(OH)_6^- + 2\ H^+ + 2\ OH^-$

$ClO_2 \rightarrow ClO_2^-$

Step 4 Combine $H^+$ and $OH^-$ to form $H_2O$; cancel $H_2O$ where possible

$2\ OH^- + 4\ H_2O + SbO_2^- \rightarrow Sb(OH)_6^- + 2\ H_2O$

$ClO_2 \rightarrow ClO_2^-$

$2\ OH^- + 2\ H_2O + SbO_2^- \rightarrow Sb(OH)_6^-$

Step 5 Balance electrically with electrons

$2\ OH^- + 2\ H_2O + SbO_2^- \rightarrow Sb(OH)_6^- + 2\ e^-$

$ClO_2 + e^- \rightarrow ClO_2^-$

Step 6 and 7 Equalize gain and loss of electrons; add half-reactions

$2\ H_2O + 2\ OH^- + SbO_2^- \rightarrow Sb(OH)_6^- + 2\ e^-$

$2\ (ClO_2 + e^- \rightarrow ClO_2^-)$

$2\ H_2O + 2\ ClO_2 + 2\ OH^- + SbO_2^- \rightarrow 2\ ClO_2^- + Sb(OH)_6^-$

(c) $Al + NO_3^- \rightarrow NH_3 + Al(OH)_4^-$ (basic solution)

Step 1 Write half-reaction equations. Balance except H and O.

$Al \rightarrow Al(OH)_4^-$

$NO_3^- \rightarrow NH_3$

Step 2 Balance H and O using $H_2O$ and $H^+$

$4\ H_2O + Al \rightarrow Al(OH)_4^- + 4\ H^+$

$9\ H^+ + NO_3^- \rightarrow NH_3 + 3\ H_2O$

Step 3 Add $OH^-$ ions to both sides (same number as $H^+$ ions)

$4\ OH^- + 4\ H_2O + Al \rightarrow Al(OH)_4^- + 4\ H^+ + 4\ OH^-$

$9\ OH^- + 9\ H^+ + NO_3^- \rightarrow NH_3 + 3\ H_2O + 9\ OH^-$

Step 4 Combine $H^+$ and $OH^-$ to form $H_2O$; cancel $H_2O$ where possible

$4\ OH^- + 4\ H_2O + Al \rightarrow Al(OH)_4^- + 4\ H_2O$

$9\ H_2O + NO_3^- \rightarrow NH_3 + 3\ H_2O + 9\ OH^-$

$4\ OH^- + Al \rightarrow Al(OH)_4^-$

$6\ H_2O + NO_3^- \rightarrow NH_3 + 9\ OH^-$

Step 5 Balance electrically with electrons

$4\ OH^- + Al \rightarrow Al(OH)_4^- + 3\ e^-$.

$6\ H_2O + NO_3^- + 8\ e^- \rightarrow NH_3 + 9\ OH^-$

Step 6 and 7 Equalize gain and loss of electrons; add half-reactions

$8\ (4\ OH^- + Al \rightarrow Al(OH)_4^- + 3\ e^-)$

$\underline{3\ (6\ H_2O + NO_3^- + 8\ e^- \rightarrow NH_3 + 9\ OH^-)}$

$8\ Al + 3\ NO_3^- + 18\ H_2O + 5\ OH^- \rightarrow 3\ NH_3 + 8\ Al(OH)_4^-$

27 $OH^-$ and 24 $e^-$ canceled from each side

(d) $P_4 \rightarrow HPO_3^{2-} + PH_3$ (basic solution)

Step 1 Write half-reaction equations. Balance except H and O.

$P_4 \rightarrow 4\ HPO_3^{2-}$

$P_4 \rightarrow 4\ PH_3$

Step 2 Balance H and O using $H_2O$ and $H^+$

$12\ H_2O + P_4 \rightarrow 4\ HPO_3^{2-} + 20\ H^+$

$12\ H^+ + P_4 \rightarrow 4\ PH_3$

Step 3 Add $OH^-$ ions to both sides (same number as $H^+$ ions)

$20\ OH^- + 12\ H_2O + P_4 \rightarrow 4\ HPO_3^{2-} + 20\ H^+ + 20\ OH^-$

$12\ OH^- + 12\ H^+ + P_4 \rightarrow 4\ PH_3 + 12\ OH^-$

Step 4 Combine $H^+$ and $OH^-$ to form $H_2O$; cancel $H_2O$ where possible

$20\ OH^- + 12\ H_2O + P_4 \rightarrow 4\ HPO_3^{2-} + 20\ H_2O$

$12\ H_2O + P_4 \rightarrow 4\ PH_3 + 12\ OH^-$

$20\ OH^- + P_4 \rightarrow 4\ HPO_3^{2-} + 8\ H_2O$

Step 5 Balance electrically with electrons

$20\ OH^- + P_4 \rightarrow 4\ HPO_3^{2-} + 8\ H_2O + 12\ e^-$

$12\ H_2O + P_4 + 12\ e^- \rightarrow 4\ PH_3 + 12\ OH^-$

Step 6 and 7 Loss and gain of electrons are equal; add half-reactions

$8\ OH^- + 4\ H_2O + 2\ P_4 \rightarrow 4\ HPO_3^{2-} + 4\ PH_3$

Divide equation by 2

$4\ OH^- + 2\ H_2O + P_4 \rightarrow 2\ HPO_3^{2-} + 2\ PH_3$

(e) $Al + OH^- \rightarrow Al(OH)_4^- + H_2$ (basic solution)

Step 1 Write half-reaction equations. Balance except H and O.

$Al \rightarrow Al(OH)_4^-$

$OH^- \rightarrow H_2$

Step 2 Balance H and O using $H_2O$ and $H^+$

$4\ H_2O + Al \rightarrow Al(OH)_4^- + 4\ H^+$

$3\ H^+ + OH^- \rightarrow H_2 + H_2O$

Step 3 Add $OH^-$ ions to both sides (same number as $H^+$ ions)

$4\ OH^- + 4\ H_2O + Al \rightarrow Al(OH)_4^- + 4\ H^+ + 4\ OH^-$

$3\ OH^- + 3\ H^+ + OH^- \rightarrow H_2 + H_2O + 3\ OH^-$

Step 4 Combine $H^+$ and $OH^-$ to form $H_2O$; cancel $H_2O$ where possible

$4\,OH^- + 4\,H_2O + Al \rightarrow Al(OH)_4^- + 4\,H_2O$

$3\,H_2O + OH^- \rightarrow H_2 + H_2O + 3\,OH^-$

$4\,OH^- + Al \rightarrow Al(OH)_4^-$

$2\,H_2O + OH^- \rightarrow H_2 + 3\,OH^-$

Step 5 Balance electrically with electrons

$4\,OH^- + Al \rightarrow Al(OH)_4^- + 3\,e^-$

$2\,H_2O + OH^- + 2\,e^- \rightarrow H_2 + 3\,OH^-$

Step 6 and 7 Equalize gain and loss of electrons; add half-reactions

$2\,(4\,OH^- + Al \rightarrow Al(OH)_4^- + 3\,e^-)$

$\underline{3\,(2\,H_2O + OH^- + 2\,e^- \rightarrow H_2 + 3\,OH^-)}$

$2\,Al + 6\,H_2O + 2\,OH^- \rightarrow 2\,Al(OH)_4^- + 3\,H_2$

$6\,OH^-$ and $6\,e^-$ canceled on each side

31. (a) $Pb + SO_4^{2-} \rightarrow PbSO_4 + 2\,e^-$

$PbO_2 + SO_4^{2-} + 4\,H^+ + 2\,e^- \rightarrow PbSO_4 + 2\,H_2O$

(b) The first reaction is oxidation ($Pb^0$ is oxidized to $Pb^{2+}$).
The second reaction is reduction ($Pb^{4+}$ is reduced to $Pb^{2+}$).

(c) The first reaction (oxidation) occurs at the anode of the battery.

32. (a) The oxidizing agent is $KMnO_4$.

(b) The reducing agent is HCl.

(c) 5 moles of electrons $5\,e^- + Mn^{7+} \rightarrow Mn^{2+}$

$$\left(\frac{5\text{ mol e}^-}{\text{mol KMnO}_4}\right)\left(\frac{6.022 \times 10^{23}\text{ e}^-}{\text{mol e}^-}\right) = 3.011 \times 10^{24}\ \frac{\text{electrons}}{\text{mol KMnO}_4}$$

33. $3\,Ag + 4\,HNO_3 \rightarrow 3\,AgNO_3 + NO + 2\,H_2O$

g Ag $\rightarrow$ mol Ag $\rightarrow$ mol NO

$$(25.0\text{ g Ag})\left(\frac{1\text{ mol}}{107.9\text{ g}}\right)\left(\frac{1\text{ mol NO}}{3\text{ mol Ag}}\right) = 0.0772\text{ mol NO}$$

34. $3\ Cl_2 + 6\ KOH \rightarrow KClO_3 + 5\ KCl + 3\ H_2O$

mol $KClO_3$ → mol $Cl_2$ → L $Cl_2$

$$(0.300\ g\ KClO_3)\left(\frac{3\ mol\ Cl_2}{1\ mol\ KClO_3}\right)\left(\frac{22.4\ L}{1\ mol}\right) = 20.2\ L\ Cl_2$$

35. $5\ H_2O_2 + 2\ KMnO_4 + 3\ H_2SO_4 \rightarrow 5\ O_2 + 2\ MnSO_4 + K_2SO_4 + 8\ H_2O$

mL $H_2O_2$ → g $H_2O_2$ → mol $H_2O_2$ → mol $KMnO_4$ → g $KMnO_4$

$$(100\ mL\ H_2O_2\ solution)\left(\frac{1.031\ g}{mL}\right)\left(\frac{9.0\ g\ H_2O_2}{100.\ g\ H_2O_2\ solution}\right)\left(\frac{1\ mol}{34.02\ g}\right)\left(\frac{2\ mol\ KMnO_4}{5\ mol\ H_2O_2}\right)\left(\frac{158.0\ g}{mol}\right)$$

$= 17\ g\ KMnO_4$

36. $Cr_2O_7^{2-} + 3\ H_3AsO_3 + 8\ H^+ \rightarrow 2\ Cr^{3+} + 3\ H_3AsO_4 + 4\ H_2O$

g $H_3AsO_3$ → mol $H_3AsO_3$ → mol $Cr_2O_7^{2-}$ → mL $Cr_2O_7^{2-}$

$$(5.00\ g\ H_3AsO_4)\left(\frac{1\ mol}{125.9\ g}\right)\left(\frac{1\ mol\ Cr_2O_7^{2-}}{3\ mol\ H_3AsO_3}\right)\left(\frac{1000\ mL}{0.200\ mol}\right) = 66.2\ mL\ of\ 0.200\ M\ K_2Cr_2O_7$$

37. $Cr_2O_7^{2-} + 6\ Fe^{2+} + 14\ H^+ \rightarrow 2\ Cr^{3+} + 6\ Fe^{3+} + 7\ H_2O$

mL $FeSO_4$ → mol $FeSO_4$ → mol $Cr_2O_7^{2-}$ → mL $Cr_2O_7^{2-}$

$$(60.0\ mL\ FeSO_4)\left(\frac{0.200\ mol}{1000\ L}\right)\left(\frac{1\ mol\ Cr_2O_7^{2-}}{6\ mol\ FeSO_4}\right)\left(\frac{1000\ mL}{0.200\ mol}\right) = 10.0\ mL\ of\ 0.200\ M\ K_2Cr_2O_7$$

38. $8\ KI + 5\ H_2SO_4 \rightarrow 4\ I_2 + H_2S + 4\ K_2SO_4 + 4\ H_2O$

g $I_2$ → mol $I_2$ → mol KI → g KI

$$(2.79\ g\ I_2)\left(\frac{1\ mol}{253.8\ g}\right)\left(\frac{8\ mol\ KI}{4\ mol\ I_2}\right)\left(\frac{160.0\ g}{mol}\right) = 3.65\ g\ KI\ in\ sample$$

$$\left(\frac{3.65\ g\ KI}{4.00\ g\ sample}\right)(100) = 91.3\%\ KI$$

39. $3\ Ag + 4\ HNO_3 \rightarrow 3\ AgNO_3 + NO + 2\ H_2O$

mol Ag → mol NO

$$(0.500\ mol\ Ag)\left(\frac{1\ mol\ NO}{3\ mol\ Ag}\right) = 0.167\ mol\ NO$$

$PV = nRT \qquad V = nRT/P$

$$P = (744 \text{ torr})\left(\frac{1 \text{ atm}}{760. \text{ torr}}\right) = 0.979 \text{ atm}$$

$$T = 301 \text{ K}$$

$$V = \frac{(0.167 \text{ mol NO})(0.0821 \text{ L atm/mol K})(301 \text{ K})}{(0.979 \text{ atm})} = 4.22 \text{ L NO}$$

40. $2\,Al + 2\,OH^- + 6\,H_2O \rightarrow 2\,Al(OH)_4^- + 3\,H_2$

    g Al → mol Al → mol $H_2$

    $$(100.0 \text{ g Al})\left(\frac{1 \text{ mol Al}}{26.98 \text{ g}}\right)\left(\frac{3 \text{ mol } H_2}{2 \text{ mol Al}}\right) = 5.560 \text{ mol } H_2$$

41. (a) $Cu^+ \rightarrow Cu^{2+}$ is an oxidation, but when electrons are gained reduction should occur.

    $Cu^+ + e^- \rightarrow Cu^0$ or $Cu^+ \rightarrow Cu^{2+} + e^-$

    (b) When $Pb^{2+}$ is reduced, it requires two individual electrons. $Pb^{2+} + 2\,e^- \rightarrow Pb^0$. An electron has only a single negative charge ($e^-$).

42. The electrons lost by the species undergoing oxidation must be gained (or attracted) by another species which then undergoes reduction.

43. $A + B^{2+} \rightarrow NR$ — $B^{2+}$ cannot take $e^-$ from A

    $A + C^+ \rightarrow NR$ — $C^+$ cannot take $e^-$ from A

    $D + 2\,C^+ \rightarrow 2\,C + D^{2+}$ — $C^+$ takes $e^-$ from D

    $B + D^{2+} \rightarrow D + B^{2+}$ — $D^{2+}$ takes $e^-$ from B

    Therefore, $B^{2+}$ is least able to attract $e^-$, then $D^{2+}$, then $C^+$, then $A^+$

44. $Sn^{4+}$ can only be an oxidizing agent.

    $Sn^{4+} + 2\,e^- \rightarrow Sn^{2+}$
    $Sn^{4+} + 4\,e^- \rightarrow Sn^0$

    $Sn^0$ can only be a reducing agent.

    $Sn^0 \rightarrow Sn^{2+} + 2\,e^-$
    $Sn^0 \rightarrow Sn^{4+} + 4\,e^-$

    $Sn^{2+}$ can be both oxidizing and reducing.

    $Sn^{2+} + 2\,e^- \rightarrow Sn^0$ (oxidizing)
    $Sn^{2+} \rightarrow Sn^{4+} + 2\,e^-$ (reducing)

45. $Mn(OH)_2$ +2
$MnF_3$ +3
$MnO_2$ +4
$K_2MnO_4$ +6
$KMnO_4$ +7

$KMnO_4$ is the best oxidizing agent of the group, since its greater positive charge (+7) makes it very attractive to electrons.

46. Equations (a) and (b) represent oxidation

(a) $Mg \rightarrow Mg^{2+} + 2\,e^-$

(b) $SO_2 \rightarrow SO_3$; $(S^{4+} \rightarrow S^{6+} + 2\,e^-)$

47. (a) $MnO_2 + 2\,Br^- + 4\,H^+ \rightarrow Mn^{2+} + Br_2 + 2\,H_2O$

(b) mL $Mn^{2+}$ → mol $Mn^{2+}$ → mol $MnO_2$ → g $MnO_2$

$$(100.0 \text{ mL } Mn^{2+})\left(\frac{0.05 \text{ mol}}{1000 \text{ mL}}\right)\left(\frac{1 \text{ mol } MnO_2}{1 \text{ mol } Mn^{2+}}\right)\left(\frac{86.94 \text{ g}}{\text{mol}}\right) = 0.4 \text{ g } MnO_2$$

(c) $$(100.0 \text{ mL } Mn^{2+})\left(\frac{0.05 \text{ mol}}{1000 \text{ mL}}\right)\left(\frac{1 \text{ mol } Br_2}{1 \text{ mol } Mn^{2+}}\right) = 0.005 \text{ mol } Br_2$$

$$PV = nRT \qquad V = \frac{nRT}{P}$$

$$V = \left(\frac{0.005 \text{ mol}}{1.4 \text{ atm}}\right)\left(\frac{0.0821 \text{ L atm}}{\text{mol K}}\right)(323 \text{ K}) = 0.09 \text{ L } Br_2 \text{ vapor}$$

48. (a) $F_2 + 2\,Cl^- \rightarrow 2\,F^- + Cl_2$

(b) $Br_2 + Cl^- \rightarrow$ NR

(c) $I_2 + Cl^- \rightarrow$ NR

(d) $Br_2 + 2\,I^- \rightarrow 2\,Br^- + I_2$

49. $Mn + 2\,HCl \rightarrow Mn^{2+} + H_2 + 2\,Cl^-$

50. $4\,Zn + NO_3^- + 10\,H^+ \rightarrow 4\,Zn^{2+} + NH_4^+ + 3\,H_2O$
See Exercise 27(a).

51.

| (1) | (2) | (3) | (4) | (5) |
|---|---|---|---|---|
| a) C oxidized | a) S oxidized | a) N oxidized | a) S oxidized | a) $O^{2-}$ oxidized |
| b) $O_2$ reduced | b) N reduced | b) Cu reduced | b) O reduced | b) $O_2^{2-}$ reduced |
| c) $O_2$, O.A. | c) $HNO_3$, O.A. | c) CuO, O.A. | c) $H_2O_2$, O.A. | c) $H_2O_2$, O.A. |
| d) $C_3H_8$, R.A. | d) $H_2S$, R.A. | d) $NH_3$, R.A. | d) $Na_2SO_3$, R.A. | d) $H_2O_2$, R.A. |
| e) $2\,⅔ \rightarrow 4$ | e) $S^{2-} \rightarrow S^0$ | e) $N^{3-} \rightarrow N_2^0$ | e) $S^{4+} \rightarrow S^{6+}$ | e) $O_2^{2-} \rightarrow O_2^0$ |
| f) $0 \rightarrow -2$ | f) $N^{5+} \rightarrow N^{2+}$ | f) $Cu^{2+} \rightarrow Cu^0$ | f) $O_2^{2-} \rightarrow O^{2-}$ | f) $O_2^{2-} \rightarrow O^{2-}$ |

O.A. = Oxidizing agent
R.A. = Reducing agent

52. $Pb + 2\,Ag^+ \rightarrow 2\,Ag + Pb^{2+}$

(a) Pb is the anode

(b) Ag is the cathode

(c) Oxidation occurs at Pb (anode)

(d) Reduction occurs at Ag (cathode)

(e) Electrons flow from the lead through the wire to the silver

(f) Positive ions flow through the salt solution towards the negatively charged strip of silver; negative ions flow toward the positively charged strip of lead.

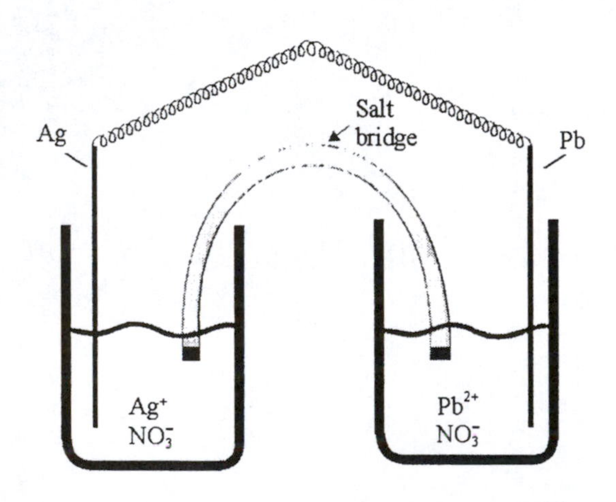

CHAPTER 18

# NUCLEAR CHEMISTRY

1. (a) Gamma radiation requires the most shielding.

   (b) Alpha radiation requires the least shielding.

2. Alpha particles are deflected less than beta particles while passing through a magnetic field, because they are much heavier (more than 7,000 times heavier) than beta particles.

3. Pairs of nuclides that would be found in the fission reaction of U-235. Any two nuclides, whose atomic numbers add up to 92 and mass numbers (in the range of 70-160) add up to 230-234. Examples include:

   ${}^{90}_{38}\text{Sr}$ and ${}^{141}_{54}\text{Xe}$ ${}^{139}_{56}\text{Ba}$ and ${}^{94}_{36}\text{Kr}$ ${}^{101}_{42}\text{Mo}$ and ${}^{131}_{50}\text{Sn}$

4. Contributions to the early history of radioactivity include:

   (a) Henri Becquerel: He discovered radioactivity.

   (b) Marie and Pierre Curie: They discovered the elements polonium and radium.

   (c) Wilhelm Roentgen: He discovered X rays and developed the tehcnique of producing them. While this was not a radioactive phenomenon, it triggered Becquerel's discovery of radioactivity.

   (d) Earnest Rutherford: He discovered alpha and beta particles, established the link between radioactivity and transmutation, and produced the first successful man-made transmutation.

   (e) Otto Hahn and Fritz Strassmann: They were first to produce nuclear fission.

5. Chemical reactions are caused by atoms or ions coming together, so are greatly influenced by temperature and concentration, which affect the number of collisions. Radioactivity is a spontaneous reaction of an individual nucleus, and is independent of such influences.

6. The term isotope is used with reference to atoms of the same element that contains different masses. For example, ${}^{12}_{6}\text{C}$ and ${}^{14}_{6}\text{C}$. The term nuclide is used in nuclear chemistry to infer any isotope of any atom.

7. $$(5 \times 10^9 \text{ years})\left(\frac{1 \text{ half-life}}{7.6 \times 10^7 \text{ years}}\right) = 70 \text{ half-lives}$$

   Even if plutonium-224 had been present in large quantities five billion years ago, no measureable amount would survive after 70 half-lives.

8.

| | charge | mass | nature of particles | penetrating power |
|---|---|---|---|---|
| Alpha | +2 | 4 amu | He nucleus | low |
| Beta | −1 | $\frac{1}{1837\ \text{amu}}$ | electron | moderate |
| Gamma | 0 | 0 | electromagnetic radiation | high |

9. Natural radioactivity is the spontaneous disintegration of those radioactive isotopes found in nature. Artificial radioactivity is the spontaneous disintegration of radioactive isotopes produced synthetically.

10. A radioactive disintegration series starts with a particular radionuclide and progresses stepwise by alpha and beta emissions to other radionuclides, ending at a stable nuclide. For example:

$$^{238}_{92}\text{U} \xrightarrow{\text{14 steps}} {}^{206}_{82}\text{Pb (stable)}$$

11. Transmutation is the conversion of one element into another by natural or artificial means. The nucleus of an atom is bombarded by various particles (alpha, beta, protons, etc.). The fast moving particles are captured by the nucleus, forming an unstable nucleus, which decays to another kind of atom. For example:

$$^{9}_{4}\text{Be} + {}^{4}_{2}\text{He} \rightarrow {}^{12}_{6}\text{C} + {}^{1}_{0}\text{n}$$

12. $$^{232}_{90}\text{Th} \xrightarrow{-\alpha} {}^{228}_{88}\text{Ra} \xrightarrow{-\beta} {}^{228}_{89}\text{Ac} \xrightarrow{-\beta} {}^{228}_{90}\text{Th} \xrightarrow{-\alpha} {}^{224}_{88}\text{Ra} \xrightarrow{-\alpha} {}^{220}_{86}\text{Rn} \xrightarrow{-\alpha} {}^{216}_{84}\text{Po} \xrightarrow{-\alpha} {}^{212}_{82}\text{Pb}$$

$$\xrightarrow{-\beta} {}^{212}_{83}\text{Bi} \xrightarrow{-\beta} {}^{212}_{84}\text{Po} \xrightarrow{-\alpha} {}^{208}_{82}\text{Pb}$$

13. $^{237}_{93}\text{Np}$ loses seven alpha particles and four beta particles.

Determination of the final product: $^{209}_{83}\text{Bi}$

$$\text{nuclear charge} = 93 - 7(2) + 4(1) = 83$$

$$\text{mass} = 237 - 7(4) = 209$$

14. Decay of bismuth-211

$$^{211}_{83}\text{Bi} \rightarrow {}^{4}_{2}\text{He} + {}^{207}_{81}\text{Tl} \qquad {}^{207}_{81}\text{Tl} \rightarrow {}^{0}_{-1}\text{e} + {}^{207}_{82}\text{Pb}$$

15. Two Germans, Otto Hahn and Fritz Strassmann, were the first scientists to report nuclear fission. The fission resulted from bombarding uranium nuclei with neutrons.

16. Natural uranium is 99+% U-238. Commercial nuclear reactors use U-235 enriched uranium as a fuel. Slow neutrons will cause the fission of U-235, but not U-238. Fast neutrons are capable of a nuclear reaction with U-238 to produce fissionable Pu-239. A breeder reactor converts nonfissionable U-238 to fissionable Pu-239, and in the process, manufactures more fuel than it consumes.

17. The fission reaction in a nuclear reactor and in an atomic bomb are essentially the same. The difference is that the fissioning is "wild" or uncontrolled in the bomb. In a nuclear reactor, the fissioning rate is controlled by means of moderators, such as graphite, to slow the neutrons and control rods of cadmium or boron to absorb some of the neutrons.

18. A certain amount of fissionable material (a critical mass) must be present before a self-supporting chain reaction can occur. Without a critical mass, too many neutrons from fissions will escape, and the reaction cannot reach a chain reaction status, unless at least one neutron is captured or every fission that occurs.

19. The mass defect is the difference between the mass of an atom and the sum of the masses of the number of protons, neutrons, and electrons in that atom. The energy equivalent of this mass defect is known as the nuclear binding energy.

20. When radioactive rays pass through normal matter, they cause that matter to become ionized (usually by knocking out electrons). Therefore, the radioactive rays are classified as ionizing radiation.

21. Some biological hazards associated with radioactivity are:

    (a) High levels of radiation can cause nausea, vomiting, diarrhea, and death. The radiation produces ionization in the cells, particularly in the nucleus of the cells.

    (b) Long-term exposure to low levels of radiation can weaken the body and cause malignant tumors.

    (c) Radiation can damage DNA molecules in the body causing mutations, which by reproduction, can be passed on to succeeding generations.

22. Strontium-90 has two characteristics that create concern. Its half-life is 28 years, so it remains active for a long period of time (disintegrating by emitting $\beta$ radiation). The other characteristic is that Sr-90 is chemically similar to calcium, so when it is present in milk Sr-90 is deposited in bone tissue along with calcium. Red blood cells are produced in the bone marrow. If the marrow is subjected to beta radiation from strontium-90, the red blood cells will be destroyed, increasing the incidence of leukemia and bone cancer.

23. A radioactive "tracer" is a radioactive material, whose presence is traced by a Geiger counter or some other detecting device. Tracers are often injected into the human body,

animals, and plants to determine chemical pathways, rates of circulation, etc. For example, use of a tracer could determine the length of time for material to travel from the root system to the leaves in a tree.

24. In living species, the ratio of carbon-14 to carbon-12 is constant due to the constant C-14/C-12 ratio in the atmosphere and food sources. When a species dies, life processes stop. The C-14/C-12 ratio decreases with time because C-14 is radioactive and decays according to its half-life, while the amount of C-12 in the species remains constant. Thus, the age of an archaeological artifact containing carbon can be calculated by comparing the C-14/C-12 ratio in the artifact with the C-14/C-12 ratio in the living species.

25. Radioactivity could be used to locate a leak in an underground pipe by using a water soluble tracer element. Dissolve the tracer in water and pass the water through the pipe. Test the ground along the path of the pipe with a Geiger counter until radioactivity from the leak is detected. Then dig.

26. The half-life of carbon-14 is 5668 years.

$$(4 \times 10^6 \text{ years})\left(\frac{1 \text{ half-life}}{5668 \text{ years}}\right) = 7 \times 10^2 \text{ half-lives}$$

700 half-lives would pass in 4 million years. Not enough C-14 would remain to allow detection with any degree of reliability. C-14 dating would not prove useful in this case.

27.

| | | Protons | Neutrons | Nucleons |
|---|---|---|---|---|
| (a) | ${}^{35}_{17}Cl$ | 17 | 18 | 35 |
| (b) | ${}^{226}_{88}Ra$ | 88 | 138 | 226 |

28.

| | | Protons | Neutrons | Nucleons |
|---|---|---|---|---|
| (a) | ${}^{235}_{92}U$ | 92 | 143 | 235 |
| (b) | ${}^{82}_{35}Br$ | 35 | 47 | 82 |

29. When a nucleus loses an alpha particle, its atomic number decreases by two, and its mass number decreases by four.

30. When a nucleus loses a beta particle, its atomic number increases by one, and its mass number remains unchanged.

31. Equations for alpha decay:

(a) ${}^{218}_{85}At \rightarrow {}^{4}_{2}He + {}^{214}_{83}Bi$

(b) ${}^{221}_{87}Fr \rightarrow {}^{4}_{2}He + {}^{217}_{85}At$

32. Equations for alpha decay:

(a) $^{192}_{78}Pt \rightarrow\ ^{4}_{2}He +\ ^{188}_{76}Os$

(b) $^{210}_{84}Po \rightarrow\ ^{4}_{2}He +\ ^{206}_{82}Pb$

33. Equations for beta decay:

(a) $^{14}_{6}C \rightarrow\ ^{0}_{-1}e +\ ^{14}_{7}N$

(b) $^{137}_{55}Cs \rightarrow\ ^{0}_{-1}e +\ ^{137}_{56}Ba$

34. Equations for beta decay:

(a) $^{239}_{93}Np \rightarrow\ ^{0}_{-1}e +\ ^{239}_{94}Pu$

(b) $^{90}_{38}Sr \rightarrow\ ^{0}_{-1}e +\ ^{90}_{39}Y$

35. $^{13}_{6}C +\ ^{1}_{0}n \rightarrow\ ^{14}_{6}C$

36. $^{30}_{15}P \rightarrow\ ^{30}_{14}Si +\ ^{0}_{+1}e$

37. (a) $^{27}_{13}Al +\ ^{4}_{2}He \rightarrow\ ^{30}_{15}P +\ ^{1}_{0}n$

(b) $^{27}_{14}Si \rightarrow\ ^{0}_{+1}e +\ ^{27}_{13}Al$

(c) $^{12}_{6}C +\ ^{2}_{1}H \rightarrow\ ^{13}_{7}N +\ ^{1}_{0}n$

(d) $^{82}_{35}Br \rightarrow\ ^{82}_{36}Kr +\ ^{0}_{-1}e$

38. (a) $^{66}_{29}Cu \rightarrow\ ^{66}_{30}Zn +\ ^{0}_{-1}e$

(b) $^{0}_{-1}e +\ ^{7}_{4}Be \rightarrow\ ^{7}_{3}Li$

(c) $^{27}_{13}Al +\ ^{4}_{2}He \rightarrow\ ^{30}_{14}Si +\ ^{1}_{1}H$

(d) $^{85}_{37}Rb +\ ^{1}_{0}n \rightarrow\ ^{82}_{35}Br +\ ^{4}_{2}He$

39. $$(112\text{ years})\left(\frac{1\text{ half-life}}{28\text{ years}}\right) = 4\text{ half-lives}$$

In 4 half-lives 1/16th $(\frac{1}{2})^4$ of the starting amount would remain.

$$\frac{1.00\text{ mg Sr-90}}{16} = 0.0625\text{ mg Sr-90 remains after 112 years.}$$

40. $\frac{240}{2} = 120; \quad \frac{120}{2} = 60; \quad \frac{60}{2} = 30;$

3 half-lives are required to reduce the count from 240 to 30 counts/min.

1980 + (3 × 28) = 2064. One eighth of the original amount Sr-90 remains. $\left[\left(\frac{1}{2}\right)^3 = \frac{1}{8}\right]$

41. (a) $^{235}_{92}U + ^{1}_{0}n \rightarrow ^{94}_{38}Br + ^{139}_{54}Xe + 3\,^{1}_{0}n + \text{energy}$

Mass loss = mass of reactants − mass of products

Mass of reactants = 235.0439 amu + 1.0087 amu = 236.0526 amu

Mass of products = 93.9154 amu + 138.9179 amu + 3(1.0087 amu) = 235.8594 amu

Mass lost = 236.0526 amu − 235.8594 amu = 0.1932 amu

$$(0.1932\ \text{amu})\left(\frac{1.000\ \text{g}}{6.022 \times 10^{23}\ \text{amu}}\right)\left(\frac{9.0 \times 10^{13}\ \text{J}}{1.00\ \text{g}}\right) = 2.9 \times 10^{-11}\ \text{J/atom U-235}$$

(b) $$\left(\frac{2.9 \times 10^{-11}\ \text{J}}{\text{atom}}\right)\left(\frac{6.022 \times 10^{23}\ \text{atoms}}{\text{mol}}\right) = 1.7 \times 10^{13}\ \text{J/mol}$$

(c) $$\left(\frac{0.1932\ \text{amu}}{236.0526\ \text{amu}}\right)(100) = 0.08185\%\ \text{mass loss}$$

42. (a) $^{1}_{1}H + ^{2}_{1}H \rightarrow ^{3}_{2}He + \text{energy}$

Mass loss = mass of reactants − mass of products

Mass of reactants = 1.00794 g/mol + 2.01410 g/mol = 3.02204 g/mol

Mass of products = 3.01603 g/mol

Mass lost =3.02204 − 3.01603 = 0.00601 g/mol

$$\left(\frac{0.00601\ \text{g}}{\text{mol}}\right)\left(\frac{9.0 \times 10^{13}\ \text{J}}{\text{g}}\right) = 5.4 \times 10^{11}\ \text{J/mol}$$

(b) $$\left(\frac{0.00601\ \text{g}}{3.02204\ \text{g}}\right)(100) = 0.199\%\ \text{mass loss}$$

43. $$(0.0100\ \text{g RaCl}_2)\left(\frac{226.0\ \text{g Ra}}{296.9\ \text{g RaCl}_2}\right)\left(\frac{\$50{,}000}{1\ \text{g Ra}}\right) = \$381$$

44. 100% to 25% requires 2 half-lives. The half-life of C-14 is 5668 years. The specimen will be the age of two half-lives:

(2)(5668 years) = 11340 years old.

45. 16.0 g → 8.0 g → 4.0 g → 2.0 g → 1.0 g → 0.50 g

16.0 g to 0.50 g requires five half-lives.

$$\frac{90 \text{ minutes}}{5 \text{ half-lives}} = 18 \text{ minutes/half-life}$$

46. (a) $^{7}_{3}Li$ is made up of 3 protons, 4 neutrons, and 3 electrons.

Calculated mass

| | | | |
|---|---|---|---|
| 3 protons | 3(1.0073 g) | = | 3.0219 g |
| 4 neutrons | 4(1.0087 g) | = | 4.0348 g |
| 3 electrons | 3(0.00055 g) | = | 0.0017 g |
| calculated mass | | | 7.0584 g |

Mass defect = calculated mass − actual mass

Mass defect = 7.0584 g − 7.0160 g = 0.0424 g/mol

(b) Binding energy

$$\left(\frac{0.0424 \text{ g}}{\text{mol}}\right)\left(\frac{9.0 \times 10^{13} \text{ J}}{\text{g}}\right) = 3.8 \times 10^{12} \text{ J/mol}$$

47. $^{235}_{92}U \rightarrow ^{207}_{82}Pb$

Mass loss: 235 − 207 = 28

Net proton loss (atomic number): 92 p − 82 p = 10 p

The mass loss is equivalent to 7 alpha particles (28/4). A loss of 7 alpha particles gives a loss of 14 protons. A decrease in the atomic number to 78 (14 protons) is due to the loss of 7 alpha particles (92 − 14 = 78). Therefore, a loss of 4 beta particles is required to increase the atomic number from 78 to 92.

The total loss = 7 alpha particles and 4 beta particles.

48. (a) Geiger counter – radiation passes through a thin glass window into a chamber filled with argon gas and containing two electrodes. Some of the argon ionizes, sending a momentary electrical impulse between the electrodes to the detector. This signal is amplified electronically and read out on a counter or as a series of clicks.

(b) Scintillation counter – radiation strikes a scintillator, which is composed of molecules that emit light in the presence of ionizing radiation. A light sensitive detector counts the flashes and converts them into a digital readout.

(c) Film badge – radiation penetrates a film holder. The silver grains in the film darken when exposed to radiation. The film is developed at regular intervals.

49. (3 days)(24 hours/day) = 72 hours

72 hr + 6 hr = 78 hr

$$\frac{78\text{ hr}}{13\ \frac{\text{hr}}{t_{0.5}}} = 6\text{ half-lives} \qquad (10\text{ mg})\left(\frac{1}{2}\right)^6 = 0.16\text{ mg remaining}$$

50. Fission is the splitting of a heavy nuclide into two or more intermediate-sized fragments with the conversion of some mass into energy. Fission occurs in nuclear reactors, or atomic bombs.

Example: $^{235}_{92}U + ^{1}_{0}n \rightarrow ^{144}_{54}Xe + ^{90}_{38}Sr + 2^{1}_{0}n$

Fusion is the process of combining two relatively small nuclei to form a single larger nucleus. Fusion occurs on the sun, or in a hydrogen bomb.

Example: $^{3}_{1}H + ^{2}_{1}H \rightarrow ^{4}_{2}He + ^{1}_{0}n$ + energy

51.

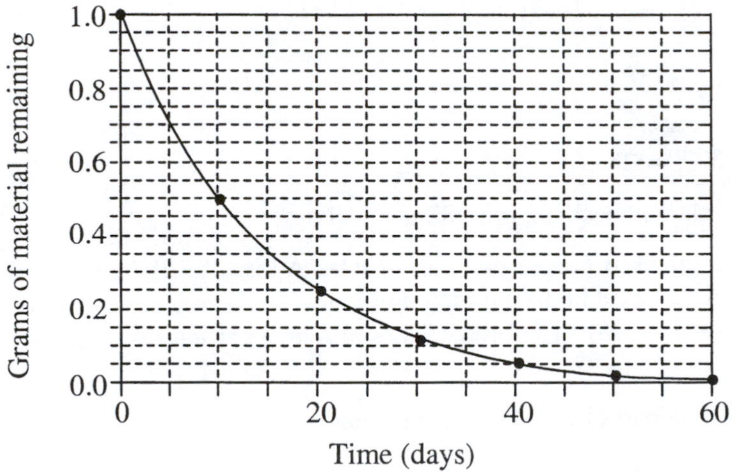

Grams of material remaining

1.0
0.8
0.6
0.4
0.2
0.0

0 20 40 60

Time (days)

The graph produces a curve for radioactive decay which never actually crosses the *x*-axis (where mass = 0), it simply approaches that point.

52. (a) $^{235}_{92}U + ^{1}_{0}n \rightarrow ^{143}_{54}Xe + 3^{1}_{0}n + ^{90}_{38}Sr$

(b) $^{235}_{92}U + ^{1}_{0}n \rightarrow ^{102}_{39}Y + 3^{1}_{0}n + ^{131}_{53}I$

(c) $^{14}_{7}N + ^{1}_{0}n \rightarrow ^{1}_{1}H + ^{14}_{6}C$

53. (a) $H_2O(l) \rightarrow H_2O(g)$

Energy$_2$: Weakest bond changes requires the least energy.

(b) $2\ H_2(g) + O_2(g) \rightarrow 2\ H_2O(g)$

Energy$_1$: medium-sized value involved in interatomic bonds.

(c) ${}^{2}_{1}H + {}^{2}_{1}H \rightarrow {}^{3}_{1}H + {}^{1}_{1}H$

Energy$_3$: Nuclear process; greatest amount of energy involved.

54. ${}^{236}_{92}U \rightarrow {}^{90}_{38}Sr + 3\,{}^{1}_{0}n + {}^{143}_{54}Xe$

55. (a) beta emission: ${}^{29}_{12}Mg \rightarrow {}^{0}_{-1}e + {}^{29}_{13}Al$

(b) alpha emission: ${}^{150}_{60}Nd \rightarrow {}^{4}_{2}He + {}^{146}_{58}Ce$

(c) positron emission: ${}^{72}_{33}As \rightarrow {}^{0}_{+1}e + {}^{72}_{32}Ge$

56. (a) ${}^{87}_{37}Rb \rightarrow {}^{0}_{-1}e + {}^{87}_{38}Sr$

(b) ${}^{87}_{38}Sr \rightarrow {}^{0}_{+1}e + {}^{87}_{37}Rb$

57.

| $t_{\frac{1}{2}}$ | 0 | 12.5 | 25.0 | 37.5 | 50.0 | 62.5 | 75.0 | 87.5 | 100. hours |
|---|---|---|---|---|---|---|---|---|---|
| Amount | 15.4 | 7.7 | 3.85 | 1.93 | 0.965 | 0.483 | 0.242 | 0.121 | 0.0605 mg |

Fraction of K-42 remaining $\dfrac{0.0605\text{ mg}}{15.4\text{ mg}} = 0.00393$ (or 0.393%)

No. After an additional eight half-lives there would be less than one microgram (0.000001 g) remaining.

$$(200\text{ hrs})\left(\frac{1\text{ half-life}}{12.5\text{ hrs}}\right) = 16\text{ half-lives}$$

$$\text{Amount remaining} = (15.4\text{ mg})\left(\frac{1}{2}\right)^{16}\left(\frac{10^3\ \mu\text{g}}{\text{mg}}\right) = 0.235\ \mu\text{g}$$

58. $$(270\text{ years})\left(\frac{1\text{ half-life}}{30\text{ years}}\right) = 9\text{ half-lives}$$

| $t_{\frac{1}{2}}$ | 0 | 30 | 60 | 90 | 120 | 150 | 180 | 210 | 240 | 270 years |
|---|---|---|---|---|---|---|---|---|---|---|
| Amount | 7680 | 3840 | 1920 | 960. | 480. | 240. | 120. | 60.0 | 30.0 | 15.0 g |

There would have been 7680 g originally

59. Element 114 would fall under lead on the periodic table. It would be a metal and would most likely form +2 and +4 ions in solution (like lead).

60. 1.00 g Co-60

(a) one half-life: $\frac{1.00\ \text{g}}{2} = 0.500\ \text{g left}$

(b) two half-lives: $\frac{0.500\ \text{g}}{2} = 0.250\ \text{g left}$

(c) four half-lives: $2^4 = 16;\ \frac{1}{16}\ \text{left} \quad \frac{1.00\ \text{g}}{16} = 0.0625\ \text{g}$

(d) ten half-lives: $2^{10} = 1024;\ \frac{1}{1024}\ \text{left} \quad \frac{1.00\ \text{g}}{1024} = 9.77 \times 10^{-4}\ \text{g}$

61. (a) ${}^{11}_{5}\text{B} \rightarrow {}^{4}_{2}\text{He} + {}^{7}_{3}\text{Li}$

(b) ${}^{88}_{38}\text{Sr} \rightarrow {}^{0}_{-1}\text{e} + {}^{88}_{39}\text{Y}$

(c) ${}^{107}_{47}\text{Ag} + {}^{1}_{0}\text{n} \rightarrow {}^{108}_{47}\text{Ag}$

(d) ${}^{41}_{19}\text{K} \rightarrow {}^{1}_{1}\text{H} + {}^{40}_{18}\text{Ar}$

(e) ${}^{116}_{51}\text{Sb} + {}^{0}_{-1}\text{e} \rightarrow {}^{116}_{50}\text{Sn}$

62. C-14 content of 1/16 of that in living plants means that four half-lives have passed. $^{14}$C half-life is 5668 years.

$$\left(\frac{5668\ \text{years}}{\text{half-life}}\right)(4\ \text{half-lives}) = 22{,}672\ \text{years} \quad (2.267 \times 10^4\ \text{years})$$

63. 1 Curie $= 3.7 \times 10^{10}$ disintegrations/sec

1 becquerel = 1 disintegration/sec

Therefore there are $3.7 \times 10^{10}$ becquerels/1 Curie

$$\left(\frac{3.7 \times 10^{10}\ \text{becquerel}}{1\ \text{Curie}}\right)(1.24\ \text{Curies}) = 4.6 \times 10^{10}\ \text{becquerels}$$

CHAPTER 19

# ORGANIC CHEMISTRY: SATURATED HYDROCARBONS

1. Two of the major reasons for the large number of organic compounds is the ability of carbon to form short or very long chains of atoms covalently bonded together and isomerism.

2. The carbon atom has only two unshared electrons, making two covalent bonds logical, but in $CH_4$, carbon forms four equivalent bonds. Promoting one 2s electron to the empty 2p orbital would make four bonds possible, but without hybridization, we could not explain the fact that all four bonds in $CH_4$ are identical, and the bond angles are equal (109.5°).

3. The first ten normal alkanes:

| | | | |
|---|---|---|---|
| methane | $CH_4$ | hexane | $C_6H_{14}$ |
| ethane | $C_2H_6$ | heptane | $C_7H_{16}$ |
| propane | $C_3H_8$ | octane | $C_8H_{18}$ |
| butane | $C_4H_{10}$ | nonane | $C_9H_{20}$ |
| pentane | $C_5H_{12}$ | decane | $C_{10}H_{22}$ |

4. (a) A molecule of ethane, $C_2H_6$, contains seven sigma bonds.

   (b) A molecule of butane, $C_4H_{10}$, contains thirteen sigma bonds.

   (c) A molecule of 2-methylpropane also contains thirteen sigma bonds.

5. Advantages: Some freons have low boiling points and therefore are excellent refrigerants. They are stable, nontoxic, nonflammable and noncorrosive.
   Disadvantages: Freons are a major factor in the destruction of the ozone layer once they get into the stratosphere.

6. *Iso* used in the word isomers means compounds having the same molecular formula but different structural formulas. *Iso* also refers to a particular isomer with a methyl branch at the end of a carbon chain. For example, isopentane

$$\begin{array}{l} CH_3\text{–}CHCH_2CH_3 \\ \quad\quad\;\; | \\ \quad\quad\; CH_3 \end{array}$$

*Sec*-: a secondary carbon atom, one that is bonded to two other carbon atoms.

$CH_3CH_2\underset{|}{C}H\,CH_3$ secondary butyl group

*sec* carbon atom

*Tert*- or *t*-: a tertiary carbon atom. A carbon atom that is bonded to three other carbon atoms as in *tert*-butyl group

$$\begin{array}{l} \quad CH_3 \\ \quad\ | \\ CH_3C- \leftarrow \textit{tert}\text{ carbon atom} \\ \quad\ | \\ \quad CH_3 \end{array}$$

7. Lewis structures:

(a) $CCl_4$

$$\begin{array}{ccc} & :\ddot{Cl}: & \\ :\ddot{Cl}: & \ddot{\underset{\cdot\cdot}{C}} & :\ddot{Cl}: \\ & :\underset{\cdot\cdot}{Cl}: & \end{array}$$

(b) $C_2Cl_6$

$$\begin{array}{cccc} & :\ddot{Cl}: & :\ddot{Cl}: & \\ :\underset{\cdot\cdot}{\ddot{Cl}}: & \ddot{\underset{\cdot\cdot}{C}} & :\ \ddot{\underset{\cdot\cdot}{C}} & :\underset{\cdot\cdot}{\ddot{Cl}}: \\ & :\underset{\cdot\cdot}{Cl}: & :\underset{\cdot\cdot}{Cl}: & \end{array}$$

(c) $CH_3CH_2CH_3$

$$\begin{array}{ccccccc} & H & & H & & H & \\ H: & \ddot{\underset{\cdot\cdot}{C}} & : & \ddot{\underset{\cdot\cdot}{C}} & : & \ddot{\underset{\cdot\cdot}{C}} & :H \\ & H & & H & & H & \end{array}$$

8. Lewis structures:

(a) $CH_4$

$$\begin{array}{ccc} & H & \\ H: & \ddot{\underset{\cdot\cdot}{C}} & :H \\ & H & \end{array}$$

(b) $C_3H_8$

$$\begin{array}{ccccccc} & H & & H & & H & \\ H: & \ddot{\underset{\cdot\cdot}{C}} & : & \ddot{\underset{\cdot\cdot}{C}} & : & \ddot{\underset{\cdot\cdot}{C}} & :H \\ & H & & H & & H & \end{array}$$

(c) $C_5H_{12}$

$$\begin{array}{ccccccccccc} & H & & H & & H & & H & & H & \\ H: & \ddot{\underset{\cdot\cdot}{C}} & : & \ddot{\underset{\cdot\cdot}{C}} & : & \ddot{\underset{\cdot\cdot}{C}} & : & \ddot{\underset{\cdot\cdot}{C}} & : & \ddot{\underset{\cdot\cdot}{C}} & :H \\ & H & & H & & H & & H & & H & \end{array}$$

9. Formulas (a) and (i) are not isomers. They are the same compound.
Formulas (b) and (c) are isomers of $C_5H_{12}$.
Formulas (f) and (h) are isomers of $C_5H_{10}$.
Formulas (d), (e), and (g) are isomers of $C_6H_{14}$.

10. Formulas (b), (e), and (f) are identical. The others are different.

11. The formulas in Exercise 9 contain the following numbers of methyl groups:
(a) 2 (b) 2 (c) 3 (d) 2 (e) 4 (f) 0 (g) 3
(h) 1 (i) 2

12. The formulas in Exercise 10 contain the following numbers of methyl groups:
(a) 4 (b) 4 (c) 4 (d) 4 (e) 4 (f) 4

13. heptane

$CH_3CH_2CH_2CH_2CH_2CH_2CH_3$

$$\begin{array}{ll} CH_3CH_2CH_2CH_2 & CHCH_3 \\ & | \\ & CH_3 \end{array}$$

$$\begin{array}{ll} CH_3CH_2CH_2 & CHCH_2CH_3 \\ & | \\ & CH_3 \end{array}$$

$$\begin{array}{ll} & CH_3 \\ & | \\ CH_3CH_2CH_2 & CCH_3 \\ & | \\ & CH_3 \end{array}$$

$$\begin{array}{lll} & CH_3 & \\ & | & \\ CH_3CH_2 & C & CH_2CH_3 \\ & | & \\ & CH_3 & \end{array}$$

$$\begin{array}{llll} & CH_3 & & \\ & | & & \\ CH_3CH_2 & CH & CHCH_3 \\ & & | & \\ & & CH_3 & \end{array}$$

$$\begin{array}{llll} CH_3 & CHCH_2 & CHCH_3 \\ & | & | \\ & CH_3 & CH_3 \end{array}$$

$$\begin{array}{lll} & CH_3 & CH_3 \\ & | & | \\ CH_3 & CH- & CCH_3 \\ & & | \\ & & CH_3 \end{array}$$

$$\begin{array}{ll} CH_3CH_2 & CHCH_2CH_3 \\ & | \\ & CH_2CH_3 \end{array}$$

14. hexane

$CH_3CH_2CH_2CH_2CH_2CH_3$

$$\begin{array}{ll} CH_3CH_2CH_2 & CHCH_3 \\ & | \\ & CH_3 \end{array}$$

$$\begin{array}{ll} CH_3CH_2 & CHCH_2CH_3 \\ & | \\ & CH_3 \end{array}$$

$$\begin{array}{ll} & CH_3 \\ & | \\ CH_3CH_2 & CCH_3 \\ & | \\ & CH_3 \end{array}$$

$$\begin{array}{lll} & CH_3 & \\ & | & \\ CH_3 & CH & CHCH_3 \\ & & | \\ & & CH_3 \end{array}$$

15. (a) $CH_2Cl_2$, one $CH_2Cl_2$

(b) $C_3H_7Br$, two $CH_3CH_2CH_2Br$ $CH_3CHBrCH_3$

(c) $C_3H_6Cl_2$, four $CH_3CH_2CHCl_2$ $CH_3CHClCH_2Cl$

$CH_2ClCH_2CH_2Cl$ $CH_3CCl_2CH_3$

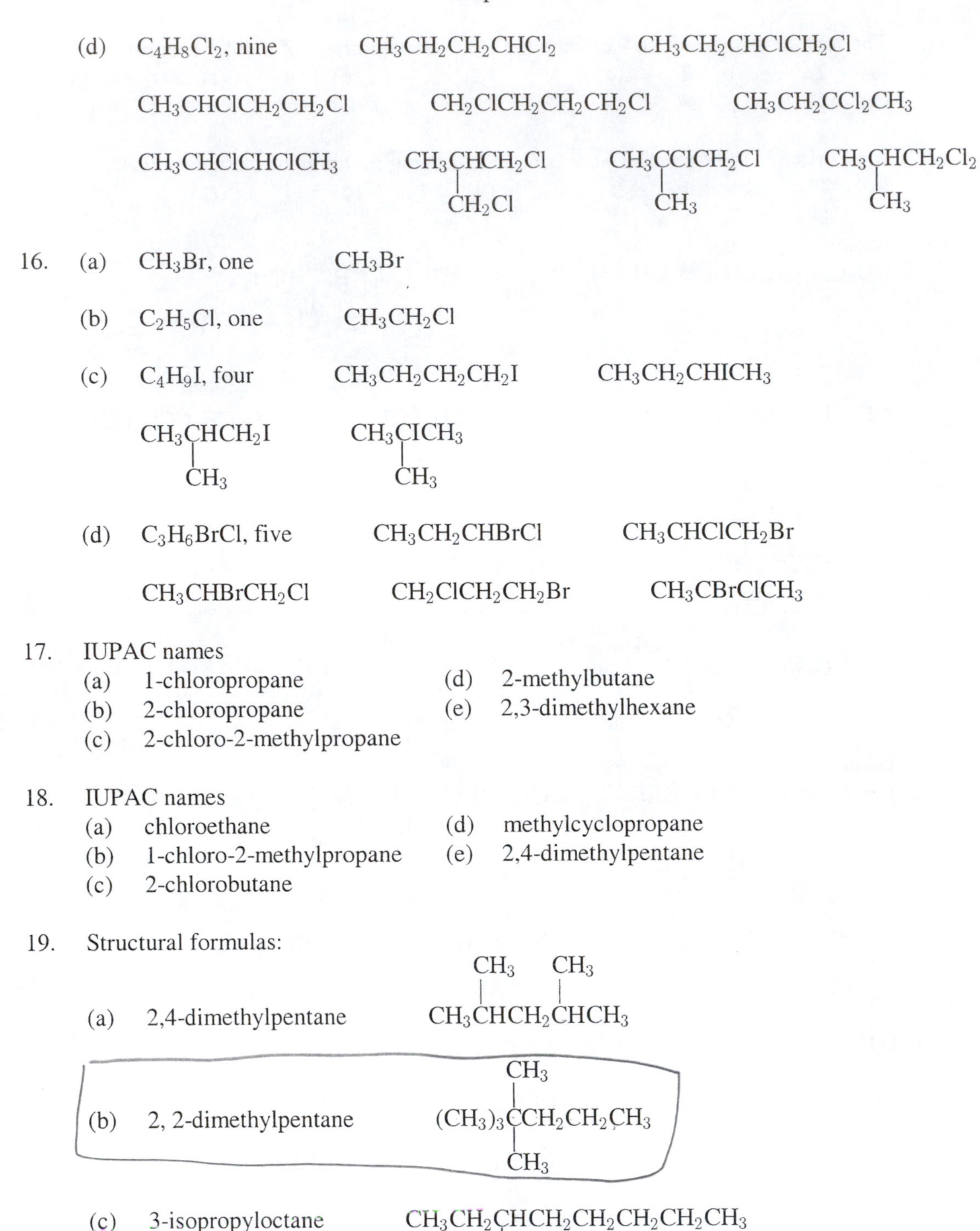

(d) $C_4H_8Cl_2$, nine $CH_3CH_2CH_2CHCl_2$ $CH_3CH_2CHClCH_2Cl$

$CH_3CHClCH_2CH_2Cl$ $CH_2ClCH_2CH_2CH_2Cl$ $CH_3CH_2CCl_2CH_3$

$CH_3CHClCHClCH_3$ $CH_3CH(CH_2Cl)CH_2Cl$ $CH_3CCl(CH_3)CH_2Cl$ $CH_3CH(CH_3)CH_2Cl_2$

16. (a) $CH_3Br$, one $CH_3Br$

(b) $C_2H_5Cl$, one $CH_3CH_2Cl$

(c) $C_4H_9I$, four $CH_3CH_2CH_2CH_2I$ $CH_3CH_2CHICH_3$

$CH_3CH(CH_3)CH_2I$ $CH_3CI(CH_3)CH_3$

(d) $C_3H_6BrCl$, five $CH_3CH_2CHBrCl$ $CH_3CHClCH_2Br$

$CH_3CHBrCH_2Cl$ $CH_2ClCH_2CH_2Br$ $CH_3CBrClCH_3$

17. IUPAC names
(a) 1-chloropropane
(b) 2-chloropropane
(c) 2-chloro-2-methylpropane
(d) 2-methylbutane
(e) 2,3-dimethylhexane

18. IUPAC names
(a) chloroethane
(b) 1-chloro-2-methylpropane
(c) 2-chlorobutane
(d) methylcyclopropane
(e) 2,4-dimethylpentane

19. Structural formulas:

(a) 2,4-dimethylpentane $CH_3CH(CH_3)CH_2CH(CH_3)CH_3$

(b) 2, 2-dimethylpentane $(CH_3)_3C(CH_3)(CH_3)CH_2CH_2CH_3$

(c) 3-isopropyloctane $CH_3CH_2CH(CH(CH_3)_2)CH_2CH_2CH_2CH_2CH_3$

(d) 5,6-diethyl-2,7-dimethyl-5-propylnonane

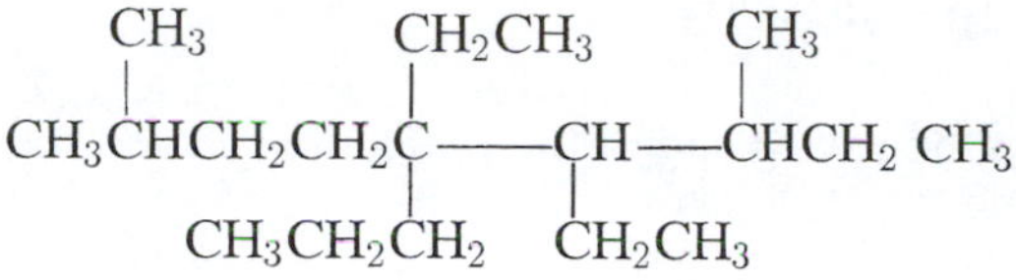

$$
\begin{array}{l}
\quad CH_3 \qquad\qquad\quad CH_2CH_3 \qquad CH_3 \\
CH_3CHCH_2CH_2C{-}{-}CH{-}{-}CHCH_2\,CH_3 \\
\qquad\qquad CH_3CH_2CH_2 \quad CH_2CH_3
\end{array}
$$

(e) 2-ethyl-1,3-dimethylcyclohexane

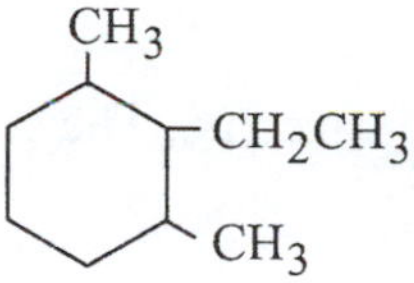

$CH_3$

$CH_2CH_3$

$CH_3$

20. (a) 4-ethyl-2-methylhexane

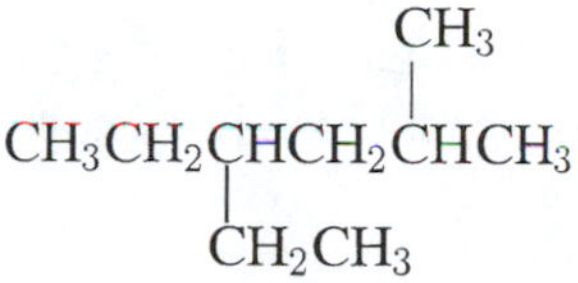

$$
\begin{array}{l}
\qquad\qquad\qquad\quad CH_3 \\
CH_3CH_2CHCH_2CHCH_3 \\
\qquad\quad CH_2CH_3
\end{array}
$$

(b) 4-*t*-butylheptane

$$
\begin{array}{l}
CH_3CH_2CH_2CHCH_2CH_2CH_3 \\
\qquad\qquad\quad C(CH_3)_3
\end{array}
$$

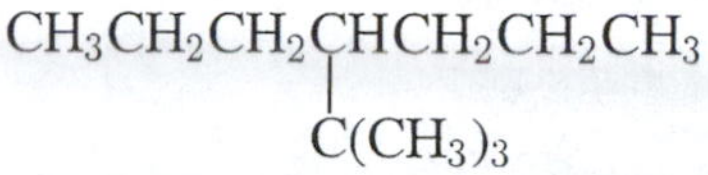

(c) 4-ethyl-7-isopropyl-2,4,8-trimethyldecane

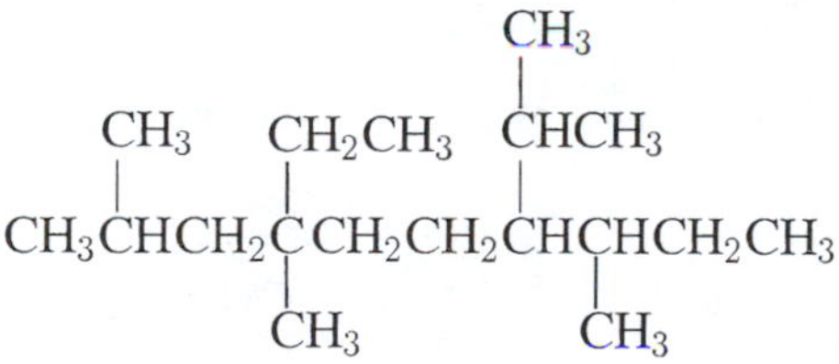

$$
\begin{array}{l}
\qquad\qquad\qquad\qquad\quad CH_3 \\
\quad CH_3 \quad\; CH_2CH_3 \;\; CHCH_3 \\
CH_3CHCH_2C\,CH_2CH_2CHCHCH_2CH_3 \\
\qquad\qquad\quad CH_3 \qquad\quad CH_3
\end{array}
$$

(d) 3-ethyl-2,2-dimethyloctane

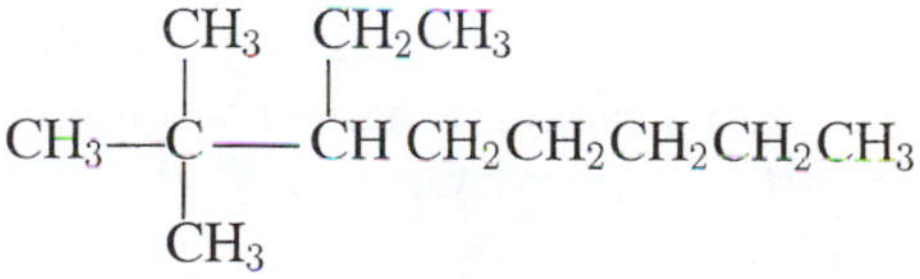

$$
\begin{array}{l}
\qquad\quad CH_3 \;\; CH_2CH_3 \\
CH_3{-}C{-}{-}CH\,CH_2CH_2CH_2CH_2CH_3 \\
\qquad\quad CH_3
\end{array}
$$

(e) 1,3-diethylcyclohexane

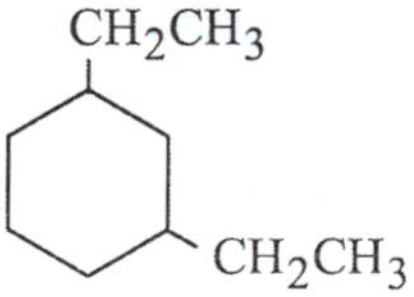

$CH_2CH_3$

$CH_2CH_3$

21. (a) 3-methylbutane

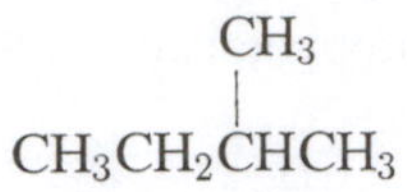

$$\begin{array}{l} \qquad\qquad\quad CH_3 \\ \qquad\qquad\quad | \\ CH_3CH_2CHCH_3 \end{array}$$

Numbering was done from the wrong end of the molecule. The correct name is 2-methylbutane.

(b) 2-ethylbutane

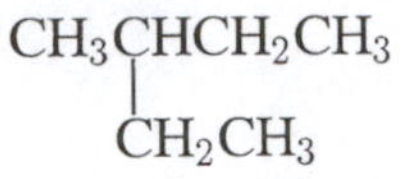

$$\begin{array}{l} CH_3CHCH_2CH_3 \\ \qquad\ | \\ \qquad CH_2CH_3 \end{array}$$

The name is not based on the longest carbon chain (5 carbons). The correct name is 3-methylpentane.

(c) 2-dimethylpentane. Each methyl group needs to be numbered. Depending on the structure, the correct name is 2,2-dimethylpentane; 2,3-dimethylpentane; or 2,4-dimethylpentane.

(d) 1,4-dimethylcyclopentane

$CH_3$

$CH_3$

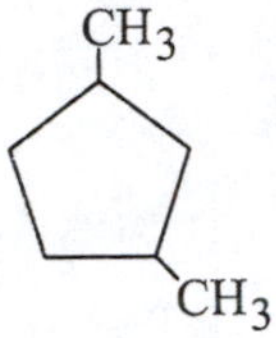

The ring was numbered in the wrong direction. The correct name is 1,3-dimethylcyclopentane.

22. (a) 3-methyl-5-ethyloctane

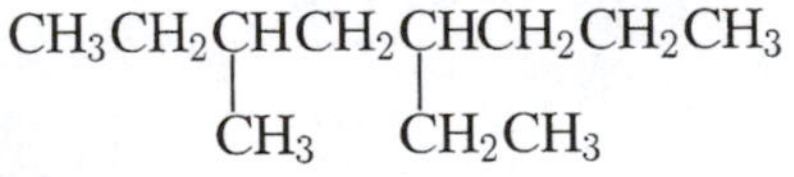

$$\begin{array}{l} CH_3CH_2CHCH_2CHCH_2CH_2CH_3 \\ \qquad\qquad\ |\qquad\quad\ | \\ \qquad\qquad CH_3\quad\ CH_2CH_3 \end{array}$$

Ethyl should be named before methyl (alphabetical order). The numbering is correct. The correct name is 5-ethyl-3-methyloctane.

(b) 3,5,5-triethylhexane

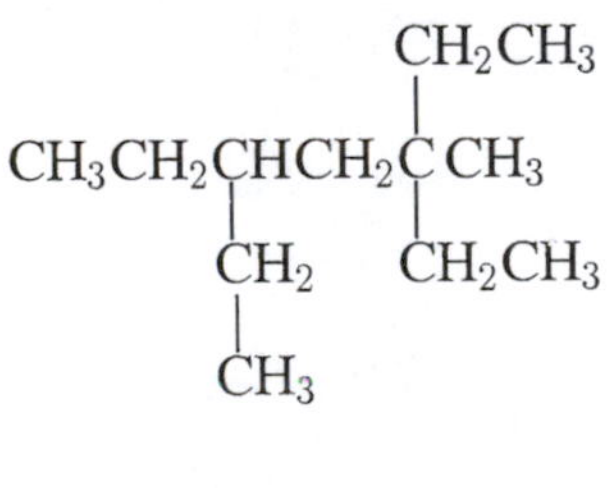

$$\begin{array}{l} \qquad\qquad\qquad\quad\ CH_2CH_3 \\ \qquad\qquad\qquad\quad\ | \\ CH_3CH_2CHCH_2C\,CH_3 \\ \qquad\qquad\ |\qquad\quad | \\ \qquad\qquad CH_2\quad CH_2CH_3 \\ \qquad\qquad\ | \\ \qquad\qquad CH_3 \end{array}$$

The name is not based on the longest carbon chain (7 carbons). The correct name is 3,5-diethyl-3-methylheptane.

(c) 4,4-dimethyl-3-ethylheptane

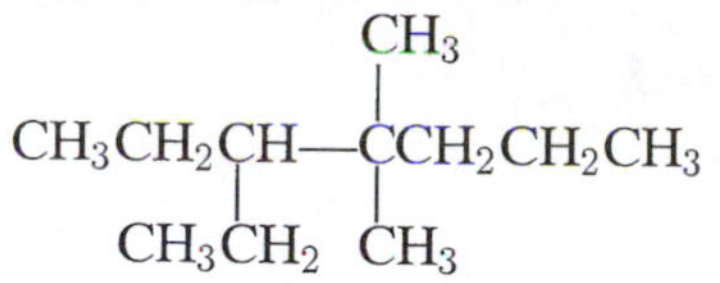

$$
\begin{array}{l}
\qquad\qquad\qquad\;\; CH_3 \\
\qquad\qquad\qquad\quad\; | \\
CH_3CH_2CH\text{—}CCH_2CH_2CH_3 \\
\qquad\quad\;\; | \qquad\;\; | \\
\quad CH_3CH_2 \;\; CH_3
\end{array}
$$

Ethyl should be named before dimethyl (alphabetical order). The correct name is 3-ethyl-4,4-dimethylheptane.

(d) 1,6-dimethylcyclohexane

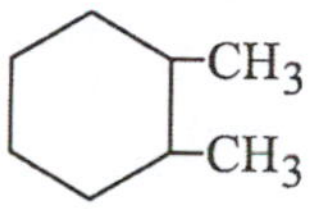

(cyclohexane ring with $-CH_3$ and $-CH_3$ on adjacent carbons)

The ring was numbered in the wrong direction. The correct name is 1,2-dimethylcyclohexane.

23. Structures for the ten dichlorosubstituted isomers of 2-methylbutane:

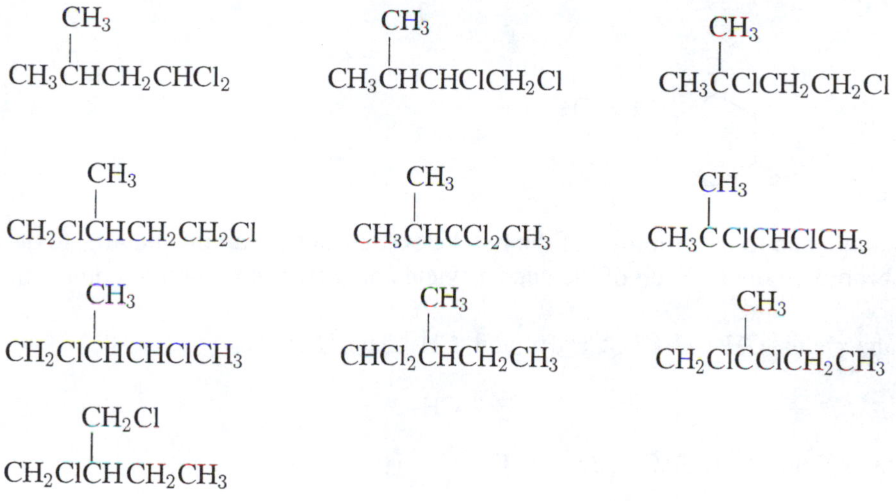

$CH_3CH(CH_3)CH_2CHCl_2$ $\qquad$ $CH_3CH(CH_3)CHClCH_2Cl$ $\qquad$ $CH_3CCl(CH_3)CH_2CH_2Cl$

$CH_2ClCH(CH_3)CH_2CH_2Cl$ $\qquad$ $CH_3CH(CH_3)CCl_2CH_3$ $\qquad$ $CH_3CCl(CH_3)CHClCH_3$

$CH_2ClCH(CH_3)CHClCH_3$ $\qquad$ $CHCl_2CH(CH_3)CH_2CH_3$ $\qquad$ $CH_2ClCCl(CH_3)CH_2CH_3$

$CH_2ClCH(CH_2Cl)CH_2CH_3$

24. $CH_3CH_2CH_2CH_2CH_2CH_2Cl$ $\qquad$ $CH_3CH_2CH_2CH_2CHClCH_3$

$CH_3CH_2CH_2CHClCH_2CH_3$

25. (a) $CH_3CH_2CH_2CH_3 + Cl_2 \xrightarrow{hv} CH_3CH_2CHClCH_3 + CH_3CH_2CH_2CH_2Cl + HCl$

(b) $2\,CH_3CH_2CH_2CH_3 + 13\,O_2 \xrightarrow{\Delta} 8\,CO_2 + 10\,H_2O$

26. (a) $CH_3CH_2CH_3 + Br_2 \xrightarrow{hv} CH_3CH_2CH_2Br + CH_3CHBrCH_3 + HBr$

(b) $CH_3CH_2CH_3 + 5\,O_2 \xrightarrow{\Delta} 3\,CO_2 + 4\,H_2O$

27. Names:

(a) 1-chloro-2ethylcyclohexane

(b) 1-chloro-3-ethyl-1-methylcyclohexane

(c) 1,4-diisopropylcyclohexane

28. Structural formulas:

(a)

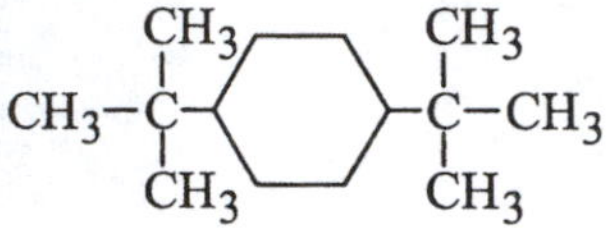

$CH_3$ $CH_3$

$CH_3-C-$ [cyclohexane ring] $-C-CH_3$

$CH_3$ $CH_3$

(b)

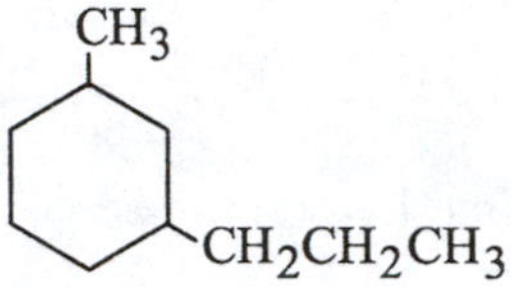

$CH_3$

[cyclohexane ring] $CH_2CH_2CH_3$

(c)

Cl

$CH_2CH_3$

$CH_3-CH_2$ [cyclohexane ring] Cl

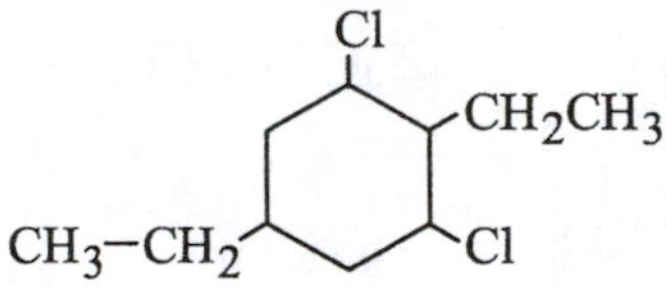

29. The sample had to be a mixture of the two butanes, *n*-butane and isobutane, to obtain four monobromo products. Each of the butanes yields only two monobromo compounds:

*n*-butane yields $CH_3CH_2CH_2CH_2Br$ and $CH_3CH_2CHBrCH_3$

isobutane yields $CH_3\underset{\displaystyle CH_3}{\underset{|}{C}}HCH_2Br$ and $CH_3\overset{\displaystyle CH_3}{\overset{|}{\underset{\displaystyle Br}{\underset{|}{C}}}}CH_3$

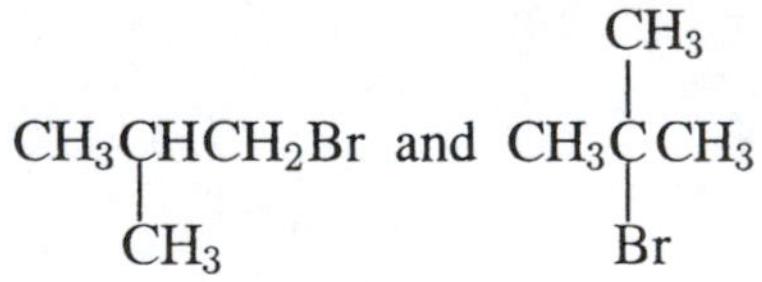

30. The formula for dodecane is $C_{12}H_{26}$.

31. $FCH_2CH_2F + Cl_2 \longrightarrow FCH_2CHClF + HCl$

32. Data: $\frac{1 \text{ gal}}{60 \text{ mi}}$; 60 mi traveled; $\frac{19 \text{ mol } C_8H_{18}}{\text{gal}}$

$$2\,C_8H_{18} + 25\,O_2 \longrightarrow 16\,CO_2 + 18\,H_2O$$

$$\frac{1 \text{ gal}}{60 \text{ mi}} \times 60 \text{ mi} = 1 \text{ gal gasoline used}$$

$$19 \text{ mol } C_8H_{18} \times \frac{16 \text{ mol } CO_2}{2 \text{ mol } C_8H_{18}} = 1.5 \times 10^2 \text{ mol } CO_2$$

$$PV = nRT \qquad V = \frac{nRT}{P}$$

$$V = \frac{1.5\times10^2 \text{ mol } CO_2 \times 0.0821 \text{ L-atm}}{1 \text{ atm} \qquad \text{mol K}} = 3.6 \times 10^3 \text{ L } CO_2$$

33. Cycloalkanes with formulas $C_5H_{10}$

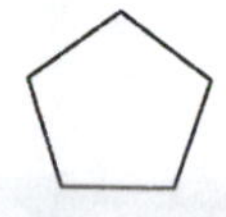

cyclopentane

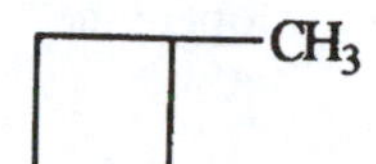


methylcyclobutane

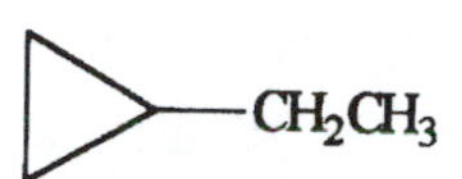


ethylcyclopropane

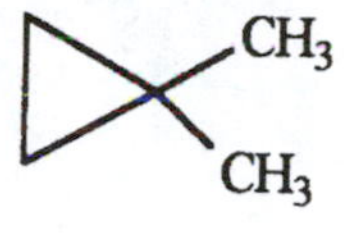


1,1-dimethylcyclopropane

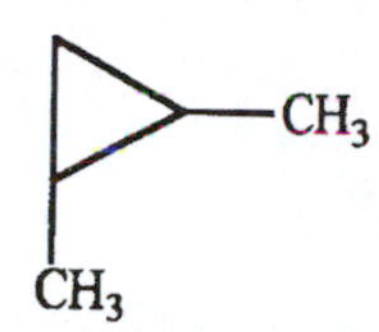


1-2-dimethylcyclopropane

34. (a) elimination
    (b) substitution
    (c) addition

35. It is not possible to distinguish hexane from 3-methylheptane based on solubility in water because both compounds are nonpolar and, thus, insoluble in water.

36. Propane is the most volatile. Volatility: propane > butane > hexane

37. (a) The compounds are isomers.
    (b) The compounds are not the same and are not isomers.
    (c) The compounds are not the same and are not isomers.
    (d) The compounds are not the same and are not isomers.

38. (a) $sp^3$ hybrid orbitals in carbon.
(b) Structural isomers are not possible because only one carbon atom is present.
(c) dichlorodifluromethane
(d) The closest classification for Freon-12 ($CF_2Cl_2$) in Table 19.1 is alkyl halide.

39. (a) $CH_3CH_2CH_2CH_2CH_2CH_2CH_2CH_2CH_2CH_2CH_3$ undecane

$CH_3CH_2CH_2CH_2CH_2CH_2CH_2CH_2CH_2CH_2CH_2CH_2CH_3$ tridecane

(b) Both compounds are alkanes. They are composed of only carbon and hydrogen and have only single bonds between the carbon atoms. Both formulas agree with the general formula for alkanes, $C_nH_{2n+2}$.

40.

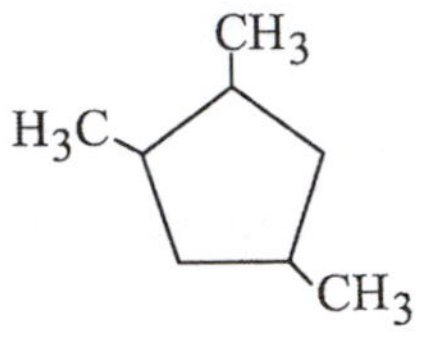

$CH_3$
$H_3C$
$CH_3$

1,3,5-trimethylcyclopentane is an incorrect name of this compound. Numbering the ring counterclockwise gives the correct name, 1,2,4-trimethylcyclopentane.

CHAPTER 20

# UNSATURATED HYDROCARBONS

1. The sigma bond in the double bond of ethene is formed by the overlap of two $sp^2$ electron orbitals and is symmetrical about a line drawn between the nuclei of the two carbon atoms. The pi bond is formed by the sidewise overlap of two p orbitals which are perpendicular to the carbon-carbon sigma bond. The pi bond consists of two electron clouds, one above and one below the plane of the carbon-carbon sigma bond.

2.

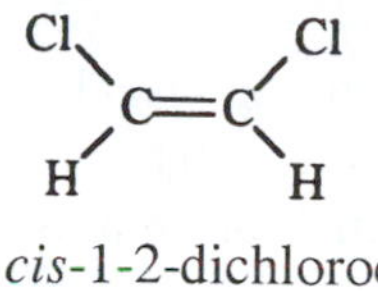

| Cl(H)C=C(Cl)H | H(Cl)C=C(Cl)H | H–CH(Cl)–CH(Cl)–H |
|---|---|---|
| *cis*-1-2-dichloroethene | *trans*-1,2-dichloroethene | 1,2-dichloroethane |

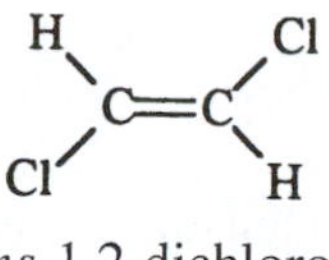

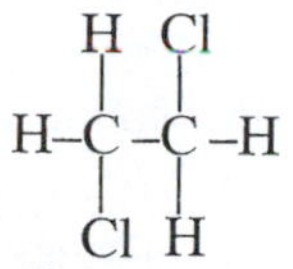

We are able to get *cis-trans* isomers of ethene because there is no rotation around the carbon-carbon double bond, so the two structures shown are not the same. With 1,2-dichloroethane, there is free rotation of the carbon-carbon single bond. In the structure shown, exchanging the chlorine on the right with either hydrogen atom on that carbon appears to make a different isomer, but rotation makes any of the three positions equivalent.

3. Rubber products deteriorate rapidly in smog-ridden areas because ozone, which is present in smog, causes oxidation of the carbon-carbon double bonds, which changes the properties of the rubber.

4. Acetylene presents two different explosion hazards:

   (a) It can form an explosive mixture with oxygen or air;
   (b) When highly compressed or liquefied, acetylene may decompose violently, either spontaneously or from a slight shock.

5. The small molecule released in an elimination reaction can be reacted with and added to the main elimination product. For example, an alkene can be prepared by release of HX from an alkyl halide. In turn, HX can be added to an alkene to form an alkyl halide.

6. During the 10-year period from 1935 to 1945, the major source of aromatic hydrocarbons shifted from coal tar to petroleum due to the rapid growth of several industries which used aromatic hydrocarbons as raw material. These industries include drugs, dyes, detergents, explosives, insecticides, plastics, and synthetic rubber. Since the raw material needs far

exceeded the aromatics available from coal tar, another source had to be found, and processes were developed to make aromatic compounds from alkanes in petroleum. World War II, which occurred during this period, put high demands on many of these industries, particularly explosives.

7. Benzene does not undergo the typical reactions of an alkene. Benzene does not decolorize bromine rapidly and it does not destroy the purple color of permanganate ions. The reactions of benzene are more like those of an alkane. Reaction of benzene with chlorine requires a catalyst. Benzene does not readily add $Cl_2$, but rather a hydrogen atom is replaced by a chlorine atom.

8. The 11-*cis* isomer of retinal combines with a protein (opsin) to form the visual pigment rhodopsin. When light is absorbed, the 11-*cis* double bond is converted to a *trans*-double bond. This process initiates the mechanism of visual excitation which our brains perceive as light or light forms.

9. (a) ethane

```
   H  H
   ..  ..
H : C : C : H
   ..  ..
   H  H
```

(b) ethene

```
H    H
..   ..
C :: C
..   ..
H    H
```

(c) ethyne

```
H : C ::: C : H
```

10. (a) propane

```
   H   H   H
   ..  ..  ..
H : C : C : C : H
   ..  ..  ..
   H   H   H
```

(b) propene

```
   H
   ..
H : C : C :: C : H
   ..   ..   ..
   H    H    H
```

(c) propyne

```
   H
   ..
H : C : C ::: C : H
   ..
   H
```

11. Isomeric iodobutenes, $C_4H_7I$

```
H        H
  \     /
   C = C
  /     \
CH3CH2    I
```

*cis*-1-iodo-1-butene

```
H        I
  \     /
   C = C
  /     \
CH3CH2    H
```

*trans*-1-iodo-1-butene

$CH_3CH_2CI{=}CH_2$
2-iodo-1-butene

$CH_3CHICH{=}CH_2$
3-iodo-1-butene

$CH_2ICH_2CH=CH_2$
4-iodo-1-butene

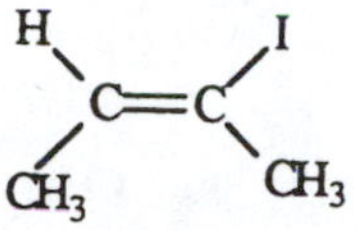

H I
C=C
$CH_3$ $CH_3$
*cis*-2-iodo-2-butene

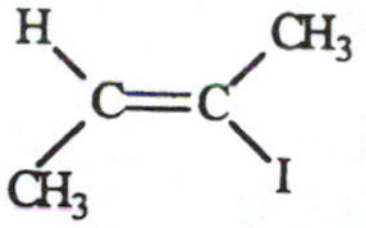

H $CH_3$
C=C
$CH_3$ I
*trans*-2-iodo-2-butene

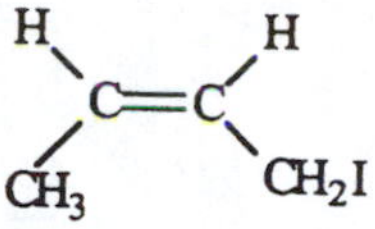

H H
C=C
$CH_3$ $CH_2I$
*cis*-1-iodo-2-butene

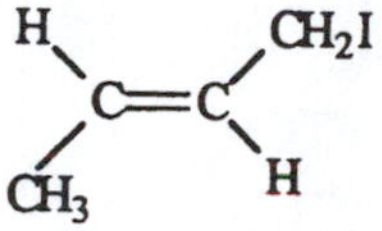

H $CH_2I$
C=C
$CH_3$ H
*trans*-1-iodo-2-butene

$CH_3$
|
$CH_3C=CHI$
1-iodo-2-methylpropene

$CH_3$
|
$CH_2IC=CH_2$
3-iodo-2-methylpropene

12. (a) $C_3H_5Cl$

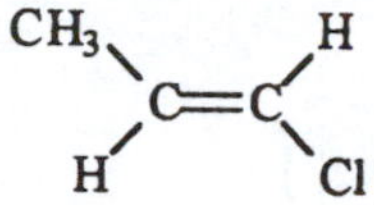

$CH_3$ H
C=C
H Cl
*trans*-1-chloropropene

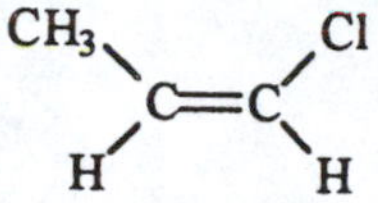

$CH_3$ Cl
C=C
H H
*cis*-1-chloropropene

$CH_3CCl=CH_2$
2-chloropropene

$CH_2ClCH=CH_2$
3-chloropropene

(b) chlorocyclopropane

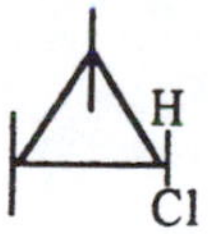

H
Cl

13. (a)

$$\begin{array}{c} CH_3 \qquad\quad CH_3 \\ | \qquad\qquad\ \ | \\ CH_3CHCH=CHCHCH_3 \end{array}$$

2,5-dimethyl-3-hexene

(b)

$CH_3(H)C=C(H)CH(CH_3)CH_3$

*cis*-4-methyl-2-pentene

(c) $CH\equiv CCH=CHCH_3$

3-penten-1-yne

(d)

$CH_3CH_2(H)C=C(H)CH_2CH_3$

*trans*-3-hexene

(e) $CH\equiv CCH(CH_3)CH_2CH_3$

3-methyl-1-pentyne

(f)

$$\begin{array}{c} CH_3 \\ | \\ CH_3CHCHCH_2CH_2CH_3 \end{array}$$

3-methyl-2-phenylhexane

14. (a)

$$\begin{array}{c} CH_2CH_3 \\ | \\ CH_2=CHCCH_2CH_3 \\ | \\ CH_3 \end{array}$$

3-ethyl-3-methyl-1-pentene

(b)

*cis,* 1,2-diphenylethene

(c) $CH\equiv CCHCH_3$ (with phenyl on C-3)

3-phenyl-1-butyne

(d)

cyclopentene

(e) $CH_3$

1-methylcyclohexene

(f) $CH(CH_3)_2$

3-isopropylcyclopentene

15. (a)

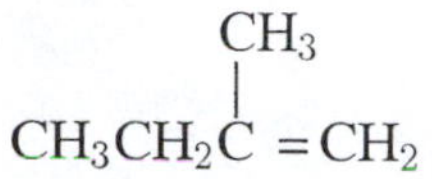

$$\begin{array}{c} \quad\quad\quad CH_3 \\ \quad\quad\quad | \\ CH_3CH_2C = CH_2 \end{array}$$

Numbering was started from the wrong end of the structure.
Correct name: 2-methyl-1-butene

(b)

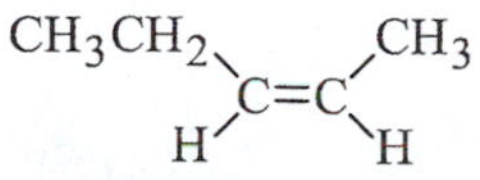

$$\begin{array}{ccc} CH_3CH_2 & & CH_3 \\ & \diagdown C{=}C \diagup & \\ & \diagup \quad\quad \diagdown & \\ H & & H \end{array}$$

Numbering was started from the wrong end of the structure.
Correct name: *cis*-2-pentene

(c)

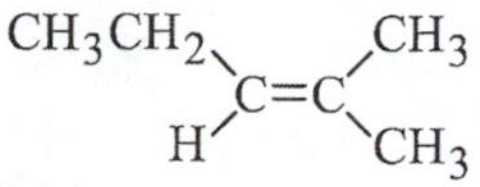

$$\begin{array}{ccc} CH_3CH_2 & & CH_3 \\ & \diagdown C{=}C \diagup & \\ & \diagup \quad\quad \diagdown & \\ H & & CH_3 \end{array}$$

This compound does not have *cis-trans* isomers. Correct name: 2-methyl-2-pentene

16. (a) $CH_2 = CHCHCH_3$ with $CH_2CH_3$ on the third carbon

$$\begin{array}{l} CH_2 = CHCHCH_3 \\ \quad\quad\quad\quad | \\ \quad\quad\quad CH_2CH_3 \end{array}$$

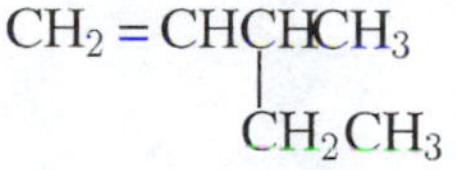

Longest chain contains five carbon atoms. Correct name: 3-methyl-1-pentene

(b)

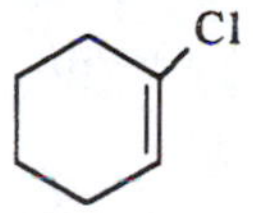

[Structure: cyclohexene ring with Cl on a double-bonded carbon]

The C = C bond in cyclohexene is numbered so that substituted groups have the smallest numbers. Correct name: 1-chlorocyclohexene

(c)

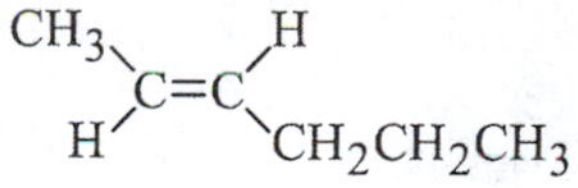

$$\begin{array}{ccc} CH_3 & & H \\ & \diagdown C{=}C \diagup & \\ & \diagup \quad\quad \diagdown & \\ H & & CH_2CH_2CH_3 \end{array}$$

Numbering was started from the wrong end of the structure.
Correct name: *trans*-2-hexene

17. (a) *trans*-3-methyl-3-hexene
    (b) 3-phenyl-1-propyne
    (c) 4,5-dibromo-2-hexyne

18. (a) *cis*-4-methyl-2-hexene
    (b) 2,3-dimethyl-2-butene
    (c) 3-isopropyl-1-pentene

19. All the hexynes, $C_6H_{10}$

$CH_3CH_2CH_2CH_2C{\equiv}CH$
1-hexyne

$CH_3CH_2CH_2C{\equiv}CCH_3$
2-hexyne

$CH_3CH_2C{\equiv}CCH_2CH_3$
3-hexyne

$CH_3CH_2CH(CH_3)C{\equiv}CH$
3-methyl-1-pentyne

$CH_3CH(CH_3)CH_2C{\equiv}CH$
4-methyl-1-pentyne

$CH_3CH(CH_3)C{\equiv}CCH_3$
4-methyl-2-pentyne

$CH_3C(CH_3)_2{-}C{\equiv}CH$
3,3-dimethyl-1-butyne

20. All the pentynes, $C_5H_8$

$CH_3CH_2CH_2C{\equiv}CH$
1-pentyne

$CH_3CH_2C{\equiv}CCH_3$
2-pentyne

$CH_3CH(CH_3)C{\equiv}CH$
3-methyl-1-butyne

21. Only structure (b) will show *cis-trans* isomers. $CH_3CH{=}CHCl$

22. Only structure (c) will show *cis-trans* isomers. $CH_2ClCH{=}CHCH_2Cl$

23. (a) $CH_3CH_2CH_2CH{=}CH_2 + Br_2 \longrightarrow CH_3CH_2CH_2CHBrCH_2Br$

(b) $CH_3CH_2C(CH_3){=}CHCH_3 + HI \longrightarrow CH_3CH_2CI(CH_3)CH_2CH_3$

(c) $CH_3CH_2CH{=}CH_2 + H_2O \xrightarrow{H^+} CH_3CH_2CH(OH)CH_3$

(d)

$$C_6H_5CH{=}CH_2 + H_2 \xrightarrow[\text{1 atm}]{\text{Pt, 25 °C}} C_6H_5CH_2CH_3$$

(e) $$CH_3CH{=}CHCH_3 + KMnO_4 \xrightarrow[\text{cold}]{H_2O} CH_3\underset{HO}{CH}\underset{OH}{CH}CH_3$$

24. (a) $$CH_3CH_2CH_2CH{=}CH_2 + H_2O \xrightarrow{H^+} CH_3CH_2CH_2\underset{OH}{CH}CH_3$$

(b) $$CH_3CH_2CH{=}CHCH_3 + HBr \longrightarrow CH_3CH_2CHBrCH_2CH_3 + CH_3CH_2CH_2CHBrCH$$

(c) $$CH_2{=}CHCl + Br_2 \longrightarrow CH_2BrCHClBr$$

(d)

$$C_6H_5CH{=}CH_2 + HCl \longrightarrow C_6H_5CHClCH_3$$

(e) $$CH_2{=}CHCH_2CH_3 + KMnO_4 \xrightarrow[\text{cold}]{H_2O} \underset{OH}{CH_2}\underset{OH}{CH}CH_2CH_3$$

25. (a) $$CH_3C{\equiv}CCH_3 + Br_2 \text{ (1 mole)} \longrightarrow CH_3CBr{=}CBrCH_3$$

(b) Two-step reaction:

$$CH{\equiv}CH + HCl \longrightarrow CH_2{=}CHCl \xrightarrow{HCl} CH_3CHCl_2$$

(c) $$CH_3CH_2CH_2C{\equiv}CH + H_2 \text{ (1 mole)} \longrightarrow CH_3CH_2CH_2CH{=}CH_2$$

26. (a) $$CH_3C{\equiv}CH + H_2 \text{ (1 mol)} \xrightarrow[\text{1 atm}]{\text{Pt, 25°C}} CH_3CH{=}CH_2$$

(b) $$CH_3C{\equiv}CCH_3 + Br_2 \text{ (2 mol)} \longrightarrow CH_3CBr_2CBr_2CH_3$$

(c) Two step reaction:

$$CH_3C{\equiv}CH + HCl \longrightarrow CH_3CCl{=}CH_2 \xrightarrow{HCl} CH_3CCl_2CH_3$$

27. When cyclohexene,  reacts with:

(a) $Br_2$, the product is 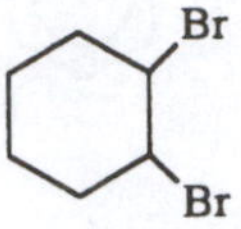 1,2-dibromocyclohexane

(b) HI, the product is 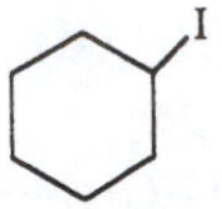 iodocyclohexane

(c) $H_2O$, $H^+$, the product is 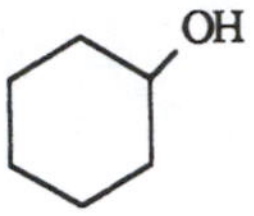 cyclohexanol

(d) $KMnO_4(aq)$, the product is 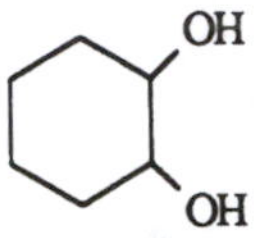 cyclohexene glycol or 1,2-dihydroxycyclohexane

28. When cyclopentene,  reacts with

(a) $Cl_2$, the product is 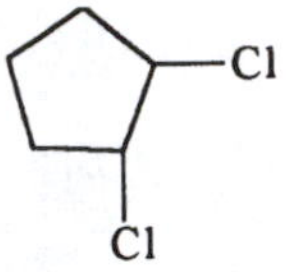 1,2-dichlorcyclopentane

(b) HBr, the product is 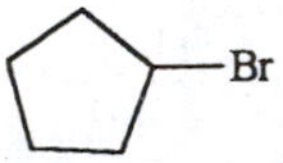 bromocyclopentane

(c) $H_2$, Pt, the product is 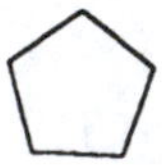 cyclopentane

(d) $H_2O$, $H^+$, the product is 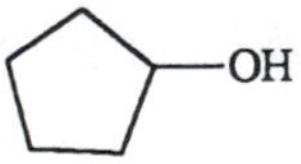 cyclopentanol

29. Reactions to convert 2-butyne to:

(a) 2,3-dibromobutane

$CH_3C{\equiv}CCH_3 + Br_2 \text{ (1 mole)} \longrightarrow CH_3CBr{=}CBrCH_3$

$$CH_3CBr=CBrCH_3 + H_2 \xrightarrow[\text{1 atm}]{\text{Pt, 25°C}} CH_3CHBrCHBrCH_3$$

(b) 2,2-dibromobutane

$$CH_3C\equiv CCH_3 + HBr \longrightarrow CH_3CH=CBrCH_3 \xrightarrow{\text{HBr}} CH_3CH_2CBr_2CH_3$$

(c) 2,2,3,3-tetrabromobutane

$$CH_3C\equiv CCH_3 + Br_2 \text{ (2 moles)} \longrightarrow CH_3CBr_2CBr_2CH_3$$

30. (a) $$CH_3C\equiv CCH_2CH_3 + Cl_2 \text{ ( 1 mole)} \longrightarrow CH_3CCl=CClCH_2CH_3$$

$$CH_3CCl=CClCH_2CH_3 + H_2 \xrightarrow[\text{1 atm}]{\text{25°C}} CH_3CHClCHClCH_2CH_3$$

(b) $$CH_3C\equiv CCH_2CH_3 + HCl \longrightarrow CH_3CCl=CHCH_2CH_3$$

$$CH_3CCl=CHCH_2CH_3 + HCl \longrightarrow CH_3CCl_2CH_2CH_2CH_3$$

Can also yield $CH_3CH_2CCl_2CH_2CH_3$ at the same time.

(c) $$CH_3C\equiv CCH_2CH_3 + 2\,Cl_2 \longrightarrow CH_3CCl_2CCl_2CH_2CH_3$$

31. (a) (b) $NH_2$ (c) COOH (d)

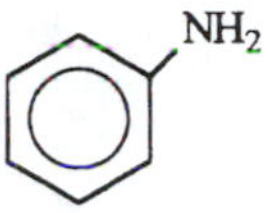

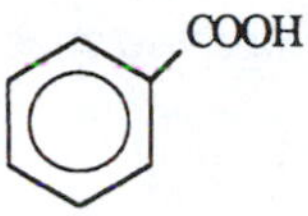

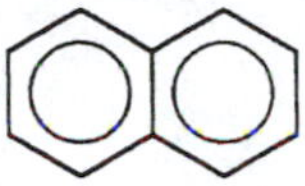

32. (a) $CH_3$ (b) $CH_2CH_3$

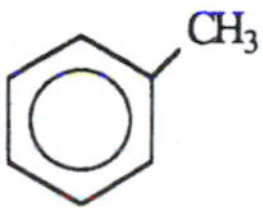

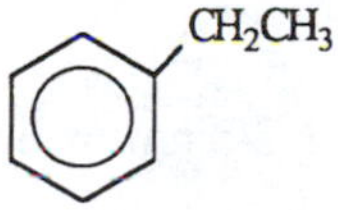

(c) OH (d)

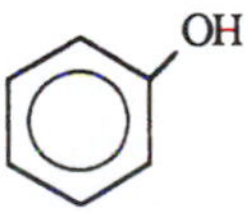

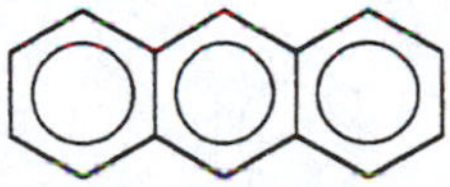

33. (a) Br Br Br

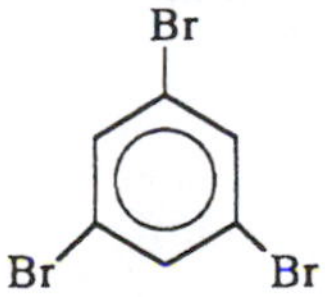

1,3,5-tribromobenzene

(b) Cl Br

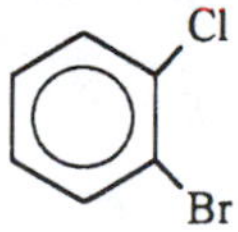

*o*-bromochlorobenzene

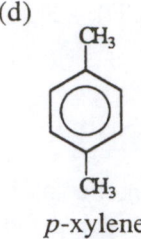

(c)

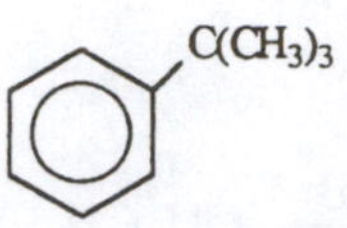

$C(CH_3)_3$

*tert*-butylbenzene

(d)

$CH_3$

$CH_3$

*p*-xylene

34. (a)

Cl

Cl $NO_2$

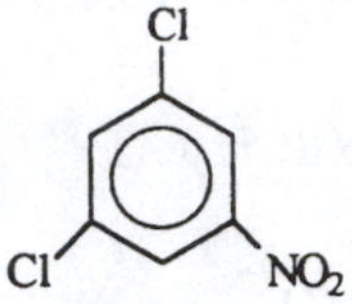

1,3-dichloro-5-nitrobenzene

(b)

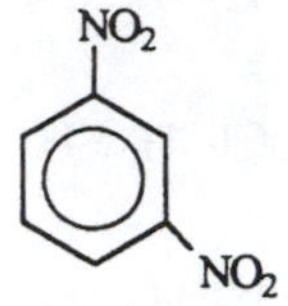

$NO_2$

$NO_2$

*m*-dinitrobenzene

(c)

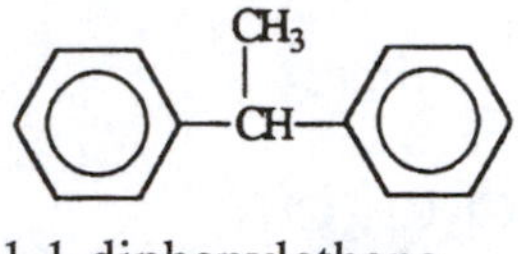

$CH_3$

CH

1,1-diphenylethane

(d)

$CH=CH_2$

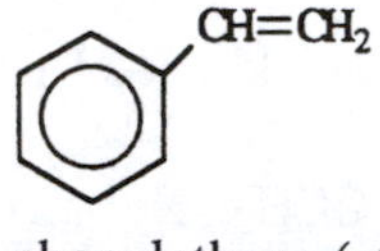

phenylethene (styrene)

35. (a) bromodichlorobenzenes

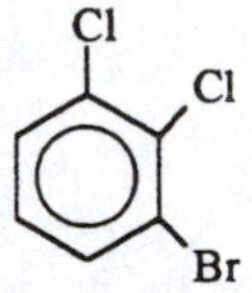

Cl Cl Br

3-bromo-1,2-dichlorobenzene

Cl Cl Br

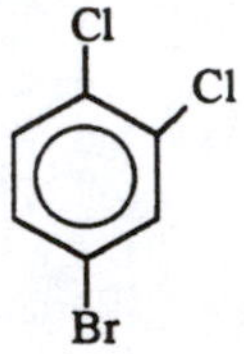

4-bromo-1,2-dichlorobenzene

Cl Br Cl

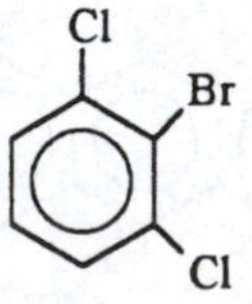

2-bromo-1,3-dichlorobenzene

Cl Cl Br

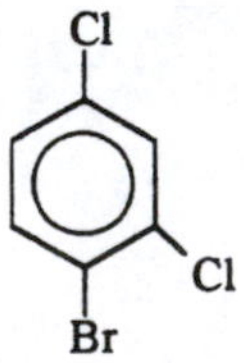

4-bromo-1,3-dichlorobenzene

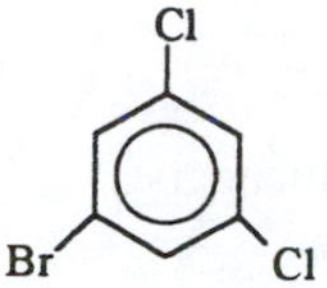

5-bromo-1,3-dichlorobenzene 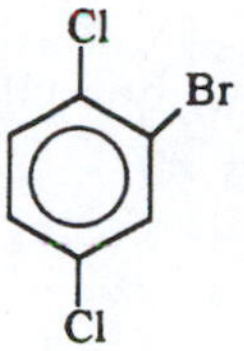　　2-bromo-1,4-dichlorobenzene

(b) The toluene derivatives of formula $C_9H_{12}$:

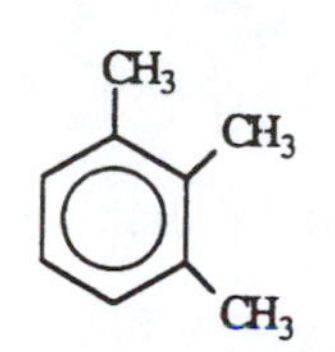

1,2,3-trimethylbenzene

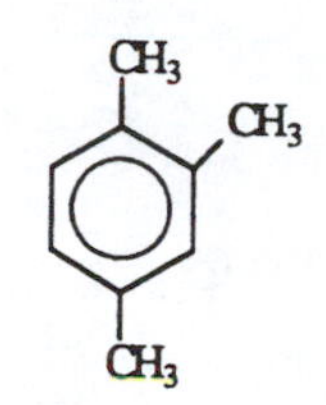

1,2,4-trimethylbenzene

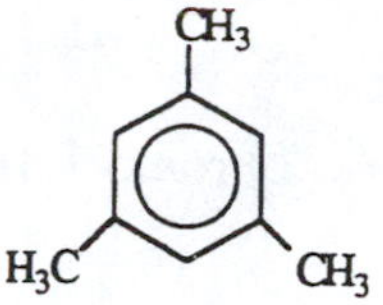

1,3,5-trimethylbenzene

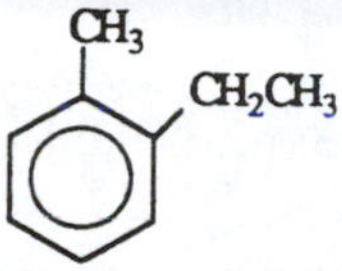

*o*-ethyltoluene

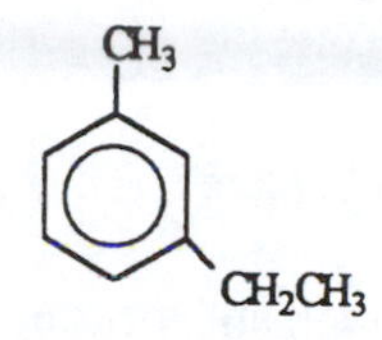

*m*-ethyltoluene

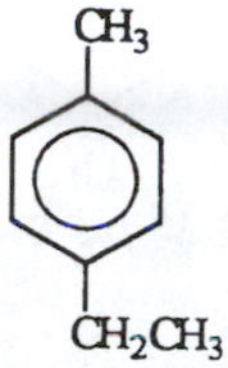

*p*-ethyltoluene

36. (a) trichlorobenzenes

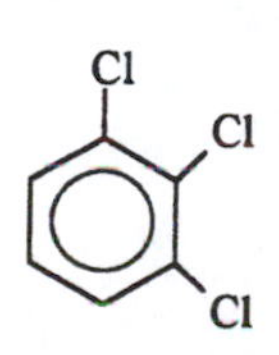

1,2,3-trichlorobenzene

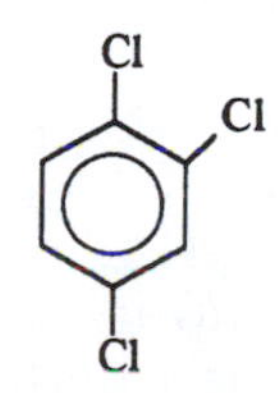

1,2,4-trichlorobenzene

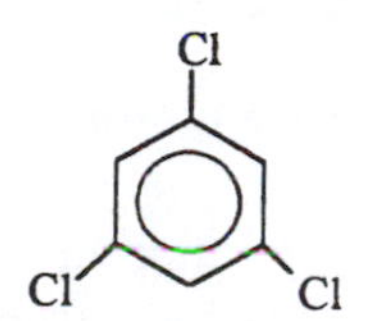

1,3,5-trichlorobenzene

(b) The benzene derivatives of formula $C_8H_{10}$:

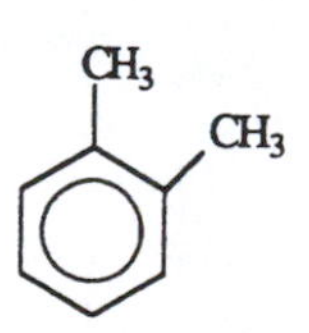

*o*-xylene or 1,2-dimethylbenzene

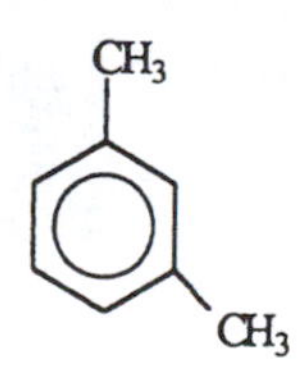

*m*-xylene or 1,3-dimethylbenzene

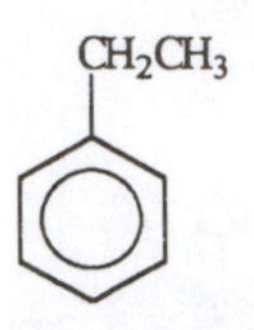


*p*-xylene or 1,4-dimethylbenzene

ethylbenzene

37. All isomers that can be written by substituting another chlorine in *o*-chlorobromobenzene.

(a)

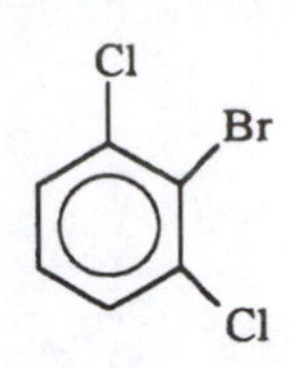


2-bromo-1,3-dichlorobenzene

(b)

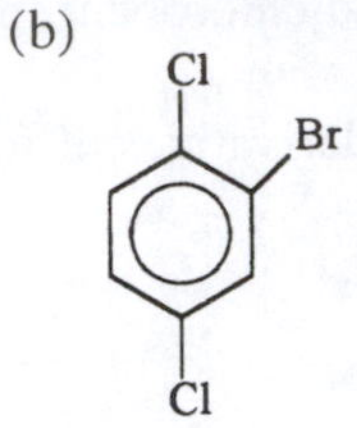


2-bromo-1,4-dichlorobenzene

(c)

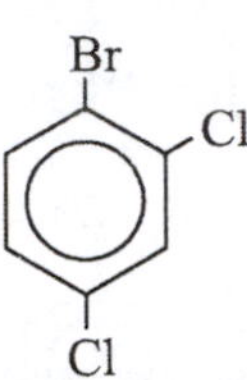


1-bromo-2,4-dichlorobenzene

(d)

1-bromo-2,3-dichlorobenzene

38. All isomers that can be written substituting a third chlorine atom on *o*-dichlorobenzene.

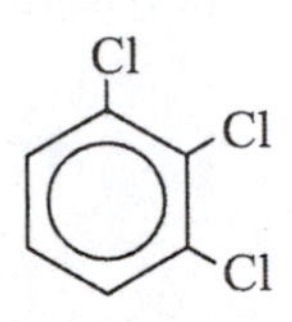


1,2,3-trichlorobenzene

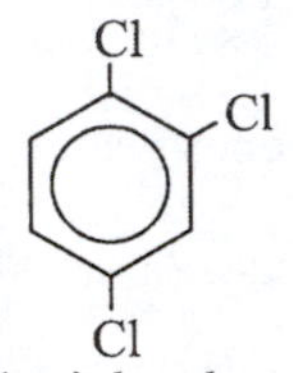


1,2,4-triclorobenzene

39. (a) *p*-chloroethylbenzene
(b) propylbenzene
(c) *m*-nitroaniline
(d) *p*-bromophenol
(e) triphenylmethane

40. (a) styrene
(b) *m*-nitrotoluene
(c) 2,4-dibromobenzoic acid
(d) isopropylbenzene
(e) 2,4,6-tribromophenol

41. (a)

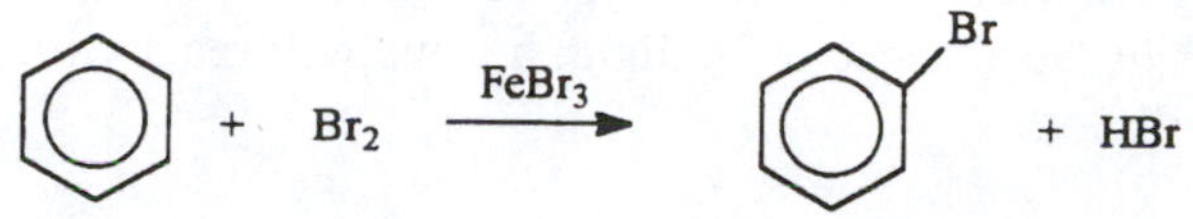

$$C_6H_6 + Br_2 \xrightarrow{FeBr_3} C_6H_5Br + HBr$$

bromobenzene

(b)

$$CH_3C_6H_4CH_3 + HNO_3 \xrightarrow{H_2SO_4} (CH_3)_2C_6H_3NO_2 + H_2O$$

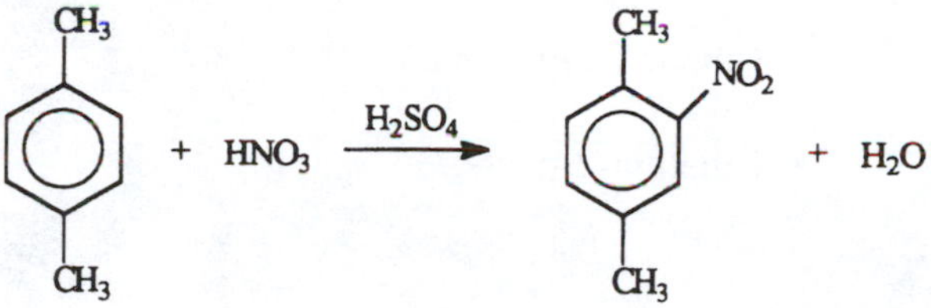

1,4-dimethyl-2-nitrobenzene

42. (a)

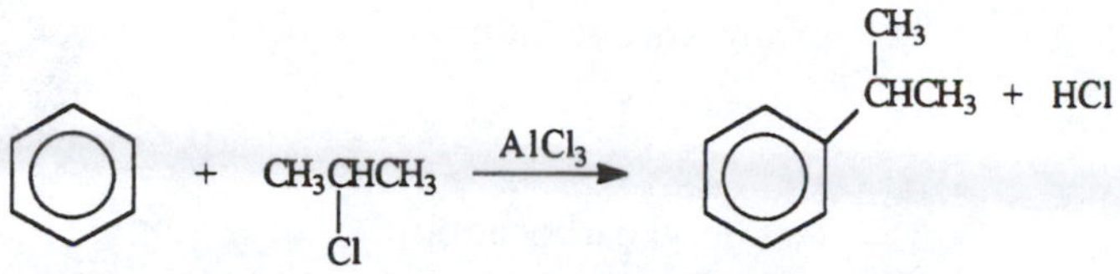

$$C_6H_6 + CH_3CHClCH_3 \xrightarrow{AlCl_3} C_6H_5CH(CH_3)CH_3 + HCl$$

isopropylbenzene

(b)

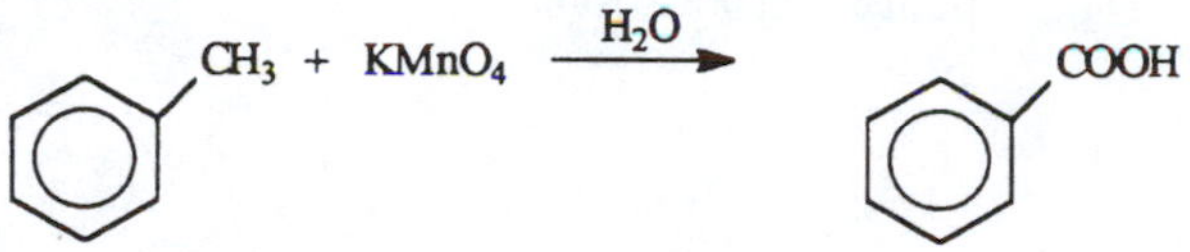

$$C_6H_5CH_3 + KMnO_4 \xrightarrow{H_2O} C_6H_5COOH$$

benzoic acid

43. When $CH_2{=}C(CH_3)CH_2CH_2CH_3$ reacts with HBr, two products are possible:

$CH_3CBr(CH_3)CH_2CH_2CH_3$ and $CH_2BrCH(CH_3)CH_2CH_2CH_3$

The first will strongly predominate. This is the product according to Markovnikov's rule and forms because the tertiary carbocation intermediate formed is more stable than a primary carbocation.

44. Two tests can be used. (1) Baeyer test—hexene will decolorize $KMnO_4$ solution; cyclohexane will not. (2) In the absence of sunlight, hexene will react with and decolorize bromine; cyclohexane will not.

45. Methylpentenes that show geometric isomerism.

$$CH_3CH{=}\overset{\overset{\large CH_3}{|}}{C}CH_2CH_3$$ 3-methyl-2-pentene

$$CH_3CH{=}CH\underset{\underset{\large CH_3}{|}}{C}HCH_3$$ 4-methyl-2-pentene

46. (a) $\overset{+}{C}H_3$ methyl carbocation

(b) $CH_3CH_2\overset{+}{C}H_2$ propyl carbocation

(c) $CH_3\overset{\overset{\large CH_3}{|}}{\underset{\underset{\large CH_3}{|}}{C^+}}$ *tert*-butyl carbocation

(d) $CH_3CH_2CH_2CH_2\overset{+}{C}H_2$ pentyl carbocation

47. (a)

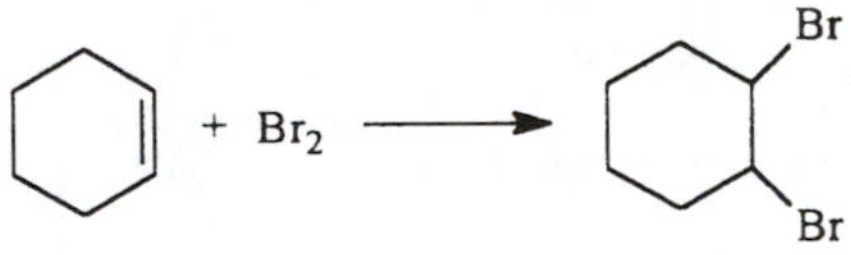

cyclohexene + $Br_2$ ⟶ 1,2-dibromocyclohexane

(b)

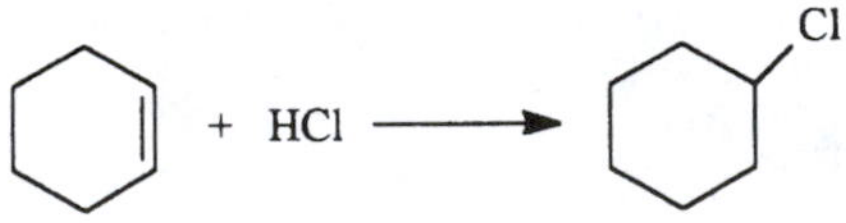

cyclohexene + HCl ⟶ chlorocyclohexane

(c)

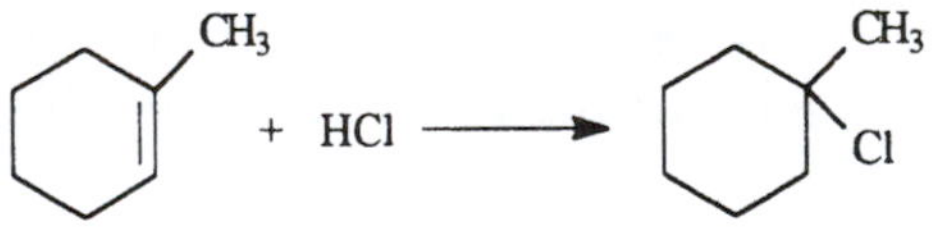

1-methylcyclohexene + HCl ⟶ 1-chloro-1-methylcyclohexane

(d)

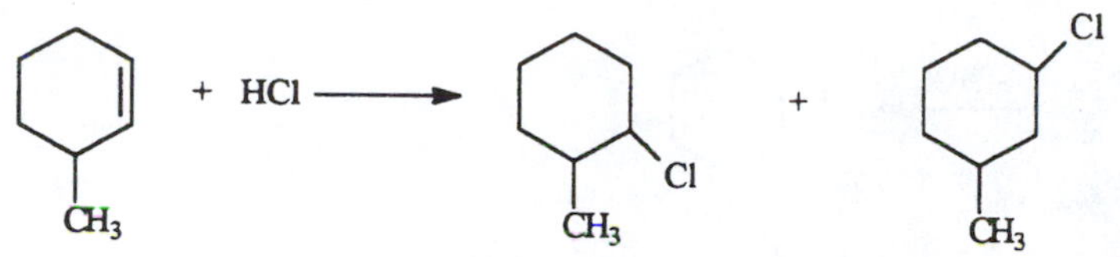

+ HCl ⟶ + 

Cl
Cl
$CH_3$
$CH_3$
$CH_3$

2-chloro-1-methylcyclohexane 3-chloro-1-methylcyclohexane

48. The reaction mechanism by which benzene is brominated in the presence of $FeBr_3$:

(a) $FeBr_3 + Br_2 \longrightarrow FeBr_4^- + Br^+$
Formation of a bromonium ion ($Br^+$), an electrophile.

(b)

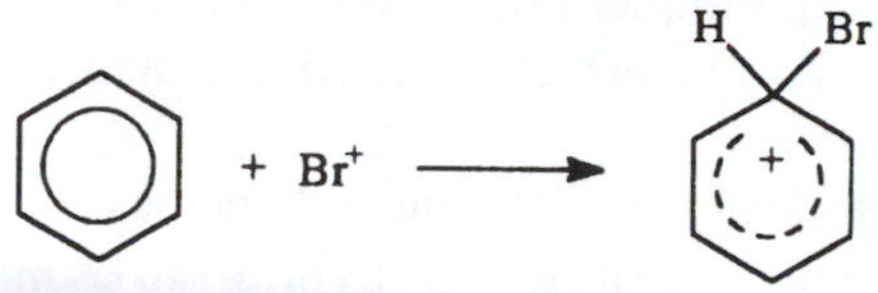

H Br

+ $Br^+$ ⟶

The bromonium ion adds to benzene forming a carbocation intermediate.

(c)

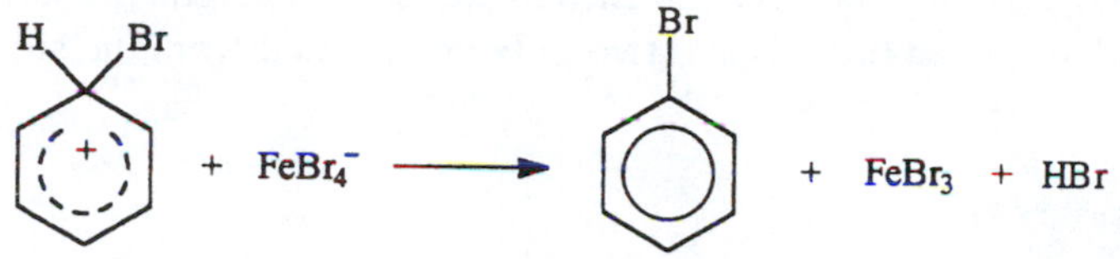

H Br Br

+ $FeBr_4^-$ ⟶ + $FeBr_3$ + HBr

A hydrogen ion is lost from the carbocation forming the product bromobenzene.

49. Chemically distinguishing between benzene, 1-hexene, and 1-hexyne

Step 1. Add $KMnO_4$ solution to a sample of each liquid. Benzene is the only one in which the $KMnO_4$ does not lose its purple color.

Step 2. To 0.5 mL samples of 1-hexene and 1-hexyne add bromine solution dropwise until there is no more color change of the bromine (from reddish-brown to colorless). 1-hexyne (with a triple bond) will decolor about twice as many drops of bromine as 1-hexene. Thus the three liquids are identified.

50. (a) $CH_3CHClCH_2CH_3 \xrightarrow{-HCl} CH_2=CHCH_2CH_3 + CH_3CH=CHCH_3$

(b) $CH_2ClCH_2CH_2CH_2CH_3 \xrightarrow{-HCl} CH_2=CHCH_2CH_2CH_3$

(c)

Cl $\xrightarrow{-HCl}$

51. Yes, there will be a color change (loss of $Br_2$ color). The fact that there is no HBr formed indicates that the reaction is not substitution but addition. Therefore, $C_4H_8$, must contain a carbon-carbon double bond. Three structures are possible.

$CH_3CH_2CH{=}CH_2$ $\quad$ $CH_3CH{=}CHCH_3$ $\quad$ $CH_3C(CH_3){=}CH_2$

52. Baeyer test: Add $KMnO_4$ solution to each sample. The $KMnO_4$ will lose its purple color with 1-heptene. There will be no reaction (no color change) with heptane.

53. The carbon-carbon bond in the three molecules is different. Ethane has a single bond between carbon atoms formed from the overlap of two $sp^3$ hybridized orbitals. Ethene has a double bond between carbon atoms. One bond is a sigma bond formed by the overlap of two $sp^2$ hybridized orbitals, the other is a pi bond formed by the sidewise overlap of two p orbitals. Ethyne has a triple bond between carbon atoms—one sigma bond formed from the overlap of two sp hybridized orbitals and two pi bonds formed from the sidewise overlap of p orbitals.

54.

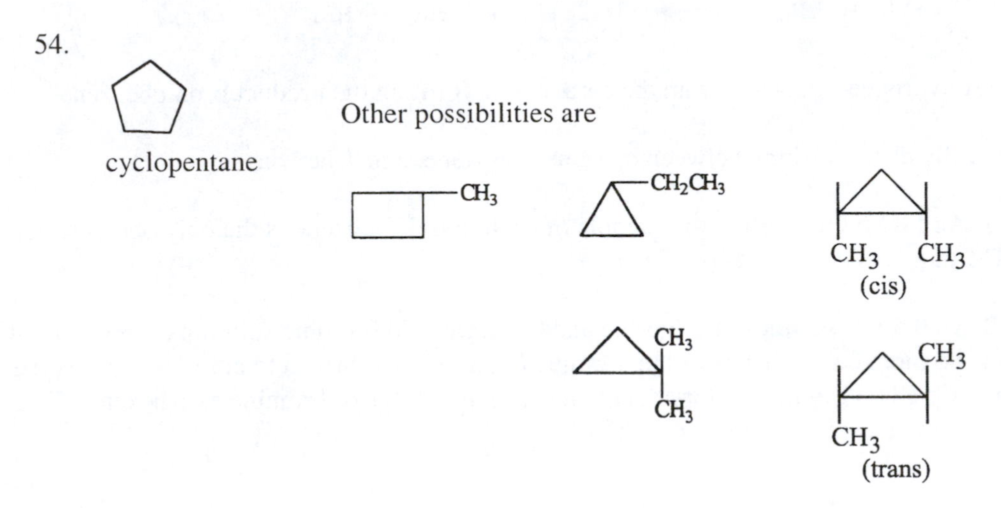

55. Benzene compounds of formula $C_8H_9Cl$

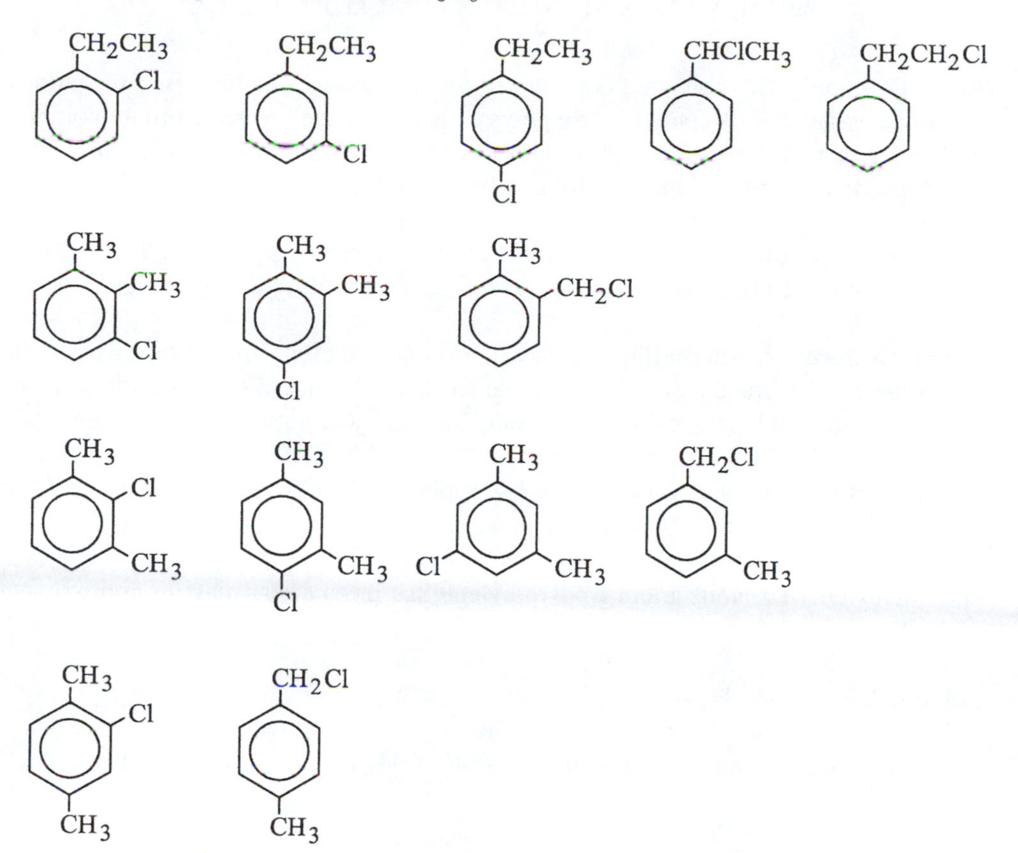

CHAPTER 21

# POLYMERS: MACROMOLECULES

1. An addition polymer is one that is produced by the successive addition of repeating monomer molecules. A condensation polymer is one that is formed from monomer molecules in a reaction which splits out water or some other simple molecule. Condensation polymerization usually involves two different monomers.

2. Those polymers which soften on reheating are thermoplastic polymers; those which set to an infusible solid and do not soften on reheating are thermosetting polymers.

3. In order to form a thermosetting polymer, there must be cross linking of polymer chains. This requires that some monomer molecules be trifunctional. If all molecules are only bi-functional, then only long chains can result, and they will not be thermosetting.

4. Vulcanization is the process of heating raw rubber with sulfur. Sulfur atoms are introduced as cross links between the polymeric chains.

5. How many ethylene units are in a polyethylene that has a molar mass of approximately 25,000?

   Polyethylene $-\!\!(CH_2-CH_2)\!\!-_n$

   Molar mass of one unit $= 2(12.01\ \text{g/mol}) + 4(1.008\ \text{g/mol}) = 28.05\ \text{g/mol}$

   $$25{,}000\ \text{g/mol} \times \frac{\text{ethylene unit}}{28.05\ \text{g/mol}} = 8.9 \times 10^2\ \text{ethylene units}$$

6. Polystyrene $\quad C_8H_8$

   $-\!\!\left[CH_2-CH(C_6H_5)\right]\!\!-_n$

   molar mass of 1 unit: $8(12.01\ \text{g/mol}) + 8(1.008\ \text{g/mol}) = 104.1\ \text{g/mol}$

   $$(3000\ \text{units})\left(\frac{104.1\ \text{g/mol}}{\text{unit}}\right) = 3 \times 10^5\ \text{g/mol} = \text{molar mass}$$

7. (a) $\left( CH_2CH \right)_n$ with Cl on CH

(b) $\left( C(CH_3)(C_6H_5) — CH(CH_3) \right)_n$

(c) $\left( CH(CH_3)—CH_2 \right)_n$

(d) $\left( C(CH_3)(C(=O)—OCH_3)—CH_2 \right)_n$

8. Structural formulas for:

(a) Saran $—\left(CH_2CCl_2\right)_n$ (b) Orlon $\left( CH_2CH(CN) \right)_n$

(c) Teflon $—\left(CF_2CF_2\right)_n$ (d) Polystyrene $\left( CH_2CH(C_6H_5) \right)_n$

9. polypropylene free radical (2 units)

Starting with RO• free radical $ROCH_2CH(CH_3)–CH_2CH(CH_3)\bullet$

10. polyethylene free radical (2 units)

Starting with RO• free radical $ROCH_2CH_2CH_2CH_2\bullet$

11. Formulas of the polymers that can be formed from:

(a) propylene $\left( CH_2—CH(CH_3) \right)_n$ (b) 2-methylpropene $\left( CH_2—C(CH_3)_2 \right)_n$

(c) 2-butene $\left( CH(CH_3)—CH(CH_3) \right)_n$

12. Formulas of the polymers that can be formed from:
(a) ethylene $-(CH_2CH_2)_n-$ (b) chloroethene $-(CH_2CHCl)_n-$

(c) 1-butene $-(CH_2-CH(CH_2CH_3))_n-$

13. Two possible ways in which vinyl chloride can polymerize to form polyvinyl chloride:

$-(CH_2CH(Cl)-CH_2CH(Cl)-CH_2CH(Cl)-CH_2CH(Cl))_n-$

$-(CH_2CH(Cl)-CH(Cl)CH_2-CH_2CH(Cl)-CH(Cl)CH_2)_n-$ (other possibilities also)

14. Two possible ways in which acrylonitrile can polymerize to form Orlon.

$-(CH_2CH(CN)-CH_2CH(CN)-CH_2CH(CN)-CH_2CH(CN))_n-$

$-(CH_2CH(CN)-CH(CN)CH_2-CH_2CH(CN)-CH(CN)CH_2)_n-$ (other possibilities also)

15. Natural rubber (all *cis*)

$-(CH_2C(CH_3)=CHCH_2-CH_2C(CH_3)=CHCH_2-CH_2C(CH_3)=CHCH_2)_n-$

16. Gutta percha (all *trans*)

$-(CH_2C(CH_3)=CHCH_2-CH_2C(CH_3)=CHCH_2-CH_2C(CH_3)=CHCH_2)_n-$

17. Yes, smog in the atmosphere contains ozone and ozone attacks natural rubber at the site of the double bond causing "age hardening" and cracking.

18. Yes, styrene-butadiene rubber contains carbon-carbon double bonds and is attacked by the ozone in smog causing "age hardening" and cracking.

19. Chemical structures for the monomers of:
    (a) natural rubber $CH_2=C(CH_3)CH=CH_2$

    (b) synthetic rubber (SBR) $C_6H_5CH=CH_2$ and $CH_2=CHCH=CH_2$

20. Chemical structures for the monomers of:
    (a) synthetic natural rubber $CH_2=C(CH_3)CH=CH_2$

    (b) neoprene rubber $CH_2=CCl\text{–}CH=CH_2$

21. Nitrile rubber (Buna N)

$$-\!\!\left(CH_2CH=CHCH_2CH_2\underset{\displaystyle CN}{\underset{|}{C}}HCH_2CH=CHCH_2CH_2CH=CHCH_2CH_2\underset{\displaystyle CN}{\underset{|}{C}}HCH_2CH=CHCH_2\right)_{\!n}$$

(other structures are possible)

22. (a) Isotactic polypropylene

$$-\!\!\left(CH_2\underset{\displaystyle CH_3}{\underset{|}{C}}HCH_2\underset{\displaystyle CH_3}{\underset{|}{C}}HCH_2\underset{\displaystyle CH_3}{\underset{|}{C}}HCH_2\underset{\displaystyle CH_3}{\underset{|}{C}}H\right)_{\!n}$$

(b) Another form of polypropylene

$$-\!\!\left(CH_2\underset{\displaystyle CH_3}{\underset{|}{C}}HCH_2\overset{\displaystyle CH_3}{\overset{|}{C}}HCH_2\underset{\displaystyle CH_3}{\underset{|}{C}}HCH_2\overset{\displaystyle CH_3}{\overset{|}{C}}H\right)_{\!n}$$

(other structures are possible)

23. Polystyrene has the highest mass percent carbon.

$$-\!\!\left(CH_2-\underset{\displaystyle C_6H_5}{\underset{|}{C}}H\right)\!\!-$$

$C_8H_8$ molar mass = 104.1 g/mol

$$\frac{96.08 \text{ g C}}{104.1 \text{ g}} \times 100 = 92.30\% \text{ C}$$

24. Polymer of $CH_2=CHC(=O)-OCH_2CH_3$

$$-\!\!\left(CH_2CH(C(=O)OCH_2CH_3)-CH_2CH(C(=O)OCH_2CH_3)-CH_2CH(C(=O)OCH_2CH_3)\right)_n$$

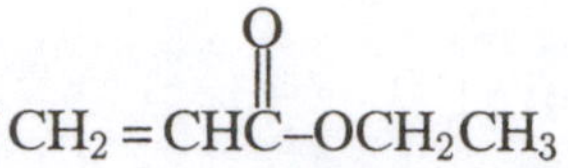

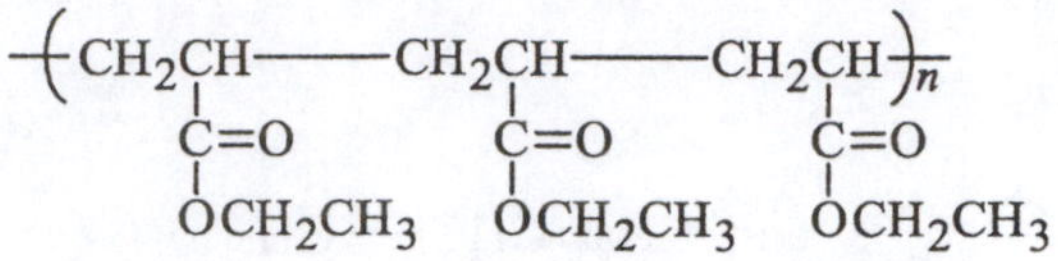

25. Cyanoacrylate ester polymer

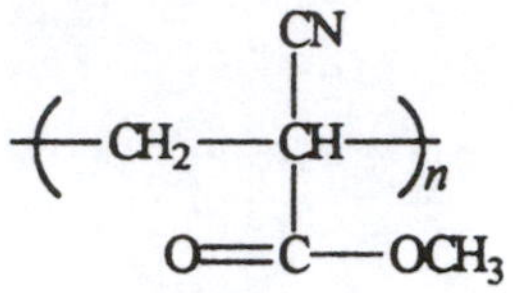

$$-\!\!\left(CH_2-C(CN)(C(=O)-OCH_3)\right)_n$$

26. (a) From 2,3-dimethy1-1,3-butadiene, $CH_2=C(H_3C)-C(CH_3)=CH_2$

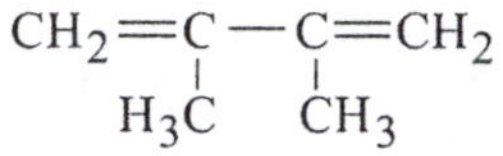

this polymer can be made $-\!\!\left(CH_2C(H_3C)=C(CH_3)CH_2\right)_n$ or $-\!\!\left(CH_2C(CH_3)=C(CH_3)CH_2\right)_n$

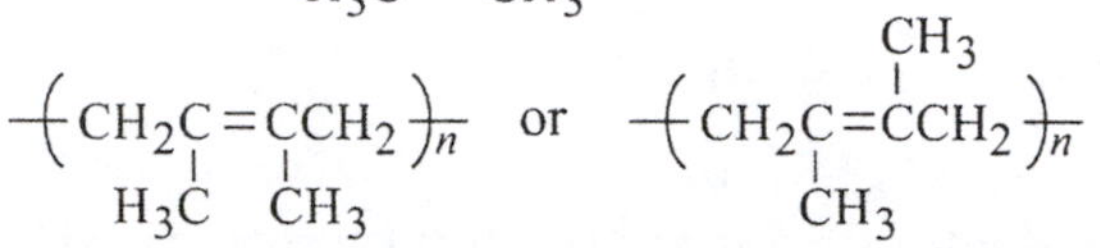

(b) If produced by the free-radical mechanism, a random mixture of *cis* and *trans* connections are made. It is possible, using catalysts, for the reaction to proceed by an ionic mechanism which will give a stereochemically controlled polymer.

27. It is easier to recycle a thermoplastic polymer because they are linear and can easily be reformed on heating. On the other hand, the cross linkages in thermosetting polymers are not easily broken or reformed.

28. $CH_2=CH_2$ ethylene $\quad CF_2=CF_2$ tetrafluoroethylene $\quad -\!\!\left(CH_2-CH_2-CF_2-CF_2-CH_2-CH_2-CF_2-CF_2\right)_n$ (many other possible structures)

29. The monomer of natural rubber is $CH_2=C(CH_3)-CH=CH_2$

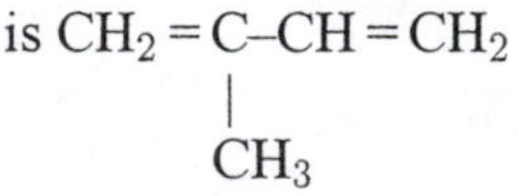

Molar mass is 68.11 g/mol

$$(250{,}000\text{ g/mol})\left(\frac{1\text{ isoprene unit}}{68.11\text{ g/mol}}\right) = 3.7\times10^3\text{ isoprene units.}$$

30. The methylmethacrylate polymers result from an addition reaction because no small molecule is split out and the polymer forms in a manner similar to that of polyethylene.

CHAPTER 22

# ALCOHOLS, ETHERS, PHENOLS, AND THIOLS

1. The question allows great freedom of choice. These shown here are very simple examples of each type.

   (a) an alkyl halide — $CH_3CH_2Cl$

   (b) a phenol — (benzene ring)–OH

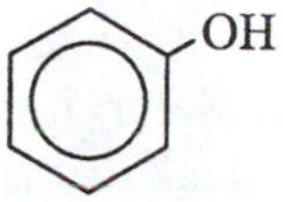

   (c) an ether — $CH_3CH_2OCH_2CH_3$

   (d) an aldehyde — $CH_3\overset{H}{\overset{|}{C}}=O$

   (e) a ketone — $CH_3\overset{O}{\overset{\|}{C}}CH_3$

   (f) a carboxylic acid — $CH_3COOH$

   (g) an ester — $CH_3\overset{O}{\overset{\|}{C}}{-}OCH_2CH_3$

   (h) a thiol — $CH_3CH_2SH$

2. Alkenes are almost never made from alcohols because the alcohols are almost always the higher value material. This is because recovering alkenes from hydrocarbon sources in an oil refinery (primarily catalytic cracking) is a relatively cheap process.

3. Isopropyl alcohol is usually used for rubbing alcohol, rather than *n*-propyl alcohol, because of price. The easy way to make the alcohol, by adding $H_2O$ to propene, yields isopropyl alcohol rather than *n*-propyl alcohol. Any process for producing *n*-propyl alcohol would be much more expensive.

4. Oxidation of primary alcohols yields aldehydes. Further oxidation yields carboxylic acids.

Examples:

$$CH_3CH_2OH \xrightarrow{[O]} CH_3\overset{\overset{\displaystyle O}{\|}}{C}\text{–}H + H_2O$$

$$CH_3\overset{\overset{\displaystyle O}{\|}}{C}\text{–}H \xrightarrow{[O]} CH_3\overset{\overset{\displaystyle O}{\|}}{C}\text{–}OH$$

5. 1,2-Ethanediol is superior to methanol as an antifreeze because of its low volatility. Methanol is much more volatile than water. If the radiator leaks gas under pressure (normally steam), it would primarily leak methanol vapor so you would soon have no antifreeze. Ethylene glycol has a lower volatility than water, so it does not present this problem.

6. Physiological effects of:

   (a) methanol: It is a poisonous liquid capable of causing blindness or death if taken internally. Exposure to methanol vapors is also very dangerous.

   (b) ethanol: It can act as a food, a drug, and a poison. The body can metabolize small amounts of ethanol to produce energy; thus it is a food. It depresses brain functions so that activities requiring skill and judgment are impaired; thus it is a drug. With very large consumption, the depression of brain function can lead to unconsciousness and death; thus it is a poison.

7. The cumene-hydroperoxide synthesis of phenol and acetone:

$$C_6H_6 + CH_3CH{=}CH_2 \xrightarrow{H_2SO_4} C_6H_5\text{–}CH(CH_3)_2 \xrightarrow{O_2} C_6H_5\text{–}C(CH_3)_2\text{–}O\text{–}O\text{–}H$$

cumene      cumene hydroperoxide

$$\xrightarrow[H_2SO_4]{\text{dilute}} C_6H_5OH + CH_3\underset{\underset{\displaystyle O}{\|}}{C}CH_3$$

phenol      acetone

8. Low molar mass ethers present two hazards. They are very volatile and their highly flammable vapors form explosive mixtures with air. They also slowly react with oxygen in the air to form unstable explosive peroxides.

9. Ethanol (molar mass = 46.07) is a liquid at room temperature because it has a significant amount of hydrogen bonding between molecules in the liquid state, and thus has a much higher boiling point than would be predicted from molar mass alone. Dimethyl ether (molar mass = 46.07) is not capable of hydrogen bonding, so has low attraction between molecules, making it a gas at room temperature.

10. Fats are a major source of energy in the body. Some infants die from SIDS due to a lack of energy because they are unable to degrade fats. One of the intermediate compounds in the degradation of fats is an alcohol, which in the process is oxidized to a ketone.

11. (a) methanol $CH_3OH$

(b) 3-methyl-1-hexanol

$$\begin{array}{c} \quad\quad\quad\;\; CH_3 \\ \quad\quad\quad\;\; | \\ CH_3CH_2CH_2CHCH_2CH_2OH \end{array}$$

(c) 1,2-propanediol

$$\begin{array}{l} CH_3CHCH_2OH \\ \quad\;\; | \\ \quad\;\; OH \end{array}$$

(d) 1-phenylethanol

$$\begin{array}{l} C_6H_5\text{–}CHCH_3 \\ \quad\quad\;\;\; | \\ \quad\quad\;\;\; OH \end{array}$$

(e) 2,3-butanediol

$$\begin{array}{l} CH_3CH\text{–}CHCH_3 \\ \quad\;\; |\quad\;\; | \\ \quad\;\; OH\; OH \end{array}$$

(f) 2-propanethiol

$$\begin{array}{l} CH_3CHCH_3 \\ \quad\;\; | \\ \quad\;\; SH \end{array}$$

12. (a) 2-butanol

$$\begin{array}{l} CH_3CHCH_2CH_3 \\ \quad\;\; | \\ \quad\;\; OH \end{array}$$

(b) 2-methyl-2-butanol

$$\begin{array}{l} \quad\;\; CH_3 \\ \quad\;\; | \\ CH_3CCH_2CH_3 \\ \quad\;\; | \\ \quad\;\; OH \end{array}$$

(c) 2-propanol

$$\begin{array}{l} CH_3CHCH_3 \\ \quad\;\; | \\ \quad\;\; OH \end{array}$$

(d) cyclopentanol

$C_5H_9\text{–}OH$

(e) 1-pentanethiol $CH_3CH_2CH_2CH_2CH_2SH$

(f) 4-ethyl-3-hexanol

$$\begin{array}{l} \qquad\qquad\;\; CH_2CH_3 \\ \qquad\qquad\quad\;\, | \\ CH_3CH_2CH\;CHCH_2CH_3 \\ \qquad\quad\;\; | \\ \qquad\quad\; OH \end{array}$$

13. Eight open chain isomeric alcohols with the formula $C_5H_{11}OH$:

1-pentanol

$CH_3CH_2CH_2CH_2CH_2OH$

2-pentanol

$$\begin{array}{l} \qquad\qquad\qquad\; OH \\ \qquad\qquad\qquad\;\; | \\ CH_3CH_2CH_2CHCH_3 \end{array}$$

3-pentanol

$$\begin{array}{l} \qquad\qquad\; OH \\ \qquad\qquad\;\; | \\ CH_3CH_2CHCH_2CH_3 \end{array}$$

3-methyl-1-butanol

$$\begin{array}{l} \quad\;\; CH_3 \\ \quad\;\;\; | \\ CH_3CHCH_2CH_2OH \end{array}$$

2-methyl-1-butanol

$$\begin{array}{l} \qquad\qquad\; CH_3 \\ \qquad\qquad\;\; | \\ CH_3CH_2CHCH_2OH \end{array}$$

2-methyl-2-butanol

$$\begin{array}{l} \qquad\qquad\; CH_3 \\ \qquad\qquad\;\; | \\ CH_3CH_2CCH_3 \\ \qquad\qquad\;\; | \\ \qquad\qquad\; OH \end{array}$$

3-methyl-2-butanol

$$\begin{array}{l} \quad\;\; CH_3 \\ \quad\;\;\; | \\ CH_3CHCHCH_3 \\ \qquad\quad\; | \\ \qquad\quad OH \end{array}$$

2,2-dimethyl-1-propanol

$$\begin{array}{l} \quad\;\; CH_3 \\ \quad\;\;\; | \\ CH_3CCH_2OH \\ \quad\;\;\; | \\ \quad\;\; CH_3 \end{array}$$

14. (a) $C_3H_8O$

| *Alcohols* | *Ethers* |
|---|---|
| $CH_3CH_2CH_2OH$ | $CH_3OCH_2CH_3$ |
| $CH_3CH(OH)CH_3$ | |

(b) $C_4H_8O$

| *Alcohols* | *Ethers* |
|---|---|
| $CH_3CH_2CH_2CH_2OH$ | $CH_3OCH_2CH_2CH_3$ |
| $CH_3CH_2CH(OH)CH_3$ | $CH_3OCH(CH_3)_2$ |
| $(CH_3)_2CHCH_2OH$ | $CH_3CH_2OCH_2CH_3$ |
| $(CH_3)_3C–OH$ | |

15. Of the eight compounds in question 13, the primary alcohols are: 1-pentanol; 3-methyl-1-butanol; 2-methyl-1-butanol; and 2,2-dimethyl-1-propanol. The secondary alcohols are: 2-pentanol; 3-pentanol; and 3-methyl-2-butanol. The tertiary alcohol is 2-methyl-2-butanol.

16. The primary alcohols in question 14 are:

$CH_3CH_2CH_2OH$ $\quad$ $CH_3CH_2CH_2CH_2OH$ $\quad$ $(CH_3)_2CHCH_2OH$

The secondary alcohols are: $\quad$ $CH_3CH(OH)CH_3$ $\quad$ $CH_3CH_2CH(OH)CH_3$

The tertiary alcohol is: $\quad$ $(CH_3)_3C{-}OH$

17. The names of the compounds are:
   - (a) 1-butanol (butyl alcohol)
   - (b) 2-propanol (isopropyl alcohol)
   - (c) 2-methyl-3-phenyl-1-propanol
   - (d) oxirane (ethylene oxide)
   - (e) 2-methyl-2-butanol
   - (f) 2-methylcyclohexanol
   - (g) 2,3-dimethyl-1,4-butanediol

18. The names of the compounds are
   - (a) ethanol (ethyl alcohol)
   - (b) 2-phenylethanol
   - (c) 3-methyl-3-pentanol
   - (d) 1-methylcyclopentanol
   - (e) 3-pentanol
   - (f) 1,2-propanediol
   - (g) 4-ethyl-2-hexanol

19. Chief product of dehydration

   (a) $CH_3\overset{CH_3|}{C}{=}CHCH_3$ $\quad$ 2-methyl-2-butene

   (b) $CH_3CH{=}CHCH_2CH_3$ $\quad$ 2-pentene

   (c)  cyclohexene

20. Chief product of dehydration

   (a) 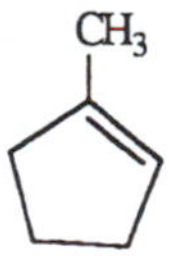 $CH_3$ 1-methylcyclopentene

   (b) $CH_3CH{=}CHCH_3$ $\quad$ 2-butene

(c) $CH_3C(CH_3){=}CHCH_2CH_3$ 2-methyl-2-pentene

21. (a) $CH_3CH_2CH_2CH(OH)CH_2CH_2CH_3$ 4-heptanol

(b) 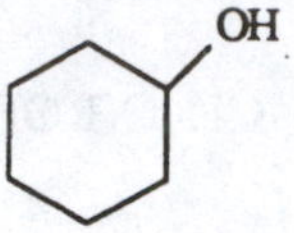 cyclohexanol

(c) 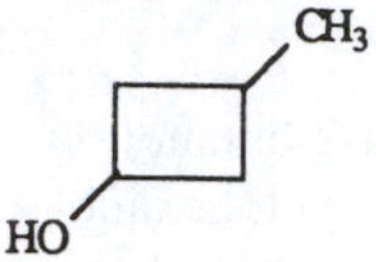 3-methylcyclobutanol

22. (a) 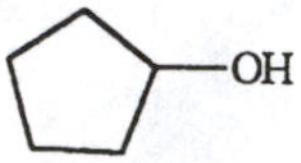 cyclopentanol

(b) 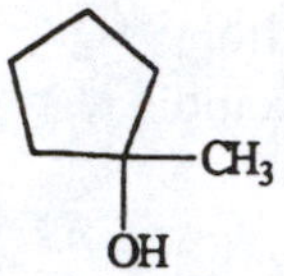 1-methylcyclopentanol

(c) $CH_3CH(CH_3)CH(OH)CH(CH_3)CH_3$ 2,4-dimethyl-3-pentanol

23. Oxidation of a primary alcohol

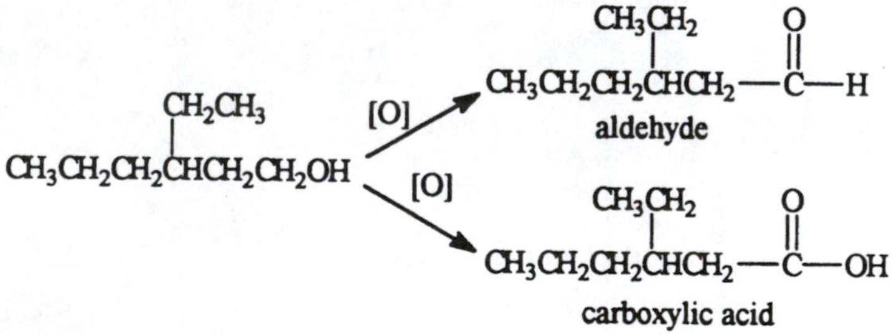

$$CH_3CH_2CH_2CH(CH_2CH_3)CH_2CH_2OH \xrightarrow{[O]} CH_3CH_2CH_2CH(CH_2CH_3)CH_2{-}C(=O){-}H$$

aldehyde

$$CH_3CH_2CH_2CH(CH_2CH_3)CH_2CH_2OH \xrightarrow{[O]} CH_3CH_2CH_2CH(CH_2CH_3)CH_2{-}C(=O){-}OH$$

carboxylic acid

24. Oxidation of a secondary alcohol

$$CH_3\underset{\displaystyle OH}{\underset{|}{C}}H\overset{\displaystyle CH_3}{\overset{|}{C}}HCH_2CH_3 \xrightarrow{[O]} CH_3\underset{\displaystyle O}{\underset{\|}{C}}\overset{\displaystyle CH_3}{\overset{|}{C}}HCH_2CH_3$$

ketone

25. (a) $CH_3CH{=}CH_2 + H_2O \xrightarrow{H^+} CH_3\underset{\displaystyle OH}{\underset{|}{C}}HCH_3$

(b) $CH_3CH_2CH{=}CH_2 + H_2O \xrightarrow{H^+} CH_3CH_2\underset{\displaystyle OH}{\underset{|}{C}}HCH_3$

(c) $CH_3CH_2CH{=}CHCH_3 + H_2O \xrightarrow{H^+} CH_3CH_2\underset{\displaystyle OH}{\underset{|}{C}}HCH_2CH_3 + CH_3CH_2CH_2\underset{\displaystyle OH}{\underset{|}{C}}HCH_3$

26. (a) $CH_3CH{=}CHCH_3 + H_2O \xrightarrow{H^+} CH_3CH_2\underset{\displaystyle OH}{\underset{|}{C}}HCH_3$

(b) $CH_3CH_2CH_2CH{=}CH_2 + H_2O \xrightarrow{H^+} CH_3CH_2CH_2\underset{\displaystyle OH}{\underset{|}{C}}HCH_3$

(c) $CH_3\underset{\displaystyle CH_3}{\underset{|}{C}}{=}CHCH_3 + H_2O \xrightarrow{H^+} CH_3\overset{\displaystyle OH}{\overset{|}{\underset{\displaystyle CH_3}{\underset{|}{C}}}}CH_2CH_3$

27. (a) $CH_3CHBrCH_3$ 2-bromopropane

(b) Br–cyclohexane (ring structure) bromocyclohexane (cyclohexyl bromide)

(c) $CH_3\overset{\displaystyle CH_3}{\overset{|}{C}}HCH_2CH_2Br$ 1-bromo-3-methylbutane

28. (a) $\underset{\displaystyle OH}{CH_3\underset{|}{C}HCH_2CH_3}$ 2-butanol

(b) $\overset{\displaystyle CH_2CH_3}{\overset{|}{\phantom{x}}}$
$CH_3\underset{\underset{\displaystyle OH}{|}}{C}H\overset{|}{C}HCH_2CH_3$ 3-ethyl-2-pentanol

(c) (cyclopentane ring)–OH cyclopentanol

29. (a) $2\,CH_3CH_2OH + H_2SO_4 \xrightarrow{140°C} \underset{\text{diethyl ether}}{CH_3CH_2OCH_2CH_3} + H_2O$

(b) $CH_3CH_2CH_2OH + H_2SO_4 \xrightarrow{180°C} \underset{\text{propene}}{CH_3CH{=}CH_2} + H_2O$

(c) $CH_3CH(OH)CH_2CH_3 \xrightarrow[H_2SO_4]{K_2Cr_2O_7} \underset{\text{2-butanone}}{CH_3\overset{\overset{\displaystyle O}{\|}}{C}CH_2CH_3} + H_2O$

(d) $CH_3CH_2\overset{\overset{\displaystyle O}{\|}}{C}\text{–}OCH_2CH_3 \xrightarrow[H_2O]{H^+} \underset{\text{propanoic acid}}{CH_3CH_2COOH} + \underset{\text{ethanol}}{CH_3CH_2OH}$

30. (a) $2\,CH_3CH_2OH + 2\,Na \longrightarrow 2\,CH_3CH_2O^-Na^+ + H_2$

(b) $CH_3CH_2CH_2CH_2OH \xrightarrow[H_2SO_4]{K_2Cr_2O_7} CH_3CH_2CH_2\overset{\overset{\displaystyle O}{\|}}{C}\text{–}H$

(c) (cyclopentane ring)–$CH{=}CH_2$ $\xrightarrow[H_2SO_4]{H_2O}$ (cyclopentane ring)–$\underset{\underset{\displaystyle OH}{|}}{C}HCH_3$

(d) $CH_3CH_2\overset{\overset{\displaystyle O}{\|}}{C}\text{–}OCH_3 + NaOH \longrightarrow CH_3CH_2\overset{\overset{\displaystyle O}{\|}}{C}\text{–}O^-Na^+ + CH_3OH$

31. (a) *o*-methylphenol

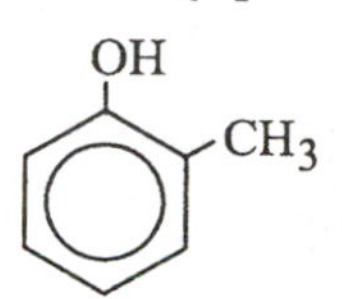


(b) *m*-dihydroxybenzene

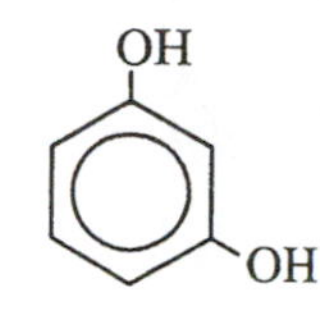


(c) 4-hydroxy-3-methoxybenzaldehyde

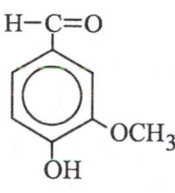

32. (a) *p*-nitrophenol

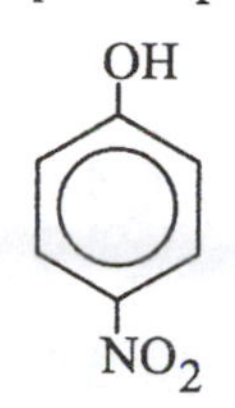


(b) 2,6-dimethylphenol

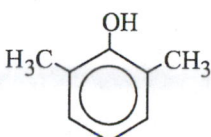

(c) *o*-dihydroxybenzene

33. (a) phenol
(b) *m*-methylphenol
(c) 2-ethyl-5-nitrophenol
(d) 4-bromo-2-chlorophenol

34. (a) *p*-dihydroxybenzene (hydroquinone)
(b) 2,4-dimethylphenol
(c) 2,4-dinitrophenol
(d) *m*-hexylphenol

35. Order of increasing solubility in water [c (lowest), a, b, d (highest)]
(c) $CH_3CH_2CH_2CH_2CH_3$
(a) $CH_3CH_2OCH_2CH_2CH_3$
(b) $CH_3CH(OH)CH_2CH_2CH_3$
(d) $CH_3CH(OH)CH(OH)CH_2CH_3$

36. Order of decreasing solubility in water: [a (highest), d, c, b (lowest)]
(a) $CH_3CH(OH)CH(OH)CH_2OH$
(d) $CH_3CH(OH)CH_2CH_2OH$
(c) $CH_3CH_2CH_2CH_2OH$
(b) $CH_3CH_2OCH_2CH_3$

37. Fourteen isomeric ethers; $C_6H_{14}O$

$CH_3OCH_2CH_2CH_2CH_2CH_3$
methyl pentyl ether
(1-methoxypentane)

$\phantom{CH_3O}CH_3$
$\phantom{CH_3O}|$
$CH_3OCHCH_2CH_2CH_3$
2-methoxypentane

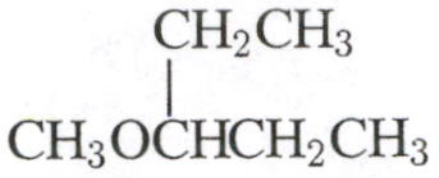

$\phantom{CH_3O}CH_2CH_3$
$\phantom{CH_3O}|$
$CH_3OCHCH_2CH_3$
3-methoxypentane

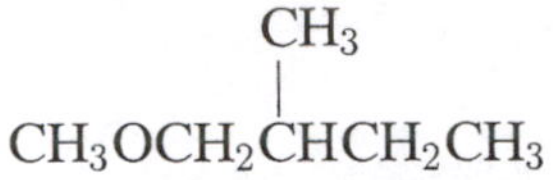

$\phantom{CH_3OCH_2}CH_3$
$\phantom{CH_3OCH_2}|$
$CH_3OCH_2CHCH_2CH_3$
1-methoxy-2-methylbutane

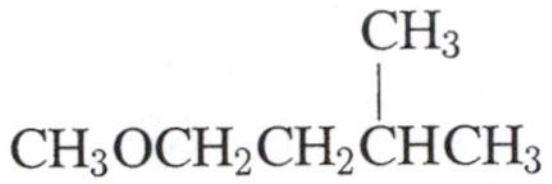

$\phantom{CH_3OCH_2CH_2}CH_3$
$\phantom{CH_3OCH_2CH_2}|$
$CH_3OCH_2CH_2CHCH_3$
1-methoxy-3-methylbutane

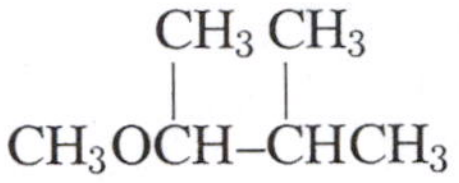

$\phantom{CH_3O}CH_3\ CH_3$
$\phantom{CH_3O}|\quad\ \ |$
$CH_3OCH{-}CHCH_3$
2-methoxy-3-methylbutane

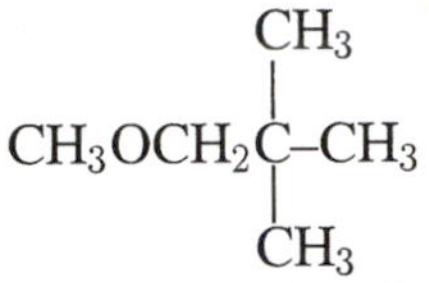

$\phantom{CH_3OCH_2}CH_3$
$\phantom{CH_3OCH_2}|$
$CH_3OCH_2C{-}CH_3$
$\phantom{CH_3OCH_2}|$
$\phantom{CH_3OCH_2}CH_3$
1-methoxy-2,2-dimethylpropane

$CH_3CH_2OCH_2CH_2CH_2CH_3$

butyl ethyl ether (1-ethoxybutane)

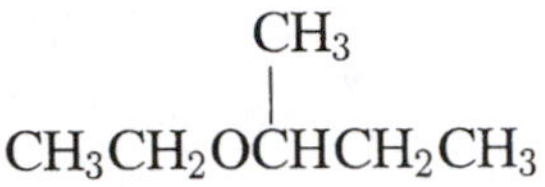

$\phantom{CH_3CH_2O}CH_3$
$\phantom{CH_3CH_2O}|$
$CH_3CH_2OCHCH_2CH_3$
*sec*-butyl ethyl ether
(2-ethoxybutane)

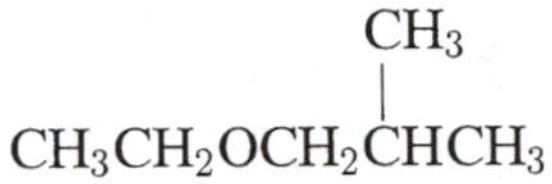

$\phantom{CH_3CH_2OCH_2}CH_3$
$\phantom{CH_3CH_2OCH_2}|$
$CH_3CH_2OCH_2CHCH_3$
ethyl isobutyl ether
(1-ethoxy-2-methylpropane)

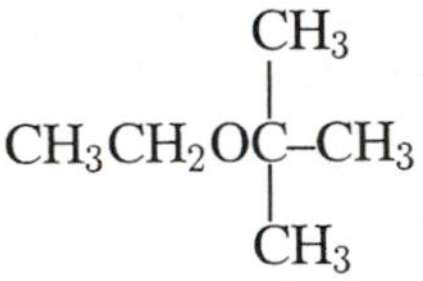

$\phantom{CH_3CH_2O}CH_3$
$\phantom{CH_3CH_2O}|$
$CH_3CH_2OC{-}CH_3$
$\phantom{CH_3CH_2O}|$
$\phantom{CH_3CH_2O}CH_3$
*t*-butyl ethyl ether
(2-ethoxy-2-methylpropane)

$CH_3CH_2CH_2OCH_2CH_2CH_3$

di-*n*-propyl ether
1-propoxypropane

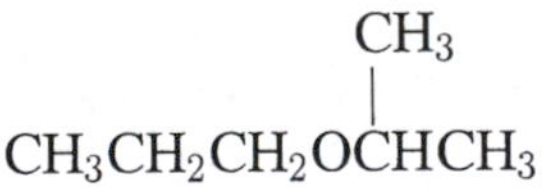

$\phantom{CH_3CH_2CH_2O}CH_3$
$\phantom{CH_3CH_2CH_2O}|$
$CH_3CH_2CH_2OCHCH_3$
isopropyl propyl ether

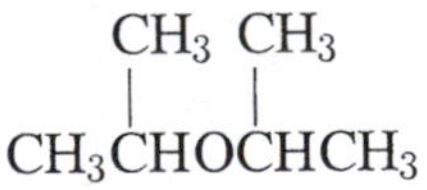

$\phantom{CH_3}CH_3\ \ CH_3$
$\phantom{CH_3}|\qquad\ |$
$CH_3CHOCHCH_3$
diisopropyl ether

38. Six isomeric ethers: $C_5H_{12}O$

$CH_3OCH_2CH_2CH_2CH_3$
butyl methyl ether
(1-methoxybutane)

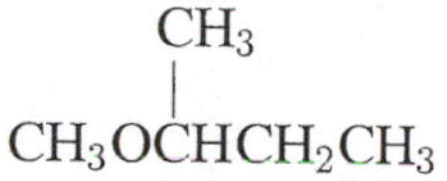

$CH_3OCH(CH_3)CH_2CH_3$ (the $CH_3$ branch is drawn on the CH carbon)
*sec*-butyl methyl ether
(2-methoxybutane)

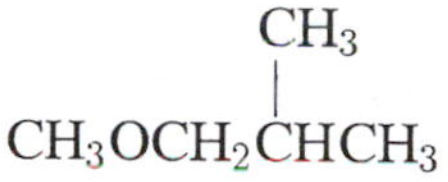

$CH_3OCH_2CH(CH_3)CH_3$ (the $CH_3$ branch is drawn on the CH carbon)
isobutyl methyl ether
(1-methoxy-2-methylpropane)

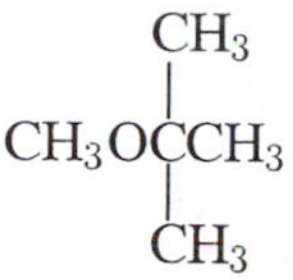

$CH_3OC(CH_3)_2CH_3$ (one $CH_3$ is drawn above the central C and one below it)
*t*-butyl methyl ether
(2-methoxy-2-methylpropane)

$CH_3CH_2OCH_2CH_2CH_3$
ethyl *n*-propyl ether
(1-ethoxypropane)

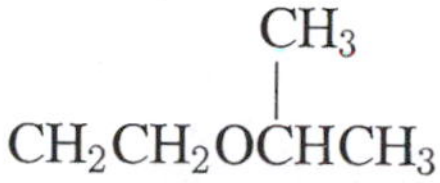

$CH_2CH_2OCH(CH_3)CH_3$ (the $CH_3$ branch is drawn on the CH carbon)
ethyl isopropyl ether
(2-ethoxypropane)

39. 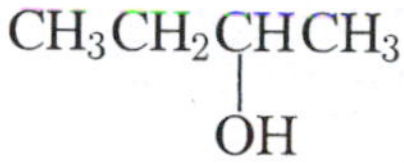 $CH_3CH_2CH(OH)CH_3 + 6\,O_2 \xrightarrow{\Delta} 4\,CO_2 + 5\,H_2O$ (the OH is drawn below the CH carbon)

40. $CH_3CH_2OCH_2CH_3 + 6\,O_2 \xrightarrow{\Delta} 4\,CO_2 + 5\,H_2O$

41. Possible combinations of reactants to make the following ethers by the Williamson synthesis:

(a) $CH_3CH_2OCH_3$ — $CH_3ONa + CH_3CH_2Cl$
or $CH_3CH_2ONa + CH_3Cl$

(b) 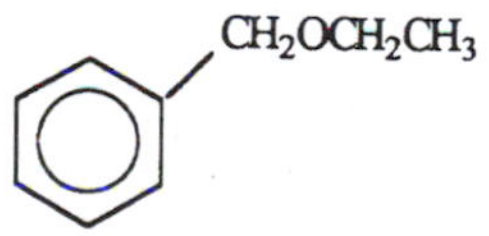 $C_6H_5CH_2OCH_2CH_3$ — $C_6H_5CH_2ONa + CH_3CH_2Cl$
or $C_6H_5CH_2Cl + CH_3CH_2ONa$

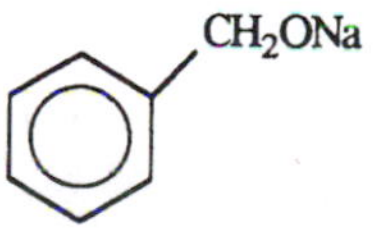

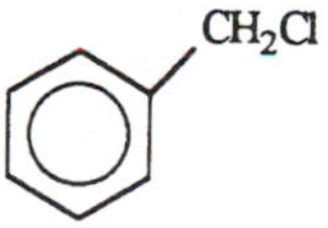

(c) $C_6H_5{-}O{-}CH_2CH_3$ $C_6H_5ONa$ + $CH_3CH_2Cl$

42. Possible combinations of reactants to make the following ethers by the Williamson synthesis:

(a) $CH_3CH_2CH_2OCH_2CH_2CH_3$ $CH_3CH_2CH_2ONa + CH_3CH_2CH_2Cl$

(b) $(CH_3)_2CHOCH_2CH_2CH_3$

$(CH_3)_2CHONa + CH_3CH_2CH_2Cl$

(c) $C_6H_5{-}O{-}C_6H_{11}$ cannot be made by the Williamson synthesis. Cannot use secondary RX.

43.

OH, OH, $CH_2(CH_2)_4CH_3$ 4-hexylresorcinol or 4-hexyl-1,3-dihydroxybenzene

44.

OH OH
*cis*-1,2-cyclopentanediol

OH, OH
*trans*-1,2-cyclopentanediol

45. Increasing acidity

$CH_2OH$    <    $H_2O$    <    HO    $CH_3$

46. The common phenolic structure in the catecholamines is catechol, *o*-dihydroxybenzene.

OH

OH

47. 1-butanol first adds a proton ($H^+$) followed by a loss of $H_2O$ to form the butyl carbocation (1° carbocation). Then a hydrogen shift occurs forming the more stable secondary carbocation followed by the loss of a proton to form the product 2-butene.

$$CH_3CH_2CH_2CH_2OH \xrightarrow{H_2SO_4} CH_3CH_2CH_2CH_2\overset{+}{O}H(H) \longrightarrow \underset{1^\circ \text{ carbocation}}{CH_3CH_2CH_2\overset{+}{C}H_2}$$

$$\longrightarrow \underset{2^\circ \text{ carbocation}}{CH_3CH_2\overset{+}{C}HCH_3} \longrightarrow \underset{\text{2-butene}}{CH_3CH{=}CHCH_3}$$

48. Methyl alcohol is converted to formaldehyde by an oxidation reaction:

$$CH_3OH \xrightarrow{[O]} H\overset{\overset{\displaystyle O}{\|}}{C}H$$

49. (a) $$CH_3\overset{\overset{\displaystyle OH}{|}}{C}HCH_3 \xrightarrow[H^+]{Cr_2O_7^{2-}} CH_3\overset{\overset{\displaystyle O}{\|}}{C}CH_3$$

(b) $$CH_3CH_2CH_2CH{=}CH_2 \xrightarrow[H_2O]{H^+} CH_3CH_2CH_2\underset{\underset{\displaystyle OH}{|}}{C}HCH_3$$

(c) $$CH_3CH_2OH + \text{Na (metal)} \longrightarrow CH_3CH_2ONa + H_2$$

(d) $$CH_3CH_2CH{=}CH_2 \xrightarrow[H_2O]{H^+} CH_3CH_2\underset{\underset{\displaystyle OH}{|}}{C}HCH_3 \xrightarrow[H^+]{Cr_2O_7^{2-}} CH_3CH_2\underset{\underset{\displaystyle O}{\|}}{C}CH_3$$

or $CH_3CH{=}CHCH_3$

(e) $$CH_3CH_2CH_2CH_2OH \xrightarrow[\Delta]{H_2SO_4} CH_3CH{=}CHCH_3 + CH_3CH_2CH{=}CH_2$$

$$\xrightarrow{HCl} CH_3CH_2\underset{\displaystyle Cl}{\underset{|}{C}}HCH_3$$

(f) $$CH_3CH_2CH_2Cl \xrightarrow{NaOH} CH_3CH_2CH_2OH \xrightarrow[H^+]{Cr_2O_7^{2-}} CH_3CH_2\overset{\displaystyle O}{\overset{\|}{C}}{-}H$$

50.

$$p\text{-}CH_3C_6H_4OH + NaOH \longrightarrow p\text{-}CH_3C_6H_4O^-Na^+ + H_2O$$

$$CH_3CH_3 + Cl_2 \xrightarrow{light} CH_3CH_2Cl + HCl$$

$$p\text{-}CH_3C_6H_4O^-Na^+ + CH_3CH_2Cl \longrightarrow p\text{-}CH_3C_6H_4OCH_2CH_3 + NaCl$$

51. A simple chemical test to distinguish between:

(a) ethanol and dimethyl ether. Ethanol will react readily with potassium dichromate and sulfuric acid to make acetaldehyde. Visibly, the orange color of the dichromate changes to green. Ethanol reacts with metallic sodium to produce hydrogen gas. Dimethyl ether does not react with either of these reagents.

(b) 1-pentanol and 1-pentene. 1-pentene will rapidly decolorize bromine as it adds to the double bond. 1-pentanol does not react.

(c) *p*-methylphenol has acidic properties so it will react with sodium hydroxide. Methoxybenzene does not have acidic properties and will not react with sodium hydroxide.

52. Differentiation between phenols and alcohols:

Phenols are acidic compounds; alcohols are not acidic.
Phenols react with NaOH to form salts; alcohols do not react with NaOH.
The OH group is bonded to a benzene ring in phenols and to a nonbenzene carbon atom in alcohols.

53. Isomers of $C_8H_{10}O$

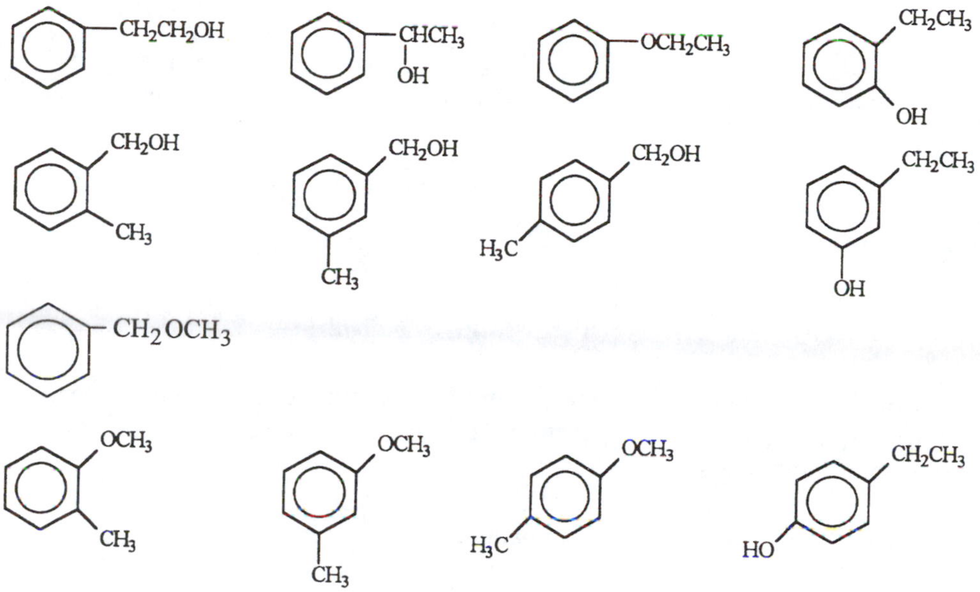

$C_6H_5-CH_2CH_2OH$ $C_6H_5-CH(OH)CH_3$ $C_6H_5-OCH_2CH_3$ (ortho $CH_2CH_3$ / OH)

(ortho $CH_2OH$ / $CH_3$) (meta $CH_2OH$ / $CH_3$) ($H_3C$ para $CH_2OH$) (meta $CH_2CH_3$ / OH)

$C_6H_5-CH_2OCH_3$

(ortho $OCH_3$ / $CH_3$) (meta $OCH_3$ / $CH_3$) ($H_3C$ para $OCH_3$) (HO para $CH_2CH_3$)

54. Only compound (a) will react with NaOH.

$$C_6H_5OH + NaOH \longrightarrow C_6H_5O^- Na^+ + H_2O$$

55. Order of increasing boiling points.

| 1-pentanol | < | 1-octanol | < | 1,2-pentanediol |
|---|---|---|---|---|
| 138°C | | 194°C | | 210°C |

All three compounds are alcohols. 1-pentanol has the lowest molar mass and hence the lowest boiling point. 1-octanol has a higher molar mass and therefore a higher boiling point than 1- pentanol. 1,2-pentanediol has two –OH groups and therefore forms more hydrogen bonds than the other two alcohols which causes its higher boiling point.

CHAPTER 23

# ALDEHYDES AND KETONES

1. (a) 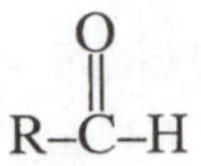 $R-\overset{\overset{\displaystyle O}{\|}}{C}-H$ aldehyde

(b) 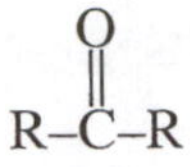 $R-\overset{\overset{\displaystyle O}{\|}}{C}-R$ ketone

(c) 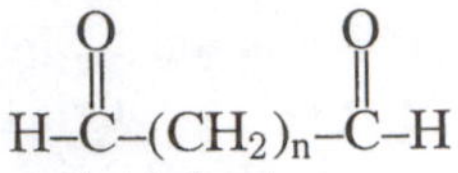 $H-\overset{\overset{\displaystyle O}{\|}}{C}-(CH_2)_n-\overset{\overset{\displaystyle O}{\|}}{C}-H$ a dialdehyde

(d) 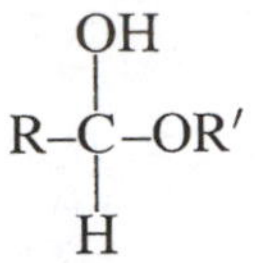 $R-\underset{\underset{\displaystyle H}{|}}{\overset{\overset{\displaystyle OH}{|}}{C}}-OR'$ a hemiacetal

(e) 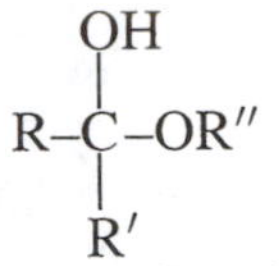 $R-\underset{\underset{\displaystyle R'}{|}}{\overset{\overset{\displaystyle OH}{|}}{C}}-OR''$ a hemiketal

(f) 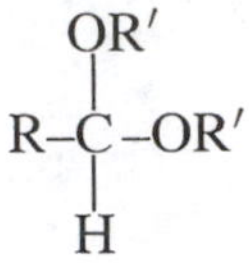 $R-\underset{\underset{\displaystyle H}{|}}{\overset{\overset{\displaystyle OR'}{|}}{C}}-OR'$ an acetal

(g) 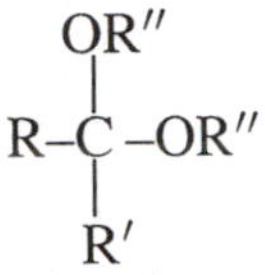 $R-\underset{\underset{\displaystyle R'}{|}}{\overset{\overset{\displaystyle OR''}{|}}{C}}-OR''$ a ketal

(h) 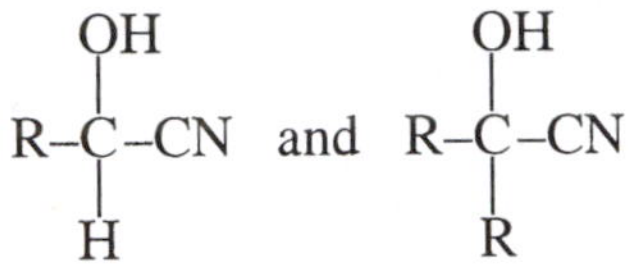 $R-\underset{\underset{\displaystyle H}{|}}{\overset{\overset{\displaystyle OH}{|}}{C}}-CN$ and $R-\underset{\underset{\displaystyle R}{|}}{\overset{\overset{\displaystyle OH}{|}}{C}}-CN$ a cyanohydrin

2. Aldehydes and ketones are unable to form hydrogen bonds to themselves because they lack a hydroxy group on the carbonyl atom. This results in a lower boiling point than alcohols of similar molar mass.

3. Propanal

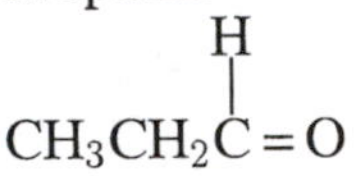

$CH_3CH_2\overset{\overset{\displaystyle H}{|}}{C}=O$

propanone

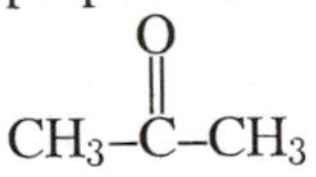

$CH_3-\overset{\overset{\displaystyle O}{\|}}{C}-CH_3$

Each has the molecular formula of $C_3H_6O$, so aldehydes and ketones appear to be isomeric with each other.

Butanal

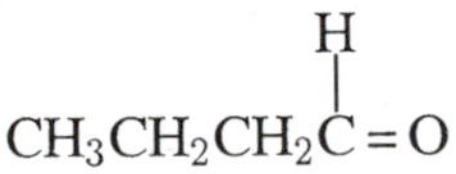

$CH_3CH_2CH_2\overset{\overset{\displaystyle H}{|}}{C}=O$

butanone

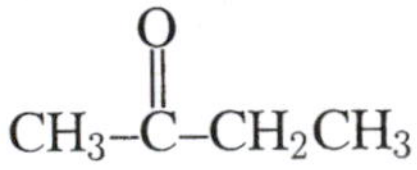

$CH_3-\overset{\overset{\displaystyle O}{\|}}{C}-CH_2CH_3$

Each has the molecular formula $C_4H_8O$, so the generalization seems to check out. The general formula for aldehydes and ketones is $C_nH_{2n}O$.

4. The strength of collagen depends on aldol condensations. After collagen is formed, aldehydes add along its length. Collagen fibers adjacent to each other undergo an aldol

condensation. The cross linking bonds between collagen strands form a strong network, giving collagen its strength.

5. The carbonyl group in high molar-mass aldehydes and ketones becomes a less significant part of the molecule and the compound behaves like a hydrocarbon. Consequently, they are not very soluble in water.

6. The electrons of the carbon atom in the carbonyl group are $sp^2$ hybridized. Therefore the double bond consists of a sigma and a pi bond.

7. Names of aldehydes.

(a) $H_2C=O$ — methanal, formaldehyde

(b) $CH_3CH(CH_3)CH_2C(=O)\text{–}H$ — 3-methylbutanal

(c) $H\text{–}C(=O)CH_2CH_2C(=O)\text{–}H$ — butanedial

(d) $C_6H_5\text{–}CH{=}CHC(=O)\text{–}H$ — 3-phenylpropenal

(e) $CH_3(H)C{=}C(H)C(=O)\text{–}H$ — *cis*-2-butenal

8. Names of aldehydes

(a) $CH_3C(=O)\text{–}H$ — ethanal, acetaldehyde

(b) $CH_3CH_2CH_2C(=O)\text{–}H$ — butanal, butyraldehyde

(c)

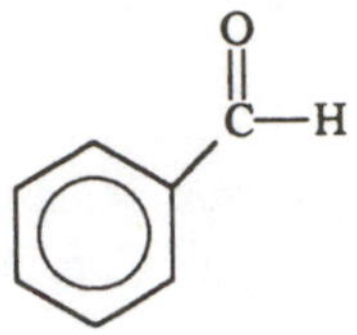

$$C_6H_5\text{–}\overset{\overset{\large O}{\|}}{C}\text{–}H$$

benzaldehyde

(d)

$$\text{2-Cl-5-}(CH_3CHCH_3)\text{-}C_6H_3\text{–}\overset{\overset{\large O}{\|}}{C}\text{–}H$$

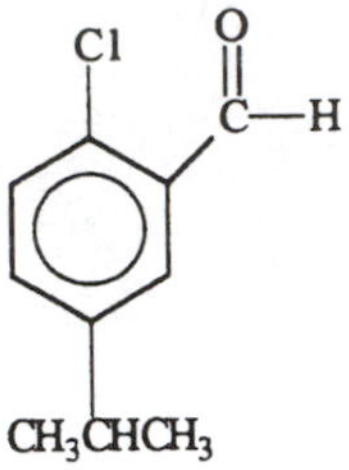

2-chloro-5-isopropylbenzaldehyde

(e) $CH_3\underset{\underset{\large OH}{|}}{C}HCH_2\overset{\overset{\large O}{\|}}{C}\text{–}H$ 3-hydroxybutanal

9. Names of ketones
   (a) propanone, acetone, dimethyl ketone
   (b) 1-phenyl-l-propanone, ethyl phenyl ketone
   (c) cyclopentanone
   (d) 4-hydroxy-4-methyl-2-pentanone

10. Names of ketones
    (a) 2-butanone, methyl ethyl ketone (MEK)
    (b) 3,3-dimethylbutanone, *t*-butyl methyl ketone
    (c) 2,5-hexanedione
    (d) 1-phenyl-2-propanone, benzyl methyl ketone

11. Structural formulas

    (a) $ClCH_2\underset{\underset{\large O}{\|}}{C}CH_2Cl$ 1,3-dichloropropanone

    (b) $CH_2 = CHCH_2\overset{\overset{\large O}{\|}}{C}\text{–}H$ 3-butenal

(c) $CH_3CH_2\overset{O}{\overset{\|}{C}}\underset{C_6H_5}{\underset{|}{C}H}CH_2CH_3$ 4-phenyl-3-hexanone

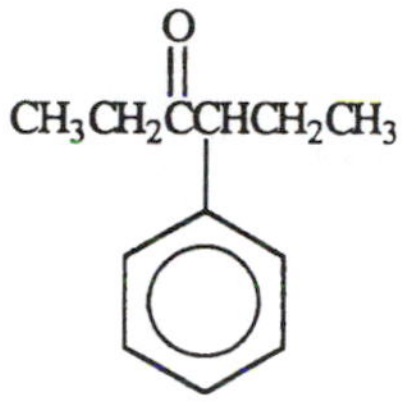

(d)  $CH_3CH_2CH_2CH_2CH_2\overset{O}{\overset{\|}{C}}\text{–}H$ hexanal

(e) 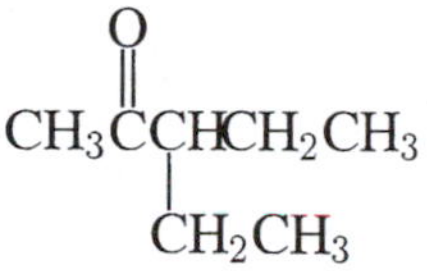 $CH_3\overset{O}{\overset{\|}{C}}\underset{CH_2CH_3}{\underset{|}{C}H}CH_2CH_3$ 3-ethyl-2-pentanone

12. Structural formulas

(a) 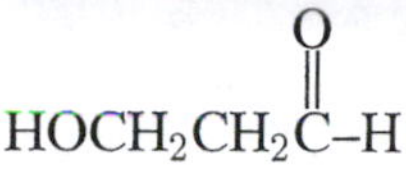 $HOCH_2CH_2\overset{O}{\overset{\|}{C}}\text{–}H$ 3-hydroxypropanal

(b) 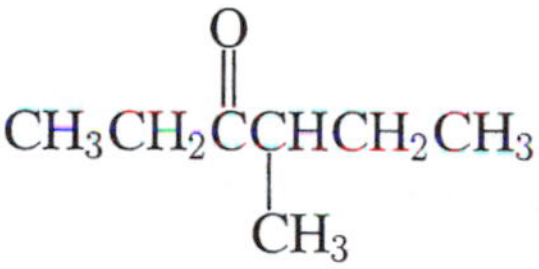 $CH_3CH_2\overset{O}{\overset{\|}{C}}\underset{CH_3}{\underset{|}{C}H}CH_2CH_3$ 4-methyl-3-hexanone

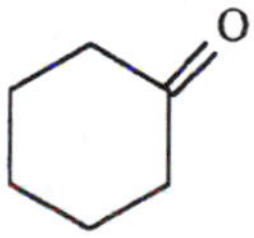

(c) cyclohexanone

(d) 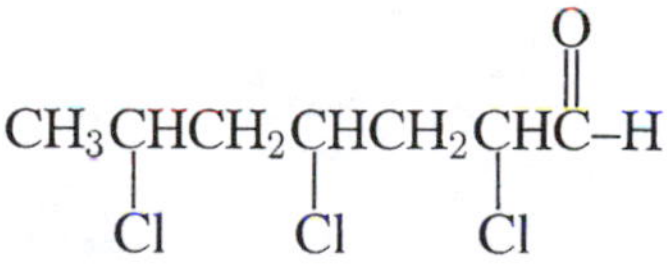 $CH_3\underset{Cl}{\underset{|}{C}H}CH_2\underset{Cl}{\underset{|}{C}H}CH_2\underset{Cl}{\underset{|}{C}H}\overset{O}{\overset{\|}{C}}\text{–}H$ 2,4,6-trichloroheptanal

(e) 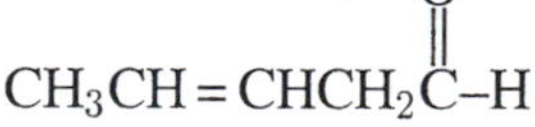 $CH_3CH=CHCH_2\overset{O}{\overset{\|}{C}}\text{–}H$ 3-pentenal

13. Higher boiling point.

(a) 2,5-hexanedione
(b) hexanal
(c) 2-pentanol
(d) butanone

14. Higher boiling point.
    (a) pentanal (c) 2-hexanone
    (b) benzyl alcohol (d) 1-butanol

15. Higher aqueous solubility.
    (a) 2,5-hexanedione (c) ethanal
    (b) propanal

16. Higher aqueous solubility.
    (a) acetaldehyde (c) propanal
    (b) 2,4-pentanedial

17. Equations for the oxidation of:

    (a) 3-pentanol

$$CH_3CH_2\underset{}{\overset{OH}{\overset{|}{C}}}HCH_2CH_3 \xrightarrow[H_2SO_4]{K_2Cr_2O_7} CH_3CH_2\overset{O}{\overset{\|}{C}}CH_2CH_3$$

same product for air ($O_2$) oxidation

    (b) 3-methyl-1-hexanol

$$CH_3CH_2CH_2\underset{CH_3}{\underset{|}{C}}HCH_2CH_2OH + O_2 \xrightarrow[\Delta]{Cu} CH_3CH_2CH_2\underset{CH_3}{\underset{|}{C}}HCH_2\overset{O}{\overset{\|}{C}}\text{–}H$$

$$CH_3CH_2CH_2\underset{CH_3}{\underset{|}{C}}HCH_2CH_2OH \xrightarrow[H_2SO_4]{K_2Cr_2O_7} CH_3CH_2CH_2\underset{CH_3}{\underset{|}{C}}HCH_2\overset{O}{\overset{\|}{C}}\text{–}H$$

$$\xrightarrow[H_2SO_4]{K_2Cr_2O_7} CH_3CH_2CH_2\underset{CH_3}{\underset{|}{C}}HCH_2COOH$$

18. Equations for the oxidation of:

    (a) 1-propanol

$$CH_3CH_2CH_2OH \xrightarrow[H_2SO_4]{K_2Cr_2O_7} CH_2CH_2\overset{O}{\overset{\|}{C}}\text{–}H \quad \text{or} \quad CH_3CH_2COOH$$

$$CH_3CH_2CH_2OH + O_2 \xrightarrow[\Delta]{Cu} CH_2CH_2\overset{\overset{\displaystyle O}{\|}}{C}\text{–}H$$

(b) $$CH_3CH\text{–}\underset{CH_3}{\overset{OH}{C}}CH_3 \;(\text{with } CH_3 \text{ on } CH) + O_2 \xrightarrow[\Delta]{Ag}$$ No reaction (3° alcohol) with either oxidizing agent

19. (a) An aldehyde group, $\text{–}\overset{\overset{\displaystyle H}{|}}{C}=O$, must be present to give a positive Tollens test.

    (b) The visible evidence for a positive Tollens test is the formation of a silver mirror on the inner walls of a test tube.

    (c) $$CH_3\overset{\overset{\displaystyle H}{|}}{C}=O + 2\,Ag^+ \xrightarrow[H_2O]{NH_3} CH_3CO\bar{O}\,\overset{+}{N}H_4 + 2\,Ag(s) \qquad \text{(silver mirror)}$$

20. (a) An aldehyde group, $\text{–}\overset{\overset{\displaystyle H}{|}}{C}=O$, must be present to give a positive Fehling test.

    (b) The visible evidence for a positive Fehling test is the formation of brick red $Cu_2O$, which precipitates during the reaction.

    (c) $$CH_3\overset{\overset{\displaystyle H}{|}}{C}=O + 2\,Cu^{2+} \xrightarrow[H_2O]{NaOH} CH_3COONa + Cu_2O(s) \qquad \text{(brick-red)}$$

21. Product(s) when each of the following is reacted with Tollens reagent:

    (a) butanal ⟶ butanoic acid + $Ag(s)$
    (b) benzaldehyde ⟶ benzoic acid + $Ag(s)$
    (c) methyl ethyl ketone ⟶ no reaction

22. Product(s) when each of the following is reacted with Fehling reagent:

    (a) propanal ⟶ propanoic acid + $Cu_2O(s)$
    (b) acetone ⟶ no reaction
    (c) 3-methylpentanal ⟶ 3-methylpentanoic acid + $Cu_2O(s)$

23. Aldol condensation

(a) butanal

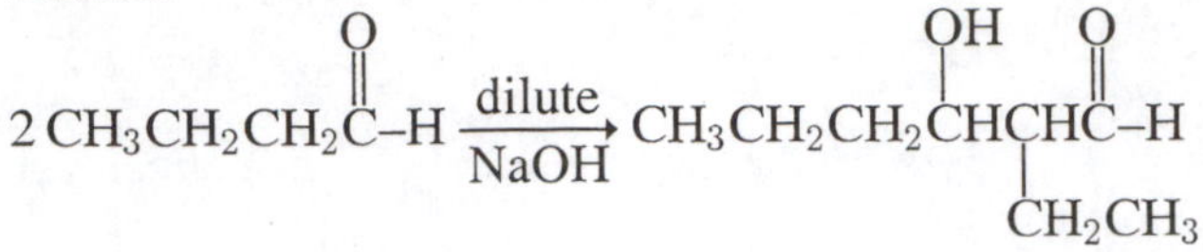

$$2\,CH_3CH_2CH_2\overset{O}{\overset{\|}{C}}\text{–}H \xrightarrow[NaOH]{dilute} CH_3CH_2CH_2\overset{OH}{\overset{|}{C}}H\underset{\underset{CH_2CH_3}{|}}{C}H\overset{O}{\overset{\|}{C}}\text{–}H$$

(b)

$$2\ C_6H_5\text{–}CH_2\overset{O}{\overset{\|}{C}}\text{—}H \xrightarrow[NaOH]{dilute} C_6H_5\text{–}CH_2\overset{OH}{\overset{|}{C}}H\text{—}\underset{\underset{C_6H_5}{|}}{C}H\overset{O}{\overset{\|}{C}}\text{—}H$$

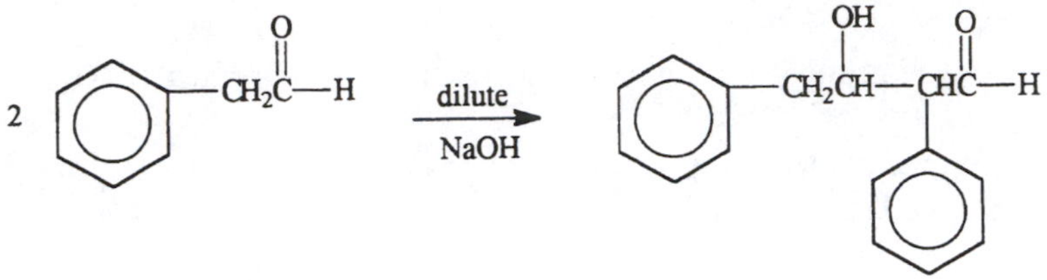

phenylethanal

24. Aldol condensation

(a) 3-pentanone

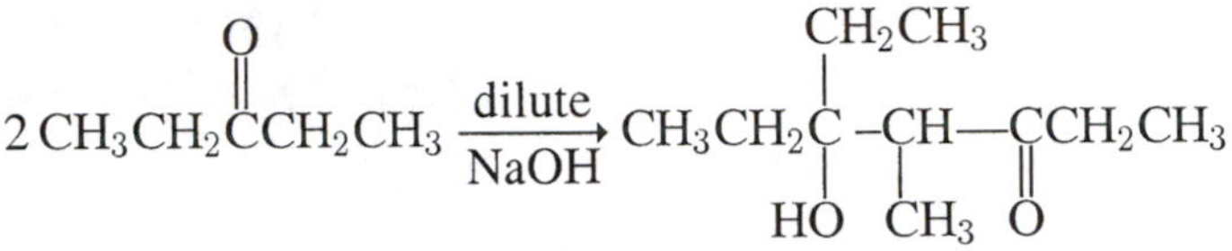

$$2\,CH_3CH_2\overset{O}{\overset{\|}{C}}CH_2CH_3 \xrightarrow[NaOH]{dilute} CH_3CH_2\underset{\underset{HO}{|}}{\overset{\overset{CH_2CH_3}{|}}{C}}\text{–}\underset{\underset{CH_3}{|}}{C}H\text{—}\underset{\underset{O}{\|}}{C}CH_2CH_3$$

(b) propanal

$$2\,CH_3CH_2\overset{\overset{H}{|}}{C}=O \xrightarrow[NaOH]{dilute} CH_3CH_2\underset{\underset{OH}{|}}{C}H\overset{\overset{CH_3}{|}}{C}H\underset{\underset{H}{|}}{C}=O$$

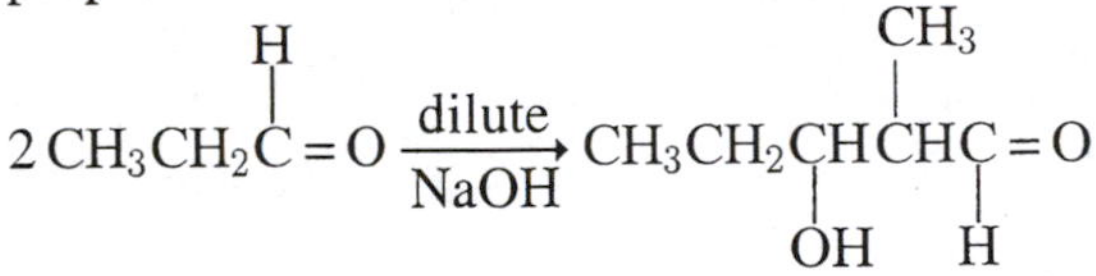

25. The completed equations are:

(a)

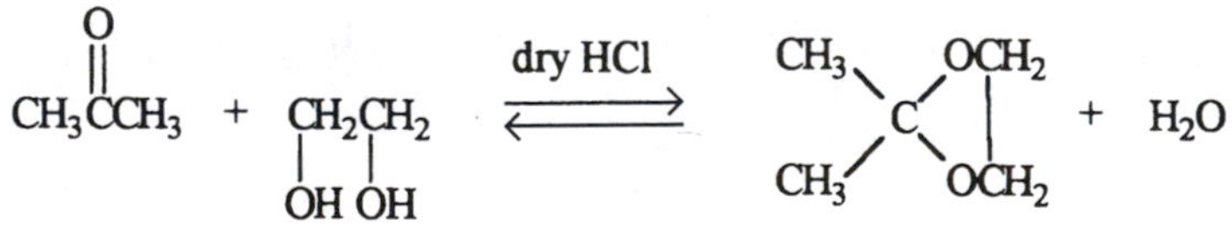

$$CH_3\overset{O}{\overset{\|}{C}}CH_3 + \underset{\underset{OH\ \ OH}{|\ \ \ \ |}}{CH_2CH_2} \overset{\text{dry HCl}}{\rightleftharpoons} (CH_3)_2C\begin{matrix}\diagup OCH_2 \\ \diagdown OCH_2\end{matrix} + H_2O$$

(b)

$$CH_3CH_2\overset{H}{\overset{|}{C}}{=}O + CH_3CH_2OH \underset{}{\overset{H^+}{\rightleftharpoons}} CH_3CH_2\overset{OH}{\overset{|}{\underset{H}{\underset{|}{C}}}}{-}OCH_2CH_3$$

(c) $$CH_3\underset{OH}{\underset{|}{C}}HCH_2CH(OCH_3)_2 \xrightarrow[H^+]{H_2O} CH_3\underset{OH}{\underset{|}{C}}HCH_2\overset{O}{\overset{\|}{C}}{-}H + 2\,CH_3OH$$

26. The completed equations are:

(a)

$$CH_3CH_2\overset{H}{\overset{|}{C}}{=}O + 2\ CH_3CH_2CH_2OH \overset{\text{dry HCl}}{\rightleftharpoons} CH_3CH_2\overset{OCH_2CH_2CH_3}{\overset{|}{\underset{H}{\underset{|}{C}}}}{-}OCH_2CH_2CH_3 + H_2O$$

acetal

(b)

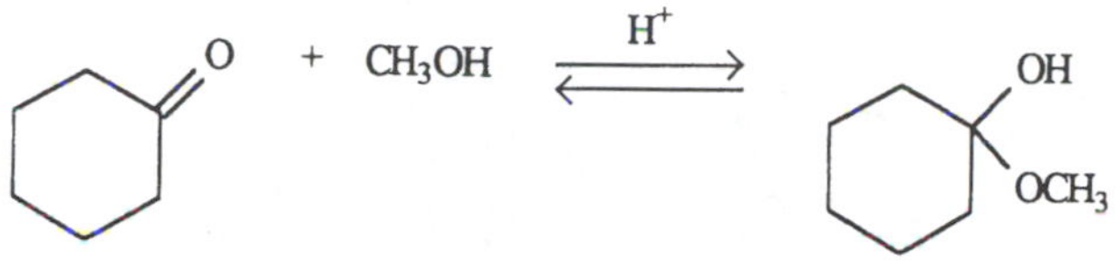

$$\text{cyclohexanone} + CH_3OH \overset{H^+}{\rightleftharpoons} \text{1-methoxycyclohexan-1-ol (ring C bearing } OH \text{ and } OCH_3)$$

(c) $$CH_3CH_2CH_2CH(OCH_3)_2 \underset{H^+}{\overset{H_2O}{\rightleftharpoons}} CH_3CH_2CH_2\overset{H}{\overset{|}{C}}{=}O + 2\ CH_3OH$$

27. Sequence of reactions:

(a) $$CH_3\underset{O}{\underset{\|}{C}}CH_3 + HCN \longrightarrow CH_3\overset{CH_3}{\overset{|}{\underset{OH}{\underset{|}{C}}}}{-}CN$$

(b) $$CH_3\overset{CH_3}{\overset{|}{\underset{OH}{\underset{|}{C}}}}{-}CN + H_2O \longrightarrow CH_3\overset{CH_3}{\overset{|}{\underset{OH}{\underset{|}{C}}}}{-}COOH$$

(c) $2\,CH_3C(CH_3)(OH){-}COOH + CH_3C(=O){-}H \xrightarrow{\text{dry HCl}} CH_3CH[O{-}C(CH_3)_2{-}COOH]_2$

28. Sequence of reactions:

(a) $C_6H_5{-}CH{=}O + HCN \xrightarrow{OH^-} C_6H_5{-}CH(OH){-}CN$

(b) $C_6H_5{-}CH(OH){-}CN + H_2O \xrightarrow{H^+} C_6H_5{-}CH(OH)COOH$

(c) $C_6H_5{-}CH(OH)COOH \xrightarrow[H_2SO_4]{K_2Cr_2O_7} C_6H_5{-}C(=O){-}COOH$

29. $HOCH_2CH_2CH{=}O \xrightarrow{H^+}$ cyclic $CH_2{-}CH_2{-}C(H)(OH){-}O$ (ring: $CH_2$—$CH_2$, $O$—$C(H)(OH)$)

30. $CH_3C(=O){-}H$ ethanal

31. Four aldol condensation products from a mixture of ethanol (E) and propanal (P) are possible.

EE PP EP PE

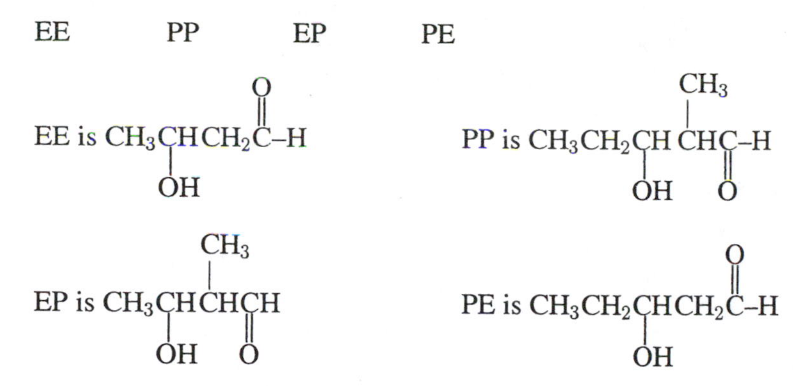


32. The secondary alcohol in compound I is oxidized to the ketone (compound II).

33. (a) $CH_3CH_2C(H){=}O$ and $CH_3C(=O)CH_3$.

The Tollens test (silver mirror) or Fehling test (red $Cu_2O$) will give positive results with propanal but not with acetone.

(b) $CH_3CH_2C(H){=}O$ and $CH_2{=}CHC(H){=}O$

Bromine will decolorize immediately with the second compound (propenal) but not with propanal.

(c) $C_6H_5$–$CH_2CH_2OH$ and $C_6H_5$–$CH(OH)CH_3$

Oxidize both compounds: 2-phenylethanol will give 2-phenylethanal and 1-phenylethanol will give methyl phenyl ketone. 1-phenylethanal will give a positive Tollens or Benedict test and methyl phenyl ketone will not give a positive test.

34. $$CH_3C(H){=}O + HCN \longrightarrow CH_3C(H)(OH){-}CN \xrightarrow[H^+]{H_2O} CH_3CH(OH){-}COOH$$

lactic acid

35. 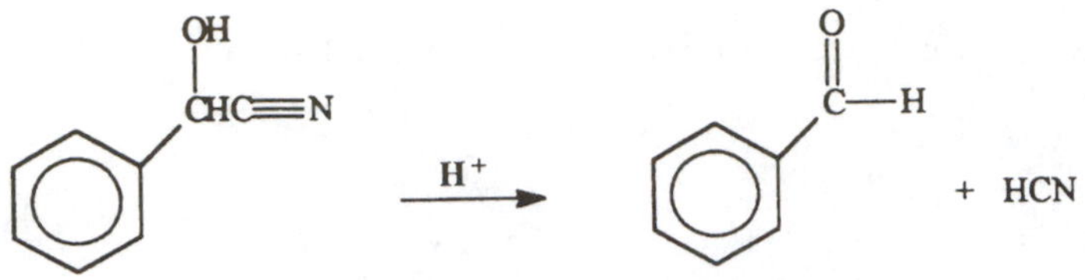 $C_6H_5CH(OH)C\equiv N \xrightarrow{H^+} C_6H_5C(=O)H + HCN$

benzaldehyde

36. Pyruvic acid is changed to lactic acid by a reduction reaction.

37. The structure of the product is $CH_2(OH)CH(OH)CH(OH)CH(OH)C(=O)CH_2OH$

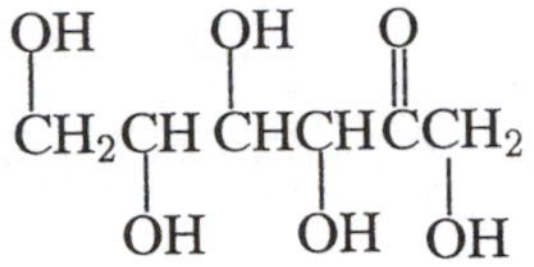

38. The alcohols which should be oxidized to give these ketones.

(a) 3-pentanone: 3-pentanol $CH_3CH_2CH(OH)CH_2CH_3$

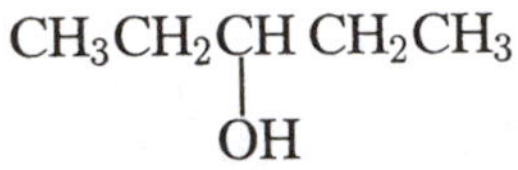

(b) methyl ethyl ketone: 2-butanol $CH_3CH(OH)CH_2CH_3$

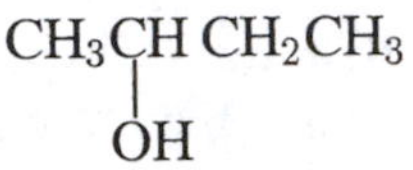

(c) 4-phenyl-2-butanone: 4-phenyl-2-butanol $C_6H_5CH_2CH_2CH(OH)CH_3$

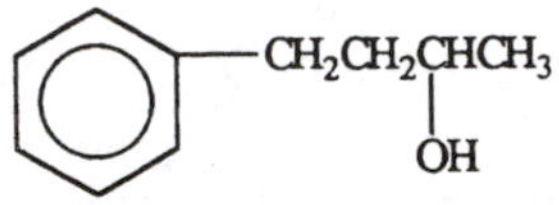

39. Isomeric aldehydes and ketones of formula $C_6H_{12}O$

*Aldehydes*

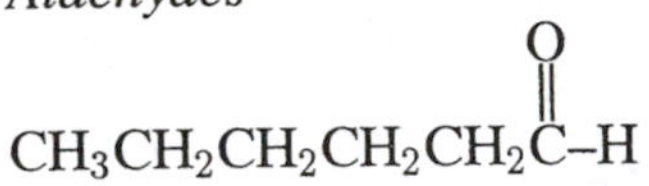

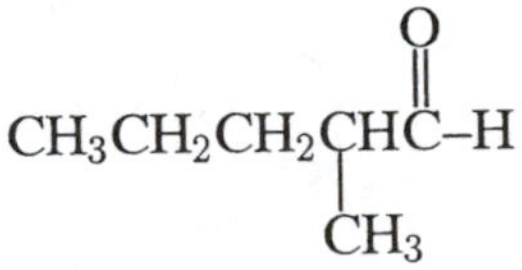

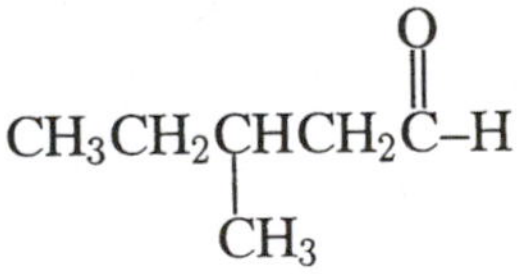

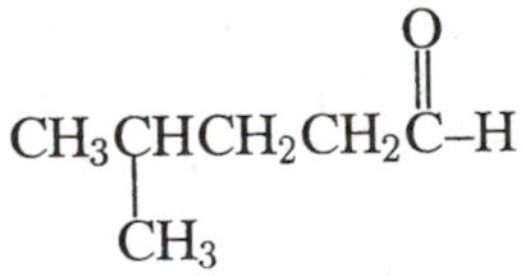

$CH_3CH_2CH_2CH_2CH_2C(=O)H$

$CH_3CH_2CH_2CH(CH_3)C(=O)H$

$CH_3CH_2CH(CH_3)CH_2C(=O)H$

$CH_3CH(CH_3)CH_2CH_2C(=O)H$

$CH_3CH_2C(CH_3)_2C(=O)H$ $\quad$ $CH_3C(CH_3)_2CH_2C(=O)H$

$CH_3CH(CH_3)CH(CH_3)C(=O)H$ $\quad$ $CH_3CH_2CH(CH_2CH_3)C(=O)H$

*Ketones*

$CH_3CH_2CH_2CH_2C(=O)CH_3$ $\quad$ $CH_3CH_2CH_2C(=O)CH_2CH_3$

$CH_3CH_2CH(CH_3)C(=O)CH_3$ $\quad$ $CH_3CH(CH_3)CH_2C(=O)CH_3$

$CH_3CH_2C(=O)CH(CH_3)CH_3$ $\quad$ $CH_3C(CH_3)_2C(=O)CH_3$

40. Benzaldehyde isomers of formula $C_9H_{10}O$

H—C=O, $CH_2CH_3$ $\quad$ H—C=O, $CH_2CH_3$ $\quad$ H—C=O, $CH_2CH_3$

H—C=O, $CH_3$, $CH_3$ $\quad$ H—C=O, $CH_3$, $H_3C$ $\quad$ H—C=O, $H_3C$, $CH_3$

H—C=O, $CH_3$, $CH_3$ $\quad$ H—C=O, $H_3C$, $CH_3$ $\quad$ H—C=O, $CH_3$, $CH_3$

41. (a) 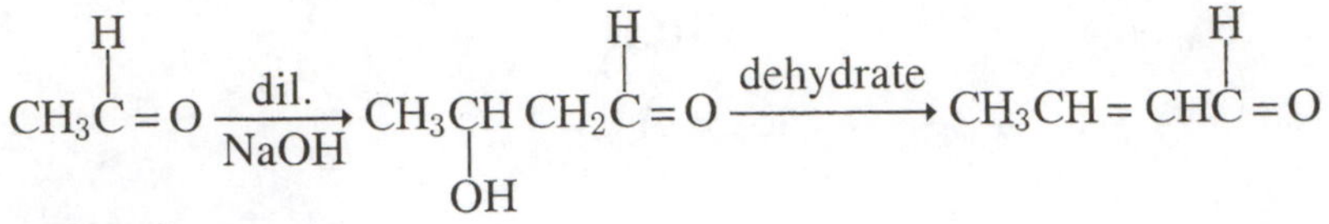 $CH_3\overset{H}{\overset{|}{C}}=O \xrightarrow[NaOH]{dil.} CH_3\underset{OH}{\underset{|}{C}}H\ CH_2\overset{H}{\overset{|}{C}}=O \xrightarrow{dehydrate} CH_3CH=CH\overset{H}{\overset{|}{C}}=O$

ethanal

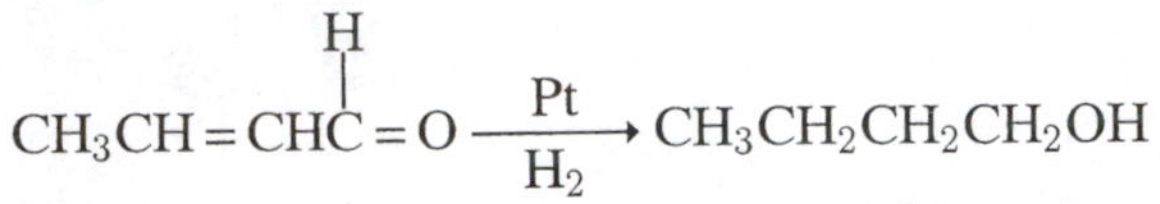

$CH_3CH=CH\overset{H}{\overset{|}{C}}=O \xrightarrow[H_2]{Pt} CH_3CH_2CH_2CH_2OH$

1 butanol

(b) 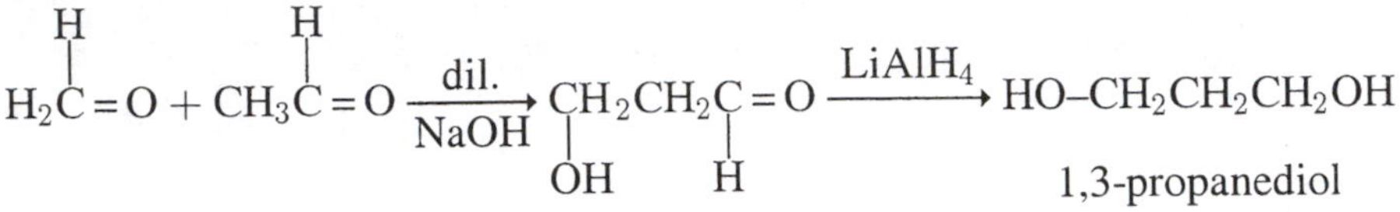 $H_2\overset{H}{\overset{|}{C}}=O + CH_3\overset{H}{\overset{|}{C}}=O \xrightarrow[NaOH]{dil.} \underset{OH}{\underset{|}{C}}H_2CH_2\underset{H}{\underset{|}{C}}=O \xrightarrow{LiAlH_4} HO\text{–}CH_2CH_2CH_2OH$

1,3-propanediol

42. 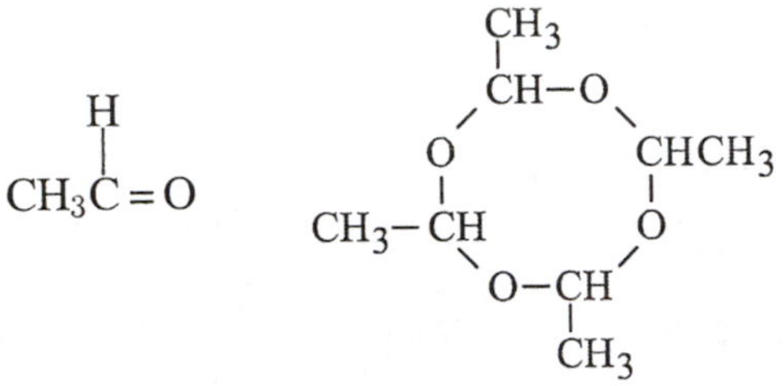 $CH_3\overset{H}{\overset{|}{C}}=O$

$$\begin{array}{ccc} & \overset{CH_3}{\overset{|}{CH}}-O & \\ O & & CHCH_3 \\ | & & | \\ CH_3-CH & & O \\ & O-\underset{CH_3}{\underset{|}{CH}} & \end{array}$$

43. In phenol, three positions, ortho, ortho, and para to the OH group are used in the reaction to form a thermosetting polymer. However, in *p*-cresol the para position is occupied by a methyl group and cannot react with formaldehyde. This leaves the *p*-cresol molecule as a bifunctional monomer, resulting in a linear, thermoplastic polymer.

CHAPTER 24

# CARBOXYLIC ACIDS AND ESTERS

1. (a) $CH_3COOH$

   (b) COOH (benzoic acid: COOH attached to a benzene ring)

   (c) $CH_3CHCOOH$ with OH on the central C: $CH_3CH(OH)COOH$

   (d) $CH_3CHCOOH$ with $NH_2$ on the central C: $CH_3CH(NH_2)COOH$

   (e) $CH_2CH_2COOH$ with Cl on the first C: $ClCH_2CH_2COOH$

   (f) COOH over COOH (bonded): $HOOC{-}COOH$

   (g) $CH_2=CHCOOH$

   (h) $CH_2\overset{O}{\overset{\|}{C}}{-}OCH_3$

   (i) $CH_3CN$

   (j) $CH_3COONa$

   (k) $CH_3\overset{O}{\overset{\|}{C}}{-}Cl$

   (l)
   $CH_2{-}O\overset{O}{\overset{\|}{C}}C_{17}H_{35}$
   $CH{-}O\overset{O}{\overset{\|}{C}}C_{17}H_{35}$
   $CH_2{-}O\overset{O}{\overset{\|}{C}}C_{17}H_{35}$

   (m) $CH_3(CH_2)_{16}COONa$

2. The butyric acid solution would be expected to have the more objectionable odor because salts normally exhibit little or no odor. Salts are ionic and therefore have low volatility. For example, dilute solutions of acetic acid (vinegar) have considerable odor, but sodium acetate does not.

3. Which has the greater solubility in water?
   (a) Propanoic acid is more soluble than methyl propanoate.
   (b) Sodium palmitate is more soluble than palmitic acid.
   (c) Sodium stearate is more soluble than barium stearate.
   (d) Sodium phenoxide is more soluble than phenol.

4. (a) The major difference between fats and oils is that fats are solids at room temperature, oils are liquids. The fatty acids in the molecules are mostly saturated in fats; more unsaturated in oils.

(b) Soaps are the sodium salts of high molar mass fatty acids. Syndets are synthetic detergents and occur in several different forms. They have cleansing action similar to soaps, but have different structural and solubility characteristics, such as being soluble in hard water. Some syndets also contain long hydrocarbon chains.

(c) Hydrolysis is the breaking apart of an ester in the presence of water to form an alcohol and a carboxylic acid. Mineral acids or digestive enzymes are used to speed the hydrolysis process. Saponification breaks the ester apart using sodium hydroxide to form an alcohol and a salt of a carboxylic acid.

5. Like a soap, a detergent contains a grease soluble component and a water soluble component. The grease soluble component of many molecules dissolves in the grease film. The water soluble components attract water, causing small droplets of grease-bearing dirt to break loose and float away.

6. The principal advantage of synthetic detergents over soap is that the syndets do not form insoluble precipitates with the ions in hard water ($Ca^{2+}$, $Mg^{2+}$, $Fe^{3+}$).

7. Medicinal effects of aspirin. Aspirin acts in the body as an antipyretic, an analgesic, and as an antiinflammatory agent.

   Risks of using aspirin. Irritation of the stomach lining; inhibit blood clotting; can prolong labor in giving birth; can bring on the development of Reyes syndrome.

8. Methyl salicylate (pain relieving liniments)
   Phenyl salicylate (protective coating for pills)

9. The most common substitutes for aspirin are acetaminophen and Ibuprofen. Acetaminophen acts as an analgesic and as an antipyretic. Ibuprofen acts as a prostaglandin inhibitor.

10. In a condensation reactoin one of the products formed is a small molecule such as water. In polyester formation between an alcohol and a carboxylic acid, water is formed in addition to the ester (polyester).

11. Names
    (a) hexanoic acid
    (b) fumaric acid
    (c) *o*-phthalic acid
    (d) stearic acid
    (e) sodium propanoate

12. Names
    (a) *o*-chlorobenzoic acid
    (b) oleic acid
    (c) *m*-toluic acid
    (d) 2-hydroxybutanoic acid
    (e) ammonium propanoate

13. Structures

(a) $CH_3(CH_2)_4COOH$ hexanoic acid

(b) $COOH$
$|$
$CH_2$
$|$
$COOH$
malonic acid

(c)

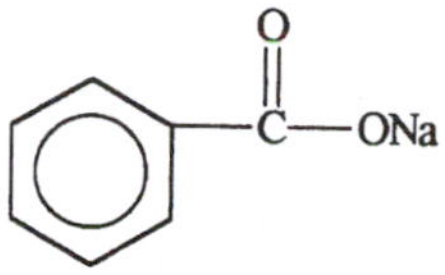

sodium benzoate

(d)

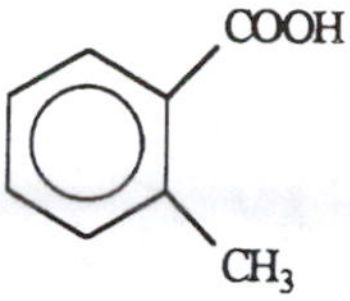

*o*-toluic acid

(e) $CH_3(CH_2)_{16}COOH$ stearic acid

(f) $CH_3CHCOOH$ (with Cl on the CH) 2-chloropropanoic acid

(g) $CH_3CH_2CH_2COOK$ potassium butyrate

14. Structures

(a) $COOH$
$|$
$COOH$
oxalic acid

(b) $CH_3CH_2CH_2CH_2COOH$ pentanoic acid

(c)

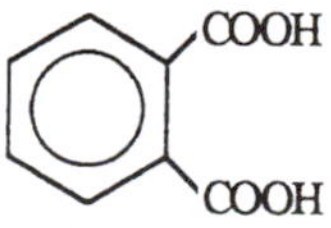

*o*-phthalic acid

(d) $CH_3CH_2CH=CHCH_2CH=CHCH_2CH=CH(CH_2)_7COOH$ linolenic acid

(e)

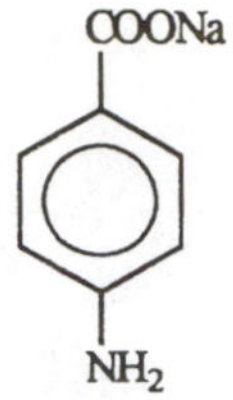

$COONa$ / $NH_2$ (para-substituted benzene ring)    sodium *p*-aminobenzoate

(f) $CH_3CH_2COONH_4$    ammonium propanoate

(g) $CH_3CH(OH)CH_2COOH$    *β*-hydroxybutyric acid

15. Increasing pH means increasing basicity

$HCl < CH_3COOH <$ $C_6H_5$–OH $< NaCl < NH_3 < NaOH$

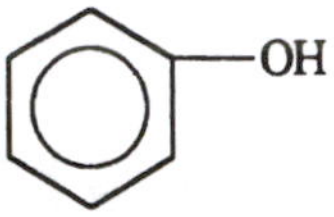

Increasing pH ⟶

16. Decreasing pH means increasing acidity

$KOH > NH_3 > KBr >$ $C_6H_5$–OH $> HCOOH > HBr$

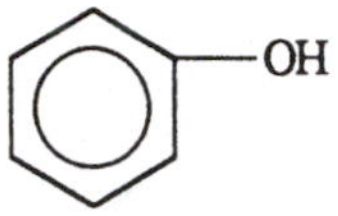

Decreasing pH ⟶

17. IUPAC and common names

| | | |
|---|---|---|
| (a) | methyl propenoate | methyl acrylate |
| (b) | ethyl butanoate | ethyl butyrate |
| (c) | phenyl-2-hydroxybenzoate | phenyl salicylate |

18. IUPAC and common names

| | | |
|---|---|---|
| (a) | methyl methanoate | methyl formate |
| (b) | propyl benzoate | propyl benzoate |
| (c) | ethyl propanoate | ethyl propionate |

19. Structural formulas

(a) $HC(=O)\text{–}OCH_3$

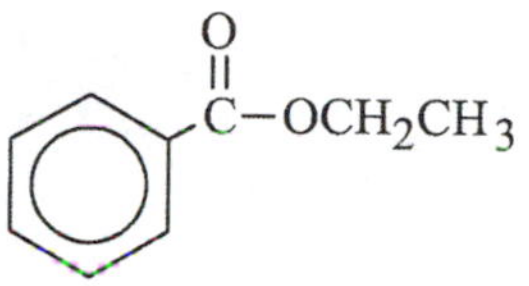

(b) $CH_3CH_2CH_2C(=O)\text{–}OCH_2CH_2CH_2CH_3$

(c) $C_6H_5\text{–}C(=O)\text{–}OCH_2CH_3$

20. Structural formulas

(a) $CH_3C(=O)\text{–}OCH_2CH_2CH_3$

(b) $C_6H_5\text{–}C(=O)\text{—}OCH_3$

(c) $CH_3CH_2CH_2CH_2CH_2C(=O)\text{–}OCH_2CH_3$

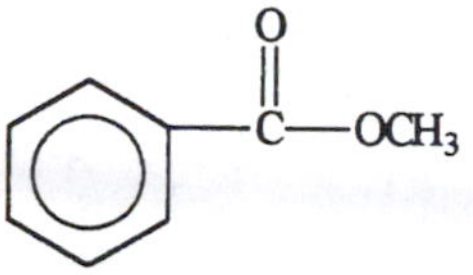

21. (a) $CH_3(CH_2)_7CH{=}CH(CH_2)_7COOH + H_2 \xrightarrow{Ni} CH_3(CH_2)_{16}COOH$

stearic acid

(b) $C_6H_5\text{–}C(=O)\text{—}H \xrightarrow[H_2SO_4]{Na_2Cr_2O_7} C_6H_5COOH$ benzoic acid

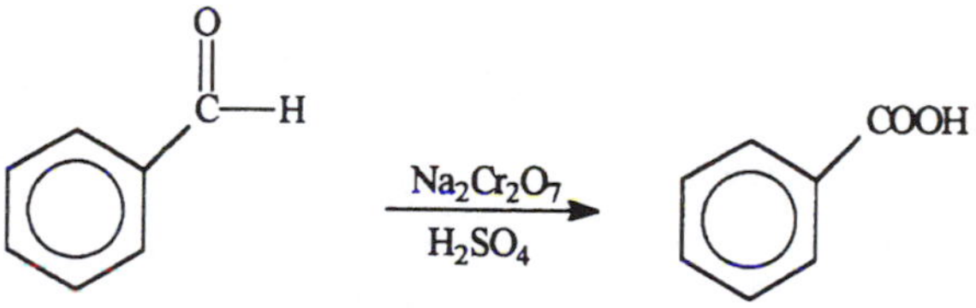

(c) $CH_3CH_2COOH + NaOH \longrightarrow CH_3CH_2COONa + H_2O$

sodium propanoate

(d) $CH_3(CH_2)_3CH_2C(=O)\text{–}OCH_2CH_2CH_3 + NaOH \xrightarrow{\Delta}$

$CH_3(CH_2)_3CH_2COONa + CH_3CH_2CH_2OH$

sodium hexanoate 1-propanol

(e)

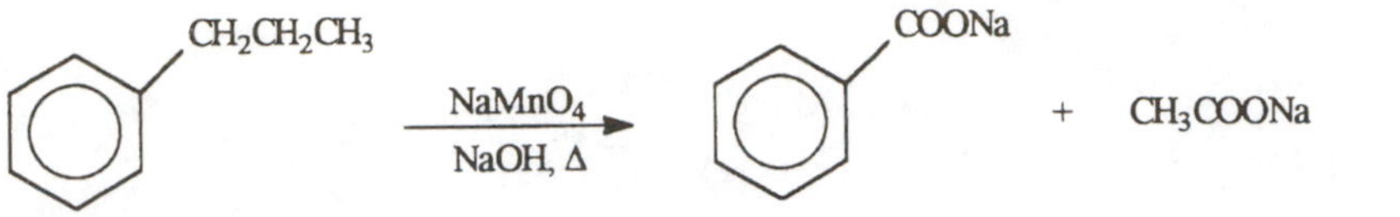

$$C_6H_5CH_2CH_2CH_3 \xrightarrow[NaOH,\ \Delta]{NaMnO_4} C_6H_5COONa + CH_3COONa$$

sodium benzoate sodium acetate

22. (a)

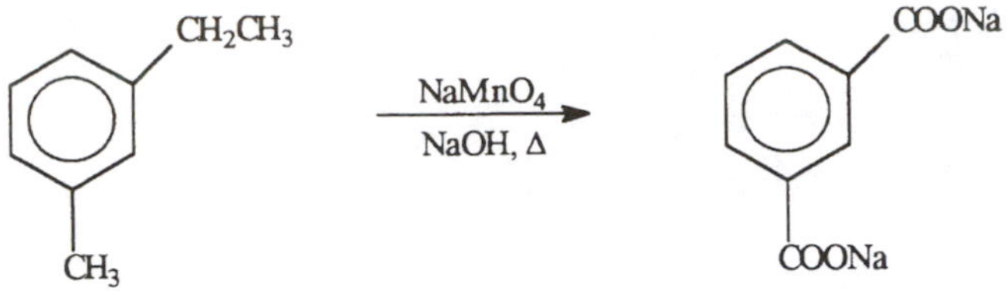

$$m\text{-}CH_3C_6H_4CH_2CH_3 \xrightarrow[NaOH,\ \Delta]{NaMnO_4} m\text{-}C_6H_4(COONa)_2$$

disodium-*m*-phthalate

(b) $HOOCCH=CHCOOH + H_2 \xrightarrow{Ni} HOOCCH_2CH_2COOH$

succinic acid

(c) $CH_3(CH_2)_7CH=CH(CH_2)_7COOH + NaOH \longrightarrow$

$CH_3(CH_2)_7CH=CH(CH_2)_7COONa + H_2O$

sodium oleate

(d) $CH_3CH_2CH_2\overset{O}{\overset{\|}{C}}\text{–}H \xrightarrow[H_2SO_4]{Na_2Cr_2O_7} CH_2CH_2CH_2COOH$

butanoic acid

(e)

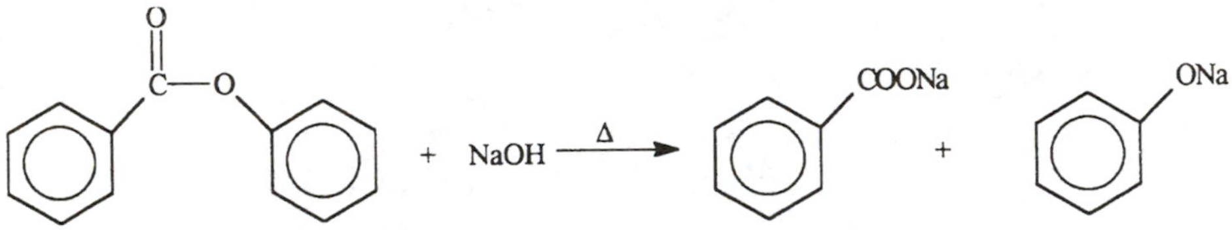

$$C_6H_5\overset{O}{\overset{\|}{C}}\text{—}OC_6H_5 + NaOH \xrightarrow{\Delta} C_6H_5COONa + C_6H_5ONa$$

sodium benzoate sodium phenolate

23. Structural formula for the ester that when hydrolyzed would yield:

(a) methanol and acetic acid $CH_3\overset{O}{\overset{\|}{C}}\text{–}OCH_3$

(b) ethanol and formic acid $H\overset{O}{\overset{\|}{C}}\text{–}OCH_2CH_3$

(c) 2-propanol and benzoic acid

$$C_6H_5-\overset{O}{\overset{\|}{C}}-O-\underset{CH_3}{\underset{|}{CH}}-CH_3$$

24. Structural formula for the ester that when hydrolyzed would yield:

(a) methanol and propanoic acid $CH_3CH_2\overset{O}{\overset{\|}{C}}-OCH_3$

(b) 1-octanol and acetic acid $CH_3\overset{O}{\overset{\|}{C}}-OCH_2(CH_2)_6CH_3$

(c) ethanol and butanoic acid $CH_3CH_2CH_2\overset{O}{\overset{\|}{C}}-OCH_2CH_3$

25. Structural formulas for the reactants that will yield the following esters:

(a) methyl palmitate $CH_3OH + CH_3(CH_2)_{14}COOH$

(b) phenyl propionate $C_6H_5OH + CH_3CH_2COOH$

(c) dimethyl succinate $CH_3OH + HOOCCH_2CH_2COOH$

26. Structural formulas for the reactants that will yield the following esters:

(a) isopropyl formate $CH_3\underset{OH}{\underset{|}{C}}HCH_3 + HCOOH$

(b) diethyl adipate $CH_3CH_2OH + HOOC(CH_2)_4COOH$

(c) benzyl benzoate $C_6H_5CH_2OH + C_6H_5COOH$

27. Structural formulas for the organic products formed in the following reactions:

(a)

$$C_6H_5\text{-}C(=O)\text{-}Cl + H_2O \longrightarrow C_6H_5\text{-}COOH + HCl$$

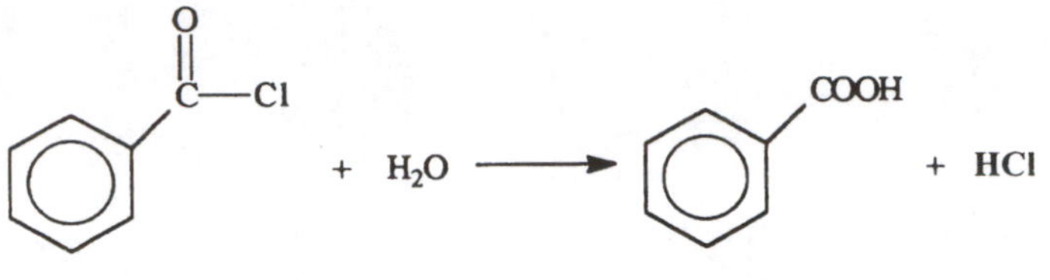

(b) $CH_2=CHCOOH + Br_2 \longrightarrow CH_2BrCHBrCOOH$

(c)

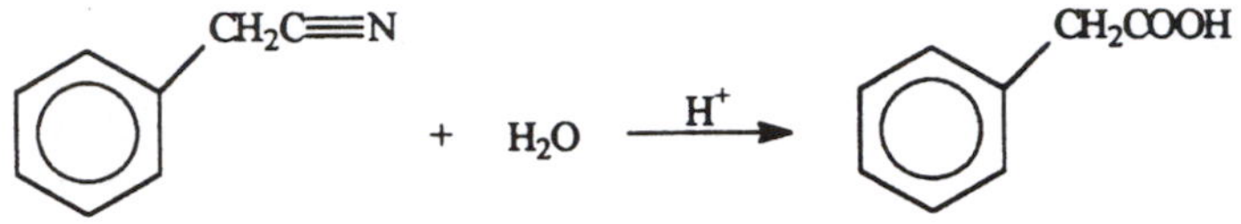

$$C_6H_5\text{-}CH_2C\equiv N + H_2O \xrightarrow{H^+} C_6H_5\text{-}CH_2COOH$$

(d) $CH_3\overset{O}{\overset{\|}{C}}\text{–}Cl + CH_3CH_2OH \longrightarrow CH_3\overset{O}{\overset{\|}{C}}\text{–}OCH_2CH_3 + HCl$

(e) $CH_3\overset{O}{\overset{\|}{C}}\text{–}Cl + 2\,NH_3 \longrightarrow CH_3\overset{O}{\overset{\|}{C}}\text{–}NH_2 + NH_4Cl$

28. Structural formulas for the organic products formed in the following reactions:

(a)

$$C_6H_5\text{-}C(=O)\text{-}Cl + 2\,NH_3 \longrightarrow C_6H_5\text{-}C(=O)\text{-}NH_2 + NH_4Cl$$

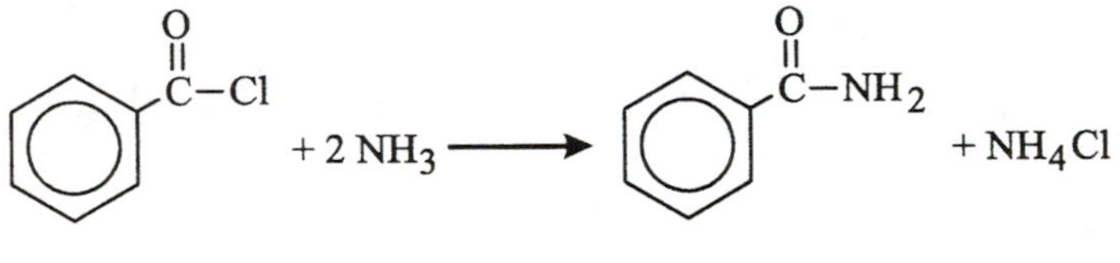

(b) $CH_3CH_2COOH + SOCl_2 \longrightarrow CH_3CH_2\overset{O}{\overset{\|}{C}}\text{–}Cl + SO_2 + HCl$

(c) $CH_3CH_2\overset{O}{\overset{\|}{C}}\text{–}Cl + CH_3OH \longrightarrow CH_3CH_2\overset{O}{\overset{\|}{C}}\text{–}OCH_3 + HCl$

(d) $CH_3CH_2CH_2C\equiv N + H_2O \xrightarrow{H^+} CH_3CH_2CH_2COOH$

(e)

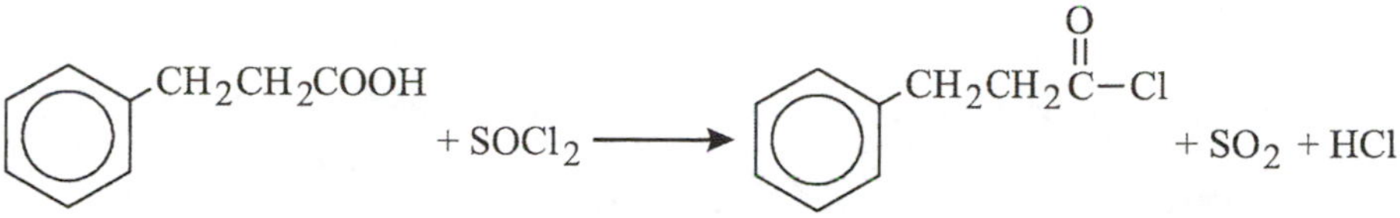

$$C_6H_5CH_2CH_2COOH + SOCl_2 \longrightarrow C_6H_5CH_2CH_2\overset{O}{\overset{\|}{C}}-Cl + SO_2 + HCl$$

29. Simple test to distinguish between:

(a) Sodium benzoate and benzoic acid: Sodium benzoate is water soluble; benzoic acid is not.

(b) Maleic acid and malonic acid. Maleic acid has a carbon-carbon double bond, it will readily add and decolorize bromine; malonic acid will not decolorize bromine.

30. Simple tests to distinguish between:

(a) Benzoic acid and ethyl benzoate; benzoic acid is an odorless solid; ethyl benzoate is a fragrant liquid.

(b) Succinic acid and fumaric acid: fumaric acid has a carbon-carbon double bond and will readily add and decolorize bromine; succinic acid will not decolorize bromine.

31. Structural formulas for the products of the following reactions:

(a) $$CH_3CH(COOH)_2 \xrightarrow{150°C} CH_3CH_2COOH + CO_2$$

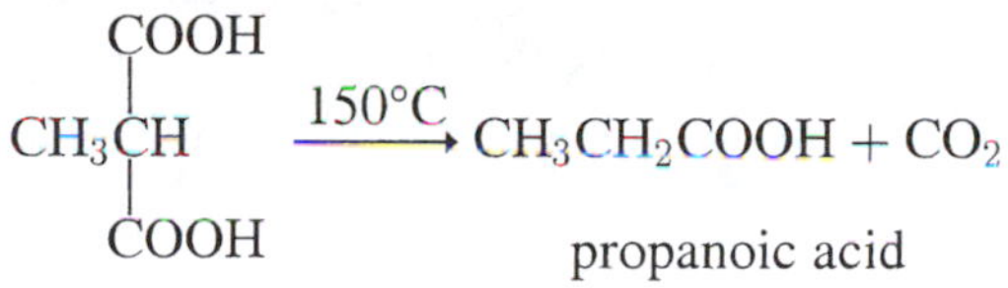

propanoic acid

(b)

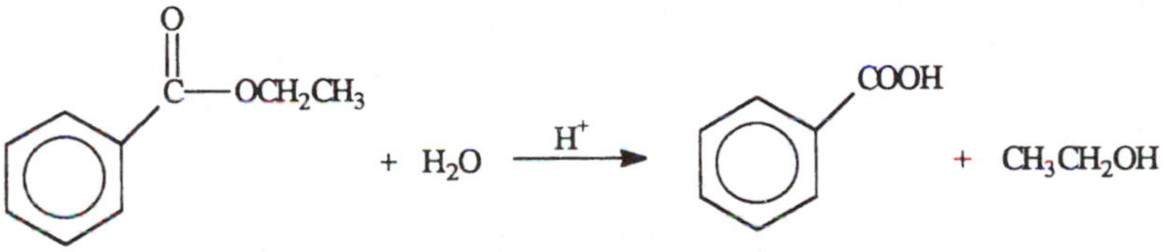

$$C_6H_5\overset{O}{\overset{\|}{C}}-OCH_2CH_3 + H_2O \xrightarrow{H^+} C_6H_5COOH + CH_3CH_2OH$$

benzoic acid

(c)

$$CH_3CH_2COOH + C_6H_5OH \xrightarrow{H^+} CH_3CH_2\overset{O}{\overset{\|}{C}}-O-C_6H_5$$

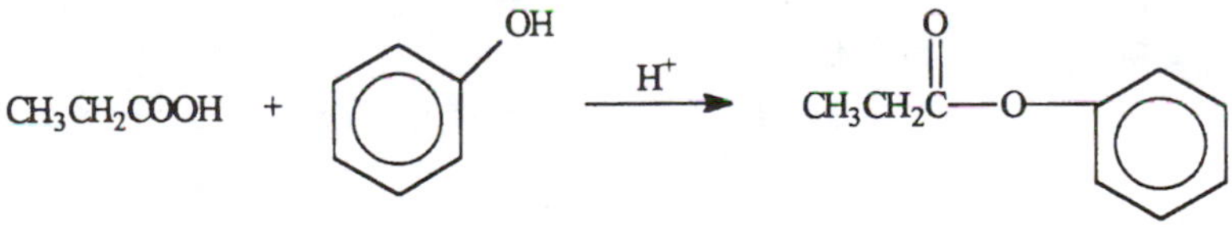

phenyl propanoate

32. (a)

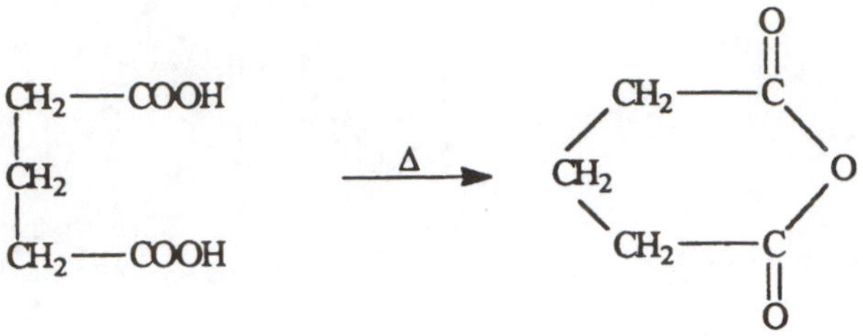

$$\begin{array}{l} CH_2\text{—}COOH \\ | \\ CH_2 \\ | \\ CH_2\text{—}COOH \end{array} \xrightarrow{\Delta} \text{cyclic: } CH_2\text{—}C(=O)\text{—}O\text{—}C(=O)\text{—}CH_2\text{—}CH_2\text{—}CH_2$$

glutaric anhydride

(b)

$$\begin{array}{l} COOH \\ | \\ CH_2 \\ | \\ COOH \end{array} + 2\,CH_3CH_2OH \xrightarrow{H^+} \begin{array}{l} COOCH_2CH_3 \\ | \\ CH_2 \\ | \\ COOCH_2CH_3 \end{array}$$

diethyl malonate

(c)

$$C_6H_5\text{—}O\text{—}\overset{O}{\overset{\|}{C}}CH_3 + H_2O \xrightarrow[\Delta]{H^+} C_6H_5OH + CH_3COOH$$

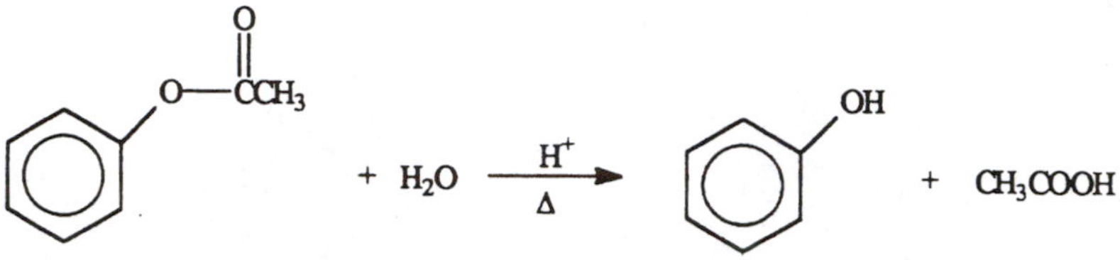

phenol acetic acid

33. Structure of *cis, cis, cis*-linolenic acid:

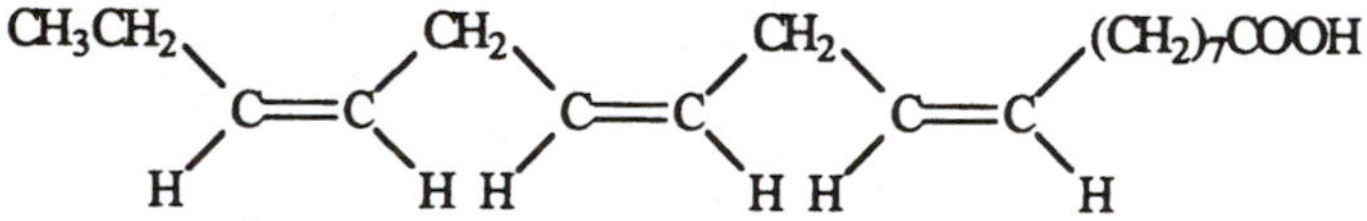

$$CH_3CH_2(H)C{=}C(H)CH_2(H)C{=}C(H)CH_2(H)C{=}C(H)(CH_2)_7COOH$$

(all double bonds cis: $CH_3CH_2$, $CH_2$, $CH_2$, $(CH_2)_7COOH$ on top; H H, H H, H on bottom)

Structure of *trans, trans, trans*-linolenic acid:

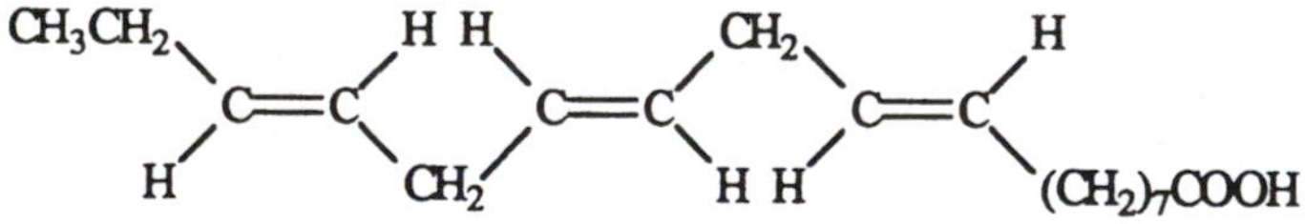

$$CH_3CH_2(H)C{=}C(H)CH_2(H)C{=}C(H)CH_2(H)C{=}C(H)(CH_2)_7COOH$$

(all double bonds trans)

34. (a) *cis*-oleic acid

$$\begin{array}{ccc} CH_3(CH_2)_7 & & (CH_2)_7COOH \\ & C{=}C & \\ H & & H \end{array}$$

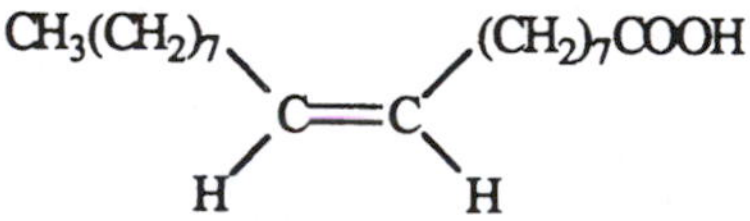

(b) *cis, cis*-linoleic acid

$$\begin{array}{c} CH_3(CH_2)_4 \quad\quad CH_2 \quad\quad (CH_2)_7COOH \\ C{=}C \quad\quad C{=}C \\ H \quad\quad H\;H \quad\quad H \end{array}$$

35. $CH_3(CH_2)_{12}COONa$ would be more useful than $CH_3(CH_2)_{12}COOH$ as a cleansing agent in soft water. Both have a long hydrocarbon chain which would dissolve in the fat, but the acid is not water soluble, whereas the sodium salt is soluble in water.

36. Sodium lauryl sulfate would be more effective as a detergent in hard water than sodium propyl sulfate because the hydrocarbon chain is only three carbons long in the latter, not long enough to dissolve grease well. Sodium lauryl sulfate is effective in both hard and soft water.

37. Only (a), hexadecyltrimethyl ammonium chloride would be a good detergent in water. It is cationic.

38. Only (c) $CH_3(CH_2)_{10}CH_2O(CH_2CH_2O)_7CH_2CH_2OH$, would be a good detergent in water. It is nonionic.

39. (a)
$$HO{-}\underset{OH}{\overset{O}{\overset{\|}{P}}}{-}OH + HOCH_2CH_3 \xrightarrow{H^+} HO{-}\underset{OH}{\overset{O}{\overset{\|}{P}}}{-}OCH_2CH_3 + H_2O$$

(b)
$$HO{-}\underset{OH}{\overset{O}{\overset{\|}{P}}}{-}OCH_2CH_3 + CH_3CH_2CH_2OH \xrightarrow{H^+} HO{-}\underset{OCH_2CH_2CH_3}{\overset{O}{\overset{\|}{P}}}{-}OCH_2CH_3 + H_2O$$

40. (a)
$$HO{-}\underset{OH}{\overset{O}{\overset{\|}{P}}}{-}OH + 3\,CH_3OH \xrightarrow{H^+} CH_3O{-}\underset{OCH_3}{\overset{O}{\overset{\|}{P}}}{-}OCH_3 + 3\,H_2O$$

(b)
$$HO{-}\underset{OH}{\overset{O}{\overset{\|}{P}}}{-}OH + CH_3OH \xrightarrow{H^+} CH_3O{-}\underset{OH}{\overset{O}{\overset{\|}{P}}}{-}OH + H_2O$$

$$CH_3O{-}\underset{OH}{\overset{O}{\overset{\|}{P}}}{-}OH + CH_3CH_2\underset{OH}{CH}\,CH_3 \xrightarrow{H^+} CH_3O{-}\underset{OH}{\overset{O}{\overset{\|}{P}}}{-}O\overset{CH_3}{CH}CH_2CH_3$$

41. Grams and moles of sodium benzoate:

$0.001 \times 1 \text{ lb} \times \frac{454 \text{ g}}{1 \text{ lb}} = 0.5 \text{ g sodium benzoate}$

$0.5 \text{ g } NaC_7H_5O_2 \times \frac{1 \text{ mol}}{145.1 \text{ g}} = 3 \times 10^{-3} \text{ mol } NaC_7H_5O_2$

42.

salicylic acid (COOH, OH on benzene ring) + acetic anhydride ($CH_3C(=O)-O-C(=O)CH_3$) → aspirin (COOH and $OCCH_3$ with C=O on benzene ring) + $CH_3COOH$

salicylic acid   acetic anhydride   aspirin (acetylsalicylic acid)   acetic acid

43. The ester is propyl propanoate $CH_3CH_2\underset{\underset{O}{\|}}{C}OCH_2CH_2CH_3$

Compound A is an acid; B is an alcohol. If B is oxidized to an acid which is the same as A, then A and B both have the same carbon structure, three carbon atoms each. The acid A must be propanoic acid, $CH_3CH_2COOH$. The alcohol can be 1-propanol or 2-propanol. Only 1-propanol can be oxidized to propanoic acid which is the same as compound A.

44. $Cl-\underset{\underset{O}{\|}}{C}-Cl$

The functional groups are acid chlorides.

The acid from which phosgene is derived is carbonic acid, $HO-\underset{\underset{O}{\|}}{C}-OH$

45.

$CH_2-O-\overset{\overset{O}{\|}}{C}(CH_2)_7CH{=}CH(CH_2)_7CH_3$

$CH-O-\overset{\overset{O}{\|}}{C}(CH_2)_7CH{=}CHCH_2CH{=}CH(CH_2)_4CH_3$

$CH_2-O-\overset{\overset{O}{\|}}{C}(CH_2)_7CH{=}CHCH_2CH{=}CH(CH_2)_4CH_3$

soybean oil

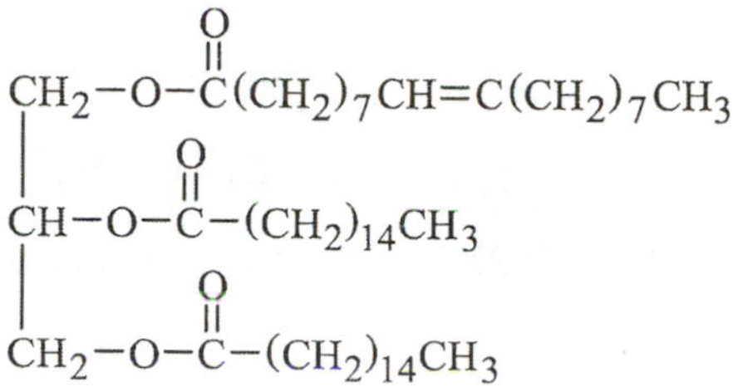

$$\begin{array}{l} \quad\quad\;\; O \\ \quad\quad\;\; \| \\ CH_2-O-C(CH_2)_7CH{=}C(CH_2)_7CH_3 \\ | \quad\quad\;\; O \\ | \quad\quad\;\; \| \\ CH-O-C-(CH_2)_{14}CH_3 \\ | \quad\quad\;\; O \\ | \quad\quad\;\; \| \\ CH_2-O-C-(CH_2)_{14}CH_3 \end{array}$$

palm oil

The bonded order of the fatty acids to glycerol may vary.

46.

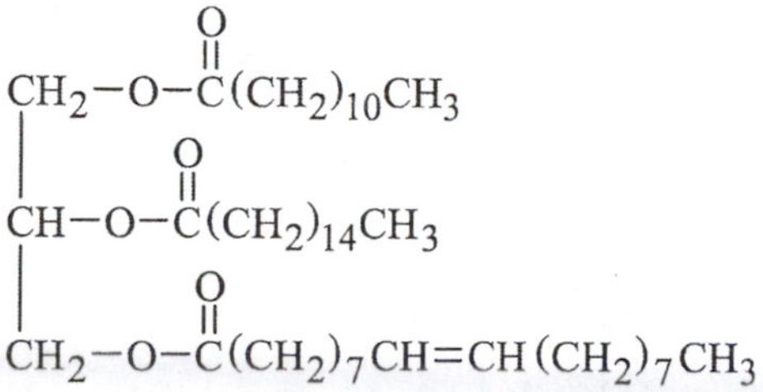

$$\begin{array}{l} \quad\quad\;\; O \\ \quad\quad\;\; \| \\ CH_2-O-C(CH_2)_{10}CH_3 \\ | \quad\quad\;\; O \\ | \quad\quad\;\; \| \\ CH-O-C(CH_2)_{14}CH_3 \\ | \quad\quad\;\; O \\ | \quad\quad\;\; \| \\ CH_2-O-C(CH_2)_7CH{=}CH(CH_2)_7CH_3 \end{array}$$

There would be two other triacylglcerols containing all three of these acids. Each of the three acids can be attached to the middle carbon of the glycerol.

47. Names and formulas of products are

(a)

| | | | |
|---|---|---|---|
| $CH_2OH$ | and | $CH_3(CH_2)_{10}COOH$ | lauric acid |
| $\vert$ $CHOH$ | | $CH_3(CH_2)_{14}COOH$ | palmitic acid |
| $\vert$ $CH_2OH$ | | $CH_3(CH_2)_7CH{=}CH(CH_2)_7COOH$ | oleic acid |

glycerol

(b)

| | | | |
|---|---|---|---|
| $CH_2OH$ | and | $CH_3(CH_2)_{10}CH_2OH$ | 1-dodecanol (lauryl alcohol) |
| $\vert$ $CHOH$ | | $CH_3(CH_2)_{14}CH_2OH$ | 1-hexadecanol (cetyl alcohol) |
| $\vert$ $CH_2OH$ | | $CH_3(CH_2)_{16}CH_2OH$ | 1-octadecanol (stearyl alcohol) |

glycerol

(c) glycerol and:

$$\begin{array}{l} CH_2OH \\ | \\ CHOH \\ | \\ CH_2OH \end{array}$$

glycerol

| | |
|---|---|
| $CH_3(CH_2)_{10}COOK$ | potassium laurate |
| $CH_3(CH_2)_{14}COOK$ | potassium palmitate |
| $CH_3(CH_2)_7CH{=}CH(CH_2)_7COOK$ | potassium oleate |

(d)

$$\begin{array}{l} CH_2{-}O{-}\overset{O}{\overset{\|}{C}}(CH_2)_{10}CH_3 \\ | \\ CH{-}O{-}\overset{O}{\overset{\|}{C}}(CH_2)_{14}CH_3 \\ | \\ CH_2{-}O{-}\overset{O}{\overset{\|}{C}}(CH_2)_{16}CH_3 \end{array}$$

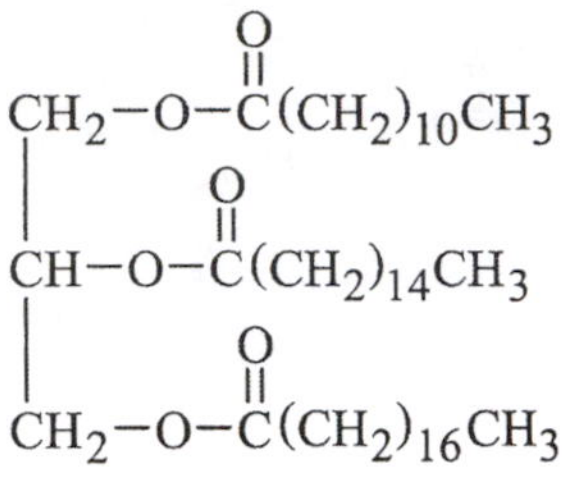

lauroylpalmitoylsteroylglycerol

48.

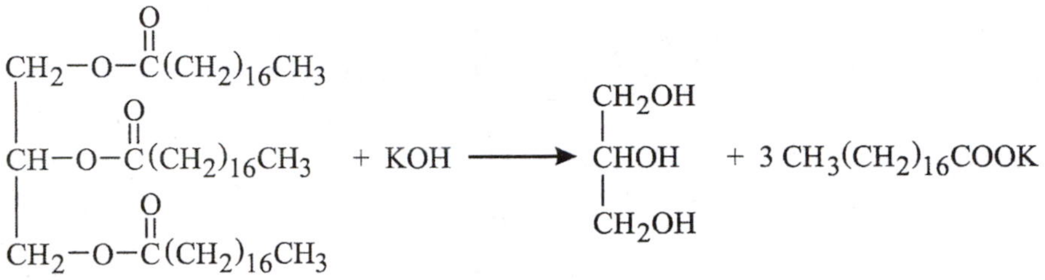

$$\begin{array}{l} CH_2{-}O{-}\overset{O}{\overset{\|}{C}}(CH_2)_{16}CH_3 \\ | \\ CH{-}O{-}\overset{O}{\overset{\|}{C}}(CH_2)_{16}CH_3 \\ | \\ CH_2{-}O{-}\overset{O}{\overset{\|}{C}}(CH_2)_{16}CH_3 \end{array} + KOH \longrightarrow \begin{array}{l} CH_2OH \\ | \\ CHOH \\ | \\ CH_2OH \end{array} + 3\ CH_3(CH_2)_{16}COOK$$

glyceryl tristearate      glycerol      potassium stearate

The solubility in water will change considerably. Glyceryl tristearate is insoluble in water but the products glycerol and potassium stearate (a salt) are soluble in water.

49. Synthesis of:

(a) acetic acid

$$CH_3CH_2OH \xrightarrow[H_2SO_4]{K_2Cr_2O_7} CH_3COOH$$

(b) ethyl acetate

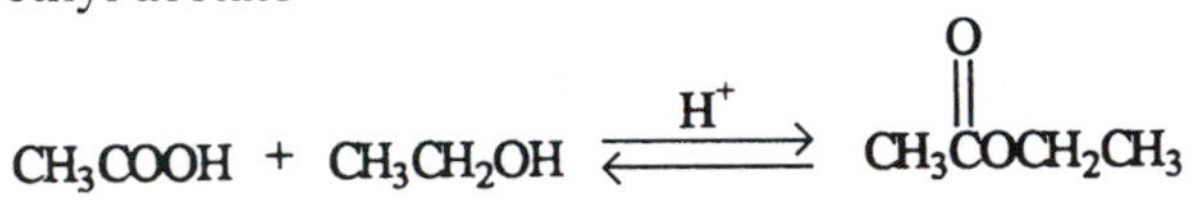

$$CH_3COOH + CH_3CH_2OH \underset{}{\overset{H^+}{\rightleftharpoons}} CH_3\overset{O}{\overset{\|}{C}}OCH_2CH_3$$

(c) $\beta$-hydroxybutyric acid

$$CH_3CH_2OH + \text{air} \xrightarrow[\Delta]{\text{Cu tube}} CH_3\overset{H}{\overset{|}{C}}=O$$

$$2\,CH_3\overset{H}{\overset{|}{C}}=O \xrightarrow[\text{NaOH}]{\text{dil.}} CH_3\underset{OH}{\underset{|}{C}H}\,CH_2\overset{H}{\overset{|}{C}}=O \xrightarrow[NH_3]{Ag_2O} CH_3\underset{OH}{\underset{|}{C}H}\,CH_2COOH$$

Aldol condensation    Tollens reagent

50. $CH_3(CH_2)_{14}\overset{O}{\overset{||}{C}}-O(CH_2)_{29}CH_3$

51.

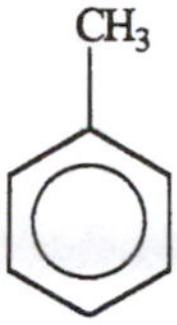

A: benzene ring with $CH_3$  B: benzene ring with $COOH$  C: benzene ring with $COOCH_3$

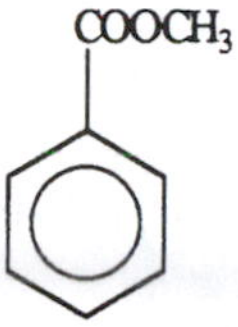

52. $CH_3(CH_2)_{12}\overset{O}{\overset{||}{C}}-O\underset{CH_3}{\underset{|}{C}H}CH_3$  isopropyl myristate

53. $C_{10}H_{12}O_2 \longrightarrow C_7H_8O + C_3H_6O_2$

A (has a benzene ring) B (alcohol) C (acid)

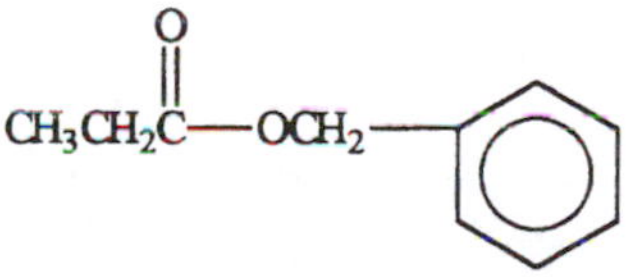

A: $CH_3CH_2\overset{O}{\overset{||}{C}}-OCH_2-C_6H_5$  B: $C_6H_5CH_2OH$  C: $CH_3CH_2COOH$

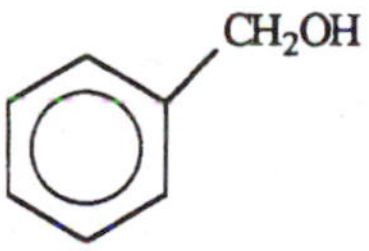

54. Smell both samples. Butanoic acid has an unpleasant rancid odor; ethyl butanoate has a pleasant odor of pineapple.

55. Each molecule of triolein requires three molecules of $H_2$ (one for each double bond)

(a) $(1.00 \text{ kg triolein})\left(\frac{10^3 \text{ g}}{\text{kg}}\right)\left(\frac{1 \text{ mol triolein}}{885.4 \text{ g triolein}}\right)\left(\frac{3 \text{ mol } H_2}{1 \text{ mol triolein}}\right)\left(\frac{22.4 \text{ L}}{1 \text{ mol } H_2}\right) = 75.9 \text{ L } H_2$

(b) $(1.00 \text{ kg triolein})\left(\frac{891.5 \text{ g tristearin}}{885.4 \text{ g triolein}}\right) = 1.01 \text{ kg tristearin}$

56. The statement is false. When methyl propanoate is hydrolyzed, propanoic acid and methanol are formed.

57. Esters of formula $C_5H_{10}O_2$

$HC(=O){-}OCH_2CH_2CH_3CH_3$ $\quad$ $HC(=O){-}OCH_2CH(CH_3)CH_3$ $\quad$ $HC(=O){-}OCH(CH_3)CH_2CH_3$

$HC(=O){-}O{-}C(CH_3)_2{-}CH_3$ $\quad$ $CH_3C(=O){-}OCH_2CH_2CH_3$ $\quad$ $CH_3C(=O){-}OCH(CH_3)CH_3$

$CH_3CH_2C(=O){-}OCH_2CH_3$ $\quad$ $CH_3CH_2CH_2C(=O){-}OCH_3$ $\quad$ $CH_3CH(CH_3)C(=O){-}OCH_3$

58. (a) Mass percent of oxygen in dacron.

$-\!\!\left(C(=O){-}C_6H_4{-}C(=O){-}OCH_2CH_2O\right)_n\!\!-$

Mass of one unit of dacron = 192.2 g
Mass of oxygen in one unit of dacron = 64.00 g

$\frac{64.00 \text{ g O}}{192.2 \text{ g}} \times 100 = 33.30\% \text{ O}$

(b) Molar mass of 105 units

$(105 \text{ units})\left(\frac{192.2 \text{ g}}{\text{unit}}\right) = 2.02 \times 10^4 \text{ g}$

59. (a) $-\!\!\left(C(=O){-}C_6H_4{-}C(=O){-}OCH_2CH_2CH_2O\right)_n\!\!-$

(b)

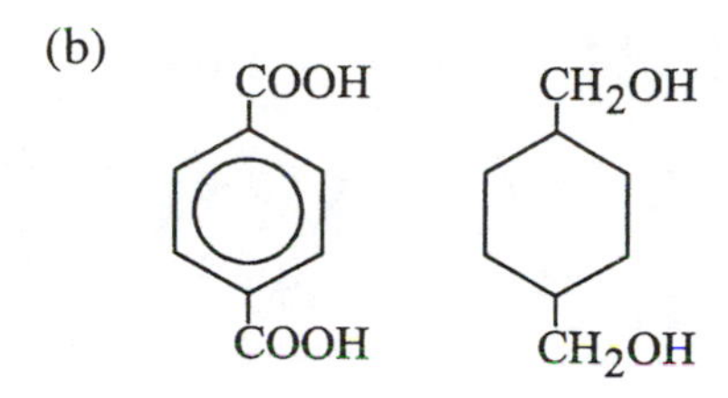


60. A glyptal polyester would most likely be thermosetling because glycerol is trifunctional and would thus allow cross linking between chains in forming the ester polymer.

CHAPTER 25

# AMIDES AND AMINES: ORGANIC NITROGEN COMPOUNDS

1. (a) $CH_3CH_2NH_2$ ethyl amine

   (b) $CH_3C(=O)-NH_2$ ethanamide

   In an amide the $NH_2$ group is bonded to a carbonyl group. In an amine the $NH_2$ group is bonded to a saturated carbon atom.

2. Amides: Unsubstituted amides (except formamide) are solids at room temperature. Many are odorless and colorless. Low molar-mass amides are water soluble. Solubility in water decreases as the molar mass increases. Amides are neutral compounds. The $NH_2$ group is capable of hydrogen bonding.

   Amines: Low molar-mass amines are flammable gases with an ammonia-like odor. Aliphatic amines up to six carbon atoms are water soluble. Many amines have a "fishy" odor and many have very foul odors. Aromatic amines occur as liquids and solids. Soluble aliphatic amines give basic solutions. Aromatic amines are less soluble in water and less basic than aliphatic amines. The $NH_2$ group is capable of hydrogen bonding.

3. Amines and alcohols of similar molar masses have approximately the same water solubility due to the fact that both of these classes of compounds can hydrogen bond with water.

4. (a) Heterocyclic compounds are those in which all the atoms in the ring are not alike.
   (b) The number of heterocyclic rings in each of the compounds is:

   (i) purine, 2 (ii) ampicillin, 2 (iii) methadone, 0 (iv) nicotine, 2

5. Functional groups in:
   (a) procaine hydrochloride: aromatic amine; ester; quaternary ammonium ion
   (b) cocaine: 3° amine; two ester groups
   (c) nicotinamide: amide; heterocyclic amine
   (d) methamphetamine: 2° amine

6. Ammonia is a toxic, basic, water soluble compound which can increase the pH of the blood and the urine and would be painful to pass through bodily tissues. However, ammonia is converted in the liver to the neutral diamide, urea, which is water soluble and is excreted in the urine.

7. (a) $NH_2$ / $CH_3$ (para-substituted benzene ring)  (b) $CH_3CH(NH_2)CH_2CH_3$  (c) $O{=}C{-}NHCH_3$ / Br (para-substituted benzene ring)

8. (a) $CH_3CH(NH_2)CH_2CHClCH_3$  (b) $C_6H_5{-}C({=}O){-}N(CH_2CH_3)_2$  (c) $NH_2$ / Cl (meta-substituted benzene ring)

9. Names
   (a) acetamide
   (b) 4-amino-2-pentanol or 4-hydroxy-2-pentanamine
   (c) *m*-methyl-N-methylbenzamide
   (d) 2,2-dimethyl-1-(N-ethyl-N-methyl) butanamine

10. Names
   (a) 3-methyl-1-butanamine
   (b) formamide
   (c) N-phenylpropanamide
   (d) ethyldiisopropylamine
   2-(N-ethyl-N-isopropyl) propanamine

11. Increasing solubility in water.

$$CH_3CH_2C({=}O){-}N(CH_3)_2 < CH_3CH_2C({=}O){-}NHCH_3 < CH_3CH_2C({=}O){-}NH_2$$

12. Increasing solubility in water.

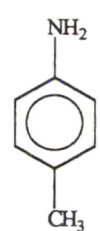

$$C_6H_5{-}C({=}O){-}NH_2 < CH_3(CH_2)_4C({=}O){-}NH_2 < CH_3C({=}O){-}NH_2$$

13. Hydrogen bonding in water.

$CH_3\overset{O}{\overset{\|}{C}}-N\begin{matrix} H\cdots\cdot :\ddot{O}(H)-H \\ H\cdots\cdot :\ddot{O}(H)-H \end{matrix}$ and $CH_3\overset{O}{\overset{\|}{C}}-\overset{H}{\underset{\cdot\cdot}{N}}-H \cdots H-\ddot{O}:-H$

14. (Several possibilities) Hydrogen bonding in water.

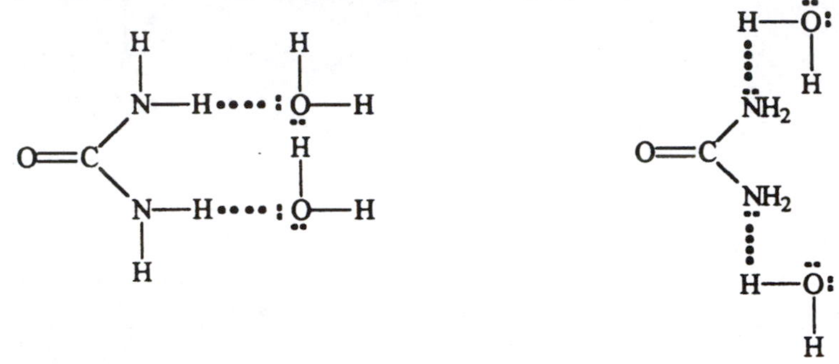


$H-\ddot{O}:(H)\cdots \ddot{N}H_2-C(=O)-\ddot{N}H_2 \cdots H-\ddot{O}:(H)$

15. (a) neither
(b) base
(c) base
(d) acid
(e) neither
(f) acid

16. (a) neither
(b) base
(c) base
(d) neither
(e) neither
(f) acid

17. Organic products

(a) $CH_3\overset{O}{\overset{\|}{C}}\text{–}OH + CH_3NH_3^+$

(b) $CH_3\overset{O}{\overset{\|}{C}}\text{–}NHCH(CH_3)_2$

(c) $H_2NCH_2CH_2CH_2CH_2COONa$

18. Organic products

(a) $CH_3\overset{O}{\overset{\|}{C}}\text{–}N(CH_2CH_3)_2$

(b) $C_6H_5CH_2COOH + CH_3\overset{+}{N}H_2CH_3$

(c) $CH_3CH(CH_3)CH_2C(=O)–NHCH_3$ (drawn with the C=O above and the $CH_3$ branch below)

19. Structures of amines with formula $C_4H_{11}N$.

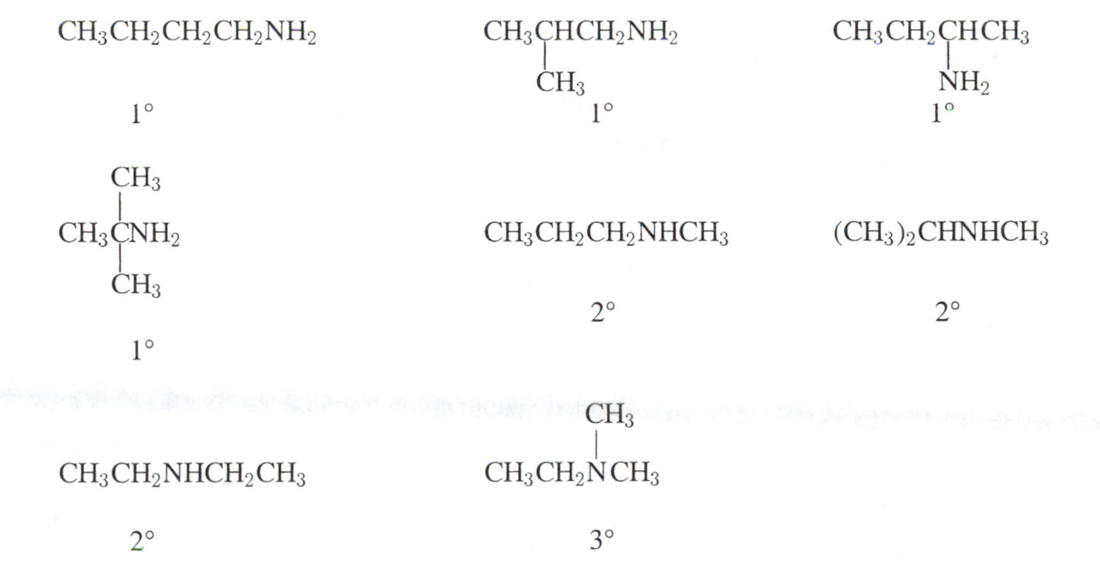


20. Structures of amines with formula $C_3H_9N$.

| | | | |
|---|---|---|---|
| $CH_3CH_2CH_2NH_2$ | $CH_3CH(NH_2)CH_3$ | $CH_3CH_2NHCH_3$ | $CH_3N(CH_3)CH_3$ |
| 1° | 1° | 2° | 3° |

21. Classification of amines
(a) primary (b) tertiary (c) primary (d) both primary

22. Classification of amines
(a) secondary (b) tertiary (c) secondary (d) tertiary

23. The triethylamine solution in 1.0 M NaOH would have the more objectionable odor because it would be in the form of the free amine, while in the acid solution the amine would form a salt that will have little or no odor.

24. The isopropylamine solution in 1.0 M KOH would have the more objectionable odor because it would be in the form of the free amine, while in the acid solution the amine would form a salt that will have little or no odor.

25. Names

(a) $CH_3NHCH_3$ dimethyl amine

(b) $CH_2CH_3$, $NH_2$ *o*-ethylaniline

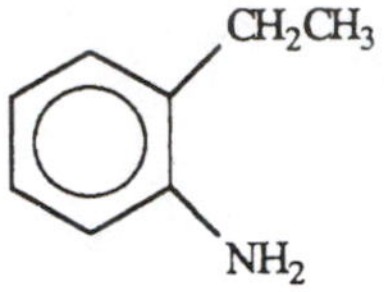

(c) $NH_2$ cyclohexanamine

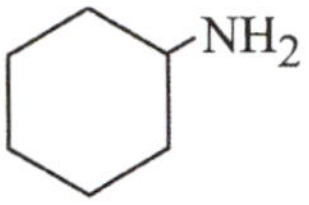

(d) $(C_2H_5)_4N^+I^-$ tetraethylammonium iodide

(e) $NH_2$, $NO_2$ *m*-nitroaniline

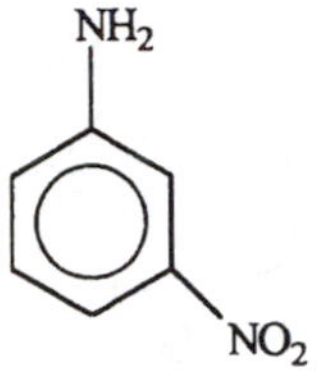

(f) $(CH_3CH_2)_2N-$ cyclohexyldiethylamine

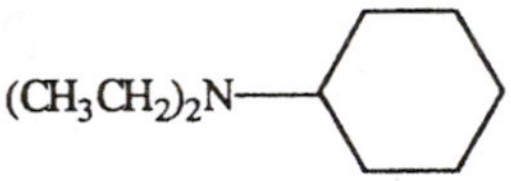

26. Names

(a)

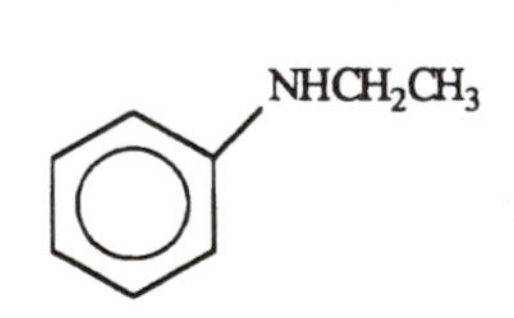


N-ethylaniline

(b) $CH_3CH_2CHCH_3$ 2-butanamine

(c) diphenylamine

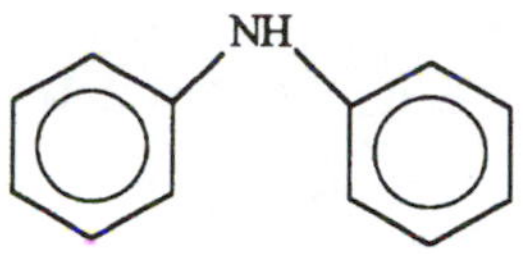

(d) $CH_3CH_2NH_3^+Br^-$ ethylammonium bromide

(e) pyridine

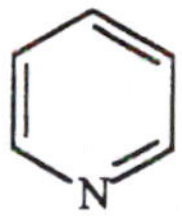

(f) $CH_3C(=O)–NHCH_2CH_3$ N-ethylacetamide

27. Structural formulas

(a) $CH_3CH_2NHCH_3$

(b) $NH_2$

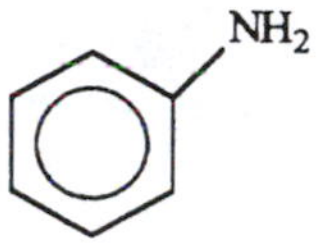

(c) $H_2NCH_2CH_2CH_2CH_2NH_2$

(d) $CH_3CH_2N(CH_3)CH(CH_3)_2$

(e)

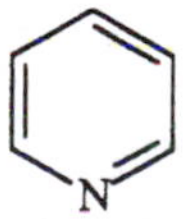

(f) $(CH_3CH_2)_3\overset{+}{N}H\ Cl^-$

28. Structural formulas

(a) $CH_3CH_2CH_2CH_2N(CH_2CH_2CH_2CH_3)CH_2CH_2CH_2CH_3$

(b) $C_6H_5\overset{+}{N}H_2CH_3\ Cl^-$

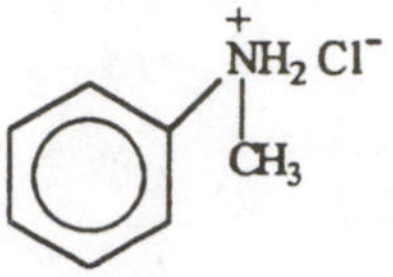

(c) $CH_3CH_2\overset{+}{N}H_3\ Cl^-$

(d) $CH_3CH_2CH_2CH(NH_2)CH_2OH$

(e) $CH_3CH_2CH_2C(CH_3)_2$—$CH(CH_3)$—NH—$C_6H_5$

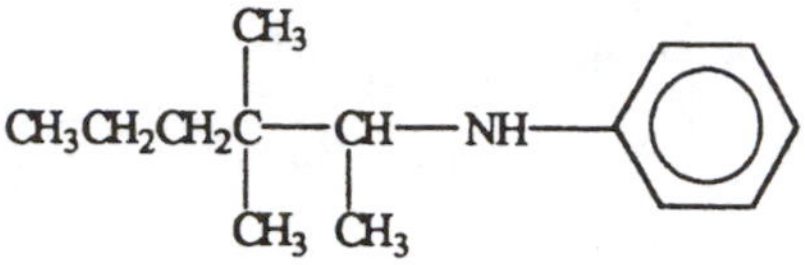

(f) $H_2NCH_2CH_2NH_2$

29. (a) $CH_3CH_2CH_2Br + NH_3 \text{ (excess)} \longrightarrow CH_3CH_2CH_2NH_2 + NH_4Br$

(b) $CH_3CH_2CH_2Br + KCN \longrightarrow CH_3CH_2CH_2CN \xrightarrow[Ni]{H_2} CH_3CH_2CH_2CH_2NH_2$

(c) $p\text{-}NO_2C_6H_4CH_2CH_3 \xrightarrow[HCl]{Sn} p\text{-}NH_3^+C_6H_4CH_2CH_3 \xrightarrow{NaOH} p\text{-}NH_2C_6H_4CH_2CH_3$

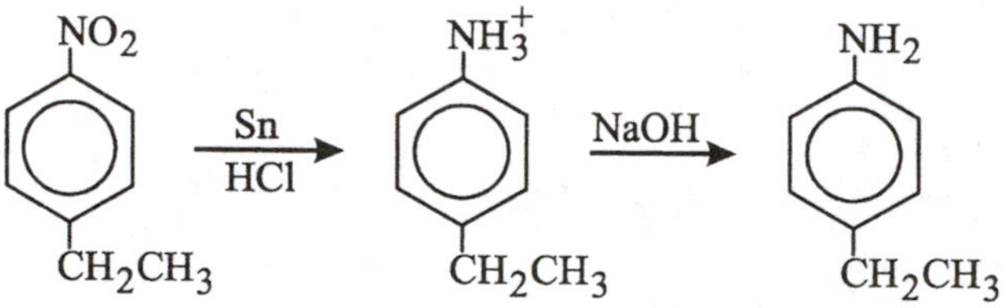

(d) $CH_3CH_2CH_2NH_2 + CH_3\overset{O}{\overset{\|}{C}}\text{–}Cl \longrightarrow CH_3\overset{O}{\overset{\|}{C}}\text{–}NHCH_2CH_2CH_3$

30. Organic products

(a) $CH_3CH_2\overset{O}{\overset{\|}{C}}\text{–}NH_2$

(b) $CH_3CH_2CH_2CH_2NH_2$

(c) $p\text{-}H_3C\text{–}C_6H_4\text{–}NH_2$

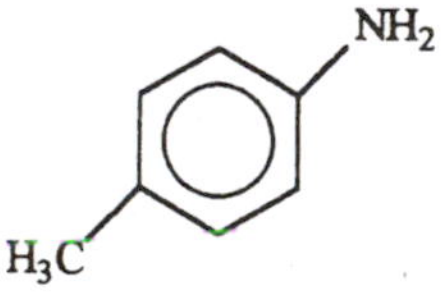

(d) N-acetylpiperidine: $C_5H_{10}N\text{–}\overset{O}{\overset{\|}{C}}\text{—}CH_3$

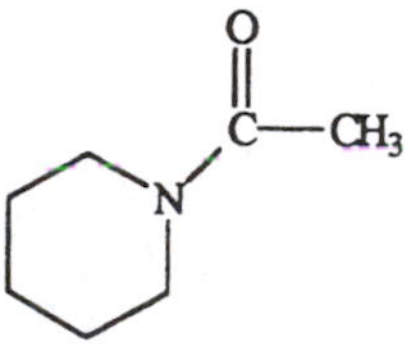

31. The main reason why trimethylamine has a lower boiling point than propylamine and ethylmethylamine is that trimethylamine cannot hydrogen bond because it has no hydrogen atoms bonded to the nitrogen atom. The other two amines do have hydrogen atom(s) bonded to the nitrogen atom and their molecules can hydrogen bond, which results in higher boiling points.

32. *Amphetamines:* stimulate the central nervous system; used to treat depression, narcolepsy, and obesity.

    *Tranquilizers* (Valium): used to modify psychotic behavior, relieve pressure and anxiety.

    *Antibacterial agents:* antibiotics

33. $\left(1.0 \times 10^{-7} \frac{\text{mol}}{\text{L}}\right)(5.0\text{ L}) = 5.0 \times 10^{-7}$ mol epinephrine

    $(5.0 \times 10^{-7}$ mol epinephrine$)$ $(183.2$ g/mol$) = 9.2 \times 10^{-5}$ g epinephrine

34.

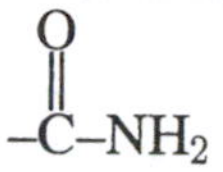

$$-\overset{\overset{\displaystyle O}{\|}}{C}-NH_2$$

amide

Some classes of biochemicals that contain an amide are:

- Antibacterial agents such as ampicillin
- B-vitamins such as niacin (nicotinamide)
- Barbiturates such as nembutal
- Tranquilizers such as valium (diazepam)

35. Lemon juice, being acidic, will react with the basic amines forming salts, which are soluble and can be washed away with water.

36.

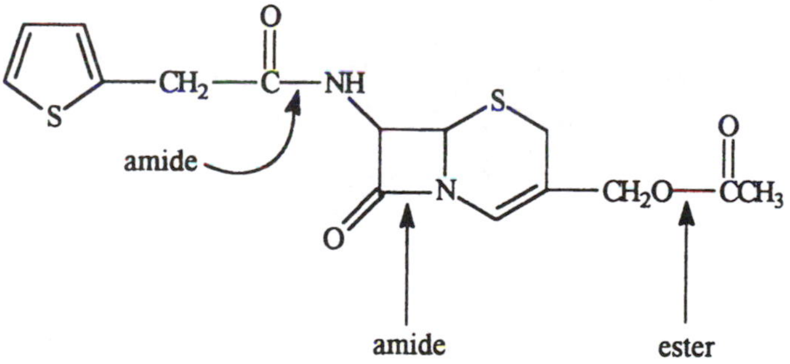

O
CH₂—C—NH
S
amide
S
O
N
CH₂O—CCH₃
O
amide
ester

37.

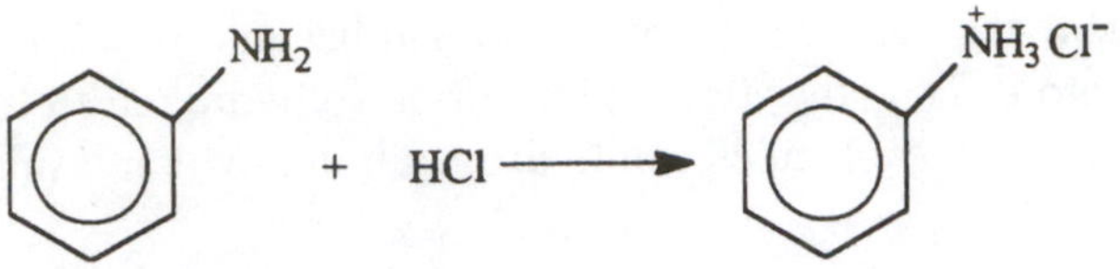

$$C_6H_5NH_2 + HCl \longrightarrow C_6H_5\overset{+}{N}H_3\,Cl^-$$

38. $H_2N{-}CH_2CH_2CH_2CH_2{-}NH_2$

Putrescine has two primary amines.

39. Drugs are given as ammonium salts because the salts are soluble in water.

40.

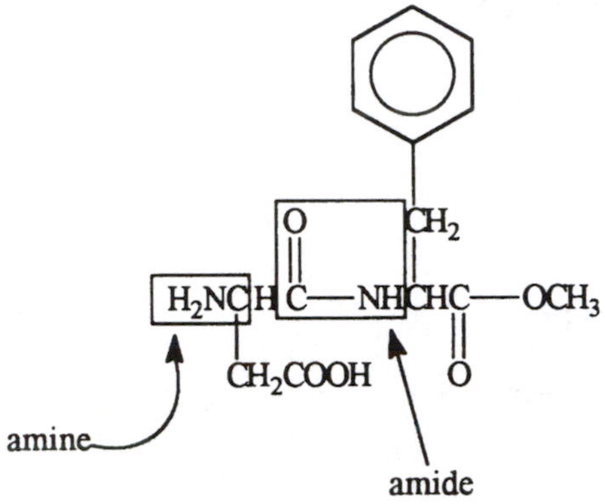

$$H_2NCH(CH_2COOH)\overset{O}{\overset{\|}{C}}{-}NHCH(CH_2C_6H_5)\overset{}{C}(=O){-}OCH_3$$

amine (boxed $H_2N$); amide (boxed $C(=O){-}NH$)

41. (a) $CH_3CH_2OH \xrightarrow[140°C]{H_2SO_4} CH_2=CH_2 \xrightarrow{Br_2} BrCH_2CH_2Br$

$\xrightarrow{KCN} NCCH_2CH_2CN \xrightarrow[Ni]{H_2} H_2NCH_2CH_2CH_2CH_2NH_2$

1,4-butanediamine

(b) $CH_3\overset{O}{\overset{\|}{C}}{-}Cl + NH_3 \longrightarrow CH_3\overset{O}{\overset{\|}{C}}{-}NH_2 \xrightarrow{LiAlH_4} CH_3CH_2NH_2$

$CH_3\overset{O}{\overset{\|}{C}}{-}Cl + CH_3NH_2 \longrightarrow CH_3\overset{O}{\overset{\|}{C}}{-}NHCH_3 \xrightarrow{LiAlH_4} CH_3CH_2NHCH_3$

$CH_3\overset{O}{\overset{\|}{C}}{-}Cl + (CH_3)_2NH \longrightarrow CH_3\overset{O}{\overset{\|}{C}}{-}N(CH_3)_2 \xrightarrow{LiAlH_4} CH_3CH_2N(CH_3)_2$

42.

$$-\!\!\left(\overset{\large O}{\overset{\|}{C}}-C_6H_4-\overset{\large O}{\overset{\|}{C}}-NH-C_6H_4-NH-\overset{\large O}{\overset{\|}{C}}-C_6H_4-\overset{\large O}{\overset{\|}{C}}-NH-C_6H_4-NH\right)_{\!n}$$

43. (a) $HO-\overset{O}{\overset{\|}{C}}-(CH_2)_8\overset{O}{\overset{\|}{C}}-OH$ and $H_2N(CH_2)_6NH_2$

(b) $HO-\overset{O}{\overset{\|}{C}}-(CH_2)_6\overset{O}{\overset{\|}{C}}-OH$ and $H_2N-C_6H_{10}-CH_2-C_6H_{10}-NH_2$

CHAPTER 26

# STEREOISOMERISM

1. A chiral carbon atom is one to which four different atoms or groups are attached and is a center of asymmetry in a molecule. In the following three compounds, the chiral carbon atoms are marked with an asterisk. (These are merely three examples; there are an infinite number of compounds which contain one chiral carbon atom.)

$$\begin{array}{ccccc} & H & & H & \\ & | & & | & \\ Cl- & \overset{*}{C} & - & C & -H \\ & | & & | & \\ & Br & & Br & \end{array} \qquad \begin{array}{ccccccc} & H & & H & & Cl & \\ & | & & | & & | & \\ H- & C & - & \overset{*}{C} & - & C & -Cl \\ & | & & | & & | & \\ & Cl & & Cl & & Cl & \end{array} \qquad CH_3\overset{*}{C}H(OH)CH_2OH$$

2. When the axes of two pieces of polaroid film are parallel, you have maximum brightness of the light passing through both. When one piece has been rotated by 90° the polaroid appears black, indicating very little light passing through.

3. A necessary and sufficient condition for a compound to show enantiomerism is that the compound not be superimposable on its mirror image.

4. Enantiomers are nonsuperimposable mirror image isomers. Diastereomers are stereoisomers that are not enantiomers (not mirror image isomers).

5. Physical properties of a pair of enantiomers

| | (+) 2-methyl-1-butanol | (−) 2-methyl-1-butanol |
|---|---|---|
| specific rotation | +5.76° | −5.76° |
| boiling point | 129°C | 129°C |
| density | 0.819 g/mL | 0.819 g/mL |

6. Most chiral molecules are stereospecific in their biological activity. Therefore a racemic mixture of a drug provides only half the bioactive material prescribed. By using a single isomer of a compound, the dosage can be cut in half and possible side effects can be avoided from its enantiomer.

7. Enantiomers have identical physical properties except their effect on polarized light, so they cannot be separated by ordinary chemical and physical means.

8. Diastereomers do not have identical physical properties, so the differences form a basis for chemical or physical separation. Differences of boiling point, freezing point, and solubilities would be most commonly used.

9. The objects that are chiral: (a) your ear; (b) a pair of pliers; (c) a coiled spring; (d) the letter b.

10. The objects that are chiral: (a) a wood screw; (c) the letter g; (d) this textbook.

11. Number of chiral carbon atoms.

(a) 1 (b) 1 (c) 2 (d) 1

12. Number of chiral carbon atoms.

(a) 0 (b) 0 (c) 3 (d) 2

13. Which compounds will show optical activity?

(a) $CH_3CH_2CH_2CH_2CH_2Cl$ no optical activity

(b) $CH_3CH_2CHClCH_2CH_3$ no optical activity

(c) $CH_3CH_2CH_2\underset{\displaystyle Cl}{\overset{\displaystyle CH_3}{C}}CH_3$ no optical activity

(d) $CH_3\overset{*}{C}HClCH_2\underset{\displaystyle CH_3}{C}HCH_3$ will show optical activity (* chiral carbon)

14. Which compounds will show optical activity?

(a) $CH_3CH_2CH_2\overset{*}{C}HClCH_3$ will show optical activity (* chiral carbon)

(b) $CH_3CH_2CH_2\underset{\displaystyle CH_3}{\overset{*}{C}}HCH_2Cl$ will show optical activity (* chiral carbon)

(c) $CH_3CH_2\overset{*}{C}HClCHCH_3$ will show optical activity (* chiral carbon)

(d) $CH_3CH_2\underset{\displaystyle CH_3}{C}ClCH_2CH_3$ no optical activity

15. Glucose, which has four chiral carbon atoms will have 16 possible stereoisomers. This can be determined from $2^n$. $2^4 = 16$.

16. Fructose, which has three chiral carbon atoms, will have 8 possible stereoisomers. This can be determined from $2^n$. $2^3 = 8$.

17. The two projection formulas (A) and (B) are the same compound, for it takes two changes to make (B) identical to (A).

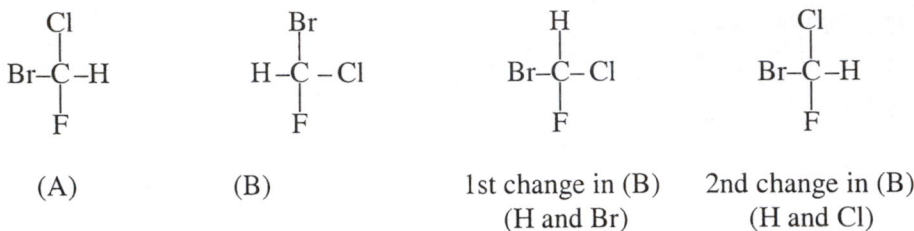

| Cl<br>\|<br>Br–C–H<br>\|<br>F | Br<br>\|<br>H–C–Cl<br>\|<br>F | H<br>\|<br>Br–C–Cl<br>\|<br>F | Cl<br>\|<br>Br–C–H<br>\|<br>F |
|---|---|---|---|
| (A) | (B) | 1st change in (B)<br>(H and Br) | 2nd change in (B)<br>(H and Cl) |

18. The two projection formulas (A) and (B) are the same compound, for it takes two changes to make (B) identical to (A).

| H<br>\|<br>Br–C–F<br>\|<br>$CH_3$ | $CH_3$<br>\|<br>Br–C–H<br>\|<br>F | H<br>\|<br>Br–C–$CH_3$<br>\|<br>F | H<br>\|<br>Br–C–F<br>\|<br>$CH_3$ |
|---|---|---|---|
| (A) | (B) | 1st change in (B)<br>(H and $CH_3$) | 2nd change in (B)<br>(F and $CH_3$) |

19. (−)-lactic acid is

$$\begin{array}{c} COOH \\ | \\ H\text{–}C\text{–}OH \\ | \\ CH_3 \end{array}$$

(+)-lactic acid is

$$\begin{array}{c} COOH \\ | \\ HO\text{–}C\text{–}H \\ | \\ CH_3 \end{array}$$

(a), (e), and (f) are (−)-lactic acid

(b), (c), and (d) are (+)-lactic acid

20. (+)-alanine is

$$\begin{array}{c} COOH \\ | \\ H_2N\text{—}\!\!+\!\!\text{—}H \\ | \\ CH_3 \end{array}$$

(−)-alanine is

$$\begin{array}{c} COOH \\ | \\ H\text{—}\!\!+\!\!\text{—}NH_2 \\ | \\ CH_3 \end{array}$$

(b), (c), and (d) are (+)-alanine

(a), (e), and (f) are (−)-alanine

All possible stereoisomers of the following compounds, with enantiomers and meso compounds labeled.

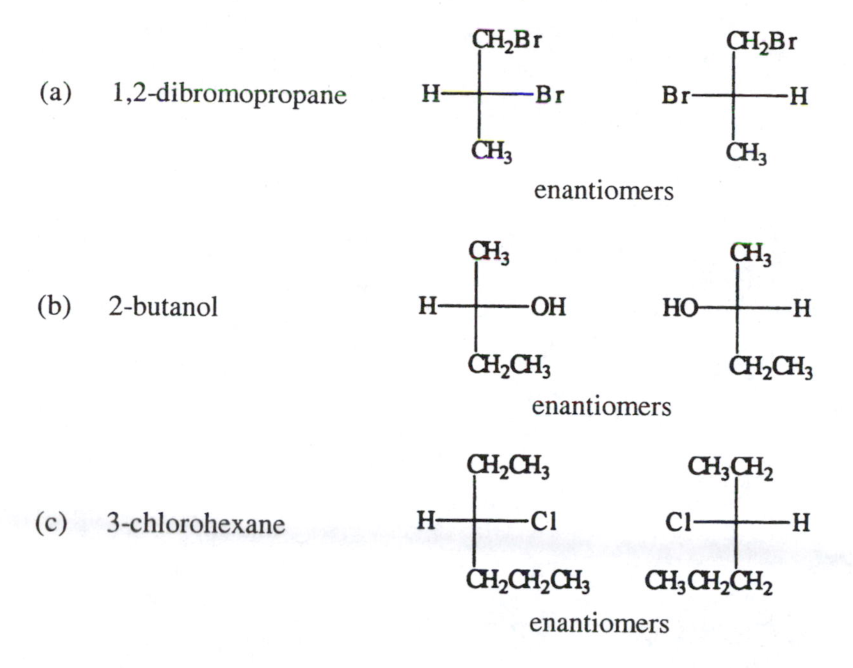


There are no meso compounds.

All possible stereoisomers of the following compounds with enantiomers and meso compounds labeled.

(a) 2,3-dichlorobutane

$CH_3$ / H—Cl / Cl—H / $CH_3$     $CH_3$ / Cl—H / H—Cl / $CH_3$     $CH_3$ / H—Cl / H—Cl / $CH_3$

enantiomers     meso

(b) 2,4-dibromopentane

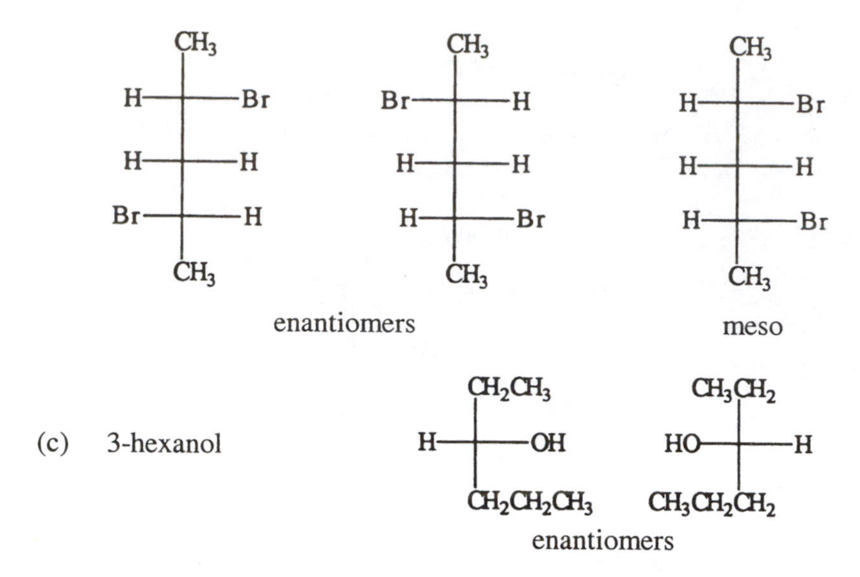


23. All the stereoisomers of 1,2,3-trihydroxybutane:

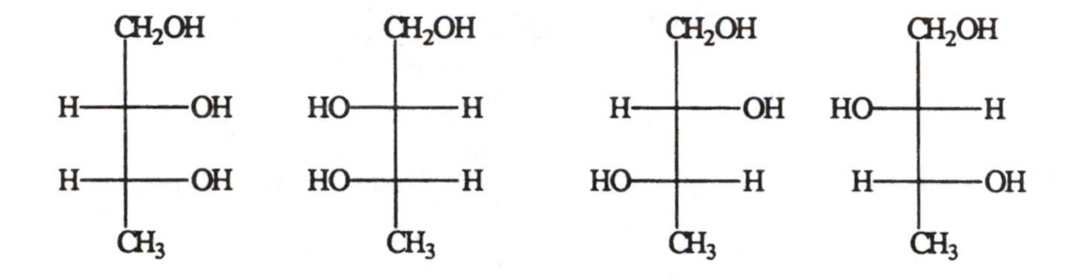

Compounds A and B, and C and D are pairs of enantiomers. There are no meso compounds. Pairs of diastereomers are A and C, A and D, B and C, and B and D.

24. All the stereoisomers of 3,4-dichloro-2-methylpentane:

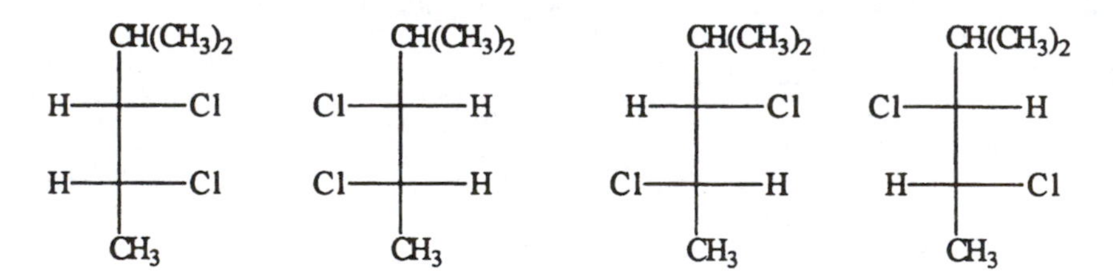

Compounds A and B, and C and D are pairs of enantiomers. There are no meso compounds. Pairs of diastereomers are A and C, A and D, B and C, and B and D.

25. The four stereoisomers of 2-hydroxy-3-pentene:

```
      CH3              CH3              CH3              CH3
      |                |                |                |
  H — C — OH      HO — C — H        H — C — OH      HO — C — H
      |                |                |                |
  H — C            H — C            H — C            H — C
      ||               ||               ||               ||
  H — C            H — C                C — H            C — H
      |                |                |                |
      CH3              CH3              CH3              CH3
      cis              cis              trans            trans
```

The two *cis* compounds are enantiomers and the two *trans* compounds are enantiomers.

26. The four stereoisomers of 2-chloro-3-hexene.

```
      CH3              CH3              CH3              CH3
      |                |                |                |
  H — C — Cl      Cl — C — H        H — C — Cl      Cl — C — H
      |                |                |                |
  H — C            H — C            H — C            H — C
      ||               ||               ||               ||
  H — C            H — C                C — H            C — H
      |                |                |                |
      CH2CH3           CH2CH3           CH2CH3           CH2CH3
      cis              cis              trans            trans
```

The two *cis* compounds are enantiomers and the two *trans* compounds are enantiomers.

27. (a)

| | | |
|---|---|---|
| $CH_3CH_2CH_2CHCl_2$<br>A | $CH_3CH_2CHClCH_2Cl$<br>B | $CH_3CHClCH_2CH_2Cl$<br>C |
| $CH_2ClCH_2CH_2CH_2Cl$<br>D | $CH_3CH_2CCl_2CH_3$<br>E | $CH_3CHClCHClCH_3$<br>F |
| $CH_3CH(CH_3)CHCl_2$<br>G | $CH_3CCl(CH_3)CH_2Cl$<br>H | $CH_2ClCH(CH_3)CH_2Cl$<br>I |

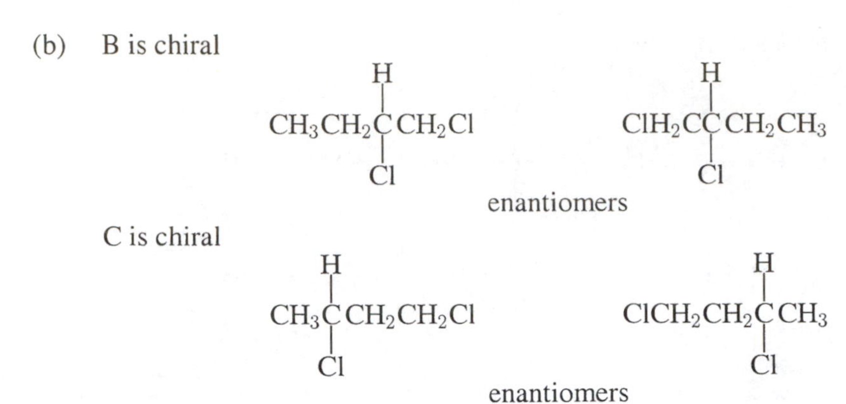


F can be both chiral and meso

$CH_3$
H—C—Cl
H—C—Cl
$CH_3$

meso

$CH_3$
H—C—Cl
Cl—C—H
$CH_3$

$CH_3$
Cl—C—H
H—C—Cl
$CH_3$

enantiomers

A, D, E, G, H, and I are achiral.

28. (a) $CH_3CH_2CHBr_2$ A

$CH_3CHBrCH_2Br$ B

$CH_3CBr_2CH_3$ C

$CH_2BrCH_2CH_2Br$ D

(b) B is chiral

$CH_2Br$
H—+—Br
$CH_3$

$CH_2Br$
Br—+—H
$CH_3$

enantiomers

A, C, and D are achiral; there are no meso compounds.

29. Assume (+)-2-bromopentane is

$$\begin{array}{c} CH_3 \\ | \\ H-C-Br \\ | \\ CH_2 \\ | \\ CH_2 \\ | \\ CH_3 \end{array}$$

All possible isomers formed when (+)-2-bromopentane is further brominated to dibromo-pentanes:

$$\begin{array}{cccc} \begin{array}{c} CH_2Br \\ | \\ H-C-Br \\ | \\ CH_2 \\ | \\ CH_2 \\ | \\ CH_3 \end{array} & \begin{array}{c} CH_3 \\ | \\ Br-C-Br \\ | \\ CH_2 \\ | \\ CH_2 \\ | \\ CH_3 \end{array} & \begin{array}{c} CH_3 \\ | \\ H-C-Br \\ | \\ H-C-Br \\ | \\ CH_2 \\ | \\ CH_3 \end{array} & \begin{array}{c} CH_3 \\ | \\ H-C-Br \\ | \\ Br-C-H \\ | \\ CH_2 \\ | \\ CH_3 \end{array} \\ \text{A} & \text{B} & \text{C} & \text{D} \end{array}$$

$$\begin{array}{ccc} \begin{array}{c} CH_3 \\ | \\ H-C-Br \\ | \\ CH_2 \\ | \\ H-C-Br \\ | \\ CH_3 \end{array} & \begin{array}{c} CH_3 \\ | \\ H-C-Br \\ | \\ CH_2 \\ | \\ Br-C-H \\ | \\ CH_3 \end{array} & \begin{array}{c} CH_3 \\ | \\ H-C-Br \\ | \\ CH_2 \\ | \\ CH_2 \\ | \\ CH_2Br \end{array} \\ \text{E} & \text{F} & \text{G} \end{array}$$

Compounds A, C, D, F, G would be optically active; B has no chiral carbon atom; E is a meso compound.

30. Assume (+)-2-chlorobutane is

$$\begin{array}{c} CH_3 \\ | \\ CH_2 \\ | \\ H—C—Cl \\ | \\ CH_3 \end{array}$$

All possible isomers formed when (+)-2-chlorobutane is further chlorinated to dichloro-butane:

| A | B | C | D | E |
|---|---|---|---|---|
| $CH_3$ | $CH_2Cl$ | $CH_3$ | $CH_3$ | $CH_3$ |
| \| | \| | \| | \| | \| |
| $CH_2$ | $CH_2$ | $CH_2$ | H—C—Cl | Cl—C—H |
| \| | \| | \| | \| | \| |
| H—C—Cl | H—C—Cl | Cl—C—Cl | H—C—Cl | H—C—Cl |
| \| | \| | \| | \| | \| |
| $CH_2Cl$ | $CH_3$ | $CH_3$ | $CH_3$ | $CH_3$ |

Compounds A, B, and E would be optically active; C does not have a chiral carbon atom; D is a meso compound.

31. Neither of the products, 1-chloropropane or 2-chloropropane, have a chiral carbon atom so neither product would rotate polarized light.

32. If 1-chlorobutane and 2-chlorobutane were obtained by chlorinating butane, and then distilled, they would be separated into the two fractions, because their boiling points are different. 1-chlorobutane has no chiral carbon, so would not be optically active. 2-chloro-butane would exist as a racemic mixture (equal quantities of enantiomers) because sub-stitution of Cl for H on carbon-2 gives equal amounts of the two enantiomers. Distillation would not separate the enantiomers because their boiling points are identical. The optical rotation of the two enantiomers of the 2-chlorobutane fraction would exactly cancel, and thus would not show optical activity.

33. Compounds (a) and (d) are meso. Make two changes on C-3 in compound (a) to prove that it is meso.

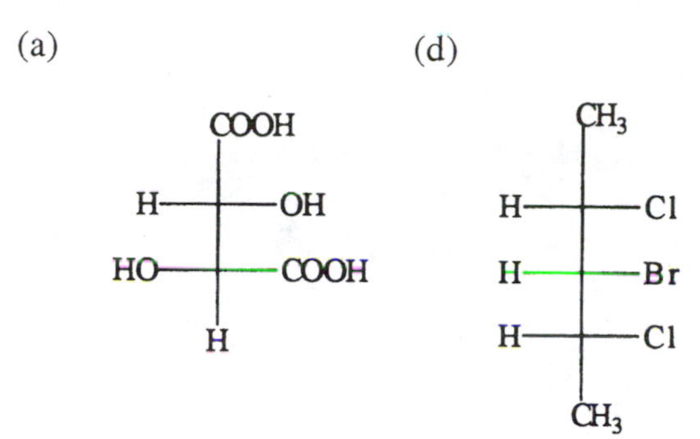


34. Compound (d) is meso.

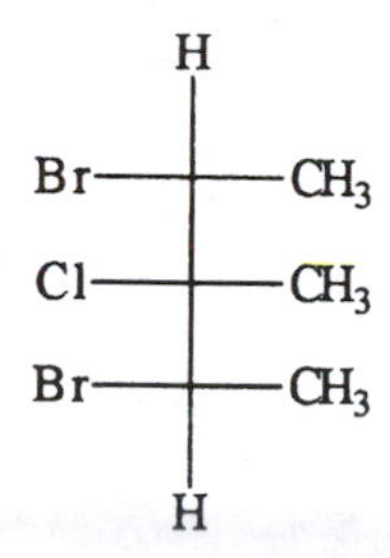


35. (b) is chiral

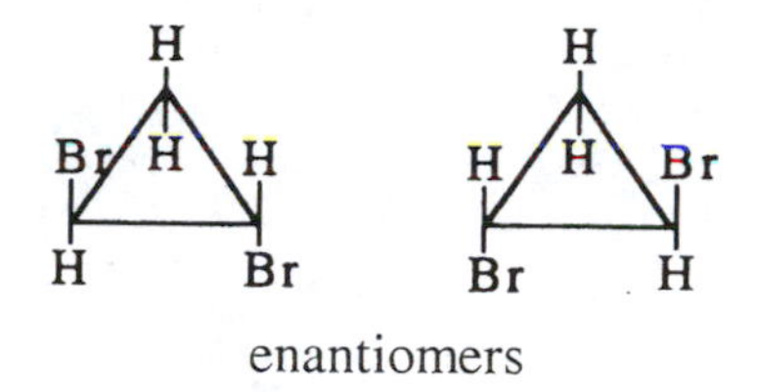

enantiomers

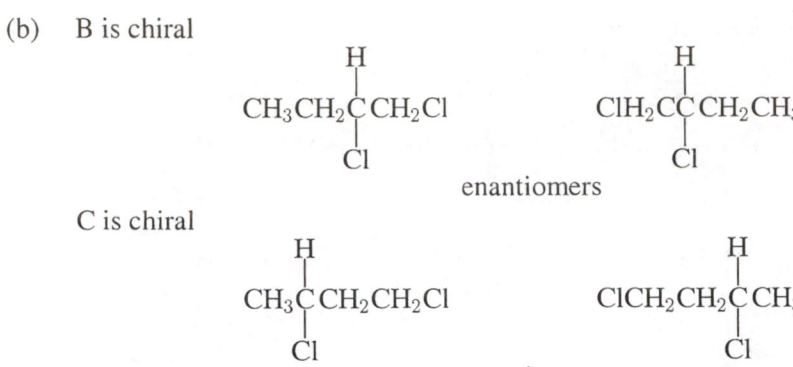

36. (c) is chiral

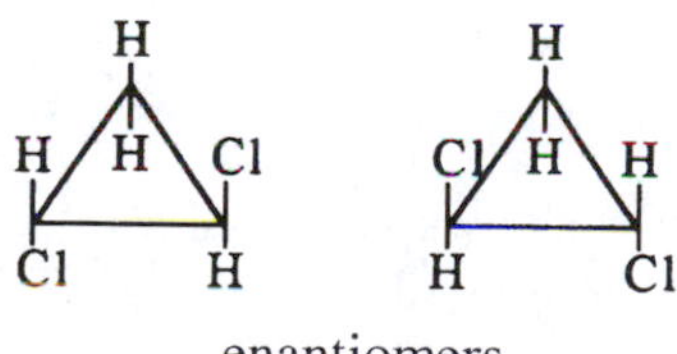

enantiomers

37. If four different groups were attached to a central carbon atom in a planar arrangement, it would not rotate polarized light because there would be a plane of symmetry in the molecule. No such plane of symmetry is possible when the four different groups are arranged in a tetrahedral structure.

38. (a) and (b)

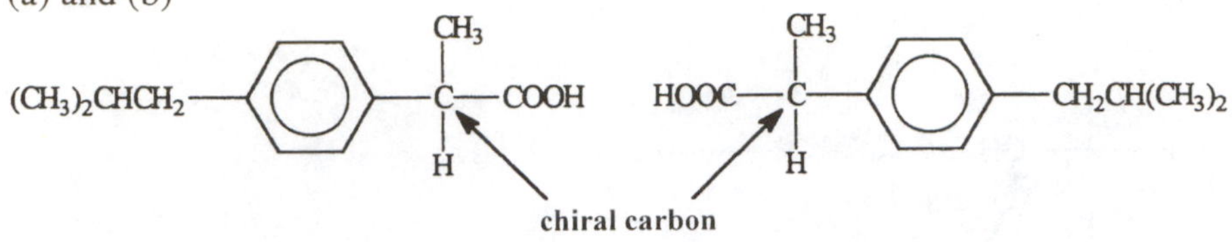

$(CH_3)_2CHCH_2$–C₆H₄–C(CH₃)(H)–COOH    HOOC–C(CH₃)(H)–C₆H₄–$CH_2CH(CH_3)_2$

chiral carbon

39. (a) A chiral primary alcohol of formula $C_5H_{12}O$.

$$CH_3CH_2\underset{}{\overset{\displaystyle CH_3 \atop |}{C}}HCH_2OH$$

(b) A compound with three primary alcohol groups is chiral, and has the formula $C_6H_{14}O_3$.

$$\underset{\displaystyle | \atop OH}{CH_2}-\underset{\displaystyle | \atop CH_2OH}{CH}-\overset{\displaystyle CH_3 \atop |}{CH}-CH_2OH$$

40. (a)

$CH_3$, H, O, H, H, H, H, H, C, H, $CH_2$

chiral carbon

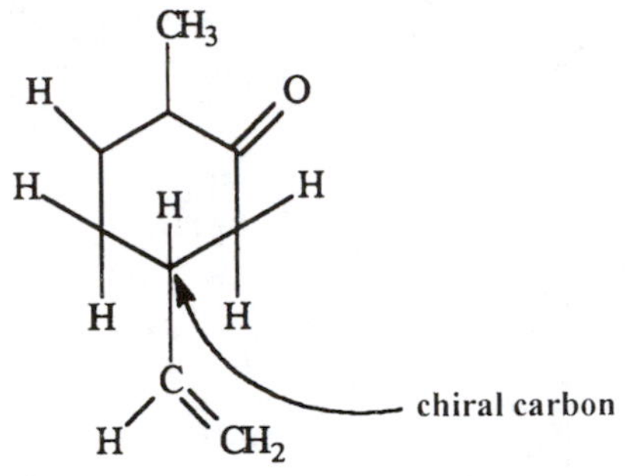

caraway

(b) The spearmint molecule is the optical isomer of the caraway molecule and differs from it in structure at the chiral carbon atom.

41. A compound of formula $C_3H_8O_2$:

(a) is chiral; contains two OH groups $CH_3\underset{\displaystyle | \atop OH}{C}HCH_2OH$

(b) is chiral; contains one OH group $CH_3{-}O{-}\underset{\displaystyle OH}{\underset{|}{C}}HCH_3$

(c) is achiral; contains two OH groups $\underset{\displaystyle OH}{\underset{|}{C}}H_2CH_2\underset{\displaystyle OH}{\underset{|}{C}}H_2$

42. Ephedrine has two chiral carbons and can have four stereoisomers. This number is calculated using $2^n$. $2^2 = 4$.

43.

$$\begin{array}{ccc} & O & \\ & \| & \\ & C{-}OH & \\ & | & \\ H_2N{-} & C & {-}H \\ & | & \\ & CH_3 & \end{array} \qquad \begin{array}{ccc} & O & \\ & \| & \\ HO{-} & C & \\ & | & \\ H{-} & C & {-}NH_2 \\ & | & \\ & CH_3 & \end{array}$$

chiral carbon

mirror image

44. (−) placed in front of a name is used to indicate the rotation of plane-polarized light to the left, and (+) for rotation to the right. Therefore, we can write (−)-methorphan for levomethorphan and (+)-methorphan for dextromethorphan. There is no obvious correlation between the structures of enantiomers and the direction in which they rotate plane polarized light.

45. Stereoisomer structures

(a) 2-bromo-3-chlorobutane

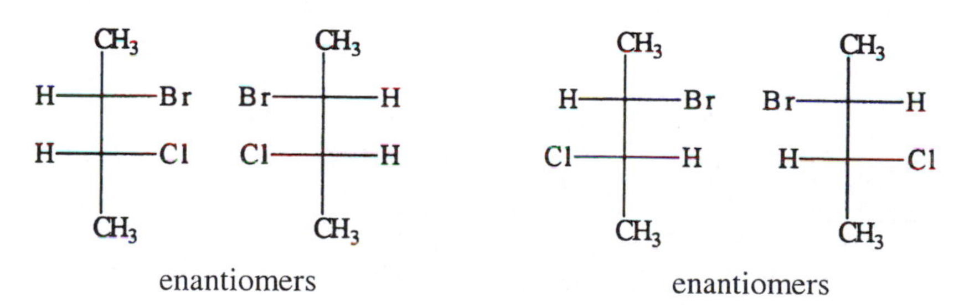

(b) 2,3,4-trichloro-l-pentanol

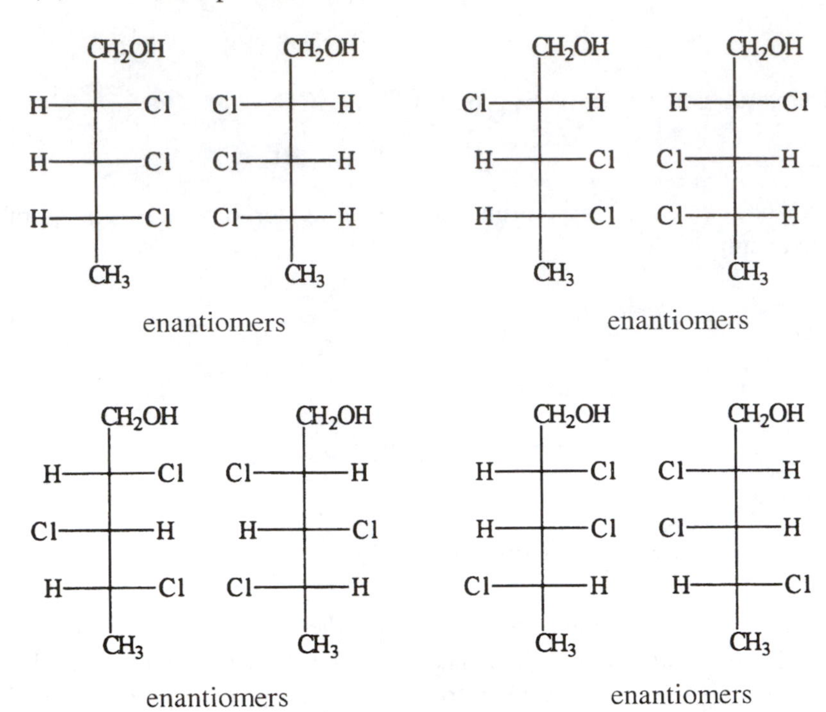


There are no meso compounds.

46. Meso structures for alkanes

(a) $C_8H_{18}$ (b) $C_9H_{20}$

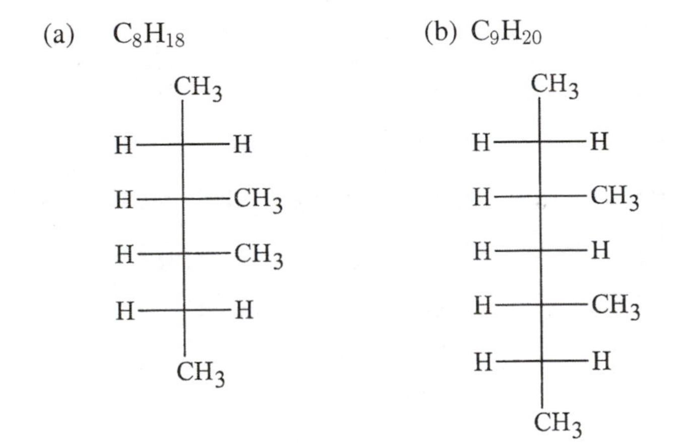

(c) $C_{10}H_{22}$

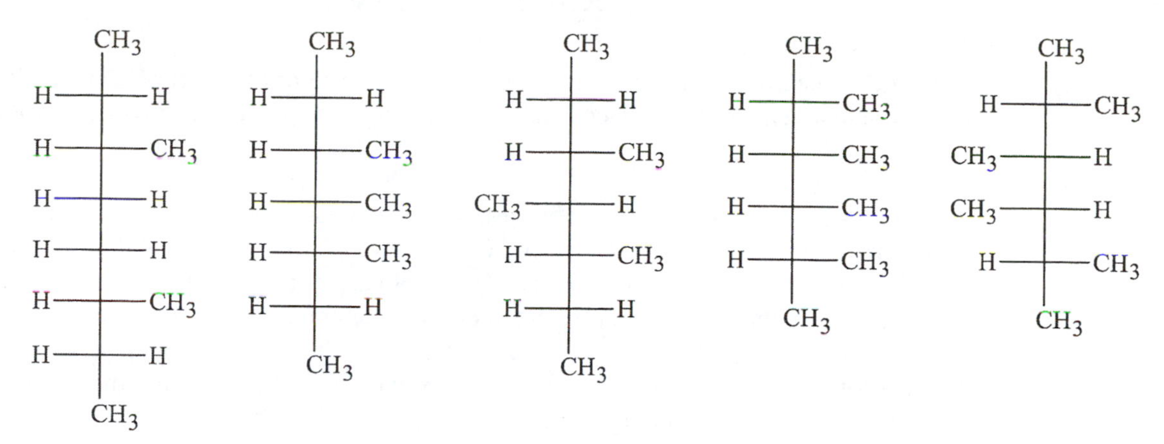

47. Optically active alcohols of $C_6H_{14}O$ (one enantiomer of each structure).

$CH_3CH_2CH_2CH_2C(H)(OH)CH_3$ $\quad$ $CH_3CH_2CH_2C(H)(OH)CH_2CH_3$

$CH_3CH_2CH_2C(CH_3)(H)CH_2OH$ $\quad$ $CH_3CH_2C(CH_3)(H)CH_2CH_2OH$

$CH_3CH_2C(CH_3)(H)—C(H)(OH)CH_3$ $\quad$ $CH_3CH_2C(CH_3)(H)—C(OH)(H)CH_3$

$CH_3CH(CH_3)CH_2C(H)(OH)CH_3$ $\quad$ $CH_3CH(CH_3)—C(H)(CH_3)CH_2OH$

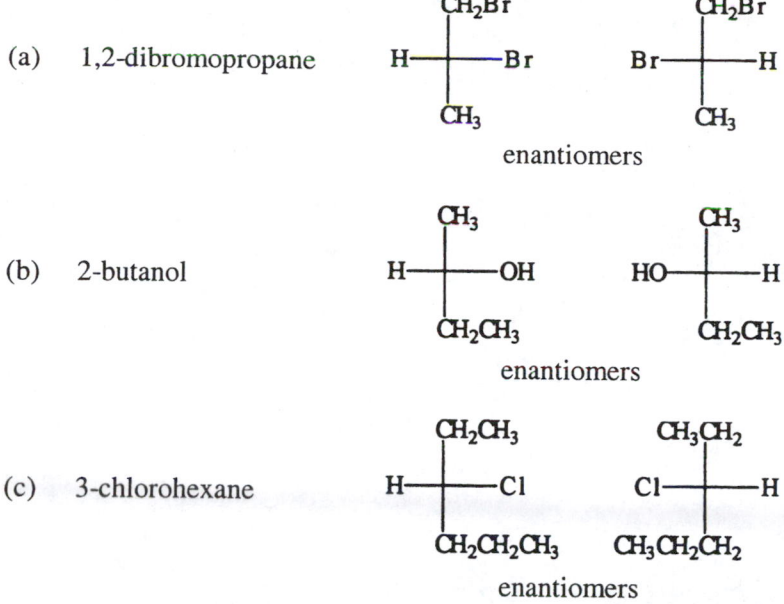

$CH_3C(CH_3)(CH_3)—C(H)(OH)CH_3$

CHAPTER 27

# CARBOHYDRATES

1. In general, the carbohydrate carbon oxidation state determines the carbon's metabolic energy content. The more oxidized a carbon is, the less energy it can provide in biological systems.

2. The notations D and L in the name of a carbohydrate specify the configuration on the last chiral carbon atom (from C-1) in the Fischer projection formula. If the –OH is written to the right of that carbon the compound is a D-carbohydrate. If the –OH is written to the left it is an L-carbohydrate. for example, D-glyceraldehyde and L-clyceraldehyde.

3. The notations (+) and (−) in the name of a carbohydrate specify whether the compound rotates the plane of polarized light to the right (+) or to the left (−).

4. Galactosemia is the inability of infants to metabolize galactose. The galactose concentration increases markedly in the blood and also appears in the urine. Galactosemia causes vomiting, diarrhea, enlargement of the liver, and often mental retardation. If not recognized a few days after birth it can lead to death.

5. There are four pairs of epimers among the D-aldohexoxes in Figure 27.1. They are: allose and altrose; glucose and mannose; gulose and idose; and galactose and talose.

6. A carbohydrate forms a five member or six member heterocyclic ring (one oxygen atom, the rest carbon atoms). If it forms a five-member ring, it is termed a furanose, after the compound furan. If it forms a six-membered ring it is termed a pyranose, after the compound pyran.

   Furan $C_4H_4O$

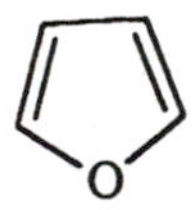

   Pyran $C_5H_6O$

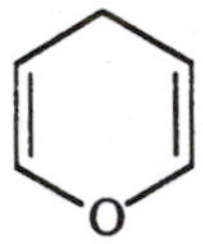

7. $\alpha$-D-glucopyranose and $\beta$-D-glucopyranose differ in the configuration at the number 1 carbon in the cyclic structure. In the open-chain structure, carbon 1 is the aldehyde group, and is not chiral. In the cyclic structure that carbon contains a hemiacetal structure, which is chiral. When the ring forms, carbon 1 can have two configurations leading to the two structures called $\alpha$ and $\beta$.

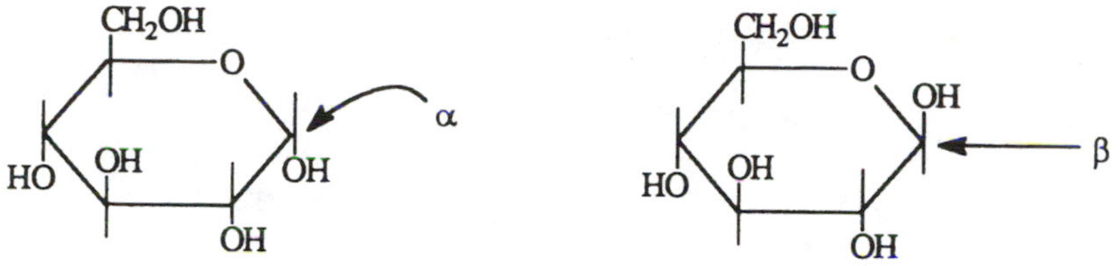

In the Haworth structure for glucose the –OH on carbon 1 is written down for the $\alpha$ structure and up for the $\beta$ structure.

8. The cyclic forms of monosaccharides are hemiacetals, because the number one carbon has an ether and an alcohol group; whereas a glycoside is an acetal which contains two ether linkages.

9. Mutarotation is the phenomenon by which the $\alpha$ or $\beta$ form of a sugar, when in solution, will undergo change to reach an equilibrium mixture, not necessarily 50%-50%, of the two forms. To achieve this equilibrium, the chain must open up and then reclose. On closing, it has the possibility of closing in either the $\alpha$ or $\beta$ form as the equilibrium mixture is achieved.

10. Major sources:
 (a) sucrose: sugar beets and sugar cane
 (b) lactose: milk
 (c) maltose: sprouting grain and partially hydrolyzed starch

11. The following parts are related to the eight D-aldohexoses shown in the text (Figure 27.1):

 (a) If each of the aldohexoses is oxidized by nitric acid to dicarboxylic acids, allose and galactose would become meso forms.

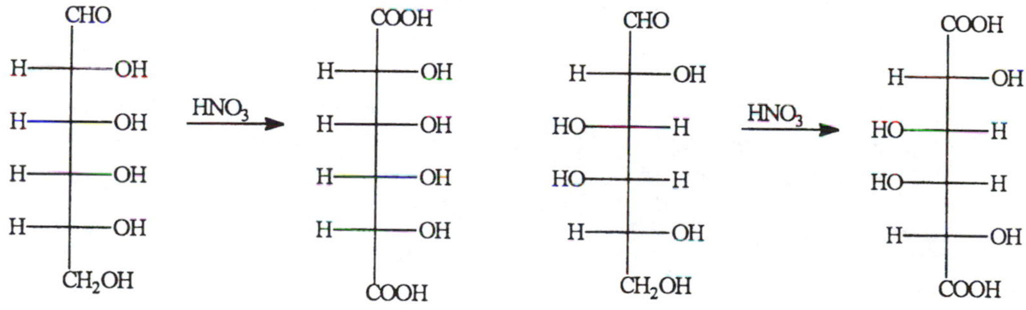

allose

galactose

(b) Names and structures of enantiomers of D-altrose and D-idose:

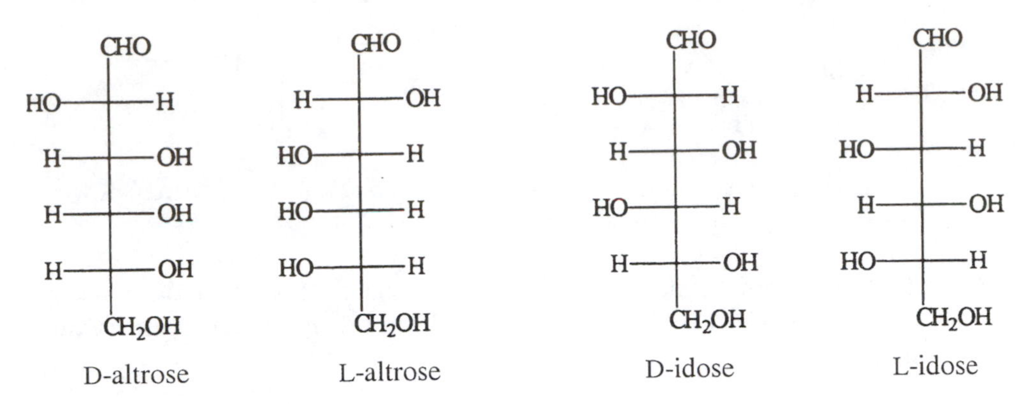


12. Invert sugar is sweeter than sucrose because it is a 50-50 mixture of fructose and glucose. Glucose is somewhat less sweet than sucrose, but fructose is much sweeter, so the mixture is sweeter.

13. Amylose is a linear polymer of D-glucopyranose units linked by $\alpha$-1,4-glycosidic bonds while cellulose is a lincar polymer of D-glucopyranose units linked by $\beta$-1,4-glycosidic bonds. Amylose molecules take the shape of a "coil" while cellulose molecules form fibers.

14. In the Benedict test, both concentrated and dilute glucose solutions will give a red precipitate of $Cu_2O$. However, during the reaction the dilute solution will appear more greenish-yellow while the concentrated solution will appear more reddish.

15. The two main components of starch are amylose and amylopectin. They are both composed of glucose units joined by $\alpha$-1,4-glycosidic linkages. The difference is that amylopectin also has branching which occurs through $\alpha$-1,6-glycosidic linkages about every 25 glucose units.

16. Some people, as they grow older, stop producing the enzyme lactase and thus lose the ability to digest lactose. This leads to lactose intolerance, a condition that produces gas and intestinal discomfort.

17. (a)

$$\begin{array}{c} CH_2OH \\ | \\ C=O \\ | \\ CH_2OH \end{array}$$

There are no chiral carbon atoms in dihydroxyacetone.

(b) If dihydroxyacetone is reacted with hydrogen in the presence of a platinum catalyst, the product will be glycerol.

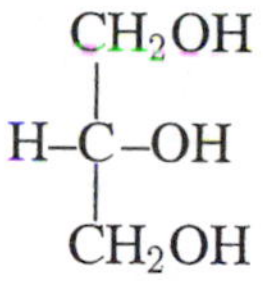

$CH_2OH$
H–C–OH
$CH_2OH$

18. (a) D-glyceraldehyde

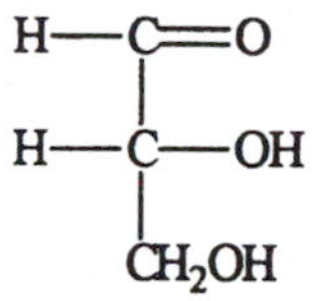

H—C=O
H—C—OH
$CH_2OH$

L-glyceraldehyde

H—C=O
HO—C—H
$CH_2OH$

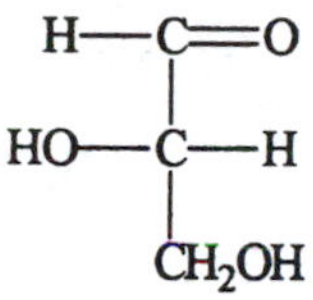

(b) If D-glyceraldehyde is reacted with hydrogen in the presence of a platinum catalyst, the product will be glycerol.

$CH_2OH$
H–C–OH
$CH_2OH$

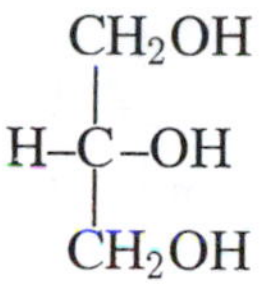

19. The structure of an epimer of D-mannose will differ from D-mannose at one chiral carbon atom. The epimer could differ at carbon 2 *or* carbon 3 *or* carbon 4 *or* carbon 5. One possible epimer is

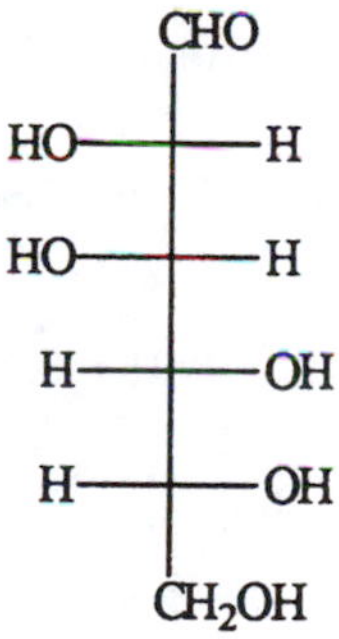

CHO
HO—H
HO—H
H—OH
H—OH
$CH_2OH$

D-mannose

CHO
HO—H
HO—H
HO—H
H—OH
$CH_2OH$

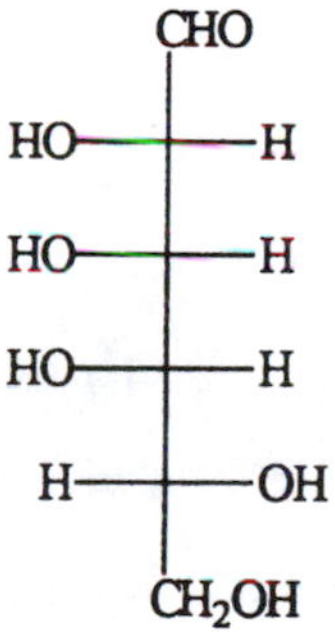

an epimer of D-mannose at carbon 4

20. The structure of an epimer of D-galactose will differ from D-galactose at one chiral carbon atom. The epimer could differ at carbon 2 *or* carbon 3 *or* carbon 4 *or* carbon 5. One possible epimer is

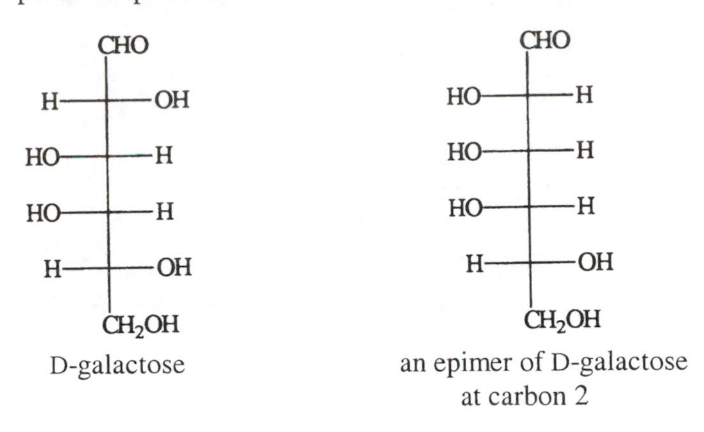


D-galactose

an epimer of D-galactose at carbon 2

21. The enantiomers are L-galactose, L-mannose and L-ribose which are mirror images of D-galactose, D-mannose and D-ribose.

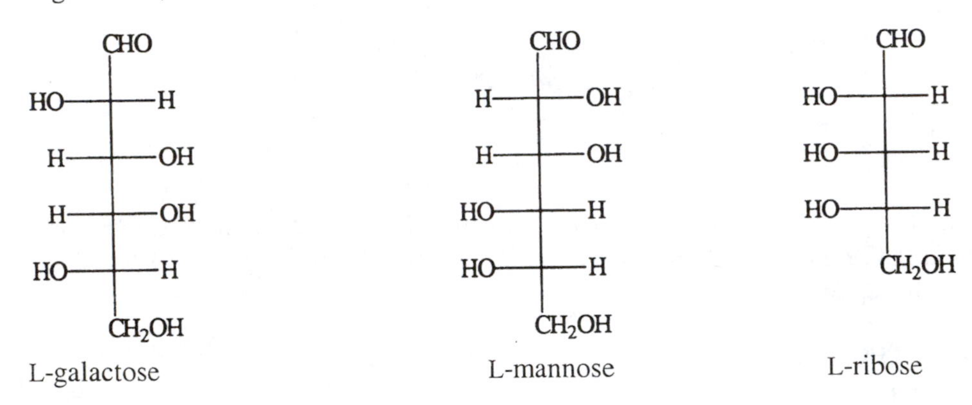


L-galactose

L-mannose

L-ribose

22. The enantiomers are L-glucose, L-fructose and L-2-deoxyribose which are mirror images of D-glucose, D-fructose and D-2-deoxyribose.

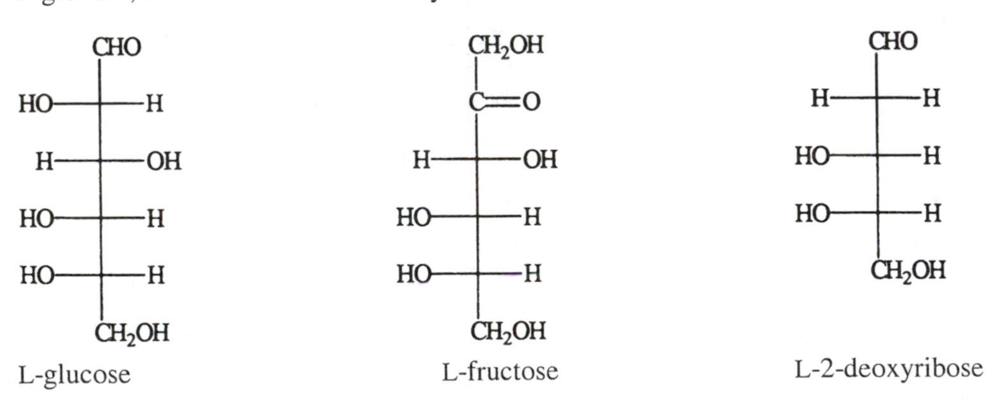


L-glucose

L-fructose

L-2-deoxyribose

23. Either the Fischer projection formula or Hawcrth formulas are satisfactory.

$\alpha$-D-glucopyranose

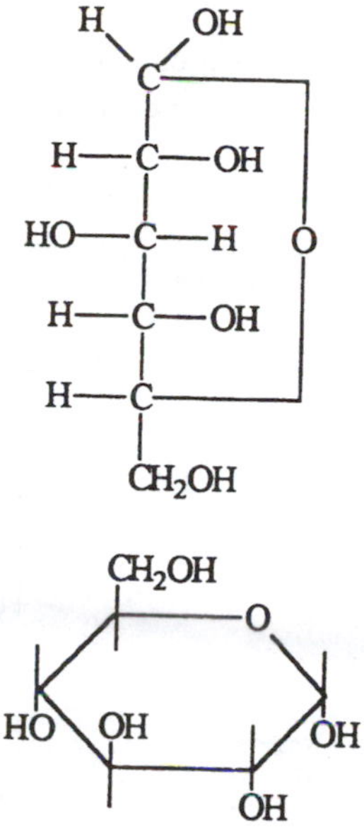

$\beta$-D-galactopyranose

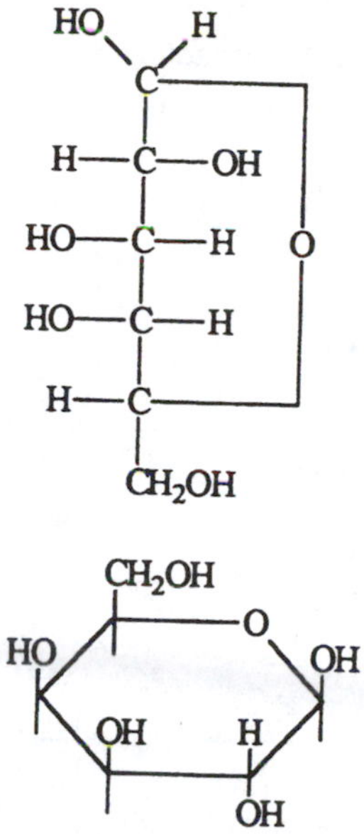

$\alpha$-D-mannopyranose

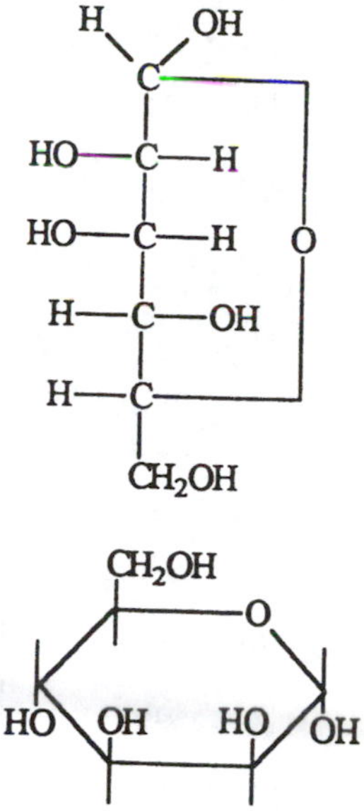

24. Either the Fischer projection formulas or Haworth formulas are satisfactory.

$\beta$-D-glucopyranose

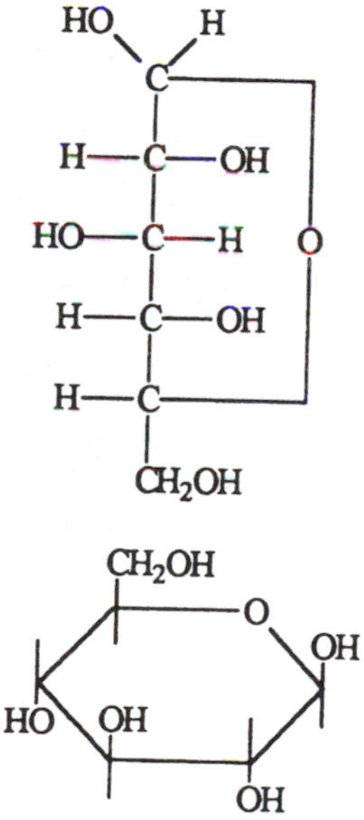

$\alpha$-D-galactopyranose

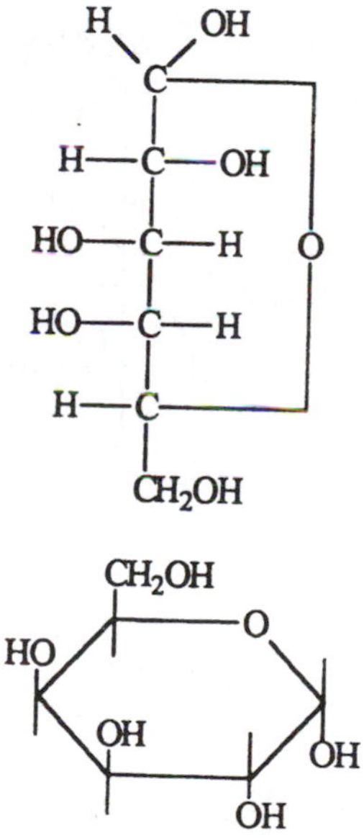

$\beta$-D-mannopyranose

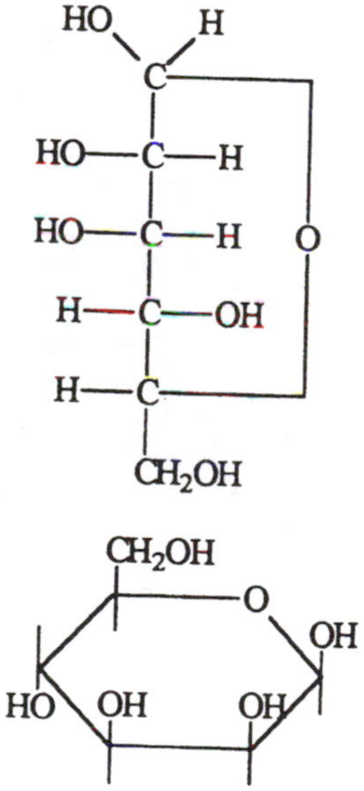

25. The glucose units in starch are connected by $\alpha$-1-4-glycosidic linkages. The human digestive system has the enzymes that catalyze the hydrolysis of starch to maltose or isomaltose and then to glucose.

26. The glucose units in cellulose are connected by $\beta$-1-4-glycosidic linkages. The human digestive system does not have the enzymes to catalyze the hydrolysis of cellulose.

27. Kiliani-Fischer synthesis of D-glucose from the proper D-tetrose (D-erythrose)

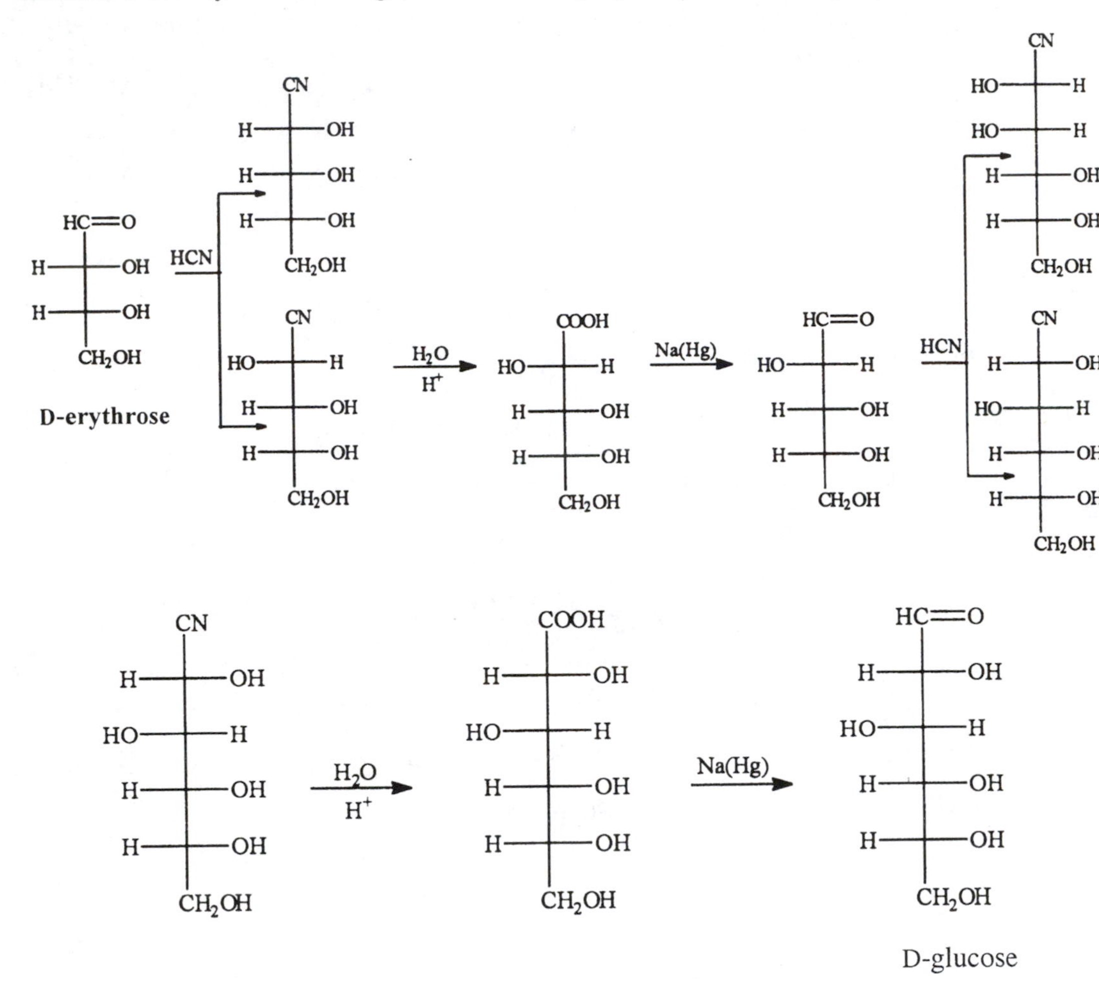

28. Kiliani-Fischer synthesis of D-ribose starts with the proper D-triose, D-glyceraldehyde.

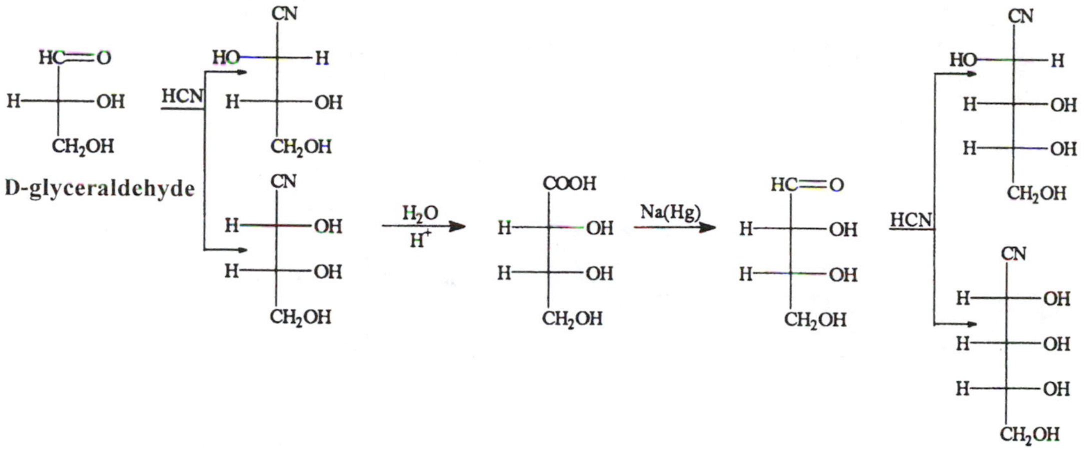

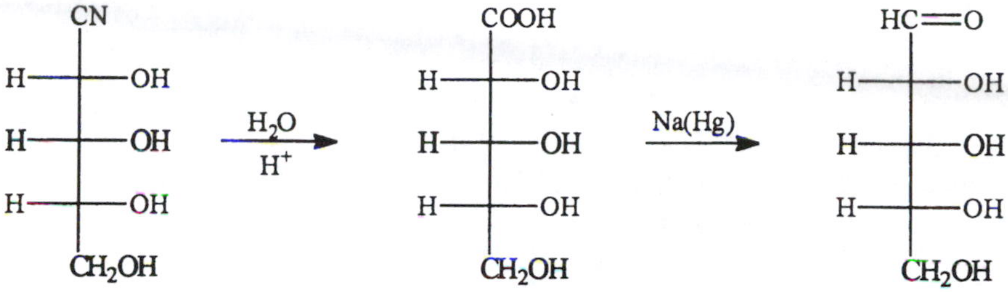

D-glyceraldehyde

D-ribose

29. Yes, D-2-deoxymannose is the same as D-2-deoxyglucose. The structures of D-mannose and D-glucose differ at carbon 2, so if the carbon 2 OH group on both molecules is changed to H, there is no difference between the two molecules.

30. No, D-2-deoxygalactose is not the same as D-2-deoxyglucose. D-galactose differs from D-glucose at carbon 4, so replacement of the carbon 2 OH with an H does not makes these two sugars identical.

31. The monosaccharide composition of:
(a) sucrose: one glucose and one fructose unit
(b) glycogen: many glucose units
(c) amylose: many glucose units
(d) maltose: two glucose units

32. The monosaccharide composition of:
(a) lactose: one glucose and one galactose unit
(b) amylopectin: many glucose units

(c) cellulose: many glucose units
(d) sucrose: one glucose and one fructose unit

33. Both cellobiose and isomaltose are disaccharides composed of two glucose units. However, in cellobiose monosaccharides are linked by a $\beta$-1,4-acetal bond while for isomaltose the linkage is $\alpha$-1,6.

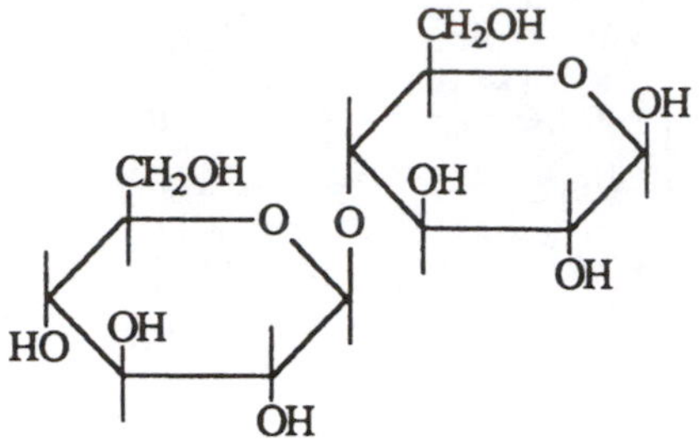

cellobiose

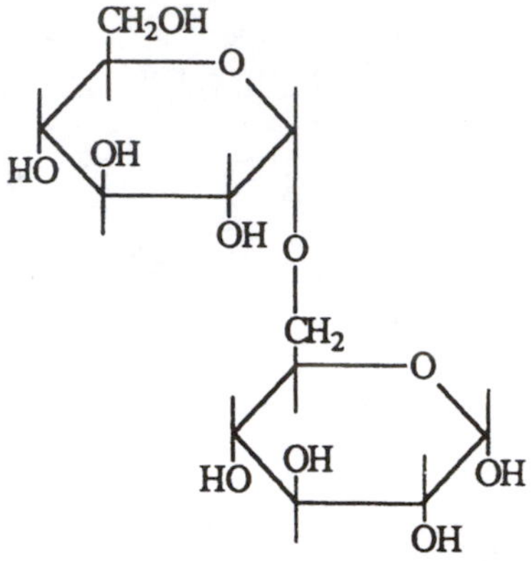

isomaltose

34. Both maltose and isomaltose are disaccharides composed of two glucose units. The glucose units in maltose are linked by an $\alpha$-1,4-glycosidic bond while the glucose units at isomaltose are linked by an $\alpha$-1,6-glycosidic bond.

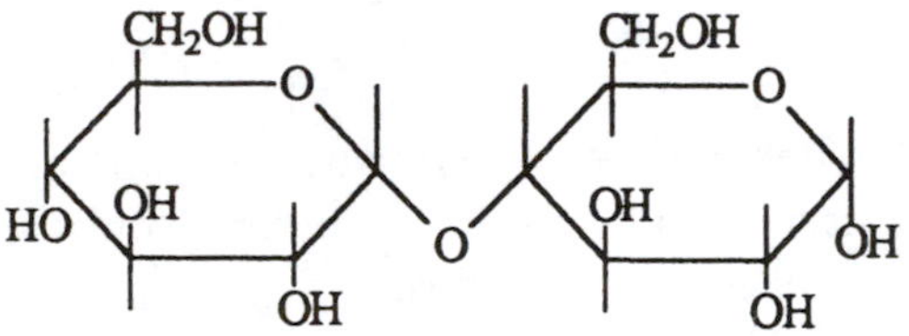

maltose

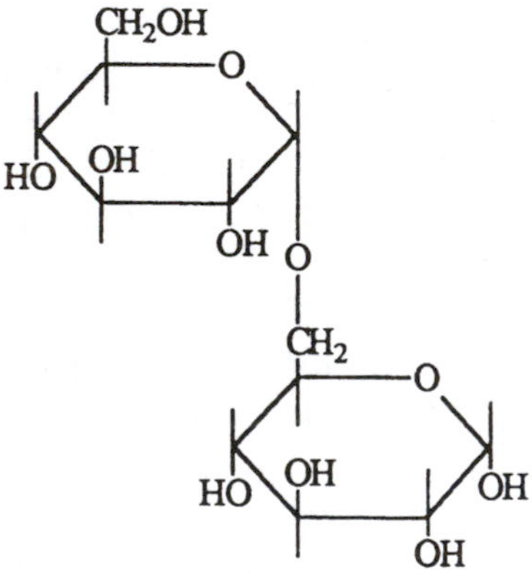

isomaltose

35. Lactose will show mutarotation; sucrose will not. The hemiacetal structure in lactose will open allowing mutarotation. Since sucrose has an acetal structure and no hemiacetal, it will not undergo mutarotation.

36. Both maltose and isomaltose will show mutarotation. Both disaccharides contain a hemiacetal structure which will open allowing mutarotation.

37.

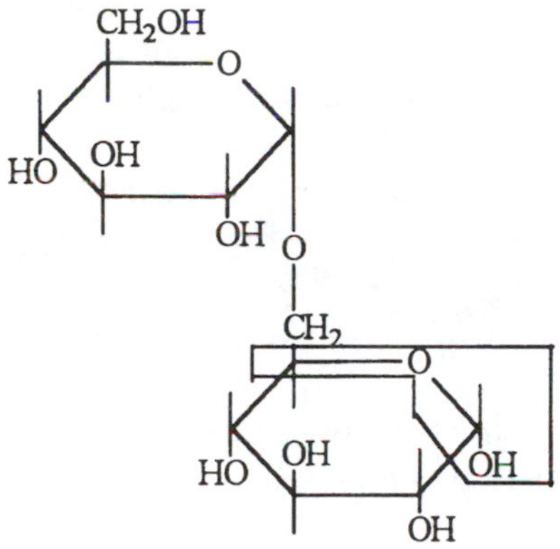

isomaltose

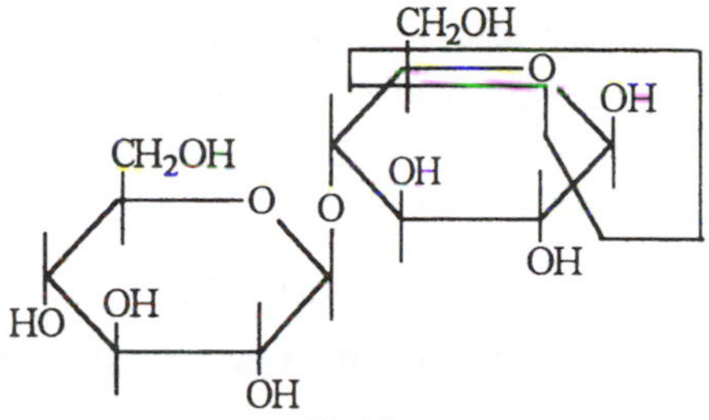

cellobiose

The circled hemiacetal structures allow these two disaccharides to be reducing sugars.

38.

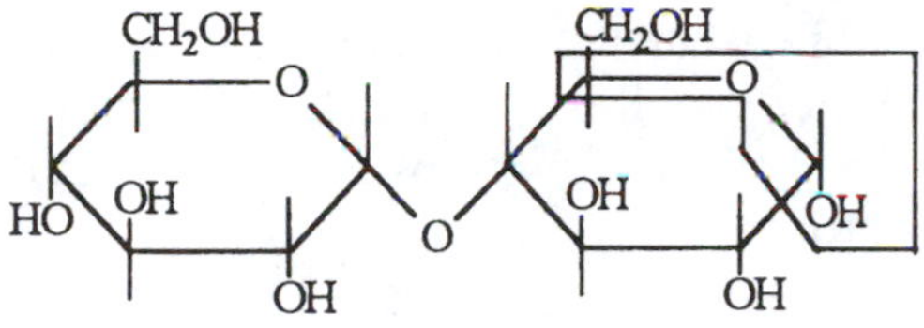

maltose

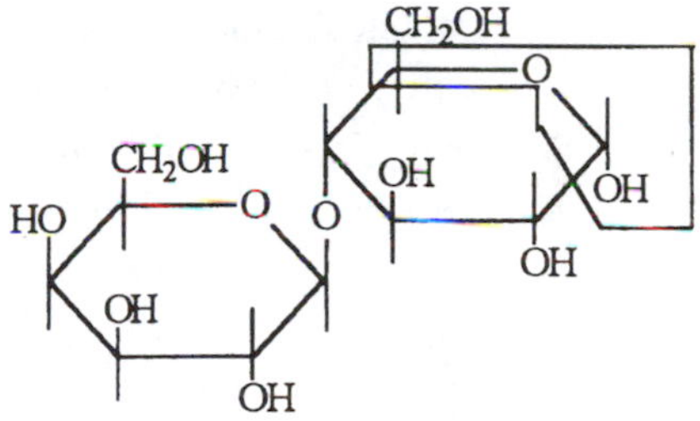

lactose

The circled hemiacetal structures allow these two disaccharides to be reducing sugars.

39. The systematic name for isomaltose is $\alpha$-D-glucopyranosyl-(1,6)-$\alpha$-D-glucopyranose. The systematic name for cellobiose is $\beta$-D-glucopyranosyl-(1,4)-$\beta$-D-glucopyranose.

40. The systematic name for maltose is $\alpha$-D-glucopyranosyl-(1,4)-$\alpha$-D-glucopyranose. The systematic name for lactose is $\beta$-D-galactopyranosyl-(1,4)-$\alpha$-D-glucopyranose.

41. The principal differences and similarities between the members of the following pairs:

(a) D-glucose and D-fructose. Glucose is an aldose; fructose is a ketose. Glucose often forms a pyranose ring structure; fructose commonly is found in a furanose structure. Both are hexoses and both are reducing sugars.

(b) Maltose and sucrose. Maltose is composed of two glucose units; sucrose is composed of one glucose and one fructose unit. Maltose is a reducing sugar; sucrose is not. Both are common disaccharides, with formulas $C_{12}H_{22}O_{11}$.

(c) Cellulose and glycogen. Cellulose is composed of glucose units linked by $\beta$-1,4-glycosidic linkages; glycogen is composed of glucose units linked by $\alpha$-1,4-glycosidic linkages. Glycogen is much more readily hydrolyzed or digested than cellulose. Both are large polymers of glucose. Glycogen is of animal origin; cellulose is from plants.

42. The principal differences and similarities between members of the following pairs:

(a) D-ribose and D-2-deoxyribose. The D-2-deoxyribose has no OH group on the number 2 carbon, only 2 hydrogen atoms. Both are five carbon sugars.

(b) Amylose and amylopectin. Amylose is a straight chain polysaccharide; amylopectin has branched chains, and more monomer units per molecule. Both are large polysaccharides composed of $\alpha$-D-glucose units.

(c) Lactose and isomaltose. These sugars are disaccharides. Lactose is composed of one galactose unit and one glucose unit while isomaltose is composed of two glucose units. The monosaccharide units in lactose are linked by a $\beta$-1,4-glycosidic bond while the units in isomaltose are linked by an $\alpha$-1,6-glycosidic bond. Both disaccharides also contain hemiacetal structures.

43.

```
H—C=O                              COOH
   |                                 |
H——+——OH                         H——+——OH
   |                                 |
HO—+——H                         HO——+——H
   |                                 |
HO—+——H    +  HNO3  ——————>     HO——+——H
   |                                 |
H——+——OH                         H——+——OH
   |                                 |
  CH2OH                            COOH
D-galactose                 mucic acid (galactaric acid)
```

44.

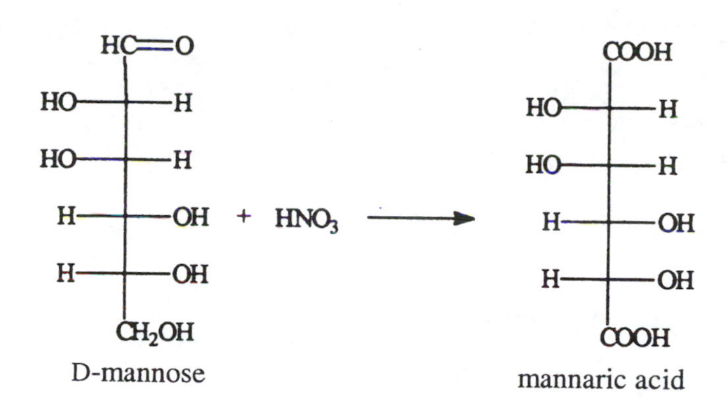


45. The formulas for the four L-ketohexoses are

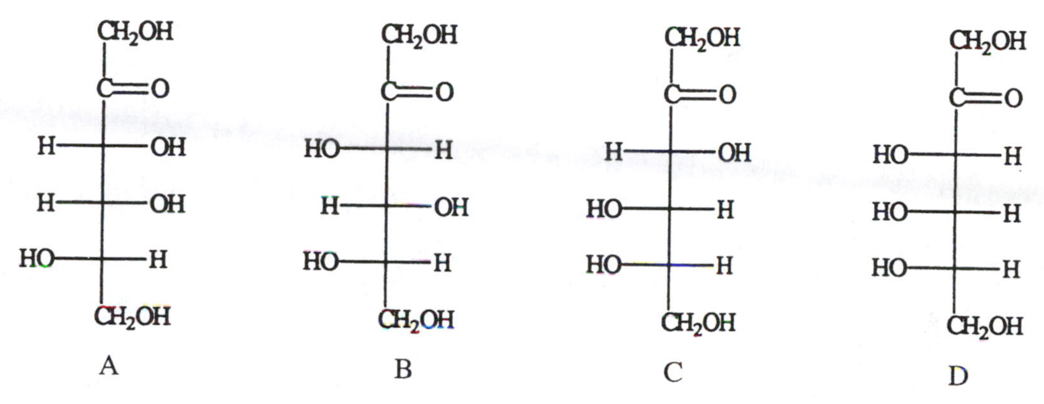


Epimers are: A and B, B and C, B and D, C and D

46. The formulas for the four L-aldopentoses are

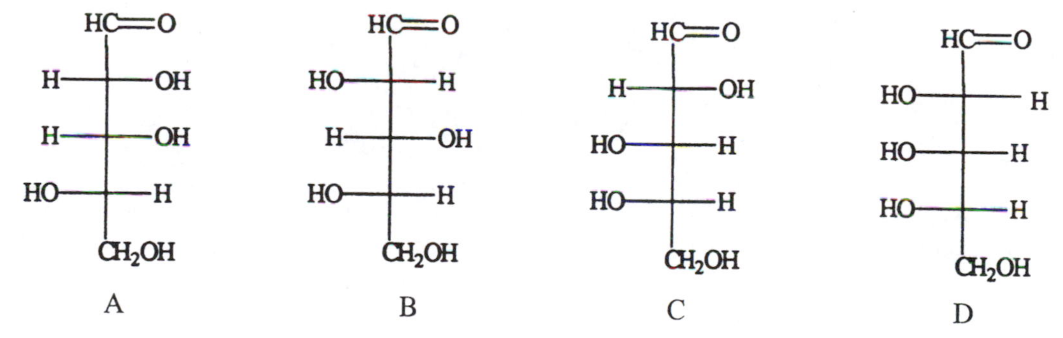


Epimers are: A and B, A and C, B and D, C and D

47. (a) β-D-mannopyranosyl-(1,4)-β-D-galactopyranose

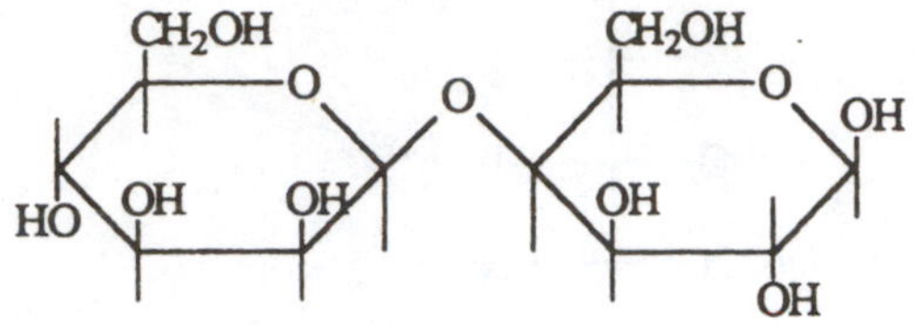

(b) β-D-galactopyranosyl-(1,6)-α-D-glucopyranose

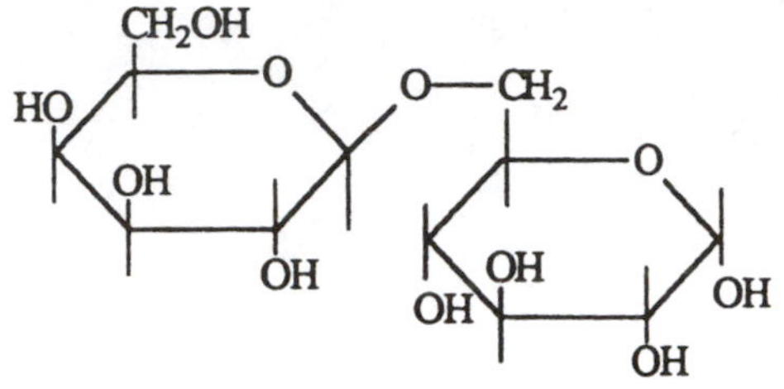

48. (a) The Haworth formula for β-D-glucopyranosyl-(1,4)-α-D-galactopyranose is

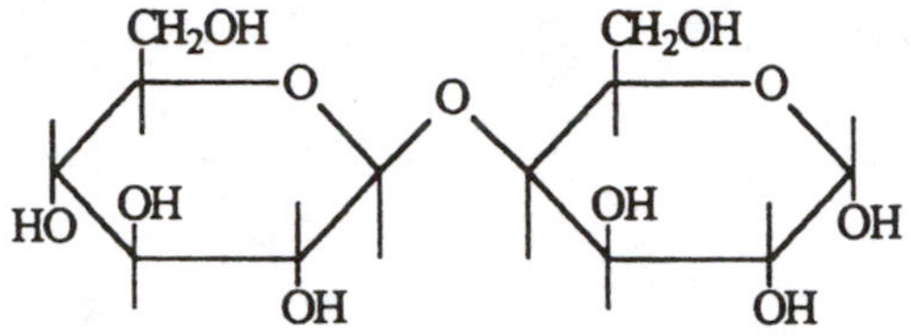

(b) The Haworth formula for β-D-galactopyranosyl-(1,6)-β-D-mannopyranose is

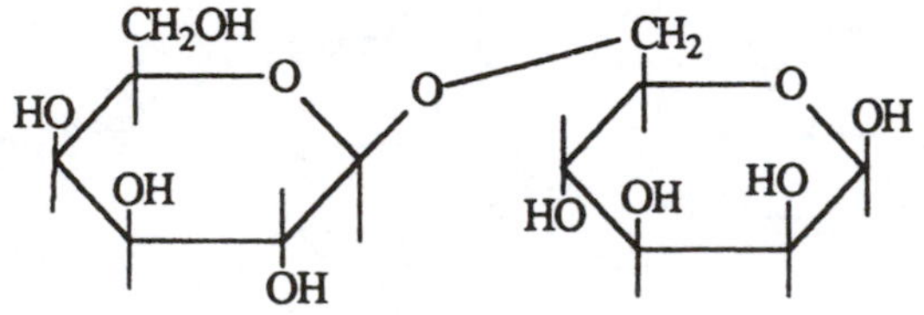

49. Glucose is called blood sugar because it is the most abundant carbohydrate in the blood and is carried by the bloodstream to all parts of the body.

50. Aspartame supplies many fewer calories than sucrose. In addition, oral bacteria cannot use aspartame as efficiently as sucrose and will form fewer dental carries.

51. The hydrolysis of glycosidic linkages is acid catalyzed. Because lemon juice contains acid it will cause sucrose to be hydrolyzed to glucose and fructose. The candy will become sweeter because fructose is much sweeter than sucrose.

52. High-fructose corn syrup is produced by breaking down some corn starch polymers to D-glucose monomers which are then converted to D-fructose, a very sweet monosaccharide.

53. (a) The sugar acid could be most easily derived from $\beta$-D-mannopyranose.

(b)

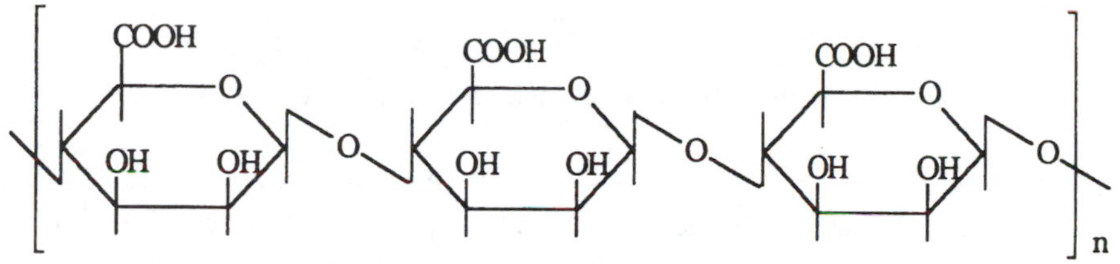

COOH O OH OH O COOH O OH OH O COOH O OH OH O n

54. Compound A must be a reducing disaccharide because it produces a reddish color in the Benedict test. Sucrose is nonreducing so compound A must be maltose.

55. (a) All starred carbons (4) in the following compound are chiral:

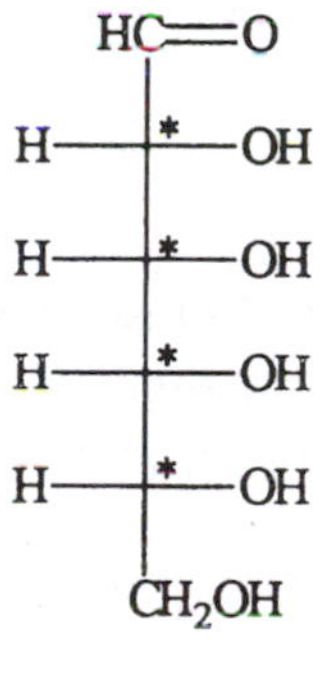

HC=O
H—*—OH
H—*—OH
H—*—OH
H—*—OH
$CH_2OH$

(b)

$CH_2OH$ O * * * * * OH OH OH OH

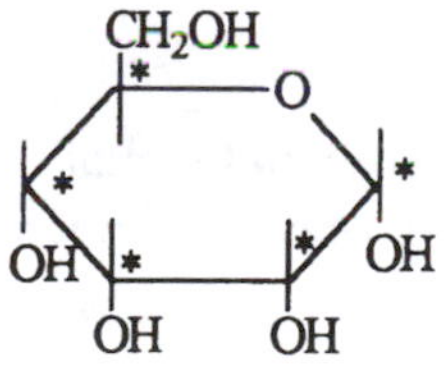

(c) All starred carbons (5) in the structure drawn in part (b) are chiral carbons.

56. If the compound (I) shown below rotates light 25° to the right, its enantiomer will rotate light 25° to the left.

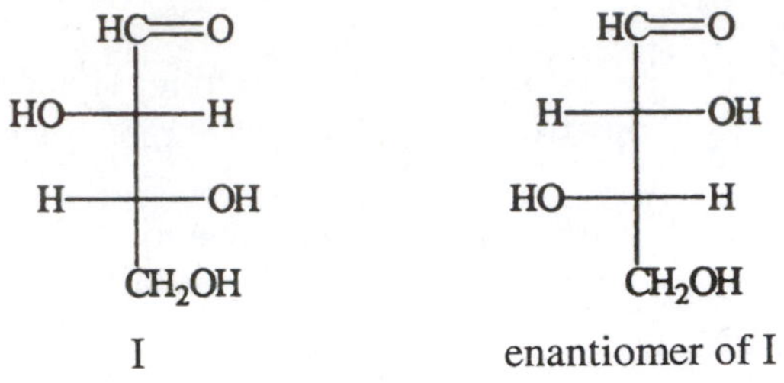

HC=O<br>HO—H<br>H—OH<br>$CH_2OH$<br>I

HC=O<br>H—OH<br>HO—H<br>$CH_2OH$<br>enantiomer of I

These compounds are not epimers because they differ at more than one chiral carbon.

57. A nonreducing disaccharide composed of two molecules of $\alpha$-D-galactopyranose can have no hemiacetal structures. Thus, the hemiacetal structure of one $\alpha$-D-galactopyranose must be used to form the glycosidic link to the hemiacetal structure of the other $\alpha$-D-galacto-pyranose unit.

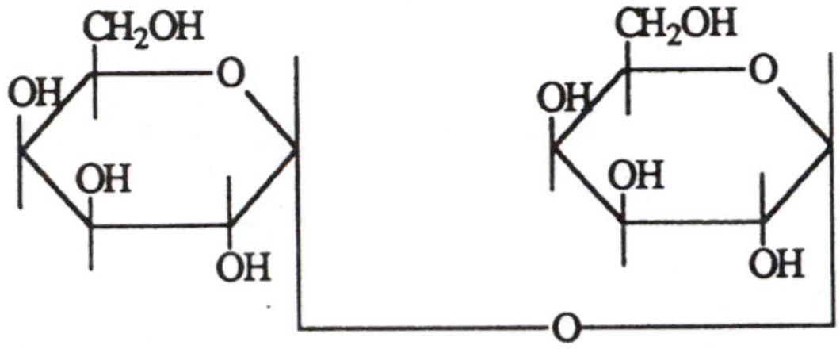

$CH_2OH$ O OH OH OH — O — $CH_2OH$ O OH OH OH

58. Cellulose, amylose, and amylopectin are polymers of glucose. Cellulose exists in the form of fibers, is not digestible by humans, and therefore remains in the digestive tract as fibers. amylose and amylopectan are digested to glucose which is dissolved into the bloodstream.

59. (a) D-galactose and D-glucose differ only at carbon 4. Thus, D-galactose must be changed at carbon 4 to be converted to D-glucose.

(b) D-galactose is an *epimer* of D-glucose.

60. No, the classmate should not be believed. Although D-glucose and D-mannose are related as epimers, it is pairs of *enantiomers* which yield equal and opposite optical rotation.

CHAPTER 28

# LIPIDS

1. The lipids, which are dissimilar substances, are arbitrarily classified as a group on the basis of their solubility in fat solvents and their insolubility in water.

2. Although caproic acid, $CH_3(CH_2)_4COOH$, has the same number of polar bonds as stearic acid, $CH_3(CH_2)_{16}COOH$, caproic acid has a shorter nonpolar hydrocarbon chain and, therefore is more water soluble.

3. Arachidonic acid is used to synthesize the eicosanoids, hormone-like substances, which include the prostaglandins, the leukotrienes, the prostacyclins and the thromboxanes.

4. The three essential fatty acids are linoleic, linolenic, and arachidonic acids. Diets lacking these fatty acids lead to impaired growth and reproduction, and skin disorders such as eczema and dermatitis.

5. Fats contain more biochemical energy than carbohydrates because (a) the carbons of fat are more reduced than those found in carbohydrates and (b) there are more reduced carbons per gram of fat than from the same mass of carbohydrate.

6. Aspirin relieves inflammation by blocking the conversion of arachidonic acid to prostaglandins.

7. Waxes serve as a protective coating because they are very hydrophobic. They do not dissolve in water and many compounds cannot pass through a wax coating.

8. A membrane lipid must be (a) partially hydrophobic to act as a barrier to water and (b) partially hydrophilic, so that the membrane can interact with water along its surface.

9. Phospholipids are mainly produced in the liver.

10. The four classes of eicosanoids are prostaglandins, prostacyclins, thromboxanes and leukotrienes.

11. In general, a membrane lipid will have both hydrophilic and hydrophobic parts. Sphingomyelin can serve as a membrane lipid because its phosphate and choline groups are hydrophilic and its two long carbon chains are hydrophobic.

```
                 OH
                 |
  O              CHCH=CH(CH2)12CH3
  ||             |
  RC—NH—CH            O                   CH3
                 |            ||                  |+
                 CH2—O—P—O—CH2CH2—N—CH3
                              |                   |
                              O⁻                  CH3
```

sphingomyelin

12. Atherosclerosis is the deposition of cholesterol and other lipids on the inner walls of the large arteries. These deposits, called plaque, accumulate, making the arterial passages narrower and narrower. Blood pressure increases as the heart works to pump sufficient blood through the restricted passages. This may lead to a heart attack, or the rough surface can lead to coronary thrombosis.

13. Dietary cholesterol is transported first to the liver where it is bound to other lipids and proteins to form the very low density lipoprotein (VLDL). This aggregate moves through the blood stream delivering lipids to various tissues. As lipids are removed, the VLDL is converted to a low density lipoprotein (LDL). Cells needing cholesterol can absorb LDL. Cholesterol is transported back to the liver by the high density lipoprotein (HDL).

14. Dietary fish oils provide fatty acids which inhibit formation of thromboxanes, compounds which participate in blood clotting.

15. HDL is a cholesterol scavenger, picking up this steroid in the serum and returning it to the liver.

16. Because the interior of a lipid bilayer is very hydrophobic, molecules with hydrophilic character only cross the lipid bilayer with difficulty. A lipid bilayer acts as a barrier to water-soluble compounds.

17. Liposomes can package drugs so that only the target organ/tissue is exposed to the drug's effects.

18. Both facilitated diffusion and active transport catalyze movement of compounds through membranes. Active transport requires an input of energy while facilitated diffusion does not.

19. All steroids possess a 17-carbon unit structure containing four fused rings known as the steroid nucleus. Many steroids could be shown; these are two common ones:

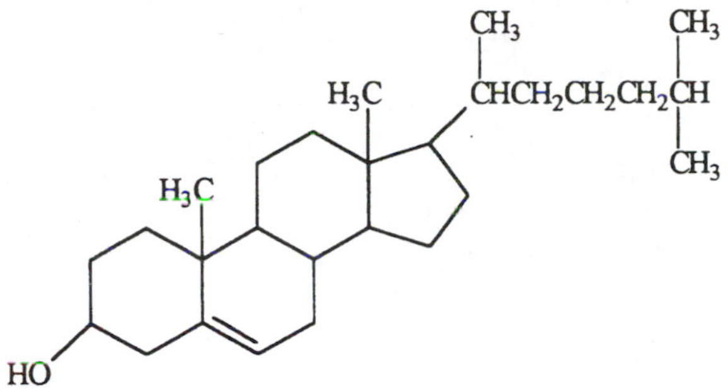

cholesterol

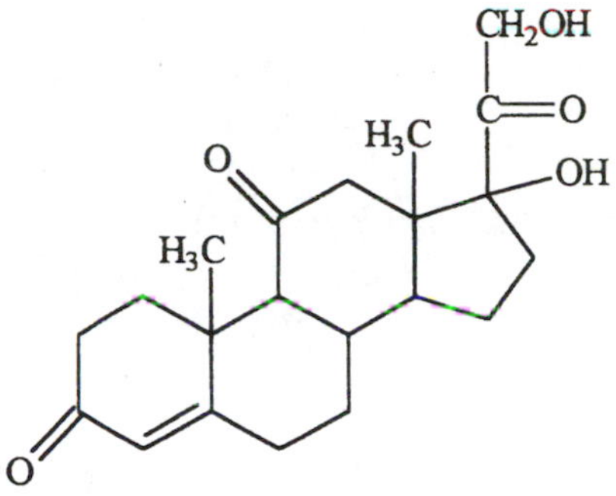

cortisone

20. Up to 80% of the total body cholesterol is metabotically produced. The statin drugs block metabolic production of cholesterol.

21. A triacylgylcerol containing one unit each of palmitic, stearic, and oleic acids:

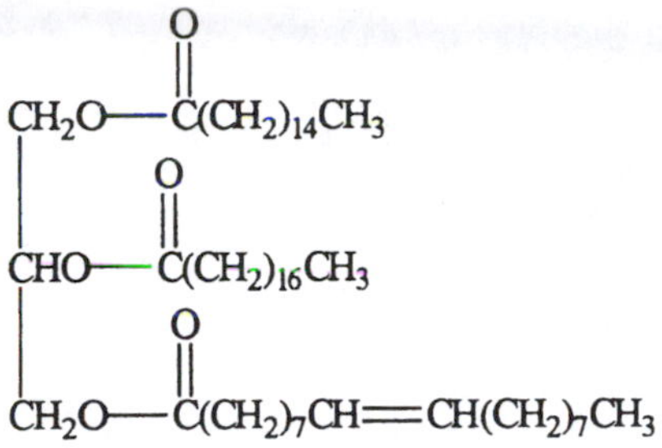

$CH_2O{-}C(=O)(CH_2)_{14}CH_3$ palmitic acid

$CHO{-}C(=O)(CH_2)_{16}CH_3$ stearic acid

$CH_2O{-}C(=O)(CH_2)_7CH{=}CH(CH_2)_7CH_3$ oleic acid

There would be two other triacylglycerols possible from these same components. Since the top and bottom attachments are equivalent, it only matters which acid is attached to the middle carbon of glycerol.

22. A triacylgylcerol containing two units of palmitic acid and one unit of oleic acid:

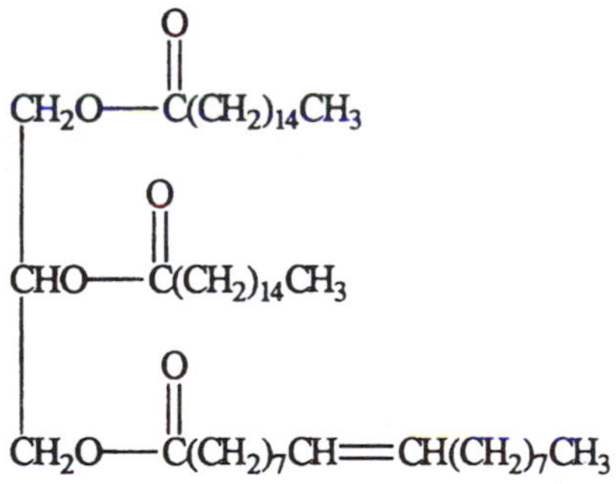

$CH_2O{-}C(=O)(CH_2)_{14}CH_3$ palmitic acid

$CHO{-}C(=O)(CH_2)_{14}CH_3$ palmitic acid

$CH_2O{-}C(=O)(CH_2)_7CH{=}CH(CH_2)_7CH_3$ oleic acid

There is one other possible triacylglycerol with the same components; this triacylglycerol would have palmitic acid units at both ends with the oleic acid unit bound to the middle carbon of glycerol.

23. No, a triacylglycerol that contains three units of lauric acid would be *less* hydrophobic than a triacylglycerol that contains three units of stearic acid. Smaller molecules tend to be less hydrophobic than larger molecules, and, a triacylglycerol with three lauric acid units is smaller than a triacylglycerol with three stearic acid units.

24. Yes, a triacylglycerol that contains three units of stearic acid would be more hydrophobic than a triacylglycerol that contains three units of myristic acid. Larger molecules tend to be more hydrophobic than smaller molecules and, a triacylglycerol with three stearic acid units is larger than a triacylglycerol with three myristic acid units.

25. Hydrolysis of a triacylglycerol will yield three fatty acids and glycerol. The products will be:

$CH_2OH$
|
$CHOH$ glycerol
|
$CH_2OH$

$CH_3(CH_2)_{14}COOH$ palmitic acid

$CH_3(CH_2)_7CH{=}CH(CH_2)_7COOH$ oleic acid

$CH_3(CH_2)_4CH{=}CHCH_2CH{=}CH(CH_2)_7COOH$ linoleic acid

26. Hydrolysis of a triacylglycerol will yield three fatty acids and glycerol. The products will be:

$CH_2OH$
|
$CHOH$ glycerol
|
$CH_2OH$

$CH_3(CH_2)_7CH{=}CH(CH_2)_7COOH$ oleic acid

$CH_3(CH_2)_{14}COOH$ palmitic acid

$CH_3(CH_2)_{16}COOH$ stearic acid

27. The phospholipid structure is:

$$\begin{array}{l} CH_2O-\overset{\overset{O}{\|}}{C}(CH_2)_{14}CH_3 \\ | \\ CHO-\overset{\overset{O}{\|}}{C}(CH_2)_{14}CH_3 \\ | \\ CH_2O-\underset{\underset{O^-}{|}}{\overset{\overset{O}{\|}}{P}}-OCH_2CH_2NH_3^+ \end{array}$$

The phosphoric acid and ethanolamine must be linked to the bottom glycerol carbon. Since a typical phospholipid contains two fatty acid units, a palmitic acid unit must be linked to the top glycerol carbon and another to the middle glycerol carbon.

28. The phospholipid structure is:

$$\begin{array}{l} CH_2O-\overset{\overset{O}{\|}}{C}(CH_2)_{16}CH_3 \\ | \\ CHO-\overset{\overset{O}{\|}}{C}(CH_2)_{16}CH_3 \\ | \\ CH_2O-\underset{\underset{O^-}{|}}{\overset{\overset{O}{\|}}{P}}-OCH_2CH_2\underset{\underset{CH_3}{|}}{\overset{\overset{CH_3}{|}}{N^+}}-CH_3 \end{array}$$

The phosphoric acid and choline units must be linked to the bottom glycerol carbon. Since a typical phospholipid contains two fatty acid units, one stearic acid unit must be linked to the top glycerol carbon and a second stearic acid unit must be linked to the middle glycerol carbon.

29. The sphingolipid structure is:

$$\begin{array}{l}
OH \\
| \\
CHCH{=}CH(CH_2)_{12}CH_3 \\
| \qquad\quad O \\
| \qquad\quad \| \\
CHNH{-}C(CH_2)_7CH{=}CH(CH_2)_7CH_3 \\
| \qquad O \qquad\qquad CH_3 \\
| \qquad \| \qquad\qquad\quad |+ \\
CH_2O{-}P{-}OCH_2CH_2N{-}CH_3 \\
\qquad\quad | \qquad\qquad\quad | \\
\qquad\quad O^- \qquad\qquad CH_3
\end{array}$$

The phosphate and choline units are linked to the bottom glycerol carbon of sphingosine (when the molecules is written as above). The oleic acid unit is then linked to the nitrogen to form the sphingolipid.

30. The sphingolipid structure is:

$$\begin{array}{l}
OH \\
| \\
CHCH{=}CH(CH_2)_{12}CH_3 \\
| \qquad\quad O \\
| \qquad\quad \| \\
CHNH{-}C(CH_2)_{16}CH_3 \\
| \qquad O \\
| \qquad \| \\
CH_2O{-}P{-}OCH_2CH_2\overset{+}{N}H_3 \\
\qquad\quad | \\
\qquad\quad O^-
\end{array}$$

The phosphate and ethanolamine units are linked to the bottom sphingosine carbon (when the molecule is written as above). The stearic acid unit is then linked to the nitrogen to form the sphingolipid.

31. The glycolipid structure is:

$$\begin{array}{l}
OH \\
| \\
CHCH{=}CH(CH_2)_{12}CH_3 \\
| \qquad\quad O \\
| \qquad\quad \| \\
CHNH{-}C(CH_2)_{14}CH_3 \\
| \\
CH_2O \\
\text{(glucose ring: } CH_2OH, O, OH, OH, OH)
\end{array}
\quad \text{or} \quad
\begin{array}{l}
OH \\
| \\
CHCH{=}CH(CH_2)_{12}CH_3 \\
| \qquad\quad O \\
| \qquad\quad \| \\
CHNH{-}C(CH_2)_{14}CH_3 \\
| \\
CH_2O \\
\text{(glucose ring: } CH_2OH, O, OH, OH, OH)
\end{array}$$

The D-glucose unit is linked to the bottom sphingosine carbon (when the molecule is written as above). The palmitic acid unit is then linked to the nitrogen to form the glycolipid.

32. The glycolipid structure is:

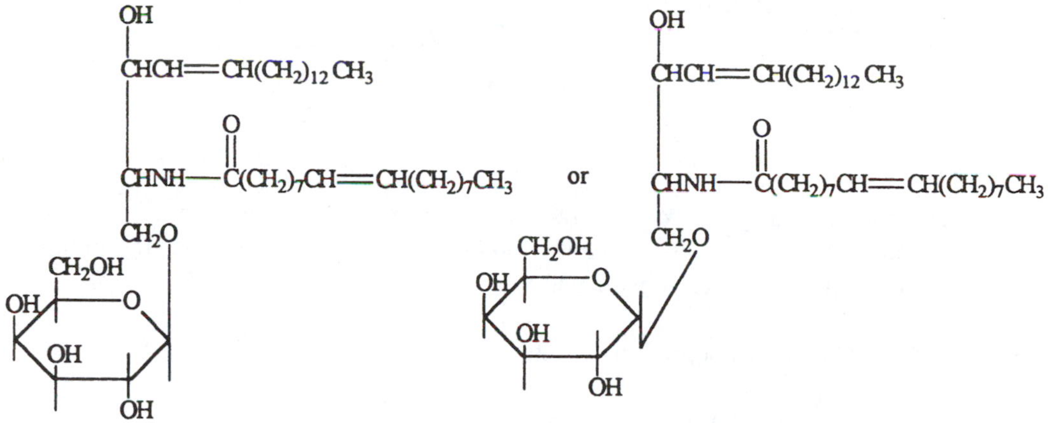

OH

$CHCH{=}CH(CH_2)_{12}CH_3$

O

$CHNH{-}C(CH_2)_7CH{=}CH(CH_2)_7CH_3$

$CH_2O$

$CH_2OH$

OH O

OH

OH

or

OH

$CHCH{=}CH(CH_2)_{12}CH_3$

O

$CHNH{-}C(CH_2)_7CH{=}CH(CH_2)_7CH_3$

$CH_2OH$ $CH_2O$

OH O

OH

OH

The D-galactose is linked to the bottom sphingosine carbon (when the molecule is written as above). The oleic acid unit is then linked to the nitrogen to form the glycolipid.

33.

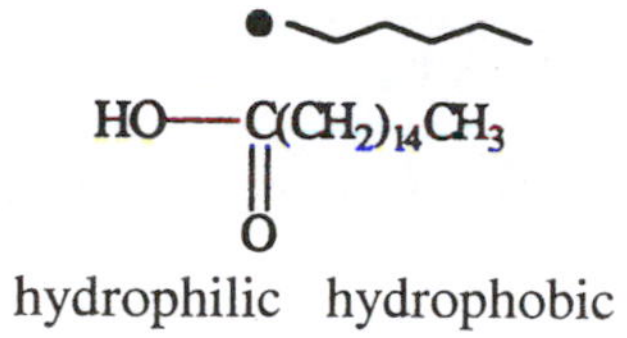

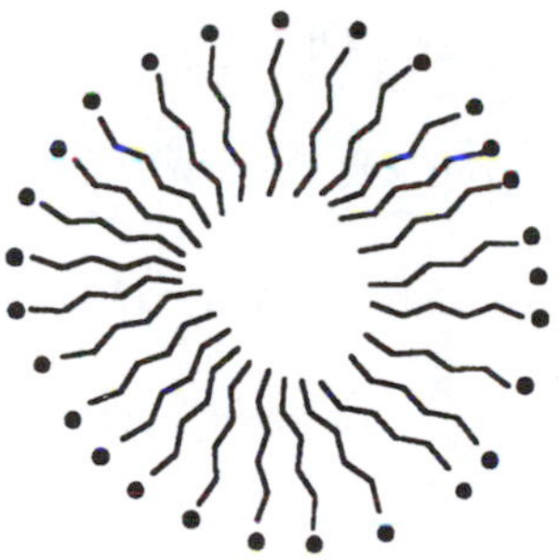

$HO{-}C(CH_2)_{14}CH_3$

O

hydrophilic hydrophobic

The micelle is a sphere with the fatty acid carboxyl groups on the hydrophilic exterior and the fatty acid alkyl chains in the hydrophobic interior. A condensed structural formula has been used to show the palmitic acid.

34.

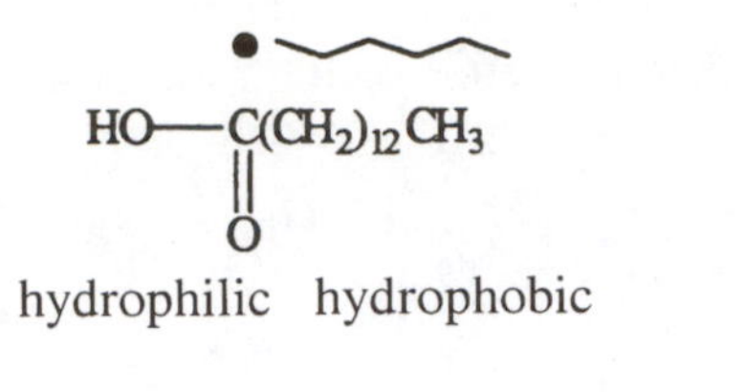


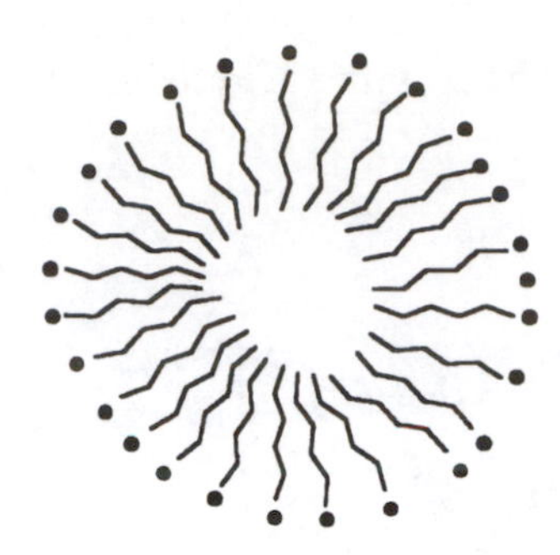

The micelle is a sphere with the fatty acid carboxyl groups on the hydrophilic exterior and the fatty acid alkyl groups in the hydrophobic interior. A condensed structural formula has been used to show the myristic acid.

35. LDL (low density lipoprotein) differs from VLDL (very low density lipoprotein) in that

(a) VLDL has a lower density than LDL.

(b) VLDL is formed in the liver while LDL is formed from VLDL as the lipoproteins circulate in the blood.

(c) VLDL is larger than LDL.

36. HDL (high density lipoprotein) differs from LDL (low density lipoprotein) in that

(a) LDL has a lower density than HDL.

(b) LDL delivers cholesterol to peripheral tissues while HDL scavenges cholesterol and returns it to the liver.

37.

$$\begin{array}{l} OH \\ | \\ CHCH{=}CH(CH_2)_{12}CH_3 \\ | \\ CH{-}NH_2 \\ | \\ CH_2OH \end{array}$$

sphingosine

$$\begin{array}{l} \qquad\qquad\quad O \\ \qquad\qquad\quad \| \\ CH_2{-}O{-}C{-}R \\ | \\ CH{-}OH \\ | \\ CH_2OH \end{array}$$

monoacylglycerol

Sphingosine is similar to the monoacyglycerol in that (a) both molecules contain a long hydrophobic chain, (b) the sphingosine amino group reacts with a fatty acid as does the secondary alcohol of the monoacylglycerol and, (c) for both compounds, the primary alcohol can react further with either acids (to form esters) or sugars (to form acetals).

38.

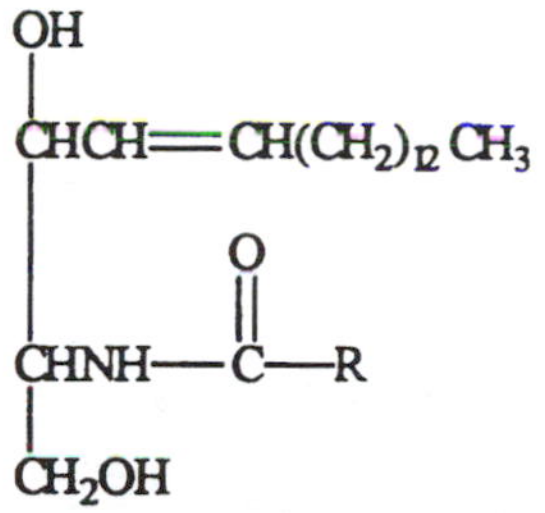

OH
|
$CHCH{=}CH(CH_2)_{12}CH_3$
|
$CHNH{-}C({=}O){-}R$
|
$CH_2OH$

sphingosine and fatty acid unit

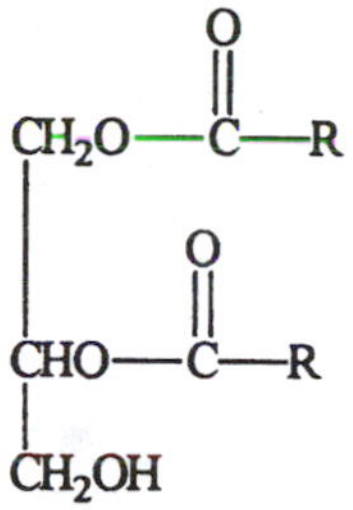

$CH_2O{-}C({=}O){-}R$
|
$CHO{-}C({=}O){-}R$
|
$CH_2OH$

diacylglycerol

(a) Both compounds have two long hydrophobic chains.

(b) For both compounds, the primary alcohol can react further with either acids (to form esters) or sugars (to form acetals).

39. Sodium ion will move from a region of high concentration to a region of low concentration as it moves from a 0.1 M solution across a membrane to a 0.001 M solution. This process does not require energy and can be accomplished by facilitated diffusion.

40. Phosphate ion will move from a region of low concentration to a region of high concentration as it moves from a 0.1 M solution across a membrane to 0.5 M solution. This process requires energy and must be accomplished by active transport.

41. Thromboxanes acts as vasoconstrictors and stimulate platelet aggregation while prostaglandins cause the redness, swelling and pain associated with tissue inflammation.

42. Thromboxanes act as vasoconstrictors and stimulate platelet aggregation while leukotrienes have been associated with many of the symptoms of an allergy attack (e.g., an asthma attack).

43. Ibuprofen (an NSAID) blocks the oxidation of arachidonic acid to form prostaglandins which, in turn, can cause inflammation, redness, and swelling.

44. Both olestra and natural fats contain fatty acid units. However, in natural fats the fatty acid units are linked to glycerol while in olestra the fatty acid units are linked to sucrose. Olestra has many of the qualities of a fat. However, it is not shaped like a natural fat and is not attacked by digestive enzymes. Thus, olestra passes through the stomach and the intestines undigested and does not provide any caloric value to the cells.

45. (a) The three essential fatty acids are:

linoleic acid $CH_3(CH_2)_4CH{=}CHCH_2CH{=}CH(CH_2)_7COOH$

linolenic acid $CH_3CH_2CH{=}CHCH_2CH{=}CHCH_2CH{=}CH(CH_2)_7COOH$

arachidonic acid $CH_3(CH_2)_4(CH{=}CHCH_2)_4CH_2CH_2COOH$

(b) A diet which is missing the essential fatty acids will lead to impaired growth and reproduction as well as skin disorders such as eczema and dermatitis.

46. This meal is changed by increasing the fat content from 10 grams to 15 grams, an increase of 5 grams. Since each gram of fat yields an average of 9.5 Cal, an increase of 5 grams equates to an increase of 47.5 Cal (5 grams $\times$ 9.5 Cal/g). Thus, this meal will now contain 234 Cal + 47.5 Cal = 282 Cal.

47. Formula for beeswax

$$CH_3(CH_2)_{14}\overset{\overset{\displaystyle O}{\|}}{C}{-}O{-}(CH_2)_{29}CH_3$$

48. $CH_3(CH_2)_7CH{=}CH(CH_2)_7COOH$
oleic acid or 9-octadeceneoic acid

CHAPTER 29

# AMINO ACIDS, POLYPEPTIDES, AND PROTEINS

1. The amino acids of proteins are called alpha amino acids because the amine group is always attached to the alpha carbon atom, that is, the carbon atom next to the carboxyl group, COOH.

$$R-\overset{\displaystyle NH_2}{\underset{\displaystyle H}{\overset{|}{\underset{|}{C}}}}-COOH$$

alpha carbon atom

2. All amino acids and proteins contain carbon, hydrogen, oxygen, and nitrogen. Sulfur is contained in some of the amino acids, and thus in most proteins.

3. Proteins from some foods are of greater nutritional value than others because they are "complete", which means they contain all eight essential amino acids, those which the human body cannot synthesize.

4. The amino acids which are essential to humans are isoleucine, leucine, lysine, methionine, phenylalanine, threonine, tryptophan, and valine.

5. Amino acids are amphoteric because the carboxyl group can react with a base to form a salt, or the amine group can react with an acid to form a salt. They are optically active because the alpha carbon is chiral, except for glycine. They commonly have the L configuration at carbon two, as in L-serine.

6. At its isoelectric point, a protein molecule must have an equal number of positive and negative charges.

7. (a) Primary structure. The number, kind, and sequence of amino acid units comprising the polypeptide chain making up a molecule.

   (b) Secondary structure. Regular three-dimensional structure held together by the hydrogen bonding between the oxygen of C=O groups and the hydrogen of the N–H groups in the polypeptide chains.

   (c) Tertiary structure. The distinctive and characteristic three-dimensional conformation or shape of a protein molecule.

(d) Quaternary structure. The three-dimensional shape formed by an aggregate of protein subunits found in some complex proteins.

8. The sulfur-containing amino acid, cysteine, has the special role in protein structure of creating disulfide bonding between polypeptide chains which helps control the shape of the molecule.

9. The major structural difference between hemoglobin and myoglobin is that hemoglobin is composed of four subunits while myoglobin only contains one. Hemoglobin's quaternary structure allows a more effective control of oxygen transport than is possible with myoglobin.

10. Both the $\alpha$-helix and $\beta$-pleated sheet are examples of secondary protein structures. The $\alpha$-helix forms a tube composed of a spiraling polypeptide chain while the $\beta$-pleated sheet forms a plane composed of polypeptide chains aligned roughly parallel to each other.

11. Hydrolysis breaks the peptide bonds, thus disrupting the primary structure of the protein. Denaturation involves alteration or disruption of the secondary, tertiary, or quaternary but not of the primary structure of proteins.

12. Amino acids containing a benzene ring give a positive xanthoproteic test (formation of yellow-colored reaction products). Among the common amino acids, these would include phenylalanine, tryptophan, and tyrosine.

13. The visible evidence observed in the:

(a) Xanthoproteic test gives a yellow-colored reaction product when a protein containing a benzene ring is reacted with concentrated nitric acid.

(b) Biuret test gives a violet color when dilute $CuSO_4$ is added to an alkaline solution of a peptide or a protein.

(c) Ninhydrin test gives a blue solution with all amino acids except proline and hydroxyproline, both of which produce a yellow solution when ninhydrin is added to an amino acid.

14. Protein column chromatography uses a column packed with polymer beads (solid phase) through which a protein solution (liquid phase) is passed. Proteins separate based on differences in how they react with the solid phase. The proteins move through the column at different rates and can be collected separately.

15. (a) Thin layer chromatography is a way of separating substances based on a differential distribution between two phases, the liquid phase and the solid phase.

(b) A strip (or sheet) is prepared with a thin coating (layer) of dried alumina or other adsorbent. A tiny spot of solution containing a mixture of amino acids is placed near the bottom of the strip. After the spot dries, the bottom edge of the strip is placed in a suitable solvent. The solvent ascends in the strip, carrying the different amino acids upwards at different rates. When the solvent front nears the top, the strip is removed from the solvent and dried.

(c) Ninhydrin is the reagent used to locate the different amino acids on the strip.

16. In ordinary electrophoresis the rate of movement of a protein depends on its charge and size. In SDS electrophoresis a detergent, sodium dodecyl sulfate, is added to the protein solution, which masks the differences in protein charges, leaving the separation primarily due to the size of the various proteins.

17. D-alanine

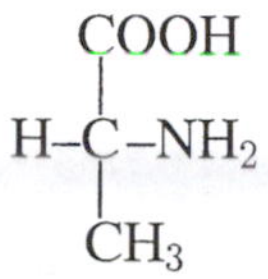

$$\begin{array}{c} COOH \\ | \\ H{-}C{-}NH_2 \\ | \\ CH_3 \end{array}$$

L-alanine
(form commonly found in proteins)

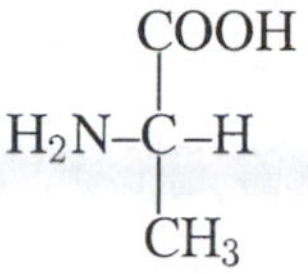

$$\begin{array}{c} COOH \\ | \\ H_2N{-}C{-}H \\ | \\ CH_3 \end{array}$$

18. D-serine

$$\begin{array}{c} COOH \\ | \\ H{-}C{-}NH_2 \\ | \\ CH_2OH \end{array}$$

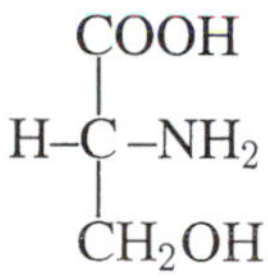

L-serine
(form commonly found in proteins)

$$\begin{array}{c} COOH \\ | \\ H_2N{-}C{-}H \\ | \\ CH_2OH \end{array}$$

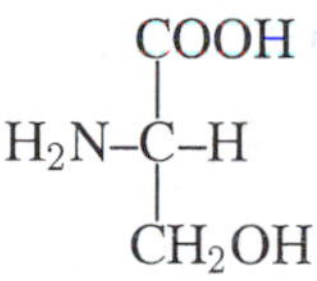

19. $(6.0 \text{ g nitrogen})\left(\frac{100. \text{ g protein}}{16 \text{ g nitrogen}}\right)\left(\frac{1}{100.0 \text{ g food product}}\right)(100) = 38\%$ protein

20. $(5.2 \text{ g nitrogen})\left(\frac{100. \text{ g protein}}{16 \text{ g nitrogen}}\right)\left(\frac{1}{250.0 \text{ g hamburger}}\right)(100) = 13\%$ protein

21. The structural formula for threonine at its isoelectric point is:

$$\begin{array}{l} CH_3{-}\underset{\displaystyle OH}{\underset{|}{CH}}{-}\underset{\displaystyle NH_3^+}{\underset{|}{CH}}{-}COO^- \end{array}$$

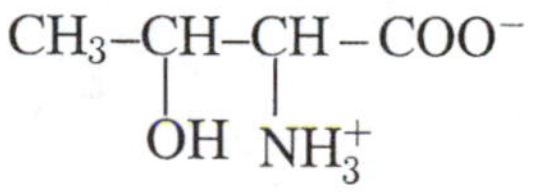

22. The structural formula for asparagine at its isoelectric point is:

$$NH_2\overset{\displaystyle O}{\overset{\|}{C}}{-}CH_2{-}\underset{\displaystyle NH_3^+}{\underset{|}{CH}}{-}COO^-$$

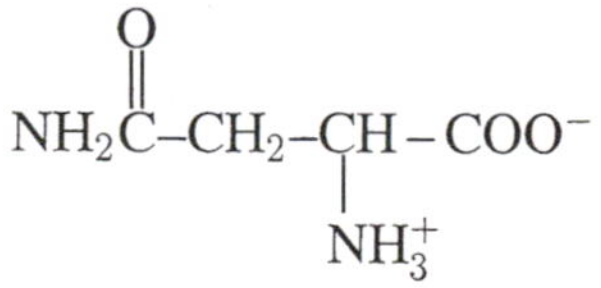

23. For phenylalanine:

(a) zwitterion formula (b) formula in 0.1 M $H_2SO_4$ (c) formula in 0.1 M NaOH

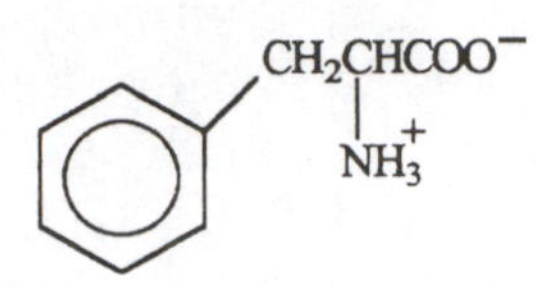


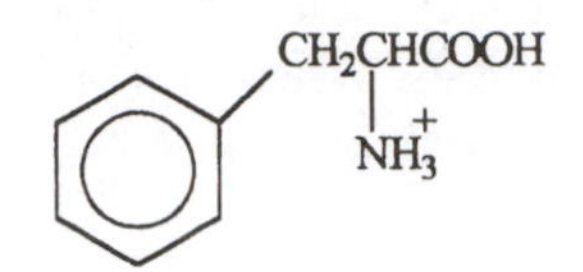


$$\text{C}_6\text{H}_5\text{-}\underset{\text{NH}_2}{\text{CH}_2\text{CHCOO}^-}$$

24. For tryptophan:

(a) zwitterion formula (b) formula in 0.1 M $H_2SO_4$ (c) formula in 0.1 M NaOH

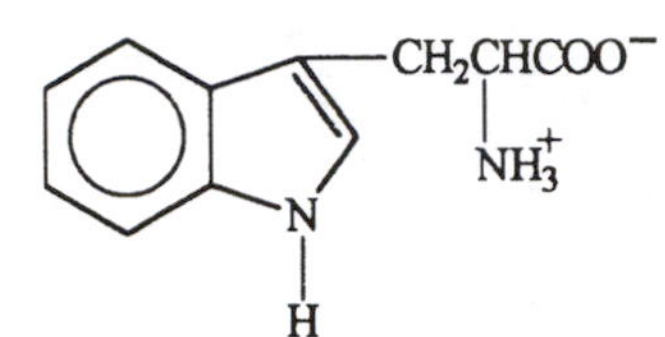


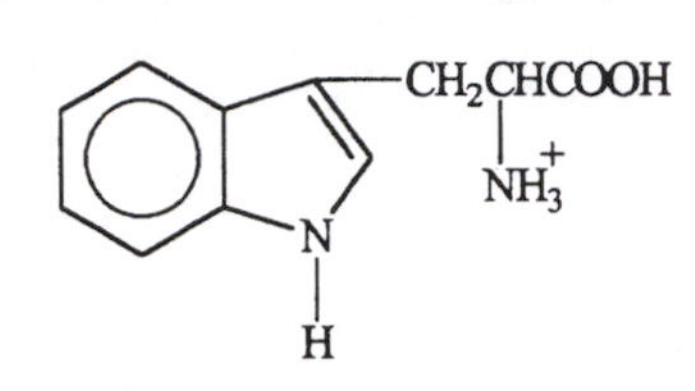


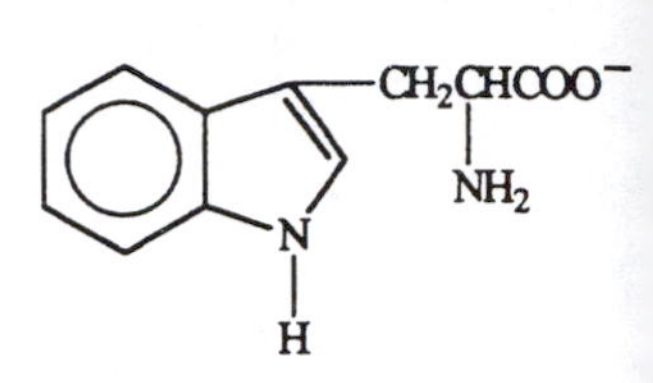


25. Ionic equations showing how alanine acts as a buffer towards:

(a) $H^+$

$$CH_3\underset{NH_3^+}{\underset{|}{CH}}COO^- + H^+ \longrightarrow CH_3\underset{NH_3^+}{\underset{|}{CH}}COOH$$

(b) $OH^-$

$$CH_3\underset{NH_3^+}{\underset{|}{CH}}COO^- + OH^- \longrightarrow CH_3\underset{NH_2}{\underset{|}{CH}}COOH + H_2O$$

26. Ionic equations showing how leucine acts as a buffer towards:

(a) $H^+$

$$(CH_3)_2CHCH_2\underset{NH_3^+}{\underset{|}{CH}}COO^- + H^+ \longrightarrow (CH_3)_2CHCH_2\underset{NH_3^+}{\underset{|}{CH}}COOH$$

(b) $OH^-$

$$(CH_3)_2CHCH_2\underset{NH_3^+}{\underset{|}{CH}}COO^- + OH^- \longrightarrow (CH_3)_2CHCH_2\underset{NH_2}{\underset{|}{CH}}COO^- + H_2O$$

27. Methionine will have the following structure at its isoelectric point:

$$CH_3SCH_2CH_2\underset{NH_3^+}{\underset{|}{CH}}COO^-$$

28. Valine will have the following structure at its isoelectric point:

$(CH_3)_2CHCH(NH_3^+)COO^-$

29. The two dipeptides containing serine and alanine:

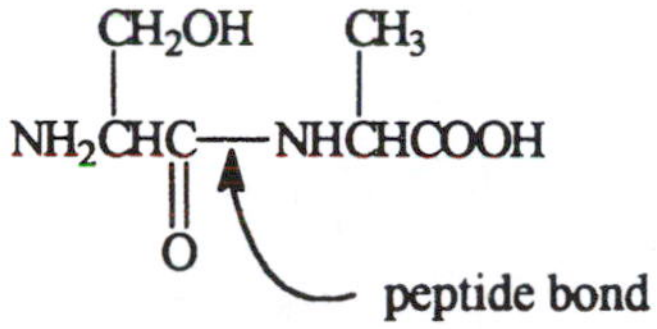

$NH_2CH(CH_2OH)C(=O)—NHCH(CH_3)COOH$ (arrow: peptide bond)

Ser-Ala

$NH_2CH(CH_3)C(=O)—NHCH(CH_2OH)COOH$ (arrow: peptide bond)

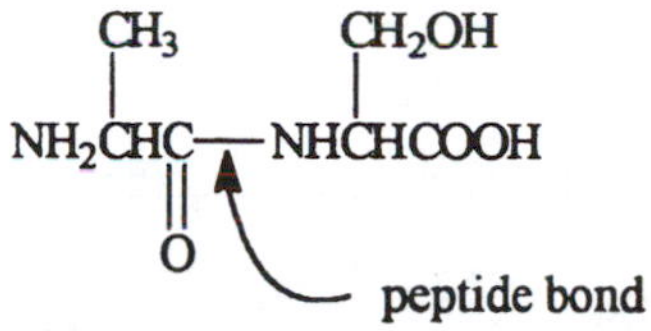

Ala-Ser

30. The two dipeptides containing glycine and threonine:

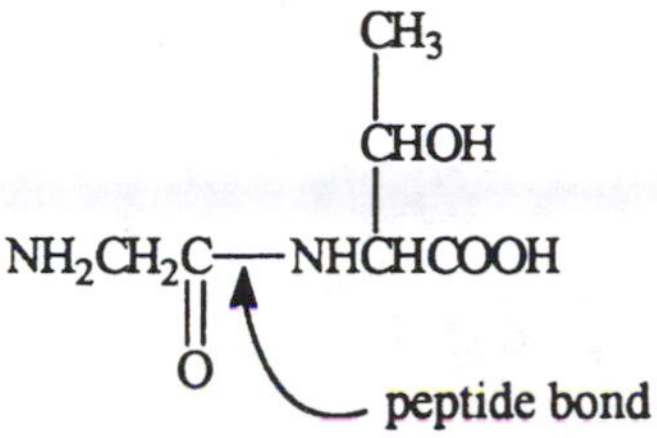

$NH_2CH_2C(=O)—NHCH(CHOHCH_3)COOH$ (arrow: peptide bond)

Gly-Thr

$NH_2CH(CHOHCH_3)C(=O)—NHCH_2COOH$ (arrow: peptide bond)

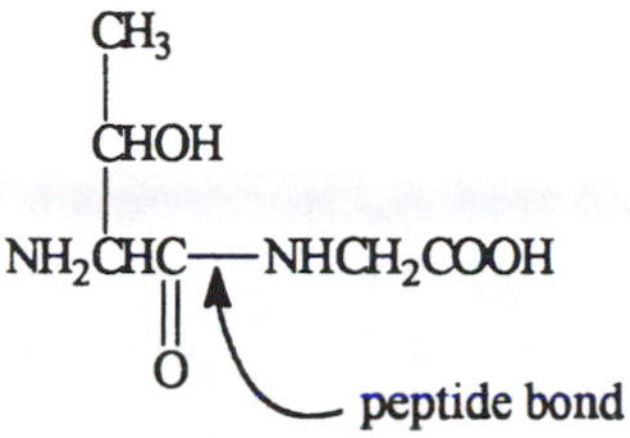

Thr-Gly

31. (a) glycylglycine $NH_2CH_2C(=O)–NHCH_2COOH$

(b) alanylglycylserine $NH_2CH(CH_3)C(=O)–NHCH_2C(=O)–NHCH(CH_2OH)COOH$

(c) glycylserylglycine $NH_2CH_2C(=O)–NHCH(CH_2OH)C(=O)–NHCH_2COOH$

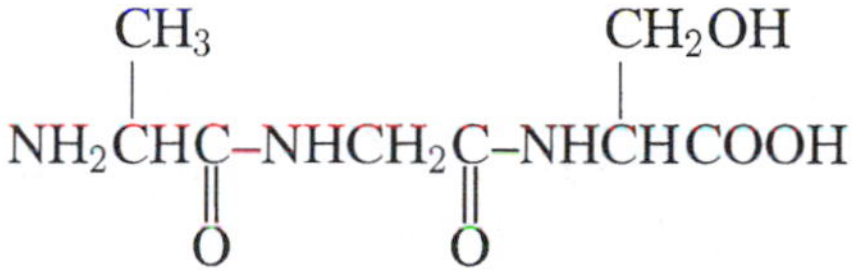

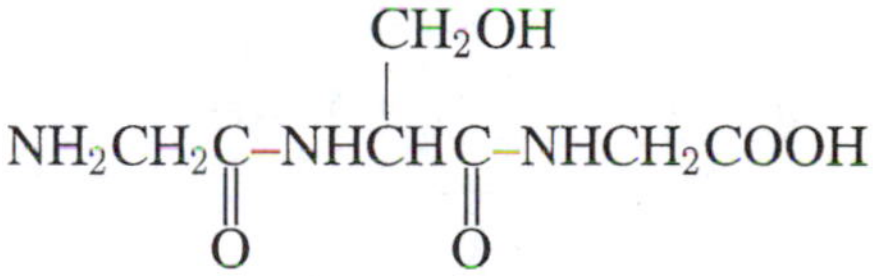

32. (a) alanylalanine $NH_2CH(CH_3)C(=O)–NHCH(CH_3)COOH$

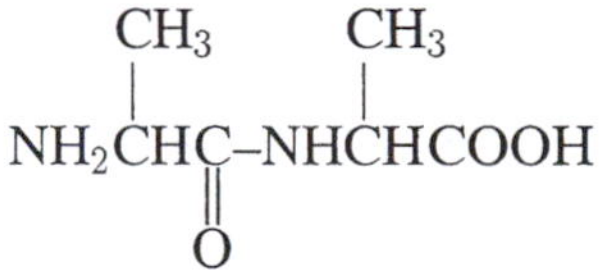

(b) serylglycylglycine

$$\begin{array}{l} \quad\;\; CH_2OH \\ \qquad | \\ NH_2CH\,\underset{\underset{O}{\|}}{C}\text{–}NHCH_2\underset{\underset{O}{\|}}{C}\text{–}NHCH_2COOH \end{array}$$

(c) serylglycylalanine

$$\begin{array}{l} \quad\;\; CH_2OH \qquad\qquad\qquad\quad CH_3 \\ \qquad | \qquad\qquad\qquad\qquad\quad\; | \\ NH_2CH\,\underset{\underset{O}{\|}}{C}\text{–}NHCH_2\underset{\underset{O}{\|}}{C}\text{–}NHCHCOOH \end{array}$$

33. All the possible tripeptides containing one unit each of glycine, phenylalanine, and leucine:

Gly-Phe-Leu Gly-Leu-Phe Phe-Gly-Leu
Phe-Leu-Gly Leu-Gly-Phe Leu-Phe-Gly

34. All the possible tripeptides containing one unit each of tyrosine, aspartic acid, and alanine:

Tyr-Asp-Ala Tyr-Ala-Asp Asp-Tyr-Ala
Asp-Ala-Tyr Ala- Tyr-Asp Ala-Asp- Tyr

35. The oxygen atom in the peptide bond and the hydrogen atom in the peptide bond participate in hydrogen bonding to hold together a $\beta$-pleated sheet:

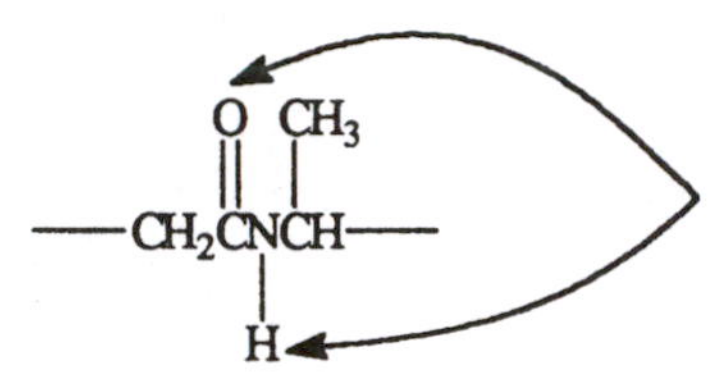


involved in hydrogen bonding

36. The oxygen atom in the peptide bond and the hydrogen bond in the peptide bond participate in hydrogen bonding to hold together an $\alpha$-helix:

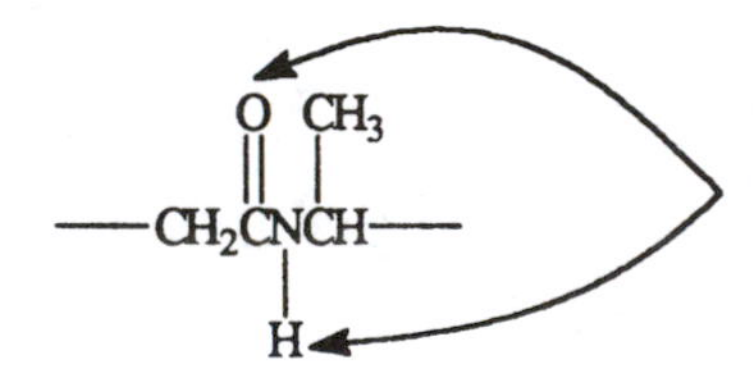


involved in hydrogen bonding

37. Tertiary protein structure is usually held together by bonds between amino acid side chains. Serine side chains will hydrogen bond to each other:

$$-CH_2O-H\cdots\cdots \underset{}{O}(H)CH_2-$$

hydrogen bond

38. Tertiary protein structure is usually held together by bonds between amino acid side chains. At pH = 7, the lysine side chain will contain a positive charge, $\overset{+}{N}H_3CH_2CH_2CH_2CH_2^-$, and the aspartic acid side chain will contain a negative charge, $^-OOCCH_2-$. These two side chains will be held together by an ionic bond:

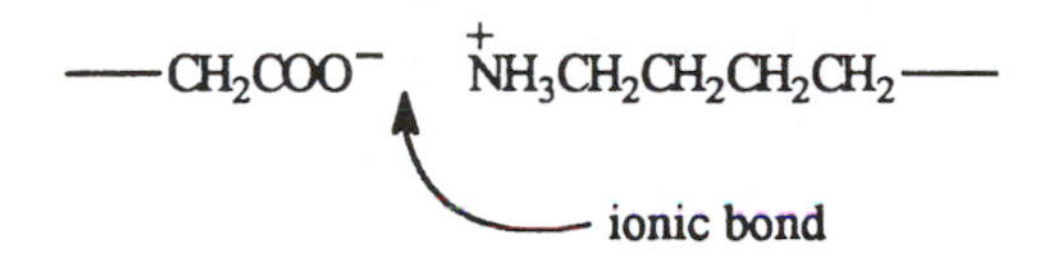


39. The tripeptide, Gly-Ala-Thr, will

    (a) react with $CuSO_4$ to give a violet color. The tripeptide has the required two peptide bonds.

    (b) not react to give a positive xanthoproteic test because there are no benzene ring compounds in this tripeptide.

    (c) react with ninhydrin to give a blue solution. (Contains the required amino acids for reaction.)

40. The tripeptide, Gly-Ser-Asp, will

    (a) react with $CuSO_4$ to give a violet color. The tripeptide has the required two peptide bonds.

    (b) not react to give a positive xanthoproteic test because there are no benzene ring amino acids in this tripeptide.

    (c) react with ninhydrin to give a blue solution. (Contains the required amino acids for reaction.)

41. Hydrolysis breaks the peptide bonds. One water molecule will react with each peptide bond, a hydrogen atom attaches to the nitrogen to complete the amino group and an –OH group attaches to the carboxyl carbon. The tripeptide, Ala-Phe-Asp, will hydrolyze to yield the following:

$$NH_2CH(CH_3)COOH + NH_2CH(CH_2C_6H_5)COOH + NH_2CH(CH_2COOH)COOH$$

42. Hydrolysis breaks the peptide bonds. One water molecule will react with each peptide bond, a hydrogen atom attaches to the nitrogen to complete the amino group and an –OH group attaches to the carboxyl carbon. The tripeptide, Ala-Glu-Tyr, will hydrolyze to yield the following:

$$NH_2CH(CH_3)COOH + NH_2CH(CH_2CH_2COOH)COOH + NH_2CH(CH_2C_6H_4OH)COOH$$

43. $$\left(\frac{1 \text{ mol Fe}}{1 \text{ mol cytochrome c}}\right)\left(\frac{55.85 \text{ g}}{\text{mol Fe}}\right)\left(\frac{100. \text{ g cytochrome c}}{0.43 \text{ g Fe}}\right) = 1.3 \times 10^4 \frac{\text{g}}{\text{mol}}$$

The molar mass of cytochrome c is $1.3 \times 10^4$ g/mol

44. $$\left(\frac{4 \text{ mol Fe}}{1 \text{ mol hemoglobin}}\right)\left(\frac{55.85 \text{ g}}{\text{mol Fe}}\right)\left(\frac{100. \text{ g hemoglobin}}{0.33 \text{ g Fe}}\right) = 6.8 \times 10^4 \frac{\text{g}}{\text{mol}}$$

The molar mass of hemoglobin is $6.8 \times 10^4$ g/mol.

45. The amino acid sequence of the heptapeptide is:

| | | | | | | |
|---|---|---|---|---|---|---|
| | | Gly | Phe | Leu | | |
| Phe | Ala | Gly | | Leu | Ala | Tyr |
| Phe | Ala | Gly | Phe | Leu | Ala | Tyr |

46. The amino acid sequence of the heptapeptide is:

Phe - Gly - Tyr

Ala - Leu - Phe

Phe - Ala - Ala

Phe - Ala - Ala - Leu - Phe - Gly - Tyr

47. The amino acid sequence of the nonapeptide is:

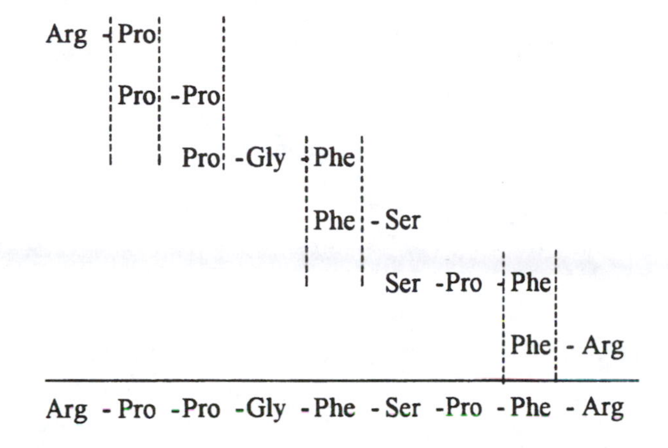


48. (a) Arginine will not migrate to either electrode in an electrolytic cell at a pH of 10.8.

(b) Arginine will migrate towards the positive electrode at pH greater than 10.8, that is, more basic than its isoelectric point.

49. Yes, all proteins have a primary structure (a sequence of amino acids linked by peptide bonds) but need not form regular, three dimensional secondary structures such as the $\alpha$-helix or $\beta$-pleated sheet.

No, all proteins must have a primary structure because, by definition, these molecules are polymers composed of a sequence of amino acids linked by peptide bonds (a primary structure).

50. A small protein like ribonuclease (or myoglobin) would have a small number of protein domains, probably only one. Small proteins (e.g., myoglobin, ribonuclease) commonly fold into one globular unit (one domain) while larger proteins will fold into more than one globular unit (more than one domain).

51. The steroisomers of threonine:

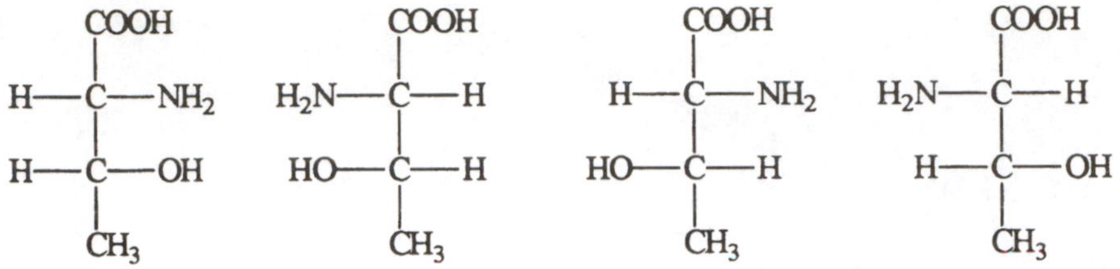

$$\begin{array}{ccc} & COOH & \\ & | & \\ H- & C & -NH_2 \\ & | & \\ H- & C & -OH \\ & | & \\ & CH_3 & \end{array} \qquad \begin{array}{ccc} & COOH & \\ & | & \\ H_2N- & C & -H \\ & | & \\ HO- & C & -H \\ & | & \\ & CH_3 & \end{array} \qquad \begin{array}{ccc} & COOH & \\ & | & \\ H- & C & -NH_2 \\ & | & \\ HO- & C & -H \\ & | & \\ & CH_3 & \end{array} \qquad \begin{array}{ccc} & COOH & \\ & | & \\ H_2N- & C & -H \\ & | & \\ H- & C & -OH \\ & | & \\ & CH_3 & \end{array}$$

52. The immunoglobulin hypervariable regions allow the body to produce millions of different immunoglobulins, each with two distinct amino acid sequences and unique antigen binding site.

53. (a) The structure of alanine at pH = 9.0 would be $NH_2\underset{\phantom{|}}{\overset{\overset{CH_3}{|}}{C}}HCOO^-$

(b) The structure of lysine at pH = 9.0 would be $\overset{+}{N}H_3CH_2CH_2CH_2CH_2\underset{\underset{NH_3^+}{|}}{C}H\,COO^-$

(c) The net charge on lysine at pH = 9.0 would be positive [see the structure of lysine in part (b)].

54. Nineteen dipeptides can be written with glycine on the N-terminal side. Another nineteen are possible with glycine on the C-terminal end. Finally, one dipeptide can be written with two glycines giving a total of thirty-nine dipeptides.

55. Vasopressin will have a higher isoelectric point than oxytocin. Vasopressin has two different amino acids as compared with oxytocin, a phenylalanine instead of an isoleucine and an arginine instead of a leucine. Thus, vasopressin has one additional basic amino acid (arginine) which will cause the vasopressin isoelectric point to be higher than the oxytocin isoelectric point.

56.

| Leucine | Alanine | Glutamic Acid |
|---|---|---|
| $(CH_3)_2CHCH_2\underset{\underset{NH_2}{\vert}}{C}HCOOH$ | $CH_3\underset{\underset{NH_2}{\vert}}{C}HCOOH$ | $HOOCCH_2CH_2\underset{\underset{NH_2}{\vert}}{C}HCOOH$ |

Glutamic acid is the only one of these three amino acids with polar bonds in its side chain. Thus, glutamic acid will be the most polar.

CHAPTER 30

# ENZYMES

1. Activation energy is the energy barrier to chemical reaction and is measured as the difference between the reactant(s) energy level and the transition state energy level. Enzymes lower the activation energy barrier and, thus, increase biochemical reaction rates.

2. Enzymes are proteins. As noted in the answer to question 1, enzymes serve as catalysts for chemical reactions in the body.

3. A coenzyme is the nonprotein part of a conjugated enzyme. An apoenzyme is the protein part of a conjugated enzyme.

4. Sucrase catalyzes the hydrolysis of sucrose; lactase catalyzes the hydrolysis of lactose; maltase catalyzes the hydrolysis of maltose.

5. The six general classes of enzymes are: (a) oxidoreductases, (b) transferases, (c) hydrolases, (d), lyases, (e) isomerases, and (f) ligases.

6. The reaction rate often can be increased (1) by increasing the reactant (substrate) concentration and (2) by raising the reaction temperature.

7. The lock-and-key hypothesis states that an enzyme active site is rigidly shaped to fit the substrate molecule(s) (as a lock fits a key). The induced-fit model envisions an active site which flexes to make a best fit with the substrate(s).

8. When substrates bind to an active site, they may be converted more easily to products because (1) the substrates are close together (proximity catalysis); (2) the substrates are oriented to best react (productive binding hypothesis); (3) when the substrates bind to the active site, they change shape to be more like products (strain hypothesis).

9. Both an enzyme substrate and an enzyme inhibitor may bind to the active site. However, a substrate will react to form product while an inhibitor will only block the active site from further reaction.

10. Pectinase is used to remove the peeling from oranges and grapefruit. The pectinase penetrates the peeling and dissolves the white stringy material that attaches the peeling to the fruit. The peeling can then be easily removed from the fruit.

11. reaction rate $= \left(\frac{0.005\text{ M}}{3.5\text{ min}}\right)\left(\frac{1\text{ min}}{60\text{ sec}}\right) = 2 \times 10^{-5}\text{ M/s}$

12. reaction rate $= \left(\frac{0.02\text{ M}}{8\text{ min}}\right)\left(\frac{1\text{ min}}{60\text{ s}}\right) = 4 \times 10^{-5}$ M/s

13. The turnover number for lysozyme shows that this enzyme converts 0.5 reactant to products every second. Thus, in 1 min,

    (0.5 reactant converted to products/s)$\left(\frac{60\text{ s}}{\text{min}}\right)$(1 min) = 30 reactants converted to products

14. The turnover number for pepsin shows that this enzyme converts 1.2 reactants to products every second. Thus, in 5 min,

    (1.2 reactants converted to products/s) $\left(\frac{60\text{ s}}{\text{min}}\right)$(5 min)(3 pepsin molecules) $= 1 \times 10^3$ reactants converted to products by three pepsin molecules

15. The lock-and-key hypothesis predicts that an enzyme active site should complement the shape of the substrate. Thus, the first enzyme would have a narrow, extended active site to fit *n*-butyl alcohol ($CH_3CH_2CH_2CH_2OH$). For the substrate, 2-methyl-2-propanol

```
      CH3
       |
CH3–C–OH
       |
      CH3
```

    the enzyme active site would need to be almost as wide as it is long.

16. The lock-and-key hypothesis predicts that an enzyme active site should complement the shape of the substrate. The first enzyme would have a narrow active site which is long enough to fit the four carbon carboxylic acid, butanoic acid ($CH_3CH_2CH_2COOH$). The second enzyme would have a narrow active site which would be shorter because it need only fit the two carbon carboxylic acid, acetic acid ($CH_3COOH$).

17. At low temperatures, fewer substrate molecules (reactants) have the energy necessary to overcome the activation energy barrier. Thus, an enzyme-catalyzed reaction rate decreases at low temperatures.

18. An enzyme-catalyzed rate decreases at high temperatures because the enzyme loses its natural shape (denatures). As an enzyme denatures it ceases to be an effective catalyst.

19. If the $V_{max}$ for an enzyme is decreased, the enzyme has lost some of its catalytic efficiency. Any process that decreases an enzyme's catalytic abilities is termed inhibition.

20. If the turnover number for an enzyme is increased, the enzyme can convert more reactants to products per unit time; the enzyme is a better catalyst. Any process that increases an enzyme's catalytic abilities is termed activation.

21. No. Enzyme A is more effective than enzyme B based on a comparison of turnover numbers (225/s for enzyme A vs. 120/s for enzyme B).

22. Yes. Enzyme B is more effective than enzyme A based on a comparison of turnover numbers ($9.8 \times 10^{-1}$/s for enzyme B vs. 0.05/s for enzyme A).

23. Glutamine is the product of this liver metabolic pathway. Thus, glutamine control of this pathway must be "feed back". Since an increase in glutamine concentration causes the metabolic pathway to slow down, this control must be feedback inhibition. Feedback inhibition means that when a large amount of product has been formed, the beginning of a process will slow down.

24. Glutamine is a starting material (reactant) for the liver metabolic pathway which produces urea. Thus, glutamine control of this pathway must be "feedforward". Since an increase in glutamine concentration causes the metabolic pathway to speed up, this control must be feedforward activation.

25. Cellulases are used to digest cellulose containing materials. Paper producers use cellulases to complete the breakdown of wood chips to wood pulp, which is used to manufacture paper products.

26. Blood clots in the brain are the major cause of strokes. Specific proteases are used to dissolve these blood clots to alleviate the cause of a stroke.

27. The turnover number measures the number of substrate molecules converted to product by one enzyme molecule under optimum conditions. Chymotrypsin can convert one glycine-containing substrate to product every twenty seconds while the enzyme can convert two hundred L-tyrosine containing substrates to products each second. Chymotrypsin is a much more efficient catalyst for the L-tyrosine-containing substrate.

28. Feedback inhibition means that when a large amount of product has been formed, the beginning of a process will slow down. Thus, feedback inhibition protects against over-production. "Feedback activation" means that when a large amount of product is formed, the beginning of the process will accelerate. A state of overproduction will disrupt the normal activity of the cell due to overproduction of products and might lead to cell death.

29.

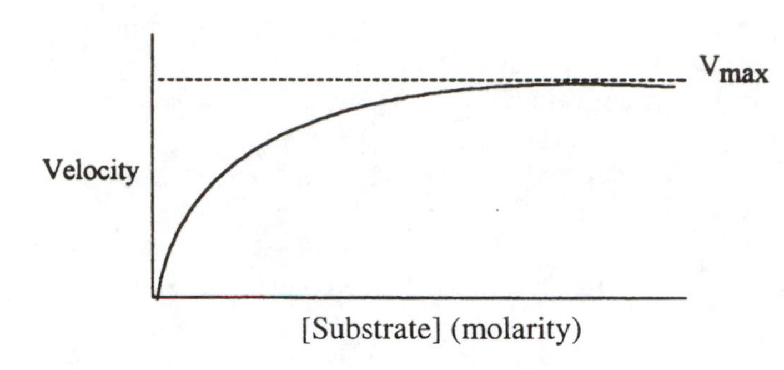


When the $V_{max}$ is reached, an increase in substrate concentration will not increase the velocity.

30. The enzyme lactase catalyzes hydrolysis of the disaccharide lactose to produce two monosaccharides, D-glucose and D-galactose.

CHAPTER 31

# NUCLEIC ACIDS AND HEREDITY

1. The five nitrogen bases found in nucleotides:

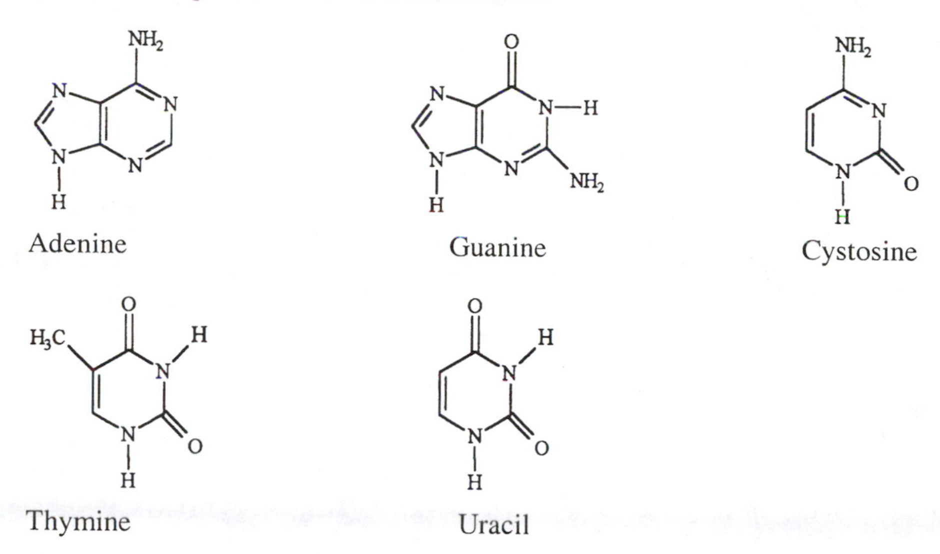


2. A nucleoside is a purine or pyrimidine base linked to a sugar molecule, usually ribose or deoxyribose. A nucleotide is a purine or pyrimidine base linked to a ribose or deoxyribose sugar which in turn is linked to a phosphate group.

3. There are three structural differences between DNA and RNA.

   (a) In RNA the sugar molecule is always ribose. In DNA, the sugar molecule is always deoxyribose, which has H instead of OH at carbon number two.

   (b) Both molecules use a mixture of four nitrogen bases. Both use cytosine, adenine, and guanine. In DNA, the fourth base is thymine. In RNA, the fourth base is uracil.

   (c) DNA exists as a double helix whereas RNA is a single strand of nucleotides.

4. The major function of ATP in the body is to store chemical energy, and to release it when called upon to carry out many of the complex reactions that are essential to most of our life processes. An equilibrium exists with ADP.

$$ATP + H_2O \underset{\text{energy storage}}{\overset{\text{energy utilization}}{\rightleftharpoons}} ADP + P_i + 35\text{ kJ}$$

5. The structure of DNA as proposed by Watson and Crick is a double-stranded helix. Each strand has a backbone of alternating phosphate and deoxyribose units. Each deoxyribose unit has one of the four nitrogen bases attached, but coming off the backbone, not part of the backbone. These nitrogen bases thus link to their complementary nitrogen base on the other strand of the double helix.

6. Complementary bases are the pairs that "fit" to hydrogen bond to each other between the two helixes of DNA. For DNA the complementary pairs are thymine with adenine, and cytosine with guanine, or T-A and C-G. For RNA it is U-A and C-G.

7. The genetic code is the sequence of nitrogen bases in a strand of DNA. The code determines the sequence of amino acids in protein molecules which will be assembled according to the code. It takes three nucleotides to make a codon, which determines one amino acid in a sequence.

8. Since there are only four different bases to make up the code, one nucleotide could only specify four possible amino acids; two nucleotides could specify 16 amino acids; 3 nucleotides could specify 64 amino acids. Since there are at least 20 amino acids needed, three nucleotides are required.

9. A brief outline of the biosynthesis of proteins:

   (a) A DNA strand produces a complementary mRNA strand which leaves the nucleus and travels to the cytoplasm where it becomes associated with a cluster of ribosomes, binding to five or more ribosomes.

   (b) With the aid of an enzyme, the proper amino acid attaches to a tRNA molecule by an ester linkage.

   (c) The amino acids are brought to the protein synthesis site by tRNA.

   (d) The initiatian of a polypeptide chain always uses the mRNA codon AUG or GUG, which ties to the tRNA anticodon UAC. This code brings N-formylmethionine for procaryotic cells. The amino group is blocked by the formyl group, leaving the carboxyl group available to react with the amino group of the next amino acid.

   (e) The next tRNA, bringing an amino acid, comes in to the mRNA and links up, anticodon to codon. The peptide linkage is then made between amino acids.

   (f) The first tRNA is ejected, and a third enters the ribosome.

   (g) The polypeptide chain terminates at a nonsense or termination-codon. The protein molecule breaks free.

10. A codon is a triplet of three nucleotides, and each codon specifies one amino acid. The cloverleaf model of transfer RNA has an anticodon loop consisting of seven unpaired nucleotides. Three of these make up the anticodon, which is complementary to, and hydrogen-bonds to the codon on mRNA.

11. The role of N-formylmethionine in procaryotic protein synthesis is to start the polypeptide chain so it goes in the right direction. It can only build from the carboxyl end. After the synthesis, the N-formylmethionine breaks loose from the protein.

12. From time to time a new trait appears in an individual that is not present in either parents or ancestors. These traits which are generally the result of genetic or chromosomal changes are called mutations.

13. A “DNA fingerprint” is a pattern of tagged DNA fragments on an electrophoretic gel which can be used to identify possible suspects.

14. With the knowledge of gene structure genetic therapists can observe DNA that has genetic defects and work towards deleting the error and replace the defect with a corrected gene. Geneticists will be able to tell in advance who is prone to getting certain diseases. They will be able to more accurately detect and predict birth defects, and in general have a better understanding of inherited diseases.

15. The letters are associated with compound names as follows:

    (a) A, adenosine
    (b) AMP, adenosine-5'-monophosphate
    (c) dADP, deoxyadenosine-5'-diphosphate
    (d) UTP, uridine-5'-triphosphate

16. The letters are associated with compound names as follows:

    (a) G, guanosine
    (b) GMP, guanosine-5'-monophosphate
    (c) dGDP, deoxyguanosine-5'-diphosphate
    (d) CTP, cytidine-5'-triphosphate

17. Structural formulas

(a) A

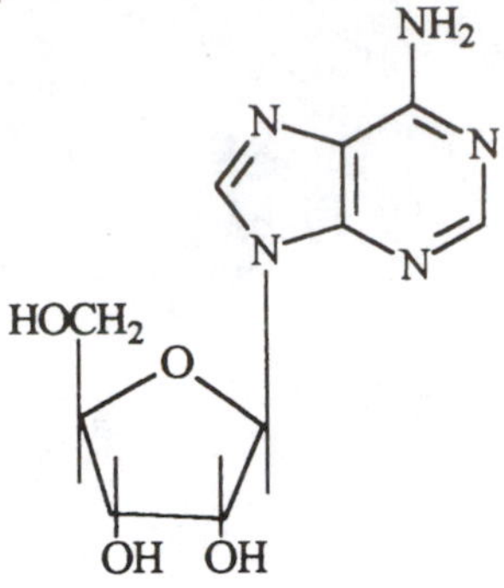

(b) AMP

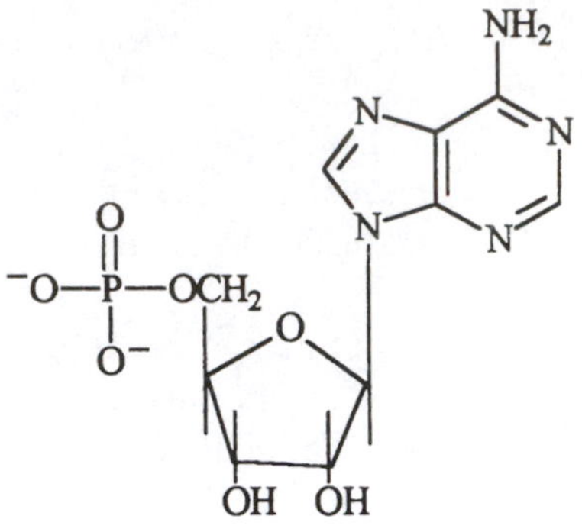

(c) CDP

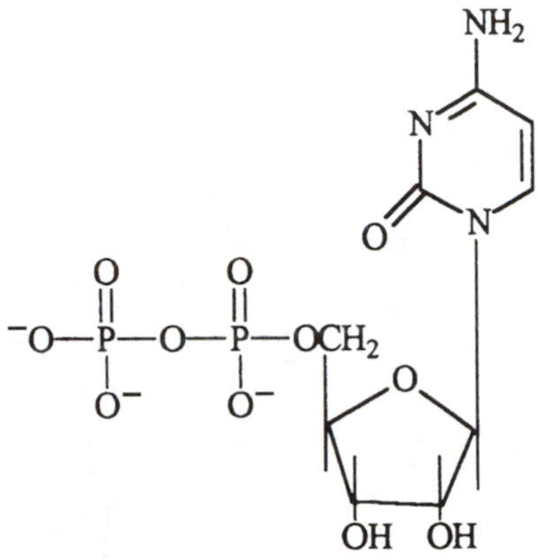

(d) dGMP

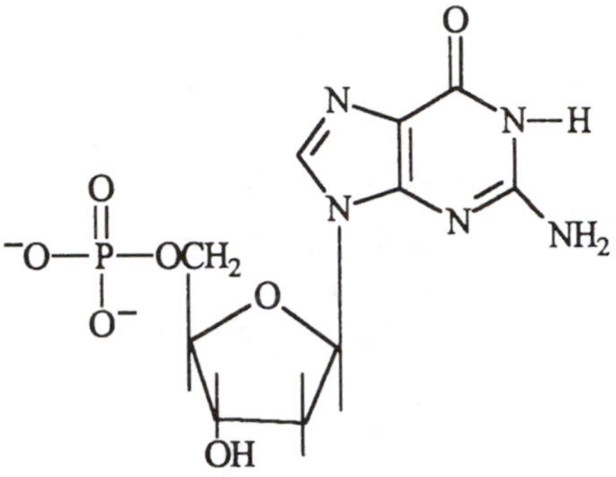

18. Structural formulas

(a) U

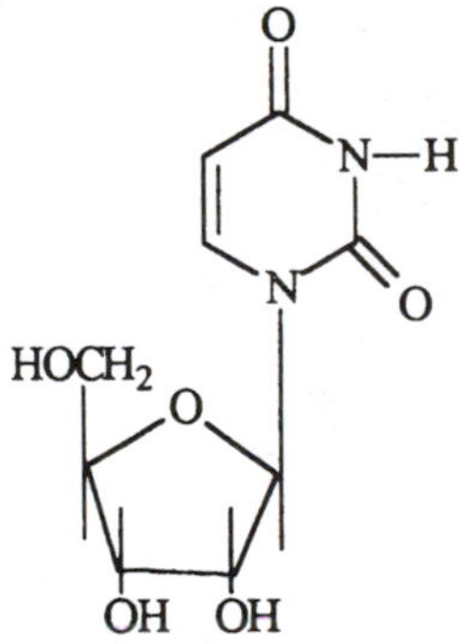

(b) UMP

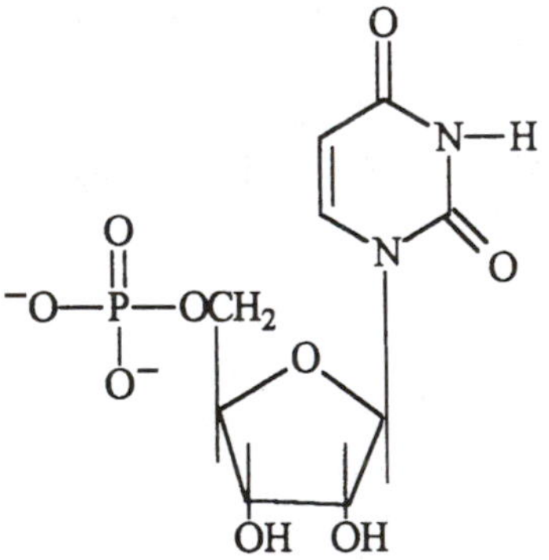

(c) CTP

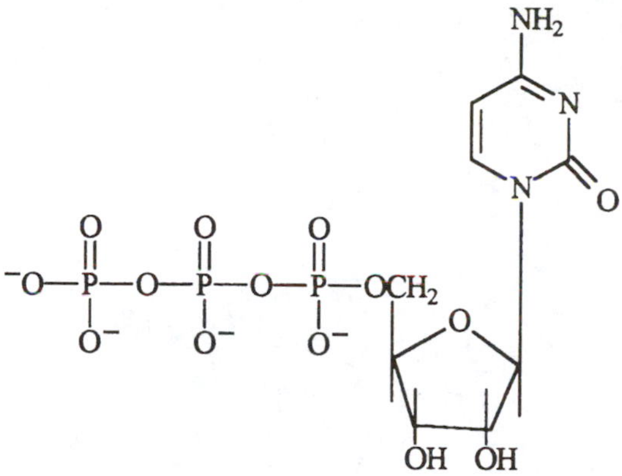

(d) dTMP

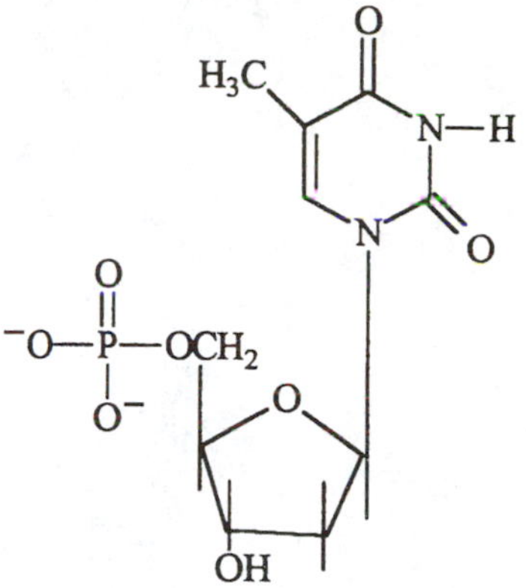

19. There are several sequences possible for the three-nucleotide, single-stranded DNA. One possible structure follows:

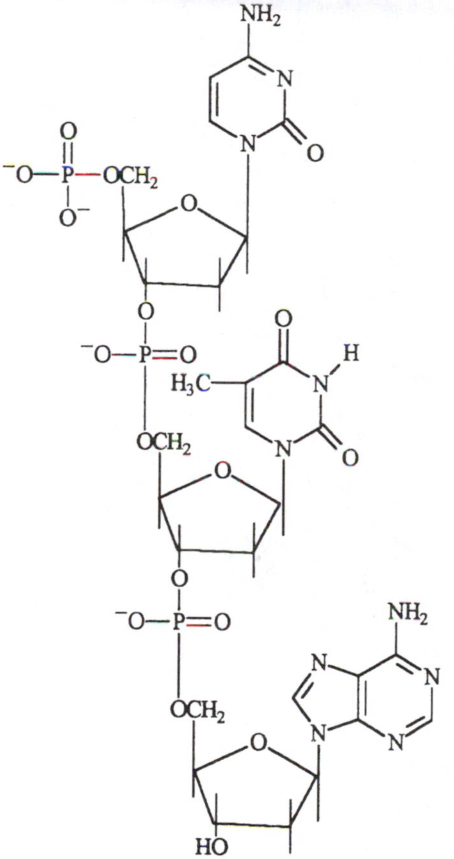

20. There are several possible sequences for the three-nucleotide, single-stranded RNA. One possible structure follows:

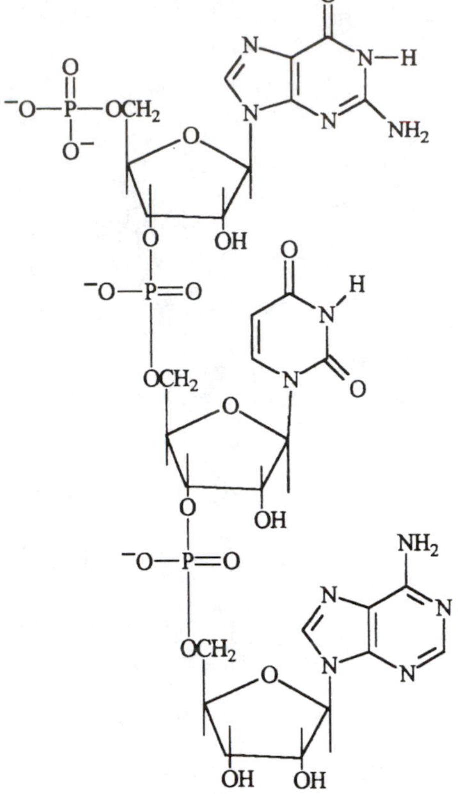

21. The hydrogen bonding between adenine and uracil:

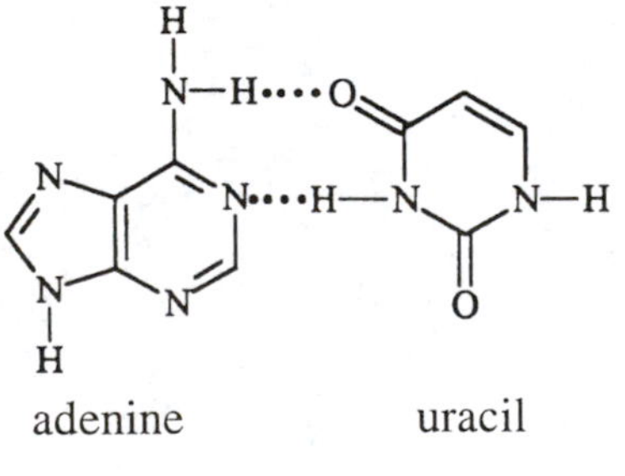

(dotted lines—hydrogen bonds)

adenine uracil

22. The hydrogen bonding between guanine and cytosine:

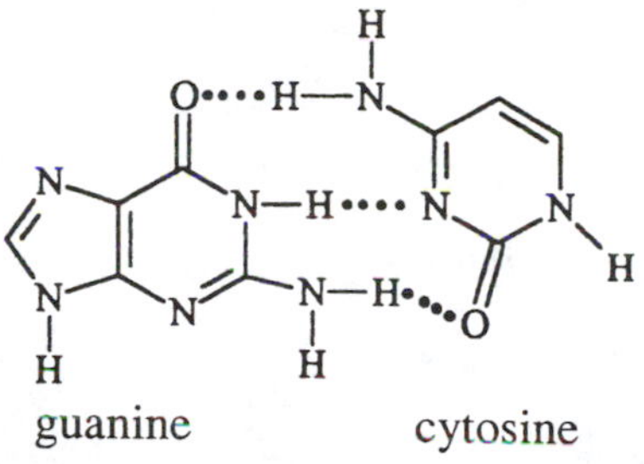

guanine cytosine

(dotted lines—hydrogen bonds)

23. RNA is responsible for moving genetic information to where it can be used for protein synthesis. There are three kinds of RNA:

    (a) Ribosomal RNA (rRNA) is found in the ribosomes where it is associated with protein roughly in the proportion 60-65% protein, 30-35% rRNA.

    (b) Messenger RNA (mRNA) carries genetic information from DNA to the ribosomes. It is a template made from DNA and carries the codons that direct the synthesis of proteins.

    (c) Transfer RNA (tRNA) is used to bring amino acids to the ribosomes for incorporation into protein molecules.

24. DNA is considered to be the genetic substance of life, because it contains the sequence of bases that carries the code for genetic characteristics.

25. Replication is the biological process of making DNA using a DNA template while the transcription process refers to the making of RNA using a DNA template.

26. Transcription is the process of making RNA using a DNA template while translation is a process for making protein using a mRNA template.

27. tRNA binds specific amino acids and brings them to the ribosome for protein synthesis. mRNA carries genetic information from DNA to the ribosome and serves as a template for protein synthesis.

28. Both mRNA and rRNA can be found in the ribosomes. rRNA serves as part of the ribosome structure while mRNA serves as a template for protein synthesis.

29. Three nucleotides are required to specify one amino acid. If there are 146 amino acid residues in the beta chain of hemoglobin, then the number of required nucleotides in mRNA is $3 \times 146 = 438$.

30. Three nucleotides are required to specify one amino acid. For the 573 amino acid residues in the phosphoglycerate kinase enzyme, the number of required nucleotides in mRNA is $3 \times 573 = 1719$.

31. For a DNA sequence, TCAATACCCGCG,

    (a) the complementary mRNA will be: AGUUAUGGGCGC.

    (b) the anticodon order in tRNA will be: UCAAUACCCGCG.

    (c) the sequence of amino acids coded by the DNA will be: Ser-Tyr-Gly-Arg

32. A segment of DNA strand consists of GCTTAGACCTGA.

    (a) The order in the complementary mRNA will be: CGAAUCUGGACU

    (b) The anticodon order in tRNA will be: GCUUAGACCUGA.

    (c) The sequence of amino acids coded by the DNA is Arg-Ile-Trp-Thr.

33. Transcription makes a polymer of nucleotides by forming phosphate ester bonds to connect the nucleotides to each other. The phosphate ester combines a phosphoric acid with an alcohol.

34. Translation makes a polymer of amino acids by forming amide bonds to connect the amino acids to each other. The amide bond combines an amine with a carboxylic acid.

35. Translation termination occurs when the ribosome reaches a "nonsense" or termination codon along the mRNA. No tRNA (in normal cells) has the anticodon to match the termination codon and so no more amino acids are added to the newly synthesized protein chain. The peptidyl-tRNA connection is broken and the protein chain is released from the ribosome.

36. Translation initiation occurs when the ribosome reaches a special AUG or GUG codon along the mRNA. Since there is commonly more than one AUG or GUG codon, the ribosome must use other information to choose the special AUG or GUG. This codon is the starting point for protein synthesis and is bound by either a special tRNA carrying N-formylmethionine (in procaryotes) or a tRNA carrying methionine (in eucaryotes).

37. mRNA codons and corresponding tRNA anticodons follow:

    (a) GUC: CAG
    (b) AGG: UCC
    (c) UUU: AAA
    (d) CCA: GGU

38. mRNA codons and corresponding tRNA anticodons follow:

(a) CGC: GCG
(b) ACA: UGU
(c) GAU: CUA
(d) UUC: AAG

39. Thymine and adenine are complementary bases in DNA; therefore each time one appears, its complement appears. There is no fixed relationship between the amount of thymine and the amount of guanine, since they are not complementary. Table 32.2 shows that the one set of bases is often far different from the other set, varying by as much as a 3:2 ratio. The complementary bases "fit" with each other with respect to H-bonding between them; also with respect to the size of the molecules.

40. In RNA the guanine content does not have to be equal to the cytosine content, because RNA is a single strand. Its complements are on the DNA template, not on the RNA. They have to be equal in DNA, because it is a double helix, with each nucleotide hydrogen-bonded to its complement.

41. A substitution mutation will change one codon. An insertion mutation will shift the sequence position of all bases following the insertion by one. Each base will take the position of its neighbor. For example, if base [C] is inserted into the sequence ... UUC ACG GCC..., every codon will change, ... U[C]U CAC GGC C... In general, every codon following an insertion is changed.

42. For the mRNA segment, UUUCAUAAG,

(a) the coded amino acids are: Phe-His-Lys.

(b) the sequence of DNA for this mRNA is: AAAGTATTC.

43. In DNA fingerprinting, DNA polymerase is used to copy very small samples of DNA so that enough will be available to separate and visualize on gel electrophoresis.

44. The Human Genome Project is a cooperative effort by leading scientific laboratories to sequence the entire human genome (approximately 3 billion base pairs).

CHAPTER 32

# NUTRITION

1. Nutrients are components of food that provide for body growth, maintenance, and repair. Food is the material we eat.

2. The "energy allowance" represents the recommended dietary energy supply needed to maintain health.

3. Marasmus is a chronic calorie deficiency while kwashiorkor represents a protein deficiency condition.

4. Candy contains primarily sucrose, a source of calories, but little or no other nutrients. Thus, candy is said to provide "empty calories."

5. The essential fatty acids have 18 or more carbon atoms per molecule and are polyunsaturated.

6. Essential fatty acids are required in the diet for normal growth and development. On a fat-free diet many deleterious physiological changes occur, including skin lesions, kidney damage, poor growth, and impaired fertility.

7. There are eight amino acids that are essential for humans. They are isoleucine, leucine, lysine, methionine, phenylalanine, threonine, tryptophan, and valine.

8. Animal proteins generally are a source of all twenty common amino acids while vegetable proteins may be deficient in one or several of the amino acids.

9. The water-soluble vitamins are C, the B vitamins, biotin and folic acid. The fat-soluble vitamins include A, D, E and K.

10. Vitamin D: functions as a regulator of calcium metabolism; Vitamin K: enables blood clotting to occur normally; Vitamin A: functions to furnish the pigment that makes vision possible. Many vitamins act as coenzymes.

11. Major elements are needed in relatively large quantities by the body while only small amounts of the trace elements are required.

12. Calcium is a major constituent of bones and teeth and is also important in nerve transmission.

13. Solid foods provide about 30-40% of the average daily water consumption while liquids yield about another 50% of the total.

14. Some common categories of food additives are: (1) nutrients; (2) preservatives; (3) anticaking agents; (4) emulsifiers; (5) thickeners; (6) flavor enhancers; (7) colors; (8) nonsugar sweeteners; (9) antimicrobials.

15. The five principal digestive juices are: saliva, gastric juice, pancreatic juice, bile and intestinal juice.

16. The U. S. Food and Drug Administration is responsible for regulating the use of food additives.

17. Enzymes present in the digestive juices are: **saliva:** salivary amylase; **gastric juice:** pepsin, and gastric lipase; **pancreatic juice,** trypsinogen, chymotrypsinogen, procarboxypeptidase, amylopsin, steapsin; **bile:** no enzymes; **intestinal juice:** sucrase, maltase, lactase, aminopeptidase, dipeptidase, nucleases, phosphatase, and intestinal lipase.

18. Chyme is the material of liquid consistency found in the stomach consisting of food particles reduced to small size resulting from gastric digestion of food.

19. The digestive function of the liver is the production of bile, one of the important digestive juices. The bile is stored in the gall bladder.

20. Intestinal mucosal cells produce many enzymes needed to complete digestion, e.g., disaccharidases, aminopeptidases, and dipeptidases.

21. The liver takes up foreign compounds which are then oxidized by enzymes on the endoplasmic reticulum.

22. GRAS means "generally recognized as safe."

23. (a) (23 g protein) + (0 g carbohydrates) + (19 g fat) = 42 g

    (b) (52 mg sodium) + (297 mg potassium) + (23 mg magnesium) + (2.5 mg iron) + (4.7 mg zinc) + (9.0 mg calcium) = 388 mg

    (c) (0.015 mg Vitamin A) + (0.092 mg thiamin) + (0.218 mg riboflavin) + (3.2 mg niacin) + (0.330 mg Vitamin $B_6$) + (0.007 mg folic acid) + (0.002 mg Vitamin $B_{12}$) = 3.9 mg

24. (a) (70 g protein) + (25 g carbohydrate) + (37 g fat) = 132 g

(b) (770 mg sodium) + (564 mg potassium) + (68 mg magnesium) + (3.5 mg iron) + (2.7 mg zinc) + (56 mg calcium) = 1464 mg

(c) (0.0565 mg Vitamin A) + (0.322 mg thiamin) + (0.408 mg riboflavin) + (29.5 mg niacin) + (1.2 mg Vitamin $B_6$) + (0.016 mg folic acid) + (0.00082 mg Vitamin $B_{12}$) = 31.5 mg

25. (a) (19 g fat) (9 kcal/g fat) = $2 \times 10^2$ kcal

(b) $\left[\frac{(2\times 10^2 \text{ kcal})}{(270 \text{ kcal})}\right](100) = 70\%$

26. (a) (37 g fat) (9 kcal/g fat) = $3 \times 10^2$ kcal

(b) $\left[\frac{(3\times 10^2 \text{ kcal})}{(728 \text{ kcal})}\right](100) = 40\%$

27. $\left[\frac{(9.0 \text{ mg})\left(\frac{1 \text{ g}}{1000 \text{ mg}}\right)}{1.3 \text{ g}}\right](100) = 0.69\%$

28. $\left[\frac{(56 \text{ mg})\left(\frac{1 \text{ g}}{1000 \text{ mg}}\right)}{1.2 \text{ g}}\right](100) = 4.7\%$

29. $\left[\frac{(2\ \mu\text{g})}{(2.0\ \mu\text{g})}\right](100) = 100\%$

30. $\left[\frac{(0.82\ \mu\text{g})}{(2.0\ \mu\text{g})}\right](100) = 41\%$

31. The three macronutrient classes are proteins, carbohydrates and lipids (fats).

32. The two micronutrient classes are minerals and vitamins.

33. The three classes of nutrients which do not commonly supply energy for the cells are (a) vitamins, (b) minerals and (c) water.

34. The three classes of nutrients which commonly supply energy for the cells are proteins, carbohydrates and lipids.

35. Starch, as a complex carbohydrate, provides dietary carbohydrate in a form which is slowly digested, enabling the body to control distribution of this energy nutrient.

36. Cellulose is important as dietary fiber. Although it is not digested, cellulose absorbs water and provides dietary bulk which helps maintain a healthy digestive tract.

37. Saliva is slightly acidic with a pH below 7.

38. Pancreatic juice is slightly basic with a pH above 7.

39. Proteins are digested in the stomach and small intestine.

40. Carbohydrates are digested in the mouth and small intestine.

41. Pancreatic amylase digests carbohydrates.

42. Pepsin digests proteins.

43. Milk contains (a) lipids, (b) minerals (calcium), (c) water and (d) carbohydrates.

44. 9 kg of body fat represents about 80,000 kcal of energy. Complete starvation for about 30 days would cause a net loss of about 9 kg of fat. Thus, this new diet is unreasonable.

45. Both vitamins and trace elements are needed in only small amounts by the body and cannot be synthesized within the body's cells. Vitamins are organic molecules and are often modified by the body. Trace elements are inorganic and, in general, are not modified before use.

46. Fat (9 kcal/g) contains more than twice as much energy as compared with carbohydrate (4 kcal/g). Thus, a tablespoon of butter will greatly increase the calorie content of a medium size baked potato.

47. (a) $\left[\frac{\text{50 fat Cal per serving}}{\text{100 Calories per serving}}\right](100) = 50\%\text{ fat}$

(b) (3 g protein per serving) (2 servings) = 6 g proteins

(c) $\left[\frac{\text{100 Calories per serving}}{\text{1000 Calories per day}}\right](100) = 10\%$

(d) $\left[\frac{\text{2 g fiber per serving}}{\text{10 g total carbohydrate per serving}}\right](100) = 20\%\text{ fiber}$

48. The foods of higher animals must be digested before they can be utilized because foods are primarily large molecules. Foods must be broken down to much smaller molecules in order to pass through the intestinal walls into the blood and lymph systems where they can be utilized in the metabolic process of animals.

49. Galactose is a component of the carbohydrate lactose which is found in milk. Thus, milk must commonly be changed to eliminate lactose from a baby's diet.

50. Nutritionists recommend less than 300 mg of dietary cholesterol per day. One large egg (270 mg cholesterol) represents almost the total daily recommended amount (about 90% of the RDA).

CHAPTER 33

# BIOENERGETICS

1. Energy is delivered to plant cells from sunlight. Sunlight is converted to chemical energy in the plant. High energy phosphate bonds are used to do the work of the cells.

   Energy is delivered to animal cells in compounds as reduced carbon atoms. The energy in reduced carbon atoms is converted to high-energy phosphate bonds, which are used to do the work of the cells.

2. Mitochondrial electron transport and oxidative phosphorylation are found in both plants and animals. Electrons are moved from NADH and $FADH_2$ to molecular oxygen to release energy. This energy allows ATP to be formed.

3. Fats and carbohydrates are good sources of cellular energy because they contain many reduced carbon atoms.

4. The most common high energy phosphate bond in the cell is the phosphate anhydride bond.

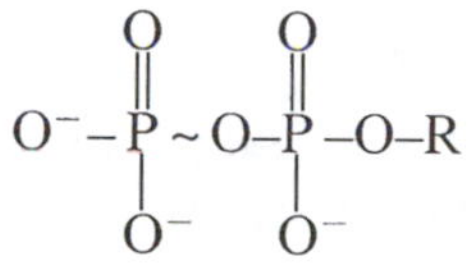

$$O^- - \overset{\overset{\displaystyle O}{\|}}{\underset{\underset{\displaystyle O^-}{|}}{P}} \sim O - \overset{\overset{\displaystyle O}{\|}}{\underset{\underset{\displaystyle O^-}{|}}{P}} - O - R$$

   It is generally associated with adenosine triphosphate (ATP).

5. ATP is known as the "common energy currency" of the cell because energy from many different catabolic processes is stored in this molecule. In turn, most anabolic pathways draw energy from this common pool.

6. An oxidation-reduction coenzyme is a reusable organic compound which helps an enzyme carry out an oxidation-reduction reaction.

7. The nicotinamide ring of $NAD^+$ becomes reduced.

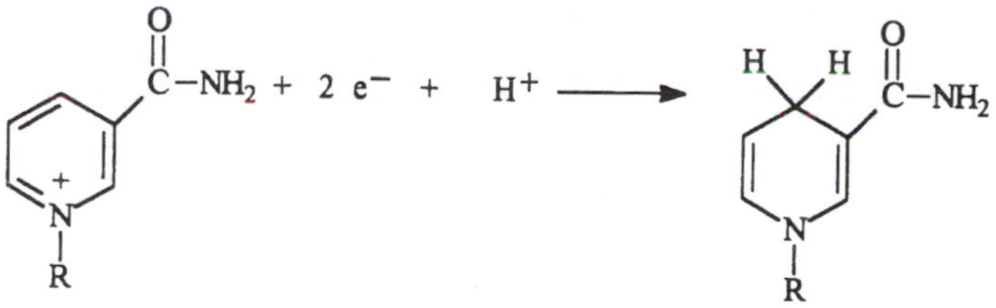

$$\text{(pyridinium ring, } N^+\text{–R, with } \overset{O}{\overset{\|}{C}}-NH_2\text{)} + 2\,e^- + H^+ \longrightarrow \text{(1,4-dihydropyridine ring, N–R, with H, H and } \overset{O}{\overset{\|}{C}}-NH_2\text{)}$$

8. Oxidative phosphorylation uses the mitochondrial electron transport system to directly produce ATP from oxidation-reduction reactions. Substrate-level phosphorylation does not use the electron transport system, but, instead involves transfer of a phosphate group from a substrate to ADP to form ATP.

9. Oxidative phosphorylation takes place in the mitochondria.

10. Both chloroplasts and mitochondria are organelles which are bounded by two membranes. In both cases they contain much folded internal membrane (which facilitates oxidation-reduction reactions).

11. Chloroplast pigments trap light to provide energy for photosynthesis.

12. The overall photosynthetic reaction in higher plants is as follows:

$$6\,CO_2 + 6\,H_2O + 2820\text{ kJ} \longrightarrow C_6H_{12}O_6 + 6\,CO_2$$

13. The oxidation state of carbon in $CH_3OH$ is $-2$. The oxidation state of carbon in $CO_2$ is $+4$. The carbon in $CH_3OH$ is in a lower energy state than in $CO_2$, and, therefore, will deliver more energy than $CO_2$ in redox reactions.

14. The average oxidation state of carbon in ethanal is $-2$. The average oxidation state of carbon in acetic acid is 0. The carbons in ethanal, being in a lower oxidation state, will deliver more energy in biological redox reactions.

15. The equation is unbalanced. An additional proton is always used when NADH is converted to $NAD^+$.

16. Both oxidized forms of the coenzymes ($NAD^+$ and FAD) are on the same side of the equation. For a reaction to occur one reduced coenzyme (NADH or $FADH_2$) must be paired with one oxidized coenzyme.

17. The conversion of glucose to carbon dioxide is catabolic because:

    (a) a larger molecule (glucose) is converted to a smaller molecule (carbon dioxide);

    (b) the carbons from glucose become oxidized as they are converted to carbon dioxide.

18. The conversion of acetate to long-chain fatty acids is anabolic because:

    (a) a smaller acetate is converted to a larger long-chain fatty acid;

(b) The carbons from acetate become reduced (on the average) as they are converted to long-chain fatty acids.

19. Procaryotic cells contain no chloroplasts.

20. Chloroplasts carry out photosynthesis, not oxidative phosphorylation.

21. The chemical changes in the mitochondria are said to be catabolic because:

(a) the mitochondrial processes convert many different molecules to smaller carbon dioxide molecules;

(b) during these chemical changes, the carbons become progressively more oxidized;

(c) as a result, ATP is produced.

22. The chemical changes in the chloroplasts are said to be anabolic because:

(a) smaller carbon dioxide molecules are converted to larger glucose molecules;

(b) as this transformation takes place, the carbons become progressively more reduced;

(c) this process requires an input of energy (from sunlight).

23. Three important characteristics of an anabolic process are: (1) simpler substances are built up into complex substances; (2) often carbons are reduced; (3) often cellular energy is consumed.

24. Three important charactcristics of a catabolic process are: (1) complex substances are broken down into simpler substances; (2) often carbons are oxidized; (3) often cellular energy is produced.

25. $2\,FADH_2 + O_2 \longrightarrow 2\,FAD + 2\,H_2O$

$$(2.38 \text{ mol } FADH_2)\left(\frac{1 \text{ mol } O_2}{2 \text{ mol } FADH_2}\right) = 1.19 \text{ mol } O_2$$

26. $2\,NADH + 2\,H^+ + O_2 \longrightarrow 2\,NAD^+ + 2\,H_2O$

$$(0.67 \text{ mol NADH})\left(\frac{1 \text{ mol } O_2}{2 \text{ mol NADH}}\right) = 0.34 \text{ mol } O_2$$

27. $NAD^+ + H^+ + 2\,e^- \longrightarrow NADH$

$$(11.75 \text{ mol } NAD^+)\left(\frac{2 \text{ mol } e^-}{1 \text{ mol } NAD^+}\right) = 23.50 \text{ mol } e^-$$

28. $FAD + 2H^{+} + 2e^{-} \longrightarrow FADH_2$
$(0.092 \text{ mol FAD})\left(\frac{2 \text{ mol e}^{-}}{1 \text{ mol FAD}}\right) = 0.18 \text{ mol e}^{-}$

29. The molecule shown contains two high energy phosphate anhydride bonds.

30. The molecule shown contains two high energy phosphate anhydride bonds.

31. $ADP + P_i + 35 \text{ kJ} \longrightarrow ATP + H_2O$
$(0.55 \text{ mol ADP})\left(\frac{35 \text{ kJ}}{\text{mol ADP}}\right) = 19 \text{ kJ}$

32. $ATP + H_2O \longrightarrow ADP + P_i + 35 \text{ kJ}$
$(1.65 \text{ mol ATP})\left(\frac{35 \text{ kJ}}{\text{mol ATP}}\right) = 58 \text{ kJ}$

33. Since the phosphorylated substrate phosphoenolpyruvate contains only one high energy phosphate bond, substrate-level phosphorylation can produce only one ATP.

34. Since the phosphorylated substrate 1,3-diphosphoglycerate contains only one high energy phosphate bond, substrate-level phosphorylation can produce only one ATP.

35. Mitochondrial electron transport and oxidative phosphorylation produce 3 ATP per NADH and 2 ATP per $FADH_2$. Thus,

$$(4 \text{ NADH})\left(\frac{3 \text{ ATP}}{\text{NADH}}\right) + (2 \text{ FADH}_2)\left(\frac{2 \text{ ATP}}{\text{FADH}_2}\right) = 16 \text{ ATP}$$

36. Mitochondrial electron transport and oxidative phosphorylation produce 3 ATP per NADH and 2 ATP per $FADH_2$. Thus,

$$(2 \text{ NADH})\left(\frac{3 \text{ ATP}}{\text{NADH}}\right) + (3 \text{ FADH}_2)\left(\frac{2 \text{ ATP}}{\text{FADH}_2}\right) = 12 \text{ ATP}$$

37. No, photosynthesis is an anabolic process by which carbon atoms are reduced. (Carbon dioxide is converted to glucose.)

38. The mitochondria carry out most of the oxidation-reduction reactions for the cell. Because these reactions involve electron movement, they are more easily controlled in a non-aqueous environment. Thus, mitochondria contain 90% membrane by mass.

39. No. Higher plant cells need mitochondria as well as chloroplasts in order to oxidize energy storage molecules to provide for the cellular energy needs.

40. Both ATP and $NAD^+$ contain a ribose which (1) is linked to an adenine at carbon one and (2) is linked to a phosphate at carbon five.

CHAPTER 34

# CARBOHYDRATE METABOLISM

1. Carbohydrates are considered energy storage molecules because they contain biologically-usable, reduced carbon atoms.

2. Glucose catabolism breaks down glucose to the smaller carbon dioxide molecule in an oxidative process. This produces energy for the cell and is catabolic. Photosynthesis produces glucose from carbon dioxide via a reductive path. Light energy is required and this is an anabolic process.

3. Each mole of glucose will yield 2820 kJ of energy when oxidized. Thus, 3 moles of glucose will provide 3 moles $\times$ 2820 kJ/mole or 8460 kJ.

4. Enzymes are needed so that biochemical reactions will proceed fast enough to keep the cell alive. Metabolism is often controlled by controlling enzyme activity.

5. Initial muscle contraction requires a readily and quickly available form of energy, ATP. Once ATP stores have been depleted, muscle cells turn to muscle glycogen as a source of chemical energy.

6. The liver replenishes muscle glucose (via the blood) by (a) releasing glucose from liver glycogen stores and (b) by converting lactic acid back into glucose.

7. The final reactions in anaerobic glucose metabolism convert NADH back to $NAD^+$. This coenzyme can then be reused to oxidize more glucose.

8. One high energy phosphate bond (in GTP) is formed directly in the citric acid cycle. In addition, one $FADH_2$ and three NADHs are produced.

9. Acetyl-CoA is an acetyl group bonded to a coenzyme A group. Coenzyme A contains adenine, ribose, diphosphate, pantothenic acid, and thioethanolamine. In the citric acid cycle, the acetyl group enters the cycle, leaving the coenzyme. In the process of the cycle, the acetyl group is oxidized to two molecules of carbon dioxide. In the process, energy is transferred, creating 11 ATP molecules and 1 GTP molecule. Acetyl-CoA serves to tie metbolism of carbohydrates, fats and certain amino acids to the citric acid cycle.

10. Hormones are chemical substances that act as control or regulatory agents in the body. Hormones are secreted by the endocrine, or ductless glands directly into the bloodstream and are transported to various parts of the body to exert specific control functions.

11. (a) The glucose concentration in blood under normal fasting conditions is about 70 to 90 mg/100 mL of blood.

    (b) The blood glucose concentrations considered as (1) hyperglycemic is 90 to 140 mg/100 mL of blood; and (2) hypoglycemic is 50 to 70 mg/100 mL of blood.

12. The renal threshold is the concentration of a substance in the blood above which the kidneys begin to excrete that substance.

13. Type II diabetes is not insulin dependent. It can often be controlled by exercise, diet, and drugs that induce insulin production. Type I diabetes is insulin dependent and requires an external source of insulin for control. Type I diabetes is a much more severe form of diabetes.

14. In gluconeogenesis glucose is synthesized from non-carbohydrate sources: lactate, amino acids and glycerol. These three substances are converted to Embden-Meyerhof pathway intermediates, which are then converted to glucose. Since the Embden-Meyerhof pathway is anaerobic, gluconeogenesis is considered an anaerobic process.

15. Glycogenolysis breaks down glycogen to form glucose.

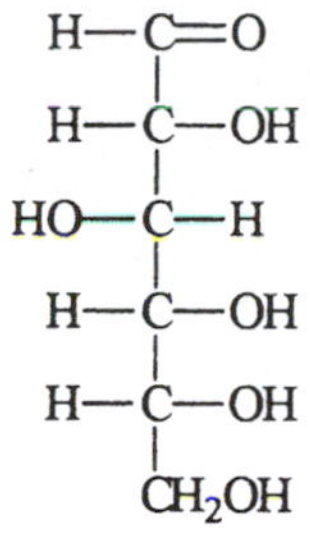

H—C=O
H—C—OH
HO—C—H
H—C—OH
H—C—OH
$CH_2OH$

D-glucose

16. Glycogenesis is the process by which glucose is used to produce glycogen.

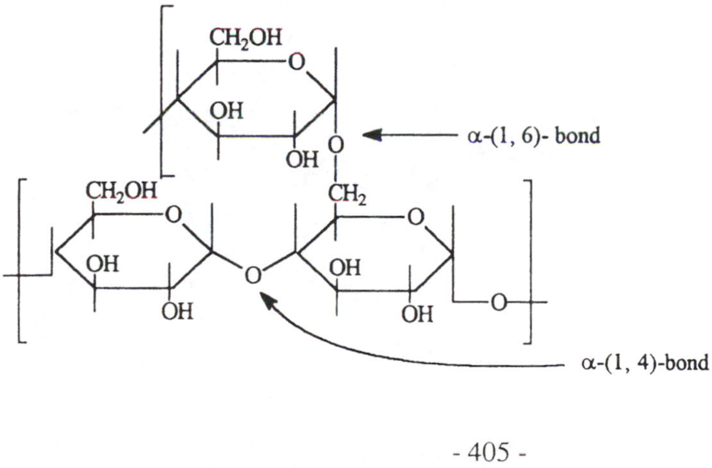

$CH_2OH$ O OH OH O

α-(1, 6)- bond

$CH_2OH$ O OH OH O $CH_2$ O OH OH O

α-(1, 4)-bond

17. Glycogenesis is anabolic because this process starts with smaller precursors (glucose molecules) and builds-up large products (glycogen molecules).

18. Glycogenolysis is catabolic because it starts with larger precursors (glycogen molecules) and forms smaller products (glucose molecules).

19. The ATP production which is associated with the citric acid cycle is oxidative phosphorylation. ATP is formed as $O_2$ is reduced during electron transport.

20. The Embden-Meyerhof pathway uses substrate-level phosphorylation to produce ATP. That is, following carbon oxidation a phosphate group is transferred from a substrate to ADP, forming ATP.

21.

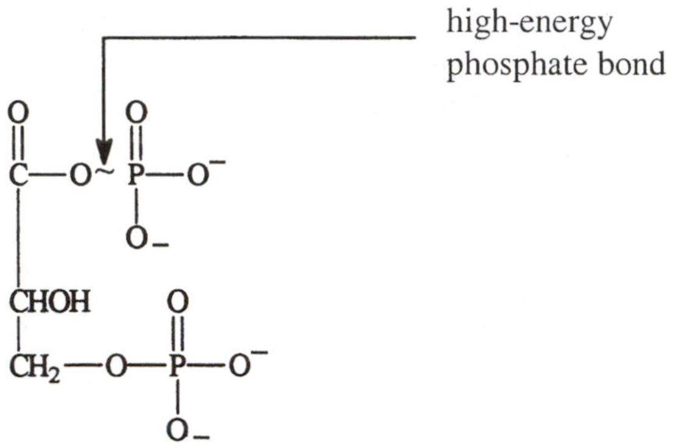

high-energy
phosphate bond

O O
C—O~P—$O^-$
$O^-$
CHOH O
$CH_2$—O—P—$O^-$
$O^-$

22.

O
C—$O^-$
high-energy
phosphate bond
O
C—O~P—$O^-$
$CH_2$ $O^-$

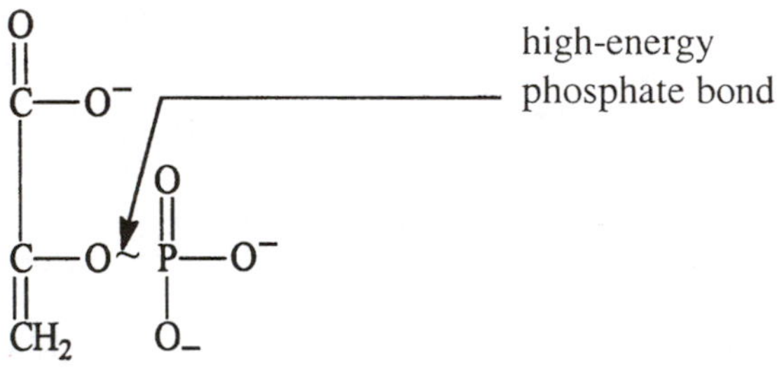

23. Although glucose is oxidized, molecular oxygen ($O_2$) is not used in the Embden-Meyerhof pathway. Thus, this pathway is anaerobic.

24. Although no molecular oxygen ($O_2$) is used in the citric acid cycle, reduced coenzymes (NADH and $FADH_2$) are produced. These reduced coenzymes must be oxidized by electron transport so that the citric acid cycle can continue. And, it is electron transport which uses molecular oxygen. Thus, the citric acid cycle depends on the presence of molecular oxygen and is considered to be aerobic.

25. Glycolysis produces 2 moles of lactate per mole of glucose. Therefore, 2.8 moles of glucose will yield 5.6 moles of lactate.

26. Glycolysis yields 2 moles of ATP per mole of glucose. Therefore, 0.85 moles of glucose will yield 1.7 moles of ATP.

27. The end products of the anaerobic catabolism of glucose in yeast cells are ethanol and carbon dioxide.

28. The end product of the anaerobic catabolism of glucose in muscle tissue is lactate.

29. The Embden-Meyerhof pathway is defined as catabolic because (a) glucose is broken down to smaller compounds, (b) the carbons of glucose are oxidized and, (c) this pathway causes production of ATP.

30. The gluconeogenesis pathway is considered to be anabolic because (a) smaller, non-carbohydrate precursors such as lactate are converted to the larger product, glucose and (b) as this conversion takes place, the carbons become progressively more reduced.

31. $$\begin{array}{l} CH_2COO^- \\ | \\ CH_2COO^- \end{array} + \underline{FAD} \longrightarrow \begin{array}{l} HC{-}COO^- \\ \parallel \\ CH{-}COO^- \end{array} + \underline{FADH_2}$$

32. $$\begin{array}{l} HO{-}CHCOO^- \\ \quad\;\; | \\ \quad\;\; CH_2COO^- \end{array} + \underline{NAD^+} \longrightarrow \begin{array}{l} O{=}C{-}COOH \\ \quad\;\; \| \\ \quad\;\; CH_2{-}COOH \end{array} + \underline{NADH} + H^+$$

33. Electron transport and oxidative phosphorylation are needed to convert the reduced coenzyme products of the citric acid cycle into oxidized coenzymes with ATP formation.

34. The glycolysis pathway uses substrate-level phosphorylation to produce ATP and, thus, does not need electron transport or oxidative phosphorylation.

35. Like vitamins, hormones are generally needed in only minute amounts.

36. Unlike vitamins which must be supplied in the diet, hormones are produced in the body.

37. The normal blood glucose concentration is 70-90 mg/100 mL of blood under fasting conditions. Hypoglycemia occurs when the blood glucose concentration drops below this range. Thus, 65 mg/100 mL of blood is hypoglycemic.

38. The normal blood glucose concentration is 70-90 mg/100 mL of blood under fasting conditions. Hyperglycemia occurs when the blood glucose concentration rises above this range. Thus, 350 mg/100 mL of blood is hyperglycemic.

39. The citric acid cycle depends on the presence of molecular oxygen. Large amounts of free oxygen were probably not available on earth until photosynthetic organisms had evolved. ($O_2$ is a product of photosynthesis.)

40. Hormones function as chemical messengers. They are chemical substances that act as control or regulatory agents in the body. Enzymes are catalysts for specific reactions, allowing these reactions to occur faster and under milder conditions than would otherwise be possible.

41. Insulin is not effective when given orally because it is a protein, and would be hydrolyzed to amino acids in the gastrointestinal tract.

42. If a large overdose of insulin was taken by accident, the blood glucose concentration would drop to a low level, probably in the hypoglycemic range. This could result in fainting, convulsions, and unconsciousness.

43. Epinephrine (adrenalin) is sometimes called the emergency or crisis hormone because it stimulates glycogenolysis, which raises the concentration of glucose in the blood, giving the body additional energy for an emergency or crisis situation.

44. $CH_3CH(OH)COO^-$

lactate

All three carbons of lactate are more reduced than the carbon in $CO_2$. Thus, lactate can provide more cellular energy upon further oxidation of the carbons.

45. A multireaction system or pathway is more efficient because (a) intermediate compounds may be used in more than one pathway and (b) energy is either produced or used more effectively when released or consumed in small increments. Energy produced in one large increment may possibly destroy the cell.

46. In aerobic catabolism, pyruvate is converted to carbon dioxide yielding many ATPs while consuming molecular oxygen. In anaerobic catabolism, pyruvate is used to recycle NADH back to $NAD^+$.

47. All these compounds contain an adenosine diphosphate group in their structure.

48. Oxidation. The average oxidation number of carbon in glucose is 0. The oxidation number of carbon in carbon dioxide is +4. Thus, the conversion of glucose to carbon dioxide is oxidation.

49. (a) Glycolysis: The anaerobic catabolic pathway for conversion of glucose to lactate.

(b) Gluconeogenesis: The metabolic pathway for the synthesis of glucose from non-carbohydrate sources.

(c) Glycogenesis: The synthesis of glycogen from glucose.

(d) Glycogenolysis: The hydrolysis, or breakdown of glycogen to glucose.

50. $$\underset{\text{glucose}}{C_6H_{12}O_6} + 2\,NAD^+ + 2\,ADP + 2\,P_i \longrightarrow \underset{\text{pyruvate}}{2\,CH_3\overset{\overset{O}{\|}}{C}{-}O^-} + 2\,NADH + 2\,ATP$$

51. (a) $CH_3\underset{\underset{OH}{|}}{C}HCOO^-$

lactate

(b) $CH_3\overset{\overset{O}{\|}}{C}COO^-$

pyruvate

(c)

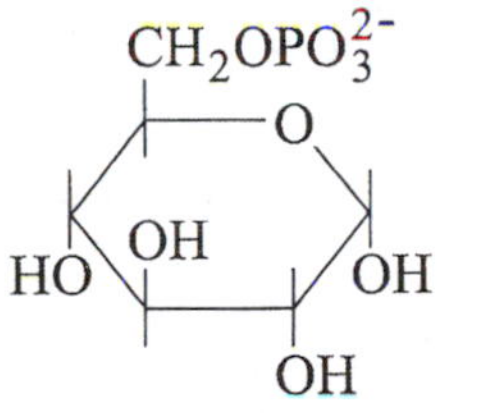

glucose-6-phosphate

(d)

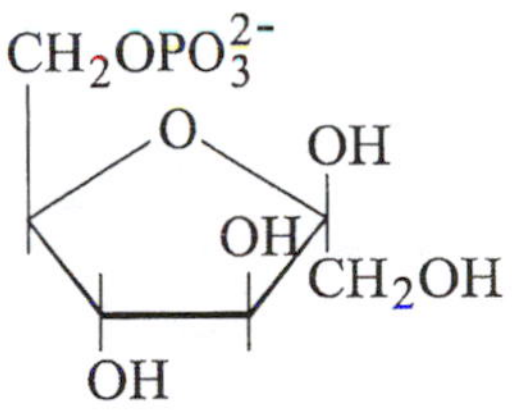

fructose-6-phosphate

(e)

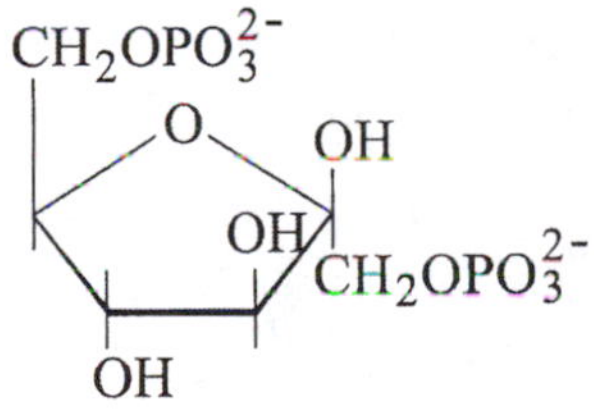

fructose-1,6-bisphosphate

(f)

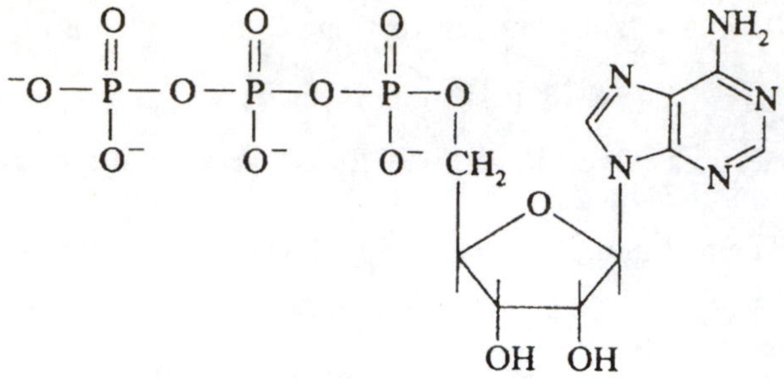

adenosine triphosphate
(ATP)

CHAPTER 35

# METABOLISM OF LIPIDS AND PROTEINS

1. A fatty acid contains many reduced carbons which can be oxidized to supply cellular energy.

2. Knoop prepared a homologous series of straight chain fatty acids with a phenyl group at one end and a carboxyl group at the other end. He fed each of these acids to test animals and analyzed the urine of the animals for metabolic products of the acid. Those animals which had been fed acids with an even number of carbons excreted phenylaceturic acid. Those animals which had been fed acids with an odd number of carbon atoms excreted hippuric acid. The experiment showed that the metabolism of the acid must occur by shortening the chain two carbon atoms at a time.

3. In the oxidation of fatty acids, the carbon atom beta to the carboxyl group is oxidized forming a $\beta$-keto acid which is then cleaved between the $\alpha$ and $\beta$ carbon atoms leaving a new fatty acid that is two carbons shorter than the original fatty acid.

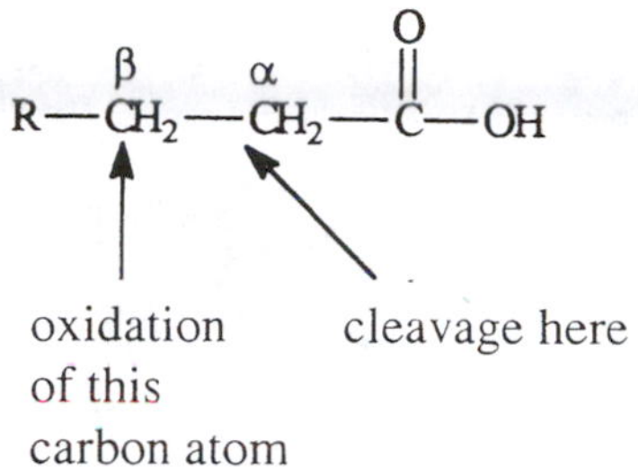

$$R-\overset{\beta}{CH_2}-\overset{\alpha}{CH_2}-\overset{\overset{\displaystyle O}{\|}}{C}-OH$$

oxidation of this carbon atom

cleavage here

4. Ketone bodies are relatively soluble in the blood stream and can be transported to energy-deficient cells in time of need. Ketone bodies supply energy via $\beta$-oxidation and the citric acid cycle.

5. Ketosis is the accelerated production of ketone bodies leading to high blood levels of these compounds with a corresponding increase in blood acidity.

6. In addition to being a food reserve, fat has two principal functions in the body. It acts as a cushion or shock absorber around internal organs, and it acts as an insulating blanket.

7. Fatty acid synthesis is not the reverse of $\beta$-oxidation. $\beta$-oxidation occurs in the mitochondria while fatty acids are synthesized in the cytoplasm. In fatty acid synthesis, the

growing molecule is linked to an acyl carrier protein while in $\beta$-oxidation the fatty acid being broken down is linked to a coenzyme A. Finally, fatty acid synthesis uses some different reactions and a different coenzyme ($NADP^+$) which are not found in $\beta$-oxidation.

8. The nitrogen cycle is the process by which nitrogen is circulated and recirculated from the atmosphere through living organisms and back to the atmosphere. Beginning with the atmosphere, nitrogen is fixed by bacterial action, combustion, or chemical fixation. In the soil nitrogen is converted to nitrates which are absorbed by higher plants and animals and converted to organic compounds. Eventually, the nitrogen is returned to the soil in the form of urea and feces or through bacterial decomposition of dead plants and animals. Some remains in the soil and the rest returns to the atmosphere as free nitrogen.

9. In the enzymatic oxidation of fatty acids from fats, the fatty acids are oxidized, ultimately forming acetyl-CoA. Acetyl-CoA enters the citric acid cycle to produce energy.

10. Soybeans enrich the soil because their roots serve as host to the symbiotic, nitrogen-fixing bacteria.

11. L-glutamic acid is involved in the majority of transaminations and is, thus, a central compound in nitrogen transfer. L-glutamic acid is also the major nitrogen source for the urea cycle.

12. The possible metabolic fates of amino acids are:

    (a) incorporation into a protein;

    (b) utilization in the synthesis of other nitrogenous compounds such as nucleic acids;

    (c) deamination to a keto acid which can be (1) utilized to synthesize other compounds, or (2) oxidized to carbon dioxide and water;

    (d) Ultimately, excess nitrogen compounds in the body are converted to uric acid (reptiles and birds) or urea (mammals) and excreted in the urine and the feces.

13. The nitrogen in L-glutamine is less basic than ammonia and, thus, is less toxic to biological organisms.

14. Nitrogen is excreted as (a) ammonia by fish and (b) uric acid by birds.

15. The structure of urea is and the structure of uric acid is

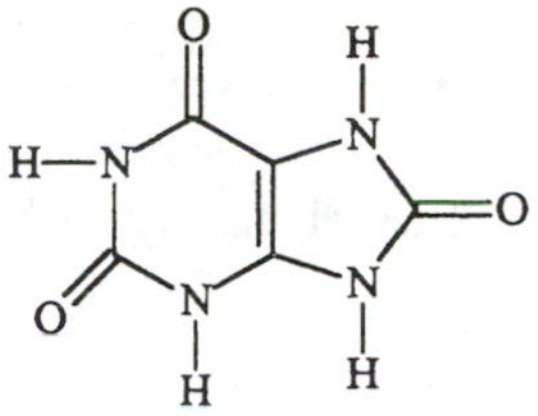

$$NH_2-\overset{\overset{\large O}{\|}}{C}-NH_2$$

16. Three ATP molecules are used for every urea produced. The production of urea is important because this molecule provides a nontoxic means for excreting nitrogen.

17. Acetyl-CoA is considered to be an important central intermediate in metabolism for the following reasons:

    (a) Complete catabolism of most energy containing nutrients is achieved by conversion to acetyl-CoA and oxidation via the citric acid cycle.

    (b) Fats and amino acids can be synthesized from acetyl-CoA.

    (c) Carbohydrates, fats and amino acids can be converted to acetyl-CoA.

    (d) Acetyl-CoA is the central intermediate in converting (1) carbohydrates to fats or amino acids, (2) fats to amino acids or, (3) amino acids to fats.

18. For a growing child the amount of nitrogen consumed is more than what is excreted. Consequently the amount of amino acids from dietary protein increases in the nitrogen pools and the child is said to be in positive nitrogen balance.

19. The two carbons that will be found in acetyl-CoA after one pass through the beta oxidation pathway by butyric acid are circled.

    $CH_3CH_2\boxed{CH_2COOH}$

20. The two carbons that will remain after two passes through the beta oxidation pathway by caproic acid are circled.

    $\boxed{CH_3CH_2}CH_2CH_2CH_2COOH$

21. $CH_3(CH_2)_{12}\overset{\overset{\beta\text{-carbon}}{\downarrow}}{C}H_2CH_2COOH$ palmitic acid

22. $CH_3(CH_2)_{10}\overset{\overset{\beta\text{-carbon}}{\downarrow}}{C}H_2CH_2COOH$ myristic acid

23. Lauric acid ($CH_3(CH_2)_{10}COOH$) has 12 carbons and will yield six molecules of acetyl-CoA.

24. Palmitic acid ($CH_3(CH_2)_{14}COOH$) has 16 carbons and will yield eight molecules of acetyl-CoA.

25. Lauric acid ($CH_3(CH_2)_{10}COOH$) will be broken down to six acetyl-CoA molecules by five passes through the beta oxidation pathway. Thus, five $FADH_2$ molecules will be produced.

26. Palmitic acid ($CH_3(CH_2)_{14}COOH$) will be broken down to eight acetyl-CoA molecules by seven passes through the beta oxidation pathway. Thus, seven NADH molecules will be produced.

27. Complete conversion of myristic acid ($CH_3(CH_2)_{12}COOH$) to seven acetyl-CoA molecules involves six passes through the beta oxidation pathway. Six NADH molecules and six $FADH_2$ molecules will be produced. Therefore:

$$(6\,\text{NADH})\left(\frac{3\,\text{ATP}}{\text{NADH}}\right) + (6\,\text{FADH}_2)\left(\frac{2\,\text{ATP}}{\text{FADH}_2}\right) = 30\,\text{ATP}$$

28. Complete conversion of lauric acid ($CH_3(CH_2)_{10}COOH$) to six acetyl-CoA molecules involves five passes through the beta oxidation pathway. Five NADH molecules and five $FADH_2$ molecules will be produced. Therefore:

$$(5\,\text{NADH})\left(\frac{3\,\text{ATP}}{\text{NADH}}\right) + (5\,\text{FADH}_2)\left(\frac{2\,\text{ATP}}{\text{FADH}_2}\right) = 25\,\text{ATP}$$

29. The acyl carrier protein (ACP) is similar to coenzyme A (CoA) in that

   (a) both molecules have –SH groups which form thioesters with carboxylic acids;

   (b) both molecules act as carriers for fatty acids during metabolism: ACP during anabolism and CoA during catabolism.

30. The acyl carrier protein (ACP) differs from coenzyme A (CoA) in that

   (a) ACP is a protein while CoA is a coenzyme;

   (b) ACP is used in fatty acid synthesis while CoA is used in the beta oxidation pathway.

31. A negative nitrogen balance means more nitrogen is excreted than ingested. A low protein diet will further decrease the amount of nitrogen ingested and should cause the nitrogen balance to become more negative.

32. A positive nitrogen balance means more nitrogen is ingested than excreted. A rapid growth spurt commonly means the body is using ingested amino acids to build new proteins. Thus, less nitrogen will be excreted and the nitrogen balance should become more positive.

33. L-leucine is a ketogenic amino acid because a diet high in L-leucine causes an increase in blood levels of the ketone body, acetoacetic acid.

34. L-aspartic acid is not a ketogenic amino acid because a diet high in L-aspartic acid caused no increase in blood levels of the ketone body $\beta$-hydroxybutyric acid. (Other ketone bodies would not increase their blood levels also.) Because L-aspartic acid is not a ketogenic amino acid, it must be a glucogenic amino acid.

35. During transamination,

    (a) L-alanine yields $CH_3C(=O)COOH$

    (b) L-serine yields $HOCH_2C(=O)COOH$

36. During transamination,

    (a) L-aspartic acid yields $HOOCCH_2C(=O)COOH$

    (b) L-phenyl alanine yields $C_6H_5{-}CH_2C(=O)COOH$

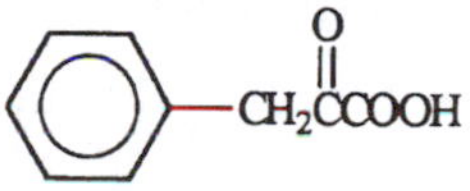

37. The portion of carbamoyl phosphate which is incorporated into urea is circled.

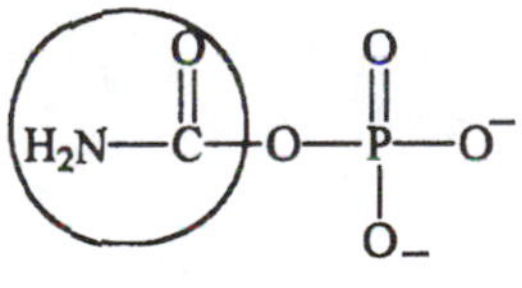

$H_2N{-}C(=O){-}O{-}P(=O)(O^-){-}O^-$ (circled portion: $H_2N{-}C(=O)$)

38. The portion of L-aspartic acid which is incorporated into urea is circled.

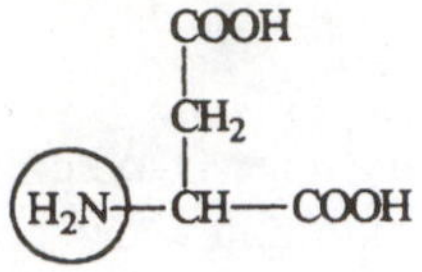

$$\begin{array}{l} \quad\;\; COOH \\ \quad\;\; | \\ \quad\;\; CH_2 \\ \quad\;\; | \\ \boxed{H_2N}-CH-COOH \end{array}$$

39. It is possible to accumulate body fat despite a low fat diet because the body is capable of synthesizing fats from nonfat starting materials.

40. $\beta$-hydroxybutyric acid is a ketone body which does not contain a ketone group.

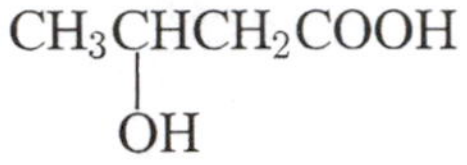

$$\begin{array}{l} CH_3\underset{\displaystyle |\atop \displaystyle OH}{C}HCH_2COOH \end{array}$$

41. Beta-oxidation is a pathway with relatively few reactions, requiring FAD, $NAD^+$, ATP, and coenzyme A. This process forms no ATP directly but depends upon mitochondrial electron transport and oxidative phosphorylation to make use of the reduced coenzymes, NADH and $FADH_2$ to form ATP.

42. In metabolism, most ATP is formed as acetyl-CoA is catabolized by the citric acid cycle, electron transport and oxidative phosphorylation.

(1) One six-carbon hexose (glucose) will be converted to two molecules of pyruvate by the Embden-Meyerhof pathway which, in turn, are converted to two acetyl-CoA molecules.

(2) A six-carbon fatty acid will yield three molecules of acetyl-CoA via the beta oxidation pathway.

The six-carbon fatty acid yields more ATP than the six carbon hexose during catabolism primarily because more acetyl-CoA molecules are formed and can feed into the citric acid cycle, electron transport and oxidative phosphorylation.

43. $$CH_3\overset{\displaystyle O\atop \displaystyle \|}{C}\text{–SCoA} + 6\,HOOCCH_2\overset{\displaystyle O\atop \displaystyle \|}{C}\text{–SCoA} + 12\,NADPH + 12\,H^+ \longrightarrow$$

$$CH_3(CH_2)_{12}COOH + 6\,CO_2 + 5\,H_2O + 7\,\text{CoA–SH} + 12\,NADP^+$$

44. 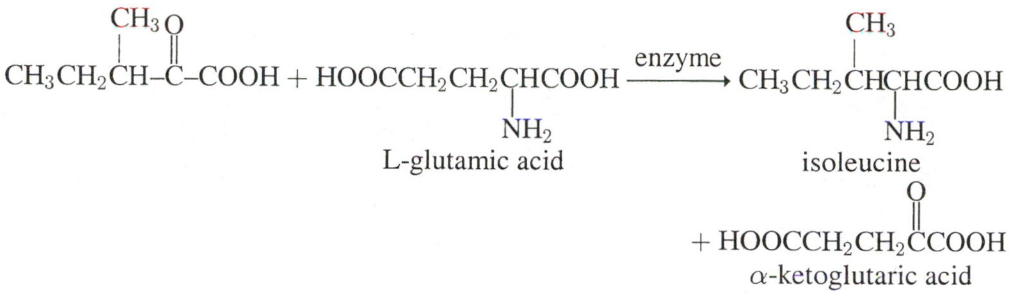 $$CH_3CH_2\overset{\overset{\Large CH_3}{|}}{C}H\text{–}\overset{\overset{\Large O}{||}}{C}\text{–}COOH + HOOCCH_2CH_2\underset{\underset{\Large NH_2}{|}}{C}HCOOH \xrightarrow{\text{enzyme}} CH_3CH_2\overset{\overset{\Large CH_3}{|}}{C}H\underset{\underset{\Large NH_2}{|}}{C}HCOOH$$

L-glutamic acid → isoleucine

$$+\ HOOCCH_2CH_2\overset{\overset{\Large O}{||}}{C}COOH$$

$\alpha$-ketoglutaric acid

45. Malonyl CoA is used in fatty acid synthesis (anabolism) but not in beta oxidation (catabolism). Thus, the lecture would probably focus on fatty acid anabolism.

46. Dehydration refers to the removal of hydrogen. In step 2 of beta oxidation,

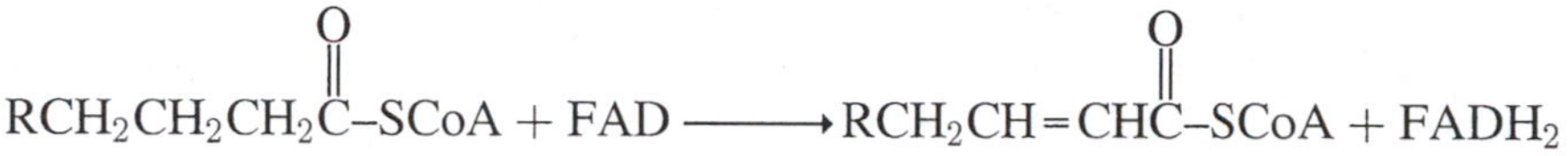

$$RCH_2CH_2CH_2\overset{\overset{\Large O}{||}}{C}\text{–}SCoA + FAD \longrightarrow RCH_2CH{=}CH\overset{\overset{\Large O}{||}}{C}\text{–}SCoA + FADH_2$$

The fatty acid carbons at the alpha and beta positions change their oxidation numbers from $-2$ to $-1$; they are oxidized.

47. $$(1.0 \text{ mol palmitic acid})\left(\frac{256.4 \text{ g}}{\text{mol palmitic acid}}\right) = 2.6 \times 10^2 \text{ g}$$

$$\left(\frac{9 \text{ kcal}}{\text{g}}\right)(2.6 \times 10^2 \text{ g}) = 2 \times 10^3 \text{ kcal}$$